Bulk Metallic Glasses

MATERIALS RESEARCH SOCIETY
SYMPOSIUM PROCEEDINGS VOLUME 554

Bulk Metallic Glasses

Symposium held December 1–3, 1998, Boston, Massachusetts, U.S.A.

EDITORS:

William L. Johnson

Keck Laboratory, California Institute of Technology
Pasadena, California, U.S.A.

Akihisa Inoue

Tohoku University
Sendai, Japan

C.T. Liu

Oak Ridge National Laboratory
Oak Ridge, Tennessee, U.S.A.

Materials Research Society
Warrendale, Pennsylvania

This symposium was partially supported by the U.S. Department of Energy and by Oak Ridge National Laboratory. The opinions, findings, conclusions, or recommendations expressed herein do not necessarily reflect the views of the U.S. government.

Single article reprints from this publication are available through University Microfilms Inc., 300 North Zeeb Road, Ann Arbor, Michigan 48106

CODEN: MRSPDH

Published by:

Materials Research Society
506 Keystone Drive
Warrendale, PA 15086
Telephone (724) 779-3003
Fax (724) 779-8313
Web site: http://www.mrs.org/

Library of Congress Cataloging-in-Publication Data

Bulk metallic glasses : symposium held December 1–3, 1998, Boston, Massachusetts,
U.S.A. / editors, William L. Johnson, Akihisa Inoue, C.T. Liu
p.cm.—(Materials Research Society symposium proceedings,
ISSN: 0272-9172; v. 554)
Includes bibliographical references and indexes.
ISBN: 1-55899-460-2
1. Metallic glasses—Congresses. 2. Bulk solids—Congresses. I. Johnson,
William L. II. Inoue, Akihisa III. Liu, C.T. IV. Series: Materials Research Society
symposium proceedings ; v. 554)

TN693.M4 B85 1999   99-048559
620.1'6—dc21

Manufactured in the United States of America

CONTENTS

*Invited Paper

*Invited Paper

vii

PART V: <u>THERMAL STABILITY, TRANSPORT,</u>
<u>AND MAGNETIC PROPERTIES</u>

PART VI: <u>MECHANICAL AND OTHER PROPERTIES I</u>

*Invited Paper

*Invited Paper

PREFACE

The papers contained in this proceedings were presented December 1–3 at Symposium MM, "Bulk Metallic Glasses," during the 1998 MRS Fall Meeting in Boston, Massachusetts. The symposium was organized in response to a growing level of interest in glass forming metallic alloys which vitrify at relatively low cooling rates (from the molten state), compared with conventional rapidly quenched metallic glasses. Owing to their exceptional resistance to crystallization, these "easy" glass-forming alloys can be cast in "bulk" form with dimensions of millimeters or centimeters. The development of such alloys over the past several years has opened the possibility for broadened studies of the kinetic and thermodynamic properties of undercooled liquids. It has also opened a spectrum of new applications for glassy metals based on the ability to fabricate three-dimensional shapes and components. Products, such as golf clubs, fabricated of bulk metallic glass, have already appeared in the commercial marketplace. It has become apparent that the development of bulk glass forming alloys has ushered in a second generation of scientific and technical development in the field of amorphous metals.

The symposium was very well attended. The organizers received over 100 submitted abstracts for the Meeting, more than could be accommodated in a three-day symposium. Of the 75 papers actually presented at the Meeting, 59 were submitted for publication and are included in this volume. One of the papers, MM5.10, was presented at the symposium as the 1998 MRS Medal Award Lecture. The organizers thank all of the participants for contributing to a highly successful symposium.

William L. Johnson
Akihisa Inoue
C.T. Liu

August 1999

ACKNOWLEDGMENTS

The organizers and participants of Symposium MM wish to acknowledge the generous support of our sponsors. These included the Alps Electric Co., Ltd., Japan; Amorphous Technologies International, U.S.A.; JEOL Ltd., Japan; Oak Ridge National Laboratory, U.S.A.; and the U.S. Department of Energy. We also wish to thank the Materials Research Society staff for their support in organizing the Meeting and in the publication of the proceedings.

MATERIALS RESEARCH SOCIETY SYMPOSIUM PROCEEDINGS

MATERIALS RESEARCH SOCIETY SYMPOSIUM PROCEEDINGS

Prior Materials Research Society Symposium Proceedings available by contacting Materials Research Society

Part I

Atomic and Electronic Structure I

OXYGEN DISTRIBUTION IN Zr-BASED METALLIC GLASSES

D.H. PING, K. HONO and A. INOUE*
National Research Institute for Metals, Sengen 1-2-1, Tsukuba 305-0047, Japan
*Institute of Materials Research, Tohoku University, Sendai 980-8577, Japan

ABSTRACT

This paper reports the atom probe analysis results of the oxygen dissolved in the as-cast amorphous and crystallized $Zr_{65}Cu_{15}Al_{10}Pd_{10}$ and $Zr_{65}Cu_{17.5}Ni_{10}Al_{7.5}$ alloys. Impurity oxygen ranging from 0.1 to 1 at.% is dissolved uniformly in the as-quenched $Zr_{65}Cu_{15}Al_{10}Pd_{10}$ and $Zr_{65}Cu_{17.5}Ni_{10}Al_{7.5}$ amorphous alloys even though the oxygen is not added intentionally. When the $Zr_{65}Cu_{15}Al_{10}Pd_{10}$ alloy is crystallized, oxygen redistribution occurs; it is rejected from the primary $Zr_2(Cu, Pd)$ crystals and partitioned in the subsequently crystallized phases. Oxygen atoms are enriched in some of the crystalline phases up to approximately 4 at.%, and virtually no oxygen is dissolved in the remaining amorphous phase. In the partially crystallized $Zr_{65}Cu_{17.5}Ni_{10}Al_{7.5}$ alloy, fine oxygen enriched particles containing ~ 15 at.%O have been detected in direct contacted with crystalline grains. This work demonstrates that oxygen redistribution occurs during the crystallization reaction, thereby influencing the kinetics of crystallization.

INTRODUCTION

Following the discovery of multi-component Zr-based amorphous alloys with wide supercooled liquid regions by Inoue et al. [1], intensive investigations have been carried out to develop new Zr-based bulk amorphous alloys which can be processed by the conventional casting method at low cooling rates [2-4]. One of these is $Zr_{65}Cu_{17.5}Ni_{10}Al_{7.5}$ alloy with a wide supercooled liquid region, $\Delta T_x = T_x - T_g = 127$ K [1], where T_x is the crystallization temperature and T_g is the glass transition temperature. It is generally known that the alloy with a wide supercooled liquid region or a high reduced glass transition temperature, T_g/T_m, where T_m is a melting temperature, has an excellent glass forming ability and the fabrication of a bulk amorphous alloy by conventional casting processes is possible. Recent investigations [5-10] reported that oxygen content in the Zr-based amorphous alloy significantly affects the glass forming ability of Zr-based amorphous alloys, but the mechanism how oxygen impurity affect the crystallization kinetics is not well established. In order to clarify this, it is essential to measure the distribution of oxygen in the metallic glasses in a microscopic scale. Atom probe field ion microscopy (APFIM) is the most suitable technique to determine the local chemical concentrations of oxygen in the amorphous alloy in a sub-nanometer scale, and we have employed this technique for characterizing oxygen dissolution and redistribution in Zr-based amorphous alloys.

$Zr_{65}Cu_{15}Al_{10}Pd_{10}$ amorphous alloy was selected for this study, because it forms nanocrystalline microstructure by crystallization [11] and is thought to be suitable for observing the crystallization event in conjunction with the oxygen redistribution by the APFIM technique. $Zr_{65}Cu_{17.5}Ni_{10}Al_{7.5}$ amorphous alloy was also studied, because this is one of the most well-studied metallic glasses with an excellent glass forming ability, and the phase formation by crystallization was most widely studied in conjunction with the oxygen impurity [9,10]. However, up to now, the microstructural evolution has not been investigated in detail with respect to the effect of impurity oxygen. To investigate the oxygen effect on the crystallization reaction, a direct observation of the oxygen distribution is desirable. In this study, we have observed the distribution and redistribution of oxygen atoms dissolved in the $Zr_{65}Cu_{15}Al_{10}Pd_{10}$ and $Zr_{65}Cu_{17.5}Ni_{10}Al_{7.5}$ metallic glasses directly by the three-dimensional atom probe (3DAP) technique.

3

EXPERIMENT

Ingots with a composition of $Zr_{65}Cu_{15}Al_{10}Pd_{10}$ and $Zr_{65}Cu_{17.5}Ni_{10}Al_{7.5}$ were prepared by arc melting the mixtures of pure metals in an argon atmosphere. Rapidly solidified ribbons of 20 μm in thickness and of 1.5 mm in width were prepared by the single roller melt spinning method in an argon atmosphere. The as-quenched sample had an amorphous structure. The ribbon were crystallized by isothermal annealing in vacuum-sealed quartz tubes. The ribbon shaped specimens were mechanically ground to square rods of approximately 20 μm × 20 μm × 8 mm and then electropolished by the micro-electropolishing technique for obtaining needle shape specimens for field ion microscope (FIM) observation. Elemental maps of alloying elements were measured by a three dimensional atom probe (3DAP). The 3DAP used in this study was equipped with CAMECA tomographic atom probe (TAP) detection system [12]. Atom probe analyses were performed at tip temperatures of about 70 K in an ultrahigh vacuum condition ($< 1 \times 10^{-8}$ Pa) with a pulse fraction (a ratio of pulse voltage to the static voltage) of 0.2 and a pulse repetition rate of 600 Hz. The microstructure was observed with a Philips CM200 transmission electron microscope (TEM) operated at 200 kV and a JEM 4000EX high resolution electron microscope (HREM) operated at 400kV. The TEM samples were prepared by mechanical grinding and ion beam thinning.

RESULTS AND DISCUSSION

Figure 1 shows a 3DAP elemental map of oxygen obtained from an as-quenched $Zr_{65}Cu_{15}Al_{10}Pd_{10}$ amorphous alloy. Atoms were corrected from a volume of approximately $12 \times 12 \times 30$ nm as shown in Fig. 1(a), and each dot corresponds to an oxygen atom. One can see that the oxygen atoms are uniformly distributed in the as-quenched alloy. The other constituent elements (Zr, Cu, Pd and Al) are also detected uniformly. Fig. 1(b) shows the corresponding concentration depth profiles determined from the analyzed volume in Fig 1(a). The overall composition (Zr = 66.5 at.%, Cu = 13.5 at.%, Pd = 11.1 at.% and Al = 8.8 at.%) determined by the 3DAP result agrees reasonably well with the nominal composition of the alloy, and the oxygen content has been determined to be approximately 0.1 at.%.

Figure 2 shows an HREM image of the $Zr_{65}Cu_{15}Al_{10}Pd_{10}$ alloy annealed at 730 K for 30 min. At this stage, the microstructure is composed of nanoscale primary crystals embedded in the amorphous matrix. The average grain size was estimated to be about 10 nm. Figure 3 shows 3DAP analysis results obtained from the specimen annealed at 730 K for 30 min. The Al elemental map from an analysis volume of $16 \times 16 \times 28$ nm (Fig. 3(a)) shows that there is a particle in which Al

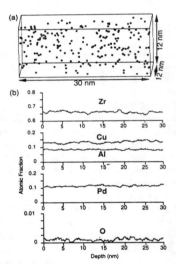

Fig. 1 (a) Oxygen mapping in the ZrCuAlPd amorphous alloy. (b) Concentration profiles obtained from the analyzed volume of (a).

atoms are depleted. The concentration changes across the particle have been determined from the inset box A as indicated in Fig. 3(b). The Al concentration of the Al depleted region is only 2 at.%, and Zr is enriched replacing for Al atoms. Oxygen is also rejected from this region. However, there are no large differences in the concentrations of Pd and Cu. Although the structures of the crystalline products have not been determined in this work, this phase is presumed to correspond to the $Zr_2(Cu, Pd)$ as proposed by Fang et al. [11] based on their X-ray diffraction results. The oxygen map in the

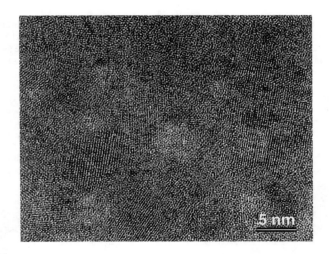

Fig. 2 HREM image of the $Zr_{65}Cu_{15}Al_{10}Pd_{10}$ alloy annealed at 730 K for 30 min.

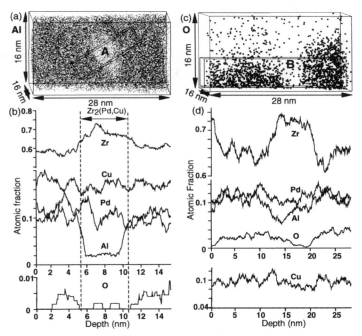

Fig. 3 3DAP analysis results of the $Zr_{65}Cu_{15}Al_{10}Pd_{10}$ alloy annealed at 730 K for 30 min. (a) Al elemental mapping revealing a $Zr_2(Pd,Cu)$ precipitate, concentration depth profiles determined from the selected box (A) are shown in (b); (c) Oxygen elemental mapping in the same analyzed volume, concentration depth profiles determined from the selected region (B) are shown in (d).

same analyzed volume is shown in Fig. 3(c). The Zr₂(Cu, Pd) phase contains little oxygen, but oxygen is enriched in two adjacent regions. In order to observe the concentration change across the oxygen enriched regions, concentration depth profiles have been calculated from the analysis volume B as shown in Fig. 3(d). The oxygen level ranges from 3 to 4 at.% in the oxygen enriched regions, suggesting that oxygen is partitioned in some crystalline phases adjacent to the primary crystals.

In order to observe the oxygen distribution in the fully crystallized microstructure, the alloy was annealed at 730 K for 60 min. Figure 4(a) shows a bright field electron micrograph and its corresponding SAED pattern of the Zr₆₅Cu₁₅Al₁₀Pd₁₀ amorphous alloy annealed at 730 K for 60 min. A large number of nanocrystals with the grain size ranging from 20 to 40 nm are observed in the micrograph. The SAED pattern shows several sharp rings, indicating that there are more than three phases in this stage, but the structures of these remain unidentified. The presence of the remaining amorphous phase is not apparent in this stage. Figure 4(b) shows 3DAP elemental map of oxygen within an analyzed volume of $17 \times 17 \times 44$ nm and 4(c) the concentration depth profiles of the

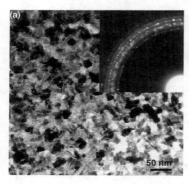

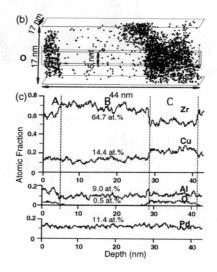

Fig. 4 (a) TEM result of the Zr₆₅Cu₁₅Al₁₀Pd₁₀ alloy annealed at 730 K for 60 min. (b) Oxygen mapping in an analysis volume $17 \times 17 \times 44$ nm obtained by 3DAP, (c) concentration depth profiles determined from the selected box $5 \times 5 \times 44$ nm as shown in (b).

alloying elements determined from the selected volume of $5 \times 5 \times 44$ nm indicated in Fig. 4(b). Each dot corresponds to a position of an oxygen atom. Oxygen distribution is not uniform within the analyzed region but enriched in some phases. The overall composition determined by this analysis is 64.7 at.%Zr, 14.4 at.%Cu, 11.4 at.%Pd, 9.0 at.%Al and 0.5at% O, which is in excellent agreement with the nominal composition of the alloy. Thus the chemical compositions determined in this analysis can be interpreted as quantitative ones. Region A is enriched in Al (~15 at.%) and the Zr concentration is lower than the average alloy composition. Although the structure has not been determined in this study, according to Fan et al. [11], possible phases after complete crystallization are Zr₂(Cu, Pd), Zr₂(Al, Pd) and Zr₃Al₂, but these were proposed based on the limited number of peaks in the x-ray diffraction results. Based on the composition determined by the 3DAP data, the phase in region A is close to Zr₂(Al, Pd) with significant substitution of Cu for Al. In this phase, approximately 4 at.% oxygen is dissolved. In region B, Zr concentration is slightly higher than those in the other two phases, i.e. ~70at.%Zr containing approximately 10at.% each of Cu, Al and Pd, thus chemically the phase is described as Zr₃AlCuPd quaternary compound, but its structure is not known.

Oxygen level in this phase is virtually zero. Region C is enriched with Al and Cu, and from the concentrations of the alloying elements, it is presumed to correspond to the $Zr_2(Cu, Pd)$ phase proposed by Fan et al. [11] with a significant substitution of Al for Cu. In this 3DAP analysis result, it should be noted that Pd concentration is uniform throughout these three phases, and it is concluded that no partitioning of Pd occurs during the crystallization reaction.

Figure 5 shows a mass spectrum of an as-quenched $Zr_{65}Cu_{17.5}Ni_{10}Al_{7.5}$ alloy. In this alloy, some of the mass peaks of Ni^{2+} and Zr^{3+} overlap and it is impossible to determine concentrations of Ni and Zr quantitatively. However, this study aimed to determine the distributions of oxygen atoms in the amorphous alloy in the course of the crystallization reaction, thus the overlapping mass peaks of some of the Zr and Ni ions do not give any influence on the oxygen distribution determined by APFIM. As shown in

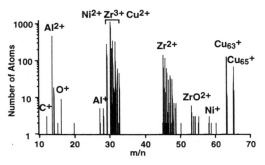

Fig. 5 mass spectrum of the as-quenched ZrCuNiAl alloy

Fig. 5, oxygen ions are detected as either O^+ or ZrO^{2+}. Figure 6 shows atom probe concentration depth profiles of an as-quenched $Zr_{65}Cu_{17.5}Ni_{10}Al_{7.5}$ alloy. Impurity oxygen atoms are uniformly distributed in the as-quenched $Zr_{65}Cu_{17.5}Ni_{10}Al_{7.5}$ amorphous alloy and its overall composition is about 1 at.%O. The other alloying elements are also homogeneously dissolved. In the partially crystallized specimen, oxygen-enriched fine particle have been detected by 3DAP as shown in Fig. 7. The oxygen content in the oxygen enriched particle has been estimated to be ~15%O with approximately 65%Zr.

As demonstrated in the two Zr-based amorphous alloys studied in this work, oxygen redistribution occurs as a result of crystallization. Oxygen is enriched in some of the crystalline phases, and this suggests a possibility of forming some metastable phases as reported by Altounian et al. [13] in Zr-Ni binary alloy. Recently, Köster et al. [14] reported that a quasicrystalline phase is formed when $Zr_{65}Cu_{17.5}Ni_{10}Al_{7.5}$ amorphous alloy is crystallized, but more recent work by Gebert et

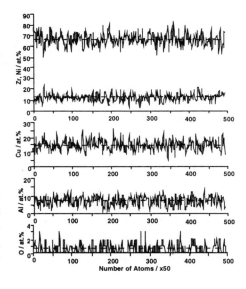

Fig. 6 Concentration depth profiles of Zr, Ni, Cu, Al and O in the as-quenched ZrCuNiAl alloy.

al. [10] on phase formation in the alloy with the same composition but with different oxygen content reported formation of only Zr_2Cu and Zr_2Ni phases. This discrepancy in the crystallization products would be originated from the different oxygen content in the starting $Zr_{65}Cu_{17.5}Ni_{10}Al_{7.5}$ amorphous alloy. In the $Zr_{65}Cu_{17.5}Ni_{10}Al_{7.5}$ amorphous alloy used in this work contained as much as 1 at.%O, although it was not intentionally added to the alloy ingot. So far, oxygen redistribution has been observed only when crystallization products are observed. As shown in Fig. 3, oxygen is not necessarily enriched in the primary crystal, but is rejected from the primary crystal. This suggests that oxides are not formed as the initial product. Thus, it is not likely that oxides themselves serve as heterogeneous nucleation sites for crystallization in these Zr-based amorphous alloys. However, as oxygen redistributes during the crystallization process and probably cause formation of some metastable phases, it is concluded that oxygen influence kinetics and phase formation of crystallization of Zr-base amorphous alloys. As oxygen is participating in phase formation, oxygen should be regarded as an alloying element in Zr-based amorphous alloys. This also suggests that controlling oxygen content in Zr-based alloy is very important to improve the thermal stability of Zr-based bulk amorphous alloy.

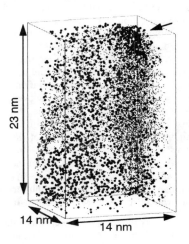

Fig. 7 3DAP elemental map of O and Ni.

CONCLUSIONS

3DAP analysis results have clearly shown that oxygen impurity up to 1 at.% is dissolved in $Zr_{65}Cu_{15}Al_{10}Pd_{10}$ and $Zr_{65}Cu_{17.5}Ni_{10}Al_{7.5}$ amorphous alloys even if oxygen is not added intentionally. When the alloy is crystallized, oxygen redistribution occurs and it is partitioned in some crystalline phases. Although the structures of these oxygen enriched phases remain unidentified, the present atom probe results suggest that oxygen plays a significant roll in crystallization of Zr-based amorphous alloy, thus controlling oxygen content in the Zr-base bulk amorphous alloy is critical for developing stable Zr-based bulk amorphous alloys.

ACKNOWLEDGMENT

This work was partly supported by the NEDO International Joint Research Grant.

REFERENCES

1. T. Zhang, A. Inoue and T. Masumoto, Mater. Trans. JIM, **32,** p. 1,005 (1991).
2. A. Peker and W. L. Johnson, Appl. Phys. Lett. **63,** p. 2,342 (1993).
3. A. Inoue, T. Zhang and T. Masumoto, Mater. Trans. JIM, **36,** P. 391 (1995).
4. A. Inoue and T. Zhang, Mater. Trans. JIM, **37,** P. 185 (1996).
5. L. Johnson, Mater. Sci. Forum **225-227,** P. 35 (1996).
6. M. Seidel, J. Eckert and L. Schultz, Mater. Sci. Forum **235-238,** P. 29 (1997).
7. X. H. Lin, W. L. Johnson and W. K. Rhim, Mater. Trans. JIM, **38,** P. 473 (1997).
8. U. Köster, J. Meinhardt, S. Roos and A. Ruedinger, Mater. Sci. Forum, **225-227,** P. 311 (1996).
9. A. Kubler, J. Eckert, A. Gebert and L. Schultz, J. Appl. Phys. **83,** P. 3,438 (1998).
10. A. Gebert, J. Eckert and L. Schultz, Acta mater. **46,** P. 5475 (1998).
11. C. Fan and A. Inoue, Mater. Trans. JIM, **38,** P. 1,040 (1997).
12. D. Blavette, B. Deconihout, A. Bostel, J. M. Sarrau, M. Bouet and A. Menand, Rev. Sci. Instrum. **64,** P. 2,911 (1993).
13. Z. Altounian, E. Batalla, J. O. Strom-Olsen and J. L. Walter, J. Appl. Phys. **61,** p. 149 (1987).
14. U. Köster, J. Meinhardt, S. Roos and H. Liebertz, Appl. Phys. Lett. **69,** p. 179 (1996).

DECOMPOSITION IN $Pd_{40}Ni_{40}P_{20}$ BULK METALLIC GLASS

M. K. MILLER*, R. B. SCHWARZ**, YI HE**,
*Metals and Ceramics Division, Oak Ridge National Laboratory, Oak Ridge, TN 37831-6376, xkm@ornl.gov
**Center for Materials Science, Los Alamos National Laboratory, CMS, Mail Stop K765, Los Alamos, NM 87545

ABSTRACT

An atom probe field ion microscope and 3-dimensional atom probe characterization of the solute distribution in a bulk $Pd_{40}Ni_{40}P_{20}$ metallic glass in the as-cast state and after annealing has been performed. Statistical analysis of the atom probe atom-by-atom data detected the presence of short range ordering in the as-cast alloy. Phase separation at the nanometer level is observed in glassy samples after annealing above the glass-transition temperature. Crystallization proceeds by phase separation into three distinct crystalline phases. Atom probe analysis of the alloy annealed for 1 h at 410°C revealed that the primary nickel phosphide phase contained significant levels of palladium, the palladium-rich Pd_3P phosphide phase contained low levels of nickel and there was a small amount of a palladium-nickel solid solution.

INTRODUCTION

The decomposition path of Pd-Ni-P glasses during crystallization is difficult to predict since there are several phases such as Pd_8P, Pd_6P, $Pd_{4.8}P$, Pd_3P, Pd_5P_2, Pd_7P_3, PdP_2 and Ni_3P, Ni_5P_2 (or Ni_7P_3), NiP_2, Ni_6P_5, NiP_2 and Ni_3P possible from the respective Pd-P and Ni-P binary phase diagrams [1,2]. Donovan et al. have shown by X-ray microanalysis in the transmission electron microscope that three crystalline phases are present after aging a $Pd_{40}Pd_{40}P_{20}$ glass for 24 h at 600°C [3]. These phases were an orthorhombic primary phosphide ($Ni_{45}Pd_{34}P_{21}$), an orthorhombic palladium-rich phosphide ($Ni_{14}Pd_{68}P_{18}$), and a face centered cubic palladium-nickel solid solution ($Ni_{59}Pd_{40}P_1$). Some previous atom probe field ion microscopy investigations have been performed on this system and the results are summarized elsewhere [4].

In this paper, the results of an atom probe field ion microscopy (APFIM) characterization of the solute distribution in a $Pd_{40}Pd_{40}P_{20}$ metallic glass that was annealed above and below the onset of crystallization temperature are reported.

EXPERIMENTAL

The ternary $Pd_{40}Ni_{40}P_{20}$ bulk metallic glass used in this investigation was prepared with a fluxing technique [5-8]. In this method, high-purity Pd, Ni (99.9% purity) and Ni_2P (99.5% purity) were placed in a fused silica tube with a high-purity dehydrated B_2O_3 flux. The silica tube was evacuated and heated to a temperature of ~1200°C which is above the melting temperature of the alloy and the B_2O_3. At this temperature, the surface oxides are reduced or dissolved in the B_2O_3 flux. The silica tubes were then quenched into water. The field ion specimens were electropolished from square bars cut from the central region of a 7 mm diameter bar to ensure that the analyzed region of the specimen was taken from the interior of the bulk specimen to eliminate the possibility of surface effects.

The thermal stability and specific heat of the alloy were measured in a Perkin-Elmer DSC-7 differential scanning calorimeter. These measurements revealed that the glass transition temperature, T_g, of this material is 303°C and the onset of crystallization, T_x, occurs at 403°C.

9

The material was characterized in the as-prepared condition and after seven annealing treatments of 48 h at 140°C, 24 h at 200°C, 2 h at 300°C, 2 h at 350°C, 0.5 h at 390°C and 1 h at 410°C and 24h at 500°C. The 140, 200 and 300°C annealing conditions are below the glass transition temperature, and the 390°C and 410°C annealing temperatures are just below and just above the onset of primary crystallization, respectively. The 500°C annealing temperature is significantly above the onset of crystallization and the time is sufficient that compositions of the phases should be close to their equilibrium values.

The material was characterized in the ORNL energy-compensated atom probe and the energy-compensated optical position-sensitive atom probe. All compositions are presented in atomic percent. In cases in which more than one atom was collected on a field evaporation pulse, the order of the atoms on that pulse was randomized to prevent bias. In these atom-by-atom statistical methods, the mixed P^+/Ni_{62}^{2+} peak was assigned to the more abundant phosphorus. The solute distribution at the atomic scale was investigated by examining the atom-by-atom data chain for ABA, ABBA, ABBBA, etc. (where B is the atom of the element of interest and A is any other atom) sequences in one dimensional columns of atoms obtained in the energy-compensated atom probe. The probability of detecting a chain containing n B atoms is given by [9] $P(n) = p^n q^2$ or $D_{exp}(n) = N P(n)$, where p is the probability of collecting a B atom, $q = 1 - p$, and N is the number of atoms in the chain. The significance of the experimental value is given by $(D_{ap}(n) - D_{exp}(n))/\sigma$, where σ was taken as $\sqrt{N} p^n q^2$. The Johnson and Klotz ordering parameter, θ, was determined from the number of AB and BB pairs in the data chain [10]. The significance is given by $(\theta - 1) / \sigma$, where σ is the standard error. In both methods, values of the significances greater than 2 or less than -2 indicate non-random behavior.

RESULTS AND DISCUSSION

No evidence of decomposition or clustering was found in the field ion micrographs or electron diffraction patterns of the materials annealed for 48 h at 140°C, 24 h at 200°C, 2 h at 300°C, 2 h at 350°C, or 0.5 h at 390°C. Representative field ion micrographs of these materials are shown in Fig. 1. These micrographs exhibited a random distribution of spots that is characteristic of an amorphous structure. With the exception of some areas in the materials annealed for 2 h at 350°C and 0.5 h at 390°C, a uniform contrast was observed indicating no extensive phase separation had occurred. In the field ion micrographs of the material annealed for 1 h at 410°C and the 24 h at 500°C, ring structures and contrast variations characteristic of the presence of crystalline phases were evident, as shown in Figs. 2 and 3, respectively. In these field ion micrographs, the dark regions correspond to the $(Ni,Pd)_3P$ phase and the brightly-imaging regions to the Pd_3P phase. Electron diffraction patterns of the field ion needles of the material aged for 1 h at 410°C showed no evidence of any remaining amorphous material. The scale of the microstructure was significantly coarser in the material aged at 500°C.

Atom probe analysis revealed that the average composition of this material was Pd- 40.4 ± 0.25 at. % Ni, 20.3 ± 0.1% P, 0.03 ± 0.01% B, 0.02 ± 0.005% Fe, 0.04 ± 0.01% O, 0.004 ± 0.002% Si+N and 0.001 ± 0.001% C. The boron concentration indicated that a small amount of boron from the B_2O_3 flux had dissolved in the alloy during preparation.

The results of the statistical analyses of the material aged below the onset of crystallization are presented in Figs. 4 and 5, for the Markov chain and the Johnson and Klotz methods, respectively. The significance of the ABA sequences for all three elements (i.e., the solute atom of interest in the data chain flanked with either of the other elements (e.g., Ni or P, Pd, Ni or P)) was positive indicating that the glass contained chemical short range ordering. The

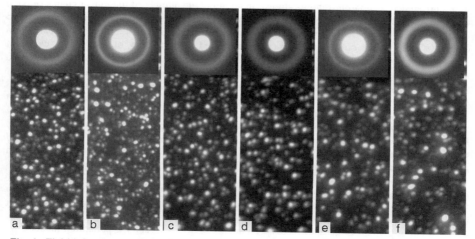

Fig. 1 Field ion micrographs and electron diffraction patterns of $Pd_{40}Ni_{40}P_{20}$ in the a) as-cast condition, and after ageing for b) 48 h at 140°C, c) 24 h at 200°C, d) 2 h at 300°C, e) 2 h at 350°C, f) and 0.5 h at 390°C.

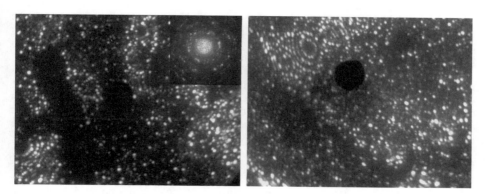

Fig. 2. Field ion micrograph and electron diffraction pattern of $Pd_{40}Ni_{40}P_{20}$ after ageing for 1 h at 410°C showing crystalline phases.

Fig. 3. Field ion micrograph of $Pd_{40}Ni_{40}P_{20}$ after ageing for 24 h at 500°C showing crystalline phases.

nickel appeared to exhibit the strongest degree of ordering followed by phosphorus and then palladium. The nickel and palladium also exhibited a slight preference for ABBA chains but the phosphorus exhibited a negative significance indicating that phosphorus has a strong preference to be surrounded by non-phosphorus atoms. In addition, the number of the longer chains (n B≥3) detected for nickel and phosphorus was significantly lower than expected in a random alloy and the phosphorus results were lower than the nickel results. Although the number of palladium chains (n Pd≥3) were consistently less than expected, the majority of the results were not statistically significantly different from a random solid solution. These results indicate that no solute clustering was occurring. In almost all cases, the Johnson and Klotz ordering parameter was statistically less than 1 indicating that the number of BB atoms in the data chain was significantly less than expected in a random alloy. Phosphorus was found to have the strongest and palladium the weakest tendency to chemical short range order.

An atom probe composition profile in the material aged for 2 h at 300°C is shown in Fig. 6. Some evidence of extremely fine scale inhomogeneities was apparent. Some regions of the material aged for 0.5 h at 390°C showed more extensive evidence of phase separation into phosphorus–enriched/palladium-depleted and phosphorus-depleted/palladium-enriched regions, as shown in Fig. 7. The compositions of the three coexisting phases after heat treatments of 1 h at 410°C and 24 h at 500°C were measured in the atom probe. Analysis of the alloy annealed for 1 h at 410°C revealed that the primary nickel phosphide had an average composition of 47.6 ± 0.3% Ni, 28.3 ± 0.3% Pd and 24.2 ± 0.3% P, the palladium-rich phosphide had an average composition of 69.8 ± 0.6% Pd, 8.6 ± 0.4% Ni and 21.6 ± 0.6% P, and the palladium-nickel solid solution had an average composition of 44.7 ± 0.4 % Pd, 39.8 ± 0.4% Ni and 15.6 ± 0.3% P. These compositions yield volume fractions of 46% for the primary nickel phosphide, 13% for the palladium-rich phosphide and 41% for the palladium-nickel solid solution. After ageing for 24 h at 500°C, the average composition of the primary phosphide was determined to be 47.4 ± 0.9% Ni, 22.2 ± 0.7 %Pd, 30.4 ± 0.8% P indicating an increase in the phosphorus and a decrease in the palladium levels. Similarly, the average composition of the palladium-rich phosphide was determined to be 72.9 ± 0.09% Pd, 23.5 ± 0.04% P 3.6 ± 0.04 Ni indicating a small increase in both palladium and phosphorus levels and a decrease in the nickel level.

The average compositions of the three phases as measured in the atom probe have been represented on an isothermal section of a ternary phase diagram in Fig. 8. In addition, the compositions which have been demonstrated previously to form bulk amorphous glass in 10 mm diameter rods [8] have been superimposed as filled circles and those compositions which do not as open circles. It is interesting to note that the phase compositions determined in the atom probe all fall just outside the range of compositions in which amorphous material is found.

CONCLUSIONS

This atom probe characterization of $Pd_{40}Ni_{40}P_{20}$ metallic glass has revealed the presence of short range ordering in the as-cast alloy. Phase separation at the nanometer level is observed in glassy samples after annealing above the glass-transition temperature. Crystallization proceeds by phase separation into three distinct crystalline phases.

ACKNOWLEDGMENTS

This research was sponsored by the Division of Materials Sciences, U. S. Department of Energy, under contract DE-AC05-96OR22464 with Lockheed Martin Energy Research Corp.,

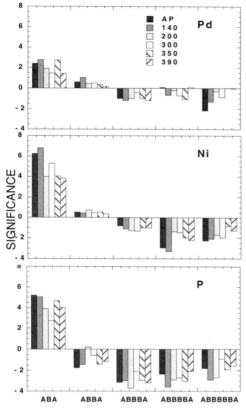

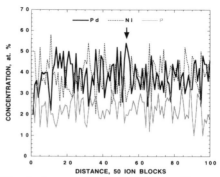

Fig. 6. Atom probe composition profiles through the $Pd_{40}Ni_{40}P_{20}$ alloy aged for 2 h at 300°C.

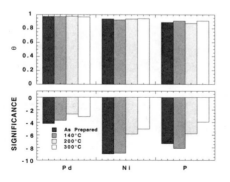

Fig. 4. Summary of the significances of the Markov chain ABA results

Fig. 5. Summary of the Johnson and Klotz results showing chemical short range order.

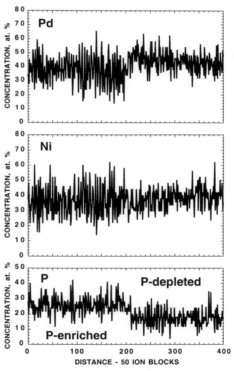

Fig. 7. Composition profiles through the $Pd_{40}Ni_{40}P_{20}$ alloy aged for 0.5 h at 390°C showing phase separation into a phosphorus-enriched and phosphorus-depleted region.

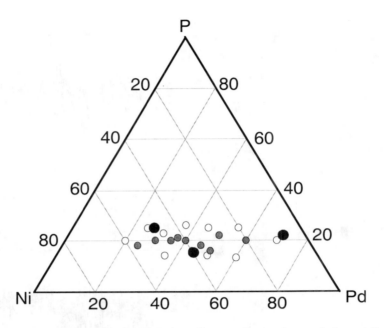

Fig. 8. Isothermal section of the Pd-Ni-P phase diagram. Gray and open circles are alloy compositions that form and do not form amorphous in 10-cm-thick sections [5,8]. Larger filled circles are the compositions of the phase formed after ageing for 1 h at 410°C.

and contract W-7405-ENG.36 with the University of California. This research was conducted utilizing the Shared Research Equipment (SHaRE) User Program facilities at Oak Ridge National Laboratory.

REFERENCES

1. I. O. Gullman, J. Less Common Metals, **11**, 157 (1966).
2. G. Hanson and K. Anderko, *Constitution of Binary Alloys*, 2nd ed. (McGraw Hill, New York, 1958)
3. P. E. Donovan, P. V. Evans and A. L. Greer, J. Mater. Sci. Lett., **5**, 951 (1986).
4. M. K. Miller, D. J. Larson, R.B. Schwarz and Yi He, Mater. Sci. Eng. **A250,** 141 (1998).
5. Yi He, R. B. Schwarz and J. I. Archuleta, Appl. Phys. Lett., **69,** 1861 (1996).
6. A. J. Drehman, A. L. Greer and D. Turnbull, Appl. Phys. Lett., **41, 716** (1982).
7. K. W. Kui, A. L. Greer and D. Turnbull, Appl. Phys. Lett., **45**, 615 (1984).
8. Yi He, T. Shen and R. B. Schwarz, Metall. Trans., in press.
9. T. T. Tsong, S. B. McLane, M. Ahmad and C. S. Wu, J. Appl. Phys., **53**, 4180 (1982).
10. C. A. Johnson and J. H. Klotz, Technometrics, **16,** 483 (1974).

A LOCAL PROBE INTO THE ATOMIC STRUCTURE OF METALLIC GLASSES USING EELS

F.M. ALAMGIR[*], Y. ITO[*], H. JAIN[*], D. B. WILLIAMS[*], R.B. SCHWARZ[**]

[*]Department of Material Science & Engr., Lehigh University, Bethlehem, PA 18015,
fma2@lehigh.edu
[**]Los Alamos National Laboratory, Los Alamos, NM 87545, USA

Abstract

Electron energy loss spectroscopy (EELS) is used to extract information on the topological arrangement of atoms around Pd in the bulk-glass-forming $Pd_{60}Ni_{20}P_{20}$. It is found that the environment around Pd in the glass is only a slight modification of the Pd crystalline structure. However, the modification is enough to allow this alloy to form a glass in bulk. In examining the differences between the structure of crystalline Pd and glassy $Pd_{60}Ni_{20}P_{20}$ it is concluded that incorporation of Ni and P into the structure frustrates the structure enough that glass formation becomes easy.

Introduction

The discovery of alloys that form bulk metallic glasses[1,2] (BMGs) has rejuvenated interest in these systems in recent years. Since the "bulk" nature of these BMGs is central to the new-found scientific interest in them, a basic question need to be answered: what governs their easy glass forming ability? This has been explained to some extent by different criteria that are based on the suppression of the nucleation of crystals[3], the suppression of the kinetics via the "confusion principle"[4], and atomic size effects of the constituent elements[5]. The experimental data in this case are profuse but structural models that tie together all the evidence are lacking.

A major reason for the lack of structural models for glass-formability and glass stability is that little information is available on the atomic structure for these glasses. Scattering experiments using neutrons, X-rays or electrons can provide total radial distribution function (RDF), but the interpretation of RDFs become difficult for BMGs, which contain many different elements, often five or more. It is much more useful to obtain a partial radial distribution function (PRDF), the RDF around a individual elements. Extended X-ray absorption fine structure (EXAFS) and anomalous X-ray scattering provides PRDFs directly but these techniques require a synchrotron light source. Only for selected elements is it possible to obtain PRDFs from neutron scattering using isotopic substitution. All these techniques give structural information with poor spatial resolution. By comparison, the electron analog of EXAFS, extended electron energy loss fine structure (EXELFS), provides a much higher spatial resolution (a sampling volume of ~50 nm^3, and a sampling area of ~ 1 nm^2 are attainable). Also the image and diffraction pattern from the sampling area may be monitored at all times during the experiment. In general, EXELFS provides supporting, and complementary structural information to EXAFS. This technique, however, is still under development.

EXELFS refers to the fine structure that appears at energies >30 eV above the ionization edge in an EELS spectrum. Incident electrons are scattered inelastically as they pass through the sample and on to an energy-loss detector, producing an EELS spectrum. If a scattering event causes an ionization event in the sample material then the ionized electron

15

wave sets up a standing wave between the central atom and a neighboring atom from which it is elastically backscattered. It does so with all its neighbors. This standing wave modulates the energy loss characteristics of the incident electrons and shows up as the fine structure (EXELFS) beyond the ionization edge of the EELS spectrum. It is obvious that this modulation function contains information on the distribution of atoms around the parent atom in reciprocal space. The Fourier transform of this function is proportional to the PRDF.

$Pd_xNi_{(80-x)}P_{20}$ is a prototypical BMG family. In a recent study[6] Egami *et al.* found using anomalous X-ray scattering the structure of $Pd_{30}Ni_{50}P_{20}$ and $Pd_{40}Ni_{40}P_{20}$ glasses to be best described by the dense random packing (DRP) model[7]. The DRP model has already been shown to work well for binary transition metal- metalloid (TM-M) glasses[8]. It was found that the total RDF for $Pd_{40}Ni_{40}P_{20}$ and $Pd_{30}Ni_{50}P_{20}$ are almost identical. However, in the Pd PRDFs for the two compositions the shapes of the well-known "split" second peak (characteristic of the DRP structure) in this case were different. If the local atomic structure in these glasses have an important bearing on the glass-formability of these alloys, then it is the PRDFs that will indicate this.

The goal of our work is to develop EXELFS for the study of BMGs using the $Pd_xNi_{(80-x)}P_{20}$ family as our case study. The viability of using EXELFS for the study of BMGs is examined in a separate paper[9]. In this paper, we present the initial results of using EXELFS in determining the structure of $Pd_{60}Ni_{20}P_{20}$. For comparison we also measured the RDF of crystalline Pd.

Experiment

The specimens are $Pd_{60}Ni_{20}P_{20}$ BMG and Pd pure metal foil (99.95%, Goodfellow metals). The glass was formed by water quenching of the melt. The details of the processing of the alloys can be found elsewhere[10]. These specimens were pre-thinned by mechanical polishing and finished with Precision Ion Polishing System (PIPS, Gatan Inc.). The samples in each case was placed on a cold stage and cooled with $N_2(l)$ to eliminate contamination buildup under the high electric field of the electron probe. The thickness of the analyzed area was in the range of 30 - 40 nm thick.

The Philips EM400 120 keV conventional transmission electron microscope (CTEM) with a single crystal LaB_6 thermionic emission gun was used. The microscope is equipped with Gatan 666 Parallel Electron Energy-Loss Spectrometer (PEELS). The spatial resolution of the electron probe used was ~ 50 nm and its effective collection semi-angle for EELS was 8 mrad. The energy dispersion of the PEELS system was set to 1 eV per channel for EXELFS analysis.

The data reduction steps involve first the removal of dark-current noise from a raw spectrum, the pre-edge background and the contribution of plural scattering by the application of the Fourier-ratio deconvolution[11]. The resulting Pd M-edges for crystalline Pd and glassy $Pd_{60}Ni_{20}P_{20}$ are shown in Fig.1.

The PRDFs were extracted using the WinXAS package[12]. The data processing using WinXAS involved the isolation of the oscillations beyond $M_{3,2}$ (Fig.1), their conversion to oscillations in k-space, correction for termination broadening by the modulation of the k-space oscillation by a Bessel function, and the Fourier transform into radial space.

Results and Discussion

In practice, a PRDF thus obtained may need further correction since a phase shift is possible for the ionized electron between its outgoing and backscattered waves. If one assumes that the phase shift to be a as a linear function of **k**, then the effect upon Fourier transform into

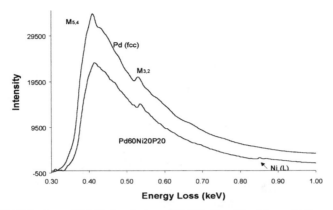

Fig1: EELS M edges of Pd (fcc) and glassy $Pd_{60}Ni_{20}P_{20}$. The $M_{5,4}$ and $M_{3,2}$ are separate ionizations. EXELFS was carried out beyond $M_{3,2}$. $Pd_{60}Ni_{20}P_{20}$ shows a small Ni L-transition.

real space is simply a constant shift in **r**. The abscissa in the RDF of Pd was thus shifted so that the first peak matches a calculated RDF of crystalline Pd and the heights of the first peaks are scaled to be equal (Fig. 2). The comparison between the calculated and phase-shifted experimental RDFs of Pd shows us that the phase shift correction assuming a linear dependence of **k** is accurate in matching the position of the second major peak at ~ 4.5 Å and its pre-peak at ~ 3.9 Å. However, the peak at 6 Å does not line up perfectly. What is more interesting is that heights of the peaks in the range of 4 Å and 6 Å are underdetermined in the experimental RDF. This may be explained by the fact that the probing electrons were sampling along a specific crystallographic direction within a single crystal grain, and therefore had "seen" more of certain radial distances and less of others. This beam orientation dependence of EXELFS in crystals has been observed by others[13]. This should not be a problem in the random structure of glass.

The same rigid phase shift was applied to the PRDF of $Pd_{60}Ni_{20}P_{20}$ and it was scaled vertically so that the height of the first peak matched that of the calculated one (Fig.3). The position of the first peak in the two cases match quite well, as should be expected since they both correspond to the Pd-Pd bond distance. The distribution of distances around this first maximum in glassy $Pd_{60}Ni_{20}P_{20}$ is wider than in Pd (fcc). This is also expected since there are Pd-Ni and Pd-P bonds within the first coordination shell of the glass, and each of these bonds may have a distribution of distances from the inherent disorder of the glass.

The second peak in the Pd (fcc) PRDF is at ~ 3.9 Å and corresponds to $\sqrt{2}*d$, where d=2.75 Å (the first peak position). This is, of course, the diagonal distance in a square plane, and characteristic of an octahedron. The following two peak positions correspond to $\sqrt{3}*d$ and $2d$ respectively.

17

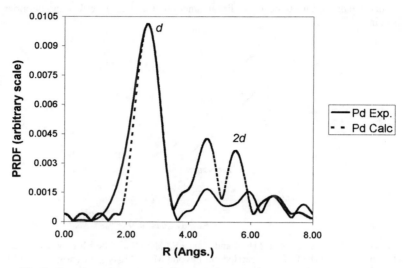

Fig. 2: Experimental and calculated RDFs of crystalline Pd. The distance d corresponds to the Pd-Pd bond distance.

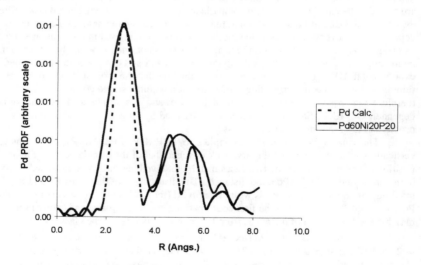

Fig 3: Pd PRDF in $Pd_{60}Ni_{20}P_{20}$ glass compared to the calculated RDF of Pd (fcc).

In the region between d and $2d$ we have an interesting correlation between the glass and Pd (fcc). The average height of broad peak between 4 Å and 6 Å in this case is close to that from the Pd (fcc) structure. Moreover, the shape of this broad peak is such that the ratio of height at the maximum (~5 Å) to that at the shoulder (~6 Å) is the same as that of the two peaks in Pd (fcc) in this region. The full-width-at-half-max (FWHM) of this broad "split" peak is slightly broader as the combined FWHM of the $\sqrt{3}*d$ and $2d$ peaks in Pd. However, in the glass, between $\sqrt{3}*d$ and $2d$ there is in the glass a broad distribution of bonds instead of the two well-defined peaks of Pd. This tells us that the atomic environment around Pd in the second coordination shell is similar as that of Pd (fcc) but with many types of bond distances. It can be suggested that it is in through such a distribution of distances that this system is able to incorporate enough "confusion" within its atomic structure and thus form a glass.

In their work Egami *et al.*[6] showed that the total RDF for $Pd_{40}Ni_{40}P_{20}$ and $Pd_{30}Ni_{50}P_{20}$ are almost identical. It is well known that the x=40 composition in the $Pd_xNi_{(80-x)}P_{20}$ family is the best glass-former. It was thus concluded in their study that the chemical substitution of Ni with Pd has minimum effect on the total atomic structure which resembles, in any case, the DRP structure commonly found in other TM-M glasses. In the same study, however, the Pd PRDFs were presented for the two compositions and, in fact, the shape of well-known "split" second peak (characteristic of the DRP structure) in this case were different. It was found that the split in peak between $\sqrt{3}*d$ and $2d$ for Pd PRDF in the $Pd_{30}Ni_{50}P_{20}$ glass is sharper than in the $Pd_{40}Ni_{40}P_{20}$ glass. It is worth investigating whether or not this splitting is correlated to glass-formability.

If the local atomic structure in these glasses have an important bearing on the glass-formability of these alloys, then the PRDFs will indicate this. For this it would be necessary to look not only at the Pd PRDFs over a composition range in this family, but also the Ni PRDFs. EXELFS of Ni is, however, still problematic[9]. We therefore have plans to carry out Ni EXAFS.

Conclusions

It is concluded from the EXELFS study of $Pd_{60}Ni_{20}P_{20}$ that the structure around Pd in this BMG is very similar to that of Pd (fcc) with slightly broader distribution of distances in the first and second coordination shells. The broader distributions can be understood from the fact that in the glass Pd-Ni and Pd-P bond distances need to be incorporated into the structure. The structure in the second coordination shell around Pd in this BMG family has been observed by others to change as Pd is substituted with Ni. The effect of structure in the second coordination shell has not been examined yet for this BMG family. EXELFS of Pd has been found, from this study to be a viable technique for this investigation.

Acknowledgement

This work was made possible support from the Basic Energy Sciences Division of the US Department of Energy (Grant no. DE-FG02-95ER45540).

References:

[1] A. Inoue, T. Zhang, T. Masumoto, Mater. Trans. JIM **31**, 177 (1990).

[2] A. Peker and W.L. Johnson, Appl. Phys. Lett. **63**, 2342 (1993).

[3] W.L. Johnson, Mater. Sci. Forum **225-227**, 35 (1996).

[4] A.L. Greer, Nature **366**, 303 (1993).

[5] R.B. Schwarz and Y. He, Proc. Int Symp Metastable, Mechanically Alloyed and Nanocrystalline Materials, LANL Publication No.LA-UR-96-1703, Los Alamos National Laboratory, Los Alamos, MN, 1996.

[6] T. Egami, W. Dmowski, Y. He, R.B. Schwarz, Met. and Mat. Trans. **29A**, 1805 (1998).

[7] J.D. Bernal, Nature **185**, 68 (1960)

[8] G.S. Cargill, J. Appl. Phys. **41**, 2248 (1970).

[9] Y. Ito, F.M. Alamgir, H. Jain, D.B. Williams, submitted for pulication in these proceedings, MRS Fall Meeting, Boston, MA, 1998.

[10] Y. He, R.B. Schwarz, J.L. Archuleta, Appl. Phys. Lett. **69**, 1861 (1996).

[11] R.F. Egerton, (1996) *Electron energy-loss spectroscopy in the electron microscope*, 2nd edn. Plenum Press, New York.

[12] T. Ressler, J. Physique IV **7**, C2 (1997)

[13] M.M. Disko, in *Analytical Electron Microscopy*, edited by R.H. Geiss (San Francisco Press, San Francisco, 1981) p. 214.

VOLUME EFFECTS IN BULK METALLIC GLASS FORMATION

A. R. YAVARI* and A. INOUE**

*LTPCM-CNRS, BP 75, Institut National Polytechnique de Grenoble, St-Martin-d'Herès Campus, 38402 France, yavari@ ltpcm.inpg.fr
**Institut for Materials Research, Tohoku University, Sendai, Japan

ABSTRACT

Atomic volume effects in amorphous alloys obtained by rapid solidification and in bulk glasses are briefly reviewed. It is recalled that at high undercoolings, the release at the growth fronts, of the volume corresponding to the reduction of the molar volume upon solidification can sharply accelerate crystal growth in the melt through viscosity reduction. A method is then proposed for estimating the volume of mixing ΔV_{mix} which is negative for elemental additions to glass-forming liquid alloy. It is argued that negative ΔV_{mix} reduces atomic mobility in easy glass-forming alloys thus allowing the suppression during cooling, of nucleation and growth of crystallites. The ZrCuNi system is used as an example for applying this reasoning. It is shown that Al or Ti addition to ZrNiCu alloys lead to strongly negative ΔV_{mix} and expected sharp drops in diffusion-controlled crystal growth kinetics in the melt.

INTRODUCTION

Atomic volume effects are important both in the theoretical understanding of amorphisation and in the experimental control of preparation and properties. In the early days of amorphisation by melt-spinning, correlations were found between glass-formability and the extent of atomic size mismatch of the constituent atoms. A proportionality between the minimum B-atom content $X_B(min)$ needed to amorphise an A matrix and the inverse of the reduced atomic volume difference $(V_A-V_B)/V_A$ was explained in terms of atomic-level stresses generated by the size differences [1,2]. It can be argued that in supersaturated solid solutions of B in A with $V_A > V_B$, beyond a certain level of compressive internal stresses, a topological instability leading to a reduced coordination number Z for example from 12 to 11 can significantly relax such stresses.

In other words, in systems with attractive interatomic A-B interactions signalled by a strongly negative solid solution heat of mixing ΔH_{mix}, but where B has little solubility within the A matrix, if the solubility existed, we would have $|\Delta H_{mix}| \gg |\Delta H_{cryst}|$ the latter being due to long-range topological order (LRO). Such solid solubility is available in the amorphous liquid-like structure with no LRO. In addition, a strong chemical short-range order within the solid solution [3] yields ΔH_{CSRO}(solid solution) < ΔH_{mix}(regular solution) and futher contributes to the energetic advantage of high CSRO compared to long-range order (crystal lattice stability).

Thus, in glass formation from liquid alloys with $V_A > V_B$ of compositions away from intermetallics such as in hypoeutectic Fe and Al-based glasses, it can be argued that with insufficient times for solute repartitioning by atomic diffusion during rapid solidification, the alloy prefers an amorphous state rather than a highly supersaturated crystalline solution.

Mat. Res. Soc. Symp. Proc. Vol. 554 © 1999 Materials Research Society

Glass formation by rapid solidification also occurs for intermetallic compositions such as Zr_2Ni, Ti_2Ni, Fe_3B and many others. These low-symmetry crystalline structures provide a local chemical short-range order very similar to that of their glasses [4]. Thus, since rapid quenching can lead to high density of quenched-in defects, it can lead to amorphisation of these intermetallic compositions.

Atomic size in the alloy is however different from that of the pure components because of charge transfer phenomena in presence of strong interactions ($\Delta H_{mix} \ll 0$). Turnbull [5] suggested that easy glass-formation (EGF) is expected in A-B alloy systems in which the repulsive branch of A-B interatomic interaction potential is softer than that of the A-A potential. Such a situation in the absence of a periodic lattice would allow A-B interatomic distances to adjust to local topology thus leading to a densification of the liquid phase. From an electronic point of view, this situation would allow better energy minimisation from interaction between conduction electrons and positive ions and the spreading of the outer electrons of B in A-atom core regions.

Easy glass-formation (EGF) has most often been obtained in liquid compositions corresponding to deep eutectics. The reduced liquidus temperature T_l of near-eutectic compositions facilitates EGF by reducing the temperature range down to the glass transition T_g that is, (T_l-T_g), through which the liquid must be cooled rapidly.
The first metallic glasses to be obtained from the liquid state in mm-thickness range were near eutectic Pd-Cu-Si and Pd-Ni-P as reported by the team of D. Turnbull at Harvard in the early 1980s [6,7]. Since 1988, A. Inoue and co-workers at Tohoku have succeeded in developing several families of metalloid-free (and metalloid-containing) bulk metallic glasses including Ln-Al-TM, Mg-Ln-(Ni,Cu), Zr-Al-TM, Pd-Ni-Cu-P, (Nd, Pr)-Fe-Al, Fe-Al-Ga-P-B-C-Si-Ge and Co-Al-Ga-P-B-Si [8,15]. In the 1990s, the team of W.L. Johnson at Cal. Tech. reported bulk glasses several mm thick in the Ti-Zr-Ni-Cu system [16]. They further succeeded in obtaining up to several cm thick glassy ingots in the Zr-Ti-TM-Be alloys at cooling rates less than 10 K/s [17]. All these systems correspond to compositions near deep-eutectic surfaces

Turnbull was first to rationalize the role of deep eutectics in terms of a reduced free-volume v_f in the liquid state. In what follows, we will show how volume and free-volume effects are major keys in understanding bulk glass formation phenomena in metallic systems.

ROLE OF THE VOLUME CHANGE ON CRYSTALLISATION $\Delta V_{L \to C}$ IN BULK GLASS FORMATION

In order to obtain a glass by cooling a liquid, nucleation and/or growth of crystalline phases must be suppressed. The difficulties of nucleating crystallites in a multicomponent liquid has recently been discussed by Desre [18]. When the nucleating crystalline phase is of a composition different from that of the melt, atomic diffusion is needed both for forming nuclei and for their growth and diffusion-controlled growth scales with $(D_a \cdot t)^{1/2}$ for atomic diffusivity D_a. Atomic diffusion in liquids is usually derived from the viscosity η using the Stokes-Einstein relation:

$$D_a \sim kT/\eta a_0 \sim v\ a_0^2\ exp(-\delta v_a/v_f)\ exp(-\Delta G_m/RT) \qquad (1)$$

where a_0 is the mean free path for diffusive jumps which is nearly equal to the atomic diameter. The second equality of eq.(1) is obtained by expressing the viscosity η in terms of the free-volume content v_f which is the liquid's equivalent of vacancies in crystals [19]. In eq.(1), v is the Debye frequency, δ is a constant near 1 and ΔG_m is the molar free energy of migration. The expression becomes identical to one for atomic diffusion by the vacancy mechanism in crystalline lattices if the probability $\exp(-\delta v_a/v_f)$ of finding a hole or free volume [20] of size $v_a = (1/6)\pi a_0^3$ on a neighboring site is replaced by the probability of finding a vacancy $\exp(-\Delta G_v)$ with free energy of formation ΔG_v. The higher the free-volume content, the faster is atomic-mobility.

We now wish to consider the effect on the free volume content V_f and atomic diffusion, of the volume change (contraction) ΔV occuring during crystallisation of the undercooled melt. Upon solidification, the volume (of metals) usually decreases by $\Delta V_{L->C}$ which is some -0.04 to -0.07·v_a. Available data indicate that the volume change -$\Delta V_{L->C}$ accompanying the melting of alloys showing EGF is much less than values of the order of 5% for pure metals [21,22]. For the early stage of crystallisation by homogeneous nucleation at frequency $I(T)$ $s^{-1}cm^{-3}$ and growth $U(T)$ cm/s, the fraction transformed X in the undercooled melt after time t at temperature T can be written as $X \sim IU^3t^4$ with the growth rate $U(T)$ given by

$$U(T) = [f(T)\, D(T)/a_0]\,[1-\exp(-\Delta H_{cryst}\,\Delta T_r\,/\,RT) \qquad (2)$$

Where ΔH_{cryst} is the enthalpy of crystallisation, $\Delta T_r = (T_f-T)/T_f$ is the reduced undercooling, $f(T)$ which is the fraction of active sites at the interface is about $0.2\,\Delta T_r$, a_0 is the interatomic distance and D is the atomic diffusion coefficient controlling growth [23]. Consider now a growth process for crystallites with higher atomic density than the melt and the release the volume change $\Delta V_{L->C}$ on crystallisation at the growth interfaces in a highly viscous undercooled liquid. This volume rejected at the growth fronts will ultimately be evacuated to the free surfaces but in so doing will accelate atomic diffusion by serving as increased free volume in eq.(1). The free volume near the growth interface can then be schematically drawn as in figure 1.

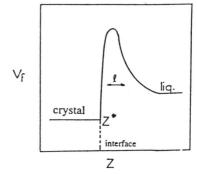

Figure 1: Schematic representation of accumulation of free-volume v_f in the undercooled melt ahead of a crystallisation front due to liberation of volume change on crystallisation ΔV_{C-L} [21].

Using the atomic diffusivity D_a from the free-volume formulation of eq.(1) in eq.(2), it can be shown that in going from a $\Delta V_{L->C}$ near zero to for example $\Delta V_{L->C} \sim$ -0.03 v_a, diffusivities and growth rates increase by several orders of magnitude usually making it impossible to obtain amorphisation via rapid solidification [4]. The densities of several bulk metallic glasses including several mm thick $Pd_{40}Cu_{30}Ni_{10}P_{20}$ and $Zr_{55}Ni_5Cu_{30}Al_{10}$ have recently been measured using helium pychnometry [24,25]. Table I from [25] shows values of density and volume per atom in the amorphous state and after crystallization (1 hour annealing 100 K above T_X) as well as the density change in going from amorphous to crystalline phase ($\Delta\rho_{A\rightarrow C}$ /ρ_A) = ($\Delta V_{A\rightarrow C}$ /V_C).

	as-cast amorphous (ρ_a) [Mg/m³]	relaxed amorphous (ρ_{ra}) [Mg/m³]	crystalline (ρ_c) [Mg/m³]	$\Delta\rho\left(-\dfrac{\rho_c-\rho_a}{\rho_a}\right)$ %
$Zr_{60}Al_{10}Cu_{30}$	6.72	—	6.74	0.30
$Zr_{60}Al_{15}Ni_{25}$	6.36	—	6.38	0.31
$Zr_{55}Al_{10}Cu_{30}Ni_{5}$	6.82	6.83	6.85	0.44
$Pd_{40}Cu_{30}Ni_{10}P_{20}$	9.27	9.28	9.31	0.54
$Zr_{55}Ti_{5}Al_{10}Cu_{20}Ni_{10}$	6.62	—	6.64	0.30
$Zr_{52.5}Ti_{5}Al_{12.5}Cu_{20}Ni_{10}$	6.52	—	6.55	0.45

Table I : Measured density changes accompanying crystallisation of bulk glass formers

It can be seen that the volume change upon crystallisation is very small. The value for $Pd_{40}Cu_{30}Ni_{10}P_{20}$, which currently holds the record for glassy thickness is the lowest reported in the literature and these low values are consistent with the theoretical expectation of increasing ease of glass formation with decreasing -$\Delta V_{L\rightarrow C}$ or -$\Delta V_{A\rightarrow C}$.

These values ($\Delta V_{A\rightarrow C}$ /$V_C \leq 0.5\%$) for bulk amorphous ingots of Table I can be compared for example, to $\Delta V_{A\rightarrow C}$ /$V_C \sim 2\%$ or more reported for Al-rich amorphous ribbons [26].

Figure 2 : Experimentally measured crystalline and amorphous mass densities versus Ti and Al content in ZrTiAlCuNi bulk glass forming alloys [25]

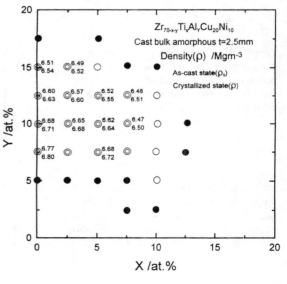

24

Some of the best bulk glass-forming alloy compositions to date have been obtained in the quinary ZrTiCuNiAl system. A study has therefore been performed on the density change $\rho_C - \rho_A$ going from the amorphous to crystalline phase $(\Delta\rho_{A\to C}/\rho_A) = (\Delta V_{A\to C}/V_C)$ in order to correlate bulk glass formability with the smallness of $\Delta V_{A\to C}$. Figure 2 shows the measured density values for amorphous ingots before and after crystallisation and figure 3 the relative density change $(\Delta\rho_{A\to C}/\rho_A)$.

Figure 3 : The density change $(\Delta\rho_{A\to C}/\rho_A) = (\Delta V_{A\to C}/V_C)$ in %, measured after crystallisation of bulk glass forming alloys of figure 2 versus Ti and Al content.

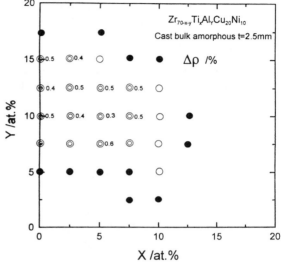

It can be seen that the smallest $\Delta\rho = 0.02$ g/cm^3 or $\Delta\rho/\rho \sim \Delta V/V = 0.3$ % is obtained for the composition $Zr_{55}Ti_5Al_{10}Cu_{20}Ni_{10}$ of Table I which is very close to optimum compositions as reported by one of the authors and by others [27,28]. In addition to their general correlation, this close association of bulk glass-formability and the smallness of the volume change $\Delta V/V$ confirms that the latter is a sensitive thermodynamic and kinetic factor in determining critical cooling rates with predictive value.

ROLE OF THE MIXING VOLUME ΔV_{MIX} IN BULK GLASS FORMATION

The volume change upon mixing or alloying of pure constituents ΔV_{mix} is usually strongly negative for attractive heteroatomic interactions ($\Delta H_{mix} \ll 0$) in good glass-forming alloys with ΔV_{mix} values of the order 5 to 20% v_a. Ramachandrarao [29] previously developped a correlation between EGF and large negative mixing volume ΔV_{mix} in the liquid based on a rough calculation of miss-match energy due to atomic size differences somewhat like the approach used by Egami and Waseda for the solid state. While it is intuitively acceptable to think that a more compact structure (strongly negative ΔV_{mix}) will have a higher viscosity and a lower diffusivity (see eq.(1)), ΔV_{mix} cannot be directly related to the hole or free volume content v_f as it corresponds to a variation in the total volume. However, it is well known that the elastic moduli (for example the bulk modulus B) of all materials depend on the interatomic distances and especially the repulsive branch of the interatomic potentials [30].

It can also be said that strongly negative ΔV_{mix} is due to charge transfer or better spreading of the outer electrons thus reducing core-core distances. The higher elastic moduli of intermetallics with highly negative ΔH_{mix} and ΔV_{mix} are consistent with this vision. Furthermore it can be said from literature on migration energies for diffusive jumps that ΔG_m in eq.(1) depends on the elastic moduli. More particularly, in order to reach the saddle point of a diffusive jump, jumping atoms must open up an activation volume v^* by straining their neighbors by $\varepsilon = v_a/Zv_a$.where Z is the number of atoms over which the strain to create the activation volume $v^* \sim v_a$ is relaxed. and following Shewmon [31],

$$\Delta G_m \propto \int B \, \varepsilon \, d\varepsilon = B/2Z^2 \tag{3}$$

It can therefore be safely said that a strongly negative ΔV_{mix} leads to higher ΔG_m values, lower atomic diffusivities D_a and lower crystal growth rates in the melt. However ΔV_m values for liquid alloys are rarely available.

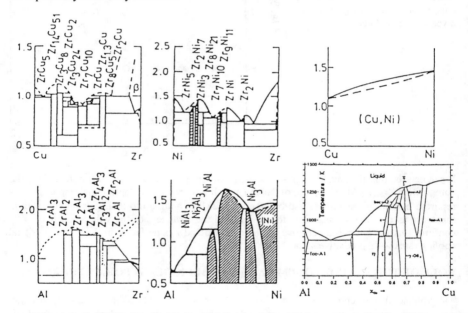

Figure 4: Zr-Ni, Zr-Cu, Cu-Ni, Zr-Al, Ni-Al and Cu-Al binary phase diagrams [32].

In the previous section we found that the volume change that accompanies crystallisation - ΔV_{C-L} of the undercooled liquid of a bulk glass-former is of the order of 0.01 or less of the atomic volume v_a. Since $-\Delta V_{mix}$ values of interest are expected to be an order of magnitude or more superior to $-\Delta V_{C-L}$, we can write:

$$\Delta V_{mix} \text{ (glass)} \sim \Delta V_{mix} \text{ (crystalline alloy)} \pm 0.01 \, v_a \tag{4}$$

and ΔV_{mix} in the crystalline state can be used to approximate that of the liquid which are generally not available. Volume changes, which through atomic mobility determine the kinetics of both nucleation and growth of crystalline phases, are consequences of the form of A-B interatomic interaction potentials and redistribution of electron densities. We can use them to reach a better understanding of bulk glass formation.

Consider first the well known bulk glass forming Zr-Ni-Cu-Al with bulk glasses forming near the composition $Zr_{60}Ni_{10}Cu_{20}Al_{10}$. Figure 4 shows the corresponding phase diagrams [32].

Zr-Cu and Zr-Ni systems are characterized by successive deep eutectics and in fact are among the best known metal-metal binary EGF alloys.

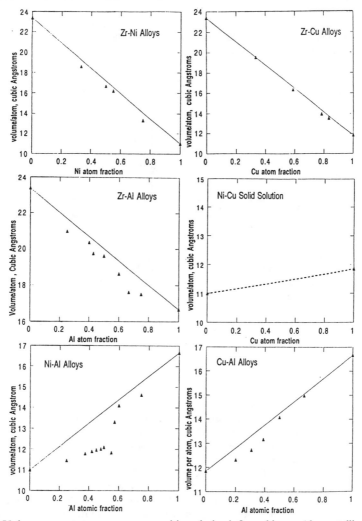

Figure 5: Volume per atom versus composition derived from binary Al crystalline phase densities. The solid line is Vegard's law with zero ΔV_{mix}

They fulfill the deep eutectic, the large difference of atomic radii and the negative heat of mixing criteria for glass formation

On the other hand, the Al binary phase diagrams are more characterized by high temperature melting intermetallics and do not correspond to easy glass formers. Why then does Al addition enhance the glass-forming ability of ZrNiCu alloys and reduce critical cooling rates in such a way as to allow maximum thicknesses to go from a few hundred μm to a cm or more ?

Figure 5 shows atomic volumes derived from available room temperature density data of intermetallic phases (and solid solutions) that appear in the Zr-Ni, Zr-Cu and Cu-Ni, Zr-Al, Ni-Al and Cu-Al phase diagrams.

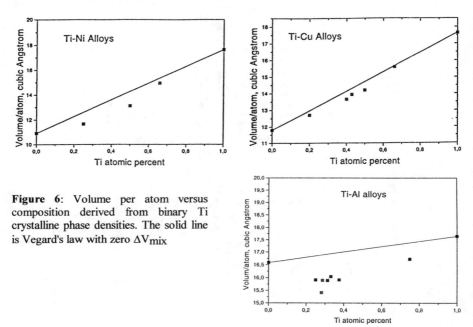

Figure 6: Volume per atom versus composition derived from binary Ti crystalline phase densities. The solid line is Vegard's law with zero ΔV_{mix}

ΔV_{mix} for these alloys can be read directly from the difference of these atomic volumes v_a with the Vegard law values (solid-lines) at the same composition. It can be seen that for Zr-Ni and Zr-Cu, $\Delta V_{mix}/v_a$ is small, with highest values of the order of a few percent. On the other hand, volume per atom in all three Al containing binary systems (Zr-Al, Ni-Al and Cu-Al) deviates strongly from Vegard's law indicating a strongly negative ΔV_{mix}. As per eq.(4), these mixing volumes are very close to those expected in the liquid state because the volume (or density) changes on crystallization ΔV_{C-G} have been measured to be less than 1% of v_a (see Table I). It can therefore be concluded that Al-addition to Zr-Ni-Cu alloys strongly reduces atomic volume and can increase the migration free energy of diffusive jumps as discussed in relation to equation 1 and thus lead to reduced atomic diffusivities, slower growth (and nucleation) of nuclei and reduced critical cooling rates for their suppression thus allowing bulk glass formation. A similar effect of Al addition has been calculated [33] in the RE-TM-Al bulk glass families such as $Nd_{70}Fe_{20}Al_{10}$.

Consider now the effect of Ti addition as in $Ti_{35}Zr_{10}Cu_{45}Ni_{10}$ [17]. The results of [17] indicate that addition of Ti to ZrCuNi alloys allows glassy thicknesses of at least 4 mm. Figure 6 shows that with Ti addition, a strong ΔV_{mix} develops just as for Al-Ni interactions. Also shown in figure 6 is the strongly negative ΔV_{mix} in Ti-Al alloys which helps explain the additional contribution of Ti-addition to bulk glass forming tendency in Al-containing bulk glass-formers such as the ZrTiAlNiCu alloys of figures 2 and 3.

In conclusion, it is found that a small value of volume change $\Delta V_{L->C}$ on solidification is essential for easy glass formation from a kinetic point of view. It also corresponds to denser packing of the liquid structure which is a particularity of deep eutectics. Experimentally measured volume changes of bulk metallic glasses during crystallisation have been measured and found to be very small and about 0.3%of the molar volume in the best known bulk glass forming alloys. From this finding, it has been possible to obtain values for the volume of mixing in binary liquid alloys ΔV_{mix} with satisfactory precision. ΔV_{mix} is then shown to be strongly negative for Al and Ti additions into ZrNiCu type bulk glass formers. It has been argued that through strongly negative ΔV_{mix} values, Ti and Al addition increase the bulk moduli and consequently the migration free energies ΔG_m for atomic diffusion, and the nucleation and growth of crystalline phases and thus strongly enhance bulk glass formability.

Acknowledgement: This work was supported by a NEDO award entitled "Nanocrystalline and Supercooled Liquid States".

References

1. T. Egami and Y. Waseda, J. Non-Cryst. Solids, **64**, p. 113 (1984)
2. D.L. Beke, H. Bakker and P.I.Loeff, Acta Metall. Matter.**39**, p. 1267 (1991)
3. A.R.Yavari, S. Gialanella, M.D. Baro and G. Le Caer, Phys. Rev. Lett.**78**, p. 4954 (1997)
4. A.R. Yavari *Amorphous Metals and Non-Equilibrium Processing*, Editions de Physique 1984, pp.31-44
5. D. Turnbull, J. Physique **35**, p. C4-1 (1974)
6. H.S. Chen and D. Turnbull, Acta Metall.**17**, 1021 (1969); Ibid **22**, p. 897 (1974)
7. H.W. Kui, A.L. Greer and D. Turnbull, Appl. Phys. Lett.**45**, p. 615 (1984)
8. A. Inoue, Proc. ISMANAM-94, ed: A.R. Yavari, Mater. Sci. Forum **179-181**, p. 691(1995)
9. A. Inoue and J.S. Gook, Mater. Trans. JIM **37**, p. 32 (1996)
10.A. Inoue Y. Shinohara and J.S. Gook, Mater. Trans. JIM **36**, p. 1427 (1995)
11.A. Inoue Sci. Rep. RITU **A42**, p. 1 (1996)
12.A. Inoue , Mater. Trans.JIM **36**, p. 866 (1995)
13.N. Nishiyama and A. Inoue, Mater. Trans.JIM **37**, p. 1531 (1996)
14.A. Inoue, T. Zhang, W. Zhang, A. Takeuchi, Mater. Trans.JIM **37**, p. 99 (1996)
15.A. Inoue, T. Zhang and A. Takeuchi, Mater. Trans.JIM **37**, p. 1731 (1996)
16.X.H. Lin and W.L. Johnson J. Appl. Phys.**78**, p. 6514 (1995)
17.W.L. Johnson, Proc. ISMANAM-95 ed: R. Schulz, Mater. Sci. Forum **225-227**, p. 35 (1996)
18.P.J. Desre, Mater. Trans. JIM **38**, p. 583 (1997)

19.F. Spaepen, Acta Metall.**25**, p. 407 (1977)

20.M.H. Cohen and D. Turnbull, J. Chem. Phys. **31**, p. 1164 (1959); Ibid **34**, p. 120 (1961)

21.A.R. Yavari, P. Hicter and P. Desre, J. Chimie Physique **79**, p. 579 (1982)

22.D. Turnbull, Scripta Metall.**11**, p.1131 (1977); Ibid **15**, p. 1039 (1981)

23.D.R. Uhlmann, J. Non-Cryst. Solids **7**, p.172 (1972)

24.A.R. Yavari, J.L. Uriarte and A. Inoue,Mater. Sci. Forum **269**, p. 533 (1998)

25.A. Inoue, T. Negishi, H.M. Kimura, T. Zhang and A.R. Yavari, Mater. Trans. JIM.**39**, p.318 (1998)

26.G.M. Dougherty, Y. He, G.J. Shiflet and S.J. Poon, Scripta Metal. Mater.**30**, p.101 (1994)

27. A. Inoue, T. Shibata and T. Zhang Mater. Trans. JIM **36**, p.1420 (1995), 28

28.L.Q. Xing and P. Ochin, Mater. Lett. **30**, p. 283 (1997)

29.P. Ramachandrarao, Z. Metallkde **71**, p. 172 (1980)

30.Ch. Kittel, *Introduction to Solid State Physics*, John Wiley & Sons, 1971

31.P.G. Shewmon, *Diffusion in Solids*, McGraw-Hill 1963

32.F.R. de Boer, R. Boom, A.R. Miedema and A.K. Niessen, *Cohesion in Metal Alloys*, North Holland 1989

33.J. L. Uriarte and A. R. Yavari, unpublished results (1997)

EXELFS OF METALLIC GLASSES

Y. ITO*, F.M. ALAMGIR*, H. JAIN*, D.B. WILLIAMS*, R.B. SCHWARZ**

*Department of Materials Science and Engineering, Lehigh University, 5 East Packer Ave. Bethlehem, PA, 18015 USA
**Los Alamos National Laboratory, Los Alamos, NM 87545, USA

ABSTRACT

The feasibility of using extended energy-loss fine structure (EXELFS) obtained from ~1 nm regions of metallic glasses to study their short-range order has been examined. Ionization edges of the metallic glasses in the electron energy-loss spectrum (EELS) have been obtained from PdNiP bulk metallic glass and Ni_2P polycrystalline powder in a transmission electron microscope. The complexity of EXELFS analysis of L- and M-ionization edges of heavy elements (Z>22, i.e. Ni and Pd) is addressed by theoretical caluculations using an *ab initio* computer code, and its results are compared with the experimental data.

INTRODUCTION

Analysis of extended energy-loss fine structure (EXELFS) in an electron energy-loss spectrum (EELS) is an ideal technique for determining the structure of amorphous materials with a high spatial resolution. It has been advanced enough in recent years such that, in some cases, the quality of the results is compatible to its X-ray analog, extended X-ray absorption fine structure (EXAFS) [1]. However, most of EXELFS analysis has been restricted to the K-ionization edge of lighter elements. For heavier elements, more complex ionization edges such as L- and M-edges have to be dealt with. The aim of our study is to obtain information on the short-range structure of bulk metallic glasses. The X-ray based techniques have a poor spatial resolution, which, therefore, cannot directly reveal, for example, the early stages of nucleation and phase separation. In principle, our electron energy-loss spectrometer is capable of acquiring the K-edge data for the light elements ($Z \leq 22$). In practice, it is limited to about $Z \leq 17$ due to the poor signal-to-noise ratio in the higher energy-loss range, beam damage, contamination and instrumental stability. Therefore, the structure around heavy elements can be determined only from their ionization edges in a lower energy-loss range. In this paper, the complexity involved with the EXELFS analysis of L- and M-edges of the heavy elements viz. Ni and Pd is addressed by comparing theoretical $\chi(k)$ functions (oscillatory component of the EXELFS) with various initial states of transition in an atomic shell. The theoretical $\chi(k)$ functions were generated by an *ab initio* EXAFS computer code called FEFF 7 [2]. Then, the combined effect of the various $\chi(k)$ functions was compared with the experimental data. Further discussion of the experimental data is presented elsewhere in these proceedings [3].

BACKGROUND

EXELFS (or EXAFS) is a quantum interference phenomenon. Due to interaction with an incoming high-energy electron (X-ray), a core electron is ejected from an atom in the specimen and is represented by an outgoing spherical wave. Weak oscillations arise from interference between the outgoing spherical wave and the elastically backscattered wave from neighboring atoms. Approximating the ejected-electron wavefunction at the scattering atom by a plane wave and assuming that multiple scattering can be neglected, the theory gives the oscillatory component as [4]

$$\chi(k) = \sum_j \frac{N_j}{r_j^2} \frac{f_j(k)}{k} \exp\left(-2r_j/\lambda_i\right) \exp\left(-2\sigma_j^2 k^2\right) \sin\left[2kr_j + \phi_j(k)\right], \qquad (1)$$

Mat. Res. Soc. Symp. Proc. Vol. 554 © 1999 Materials Research Society

where r_j is the radius of a particular shell of neighboring atoms, N_j is the number of atoms in shell j and N_j/r_j^2 (as a function of r) is partial radial distribution function (RDF). Therefore, $\overline{\chi}(k) = |RDF|$ can be obtained as the modulus of the Fourier transform of (1).

The FEFF 7 code was used for theoretical calculations of the $\chi(k)$ functions for Pd M_1-, M_2-, M_3-, M_4-, and M_5-edges of a Pd crystal (fcc) and Ni L_1-, L_2-, L_3-edges of a Ni_3P crystal (tetragonal).

EXPERIMENTAL

Five specimens were investigated in this study: two $Pd_{80-x}Ni_xP_{20}$ (x=20 and 40) bulk metallic glasses, a polycrystalline Pd metal foil (99.95%, Goodfellow metals) and Ni_2P polycrystalline powder. These specimens were thinned first by mechanical polishing and then to electron transparent thickness by an ion-beam using a Precision Ion Polishing System (PIPS, Gatan Inc.). The thickness of the analyzed area was in the range of 30 - 40 nm.

A Philips EM400 120 keV transmission electron microscope (TEM) with a LaB_6 thermionic emission gun, and a VG HB501 dedicated scanning transmission electron microscope (STEM) with a cold field-emission gun were used in this investigation. Both microscopes were equipped with Gatan 666 Parallel Electron Energy-Loss Spectrometers (PEELS). The former microscope acquired EXELFS with a spatial resolution of ~50 nm and its effective collection semi-angle for EELS was 8 mrad. The latter microscope acquired EXELFS with a spatial resolution of ~1 nm and its effective collection semi-angle for EELS was 18 - 20 mrad.

DATA ANALYSIS

The iterative gain averaging method [5] was used to minimise the gain variations in the photodiode array of the PEELS system. After the removal of the dark-current noise from a raw spectrum, the pre-edge background was subtracted by using a pre-edge curve fit to a power-law model. The contribution of plural scattering was removed by applying the Fourier-ratio deconvolution [4] (a software routine in Gatan, EL/P 3.0).

The WinXAS package [6] was used for data processing. The program performs the following steps:
1. Conversion of the scale of $\chi(E)$ from energy to k-space, $\chi(k)$.
The magnitude, k, of the wavevector of the ejected electron is given by

$$k = 2\pi/\lambda \cong \left[2m_0E_{kin}\right]^{1/2} / \hbar = 5.123\left(E - E^0\right)^{1/2}, \tag{2}$$

where E_{kin} is the kinetic energy of the ejected inner-shell electron and E^0 is the energy corresponding to $E_{kin} = 0$. Here, the maximum of the ionization edge was taken as E^0.
2. Isolation of the oscillatory component of the EXELFS ($\chi(E)$ function) by using n-spline fitting over the k range of interest.
3. Correction for k-dependence of backscattering by multiplying with k^n where n=1, 2 or 3.
4. Truncation of $\chi(k)*k^n$ by multiplying with a window function.
5. Fourier transform of $\chi(k)*k^n$ to give a raw RDF (phase shift uncorrected),
$$\left|FT\left(\chi(k)*k^n\right)\right| = |RDF|.$$
6. Correction for phase shifts to convert $|RDF|$ peak positions into interatomic distances.

RESULTS AND DISCUSSION

Pd M-edges

Fig. 1 shows a typical EEL spectrum of a Pd M-edge of $Pd_{40}Ni_{40}P_{20}$ bulk metallic glass. M_1-, M_2-, M_3-, M_4-, and M_5-edges are located at ~670, 559, 531, 340 and 335 eV, respectively. The maximum of the intensity is located at ~70 eV after the threshould of the M_4-, and M_5-

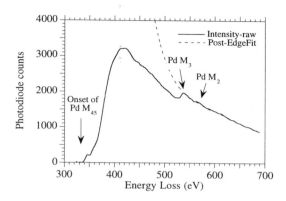

Fig. 1 Typical EEL spectrum of a Pd M-edge of $Pd_{40}Ni_{40}P_{20}$ bulk metallic glass.

edges. The EXELFS oscillations are clearly visible beyond the intensity maximum and extend beyond the M_2- and M_3-edges. Since the M_1-edge step was not recognizable and, beyond the M_1-edge, all the transitions, i.e. M_1-, M_2-, M_3-, M_4- and M_5-edges contribute to the EXELFS, the experimental analysis was carried out on the EXELFS beyond the M_2- and M_3-edges.

To examine the combined effect of M_1-, M_2-, M_3-, M_4-, and M_5-edges on the Fourier transform (FT) of the $(\chi(k)*k)$-function, the M-edges of Pd (fcc) metal were theoretically examined, using the FEFF 7 code. Although the accuracy of the results of M-edges may be less than that of K-edge, useful results are still expected. Fig. 2 shows calculated $\chi(E)$ functions for each single ionization edge, i.e. M_1-, M_2-, M_3-, M_4-, and M_5-edges, multiplied by appropriate weighting factors [7]. The dominant contribution from the M_4-, and M_5 transitions is clearly shown. Therefore, the total contribution of all the $\chi(E)$ functions is very similar to those of M_4-, and M_5-edges but with some minor deviations due to the small contributions from the M_1-, M_2-, M_3-edges as demonstrated in Fig. 3. In general all the curves show 1st, 2nd and 3rd nearst neighbor (nn) peaks at almost the same radial distances. Differences between $FT(\chi(k)*k)$ of the M_5-edge and the sum of all the edges are a slight shift of the 1st nn peak and the appearance of a shoulder peak at ~3 Å in the FT of the latter. The $FT(\chi(k)*k)$ of M_5-edge is almost identical to that obtained from the K-edge (not shown). Thus, the result obtained from the M_5-edge can be used for structural analysis and may be regarded as the true RDF. It should be also noted that, unlike the Ni L-edge (discussed later), the FT of the Pd M-edge is rather insensitive to the choice of the range of k, which adds an extra reliability to the calculation.

The above results may be compared with experimental results from polycrystalline Pd and $Pd_{60}Ni_{20}P_{20}$ bulk metallic glasses in Fig. 4. Generally the theoretical prediction and the experimental results agree well with each other. Also, good agreement with the literature [7] was obtained despite the fact that, ideally, the FT range should be to higher k values to reduce distortion caused by the combined effect of the M_{23} and M_1 EXELFS oscillations. The 2nd nn peak is split, which is a characteristic of random close packing of hard spheres in a metallic structure. Further details of these observations are discussed in another paper [3].

Ni L-edges

As an example, a typical EEL spectrum of an Ni L-edge from Ni_2P polycrystalline powder is shown in Fig. 5. L_1-, L_2- and L_3-edges are located at ~1007, 871 and 853 eV, respectively.

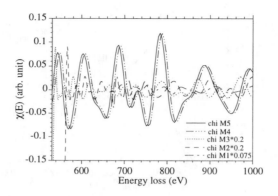

Fig. 2 Calculated $\chi(E)$ functions for M_1-, M_2-, M_3-, M_4-, and M_5-edges of Pd (fcc).

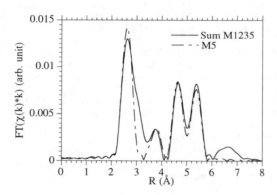

Fig. 3 Fourier transform of $(\chi(k)*k)$ of M_5- and the sum of the M_1-, M_2-, M_3- and M_5-edges of Pd (fcc).

Unlike the Pd M-edge EEL spectrum, the Ni L_2- and L_3-edge peaks are distinctively sharp. They are equivalent to white lines, well known in X-ray absorption spectroscopy. The L_2-edge appears 18 eV above the L_3 peak, due to spin-orbit splitting.

The same theoretical analysis as for the Pd M-edges was applied to the Ni L-edges of Ni_3P, which was chosen because of its the similarity to the $Ni_{80}P_{20}$ glass regarding short-range order [8]. Fig. 6 shows that the analysis of the Ni L-edge EXELFS is sensitive to the choice of the range of k for the FT performed. It is complicaed by the interference between L_2- and L_3-EXELFS. The former effect is observed as the extinction of the first major peak in the L_3 (2, 20) from the other two FTs. The latter effect is responsible for the reduction of the first peak in L_3 (6.4, 17.4). Thus, for Ni L-edges, direct FT of the sum of L-edges above the L_1-edge provides a severely distorted RDF.

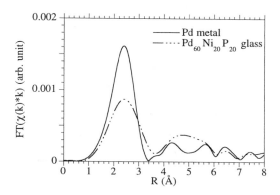

Fig.4 Fourier transform of $(\chi(k)*k)$ obtained experimentally from Pd metal and $Pd_{60}Ni_{20}P_{20}$ bulk metallic glasses (without phase shift correction).

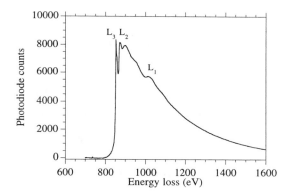

Fig. 5 Ni L-edges of Ni_2P polcrystalline powder

SUMMARY AND FUTURE WORK

The theoretical study shows that, for the Pd M-edge, the FT of the sum of all M-edges is in good agreement with the true partial RDF. For Pd metal, the FT of the experimental M-edge EXELFS agrees with the theoretical EXELFS. However, the analysis of the Ni L-edges is very sensitive to the selection of the range of the FT and affected by the interaction between L_2 and L_3 EXELFS. The FTs of experimental EXELFS agree better with the FT of the $(\chi(k)*k)_{sum}$ than those of a single ionization edge. To improve the analysis of theses edges, a more accurate structural model for the theoretical calculations, optimizing parameters for both theortical and experimental analysis, deconvolution of overlapped edges [9], comparison with near-edge structures and systematic compositional changes will be examined in future.

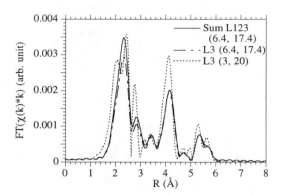

Fig. 6 Theoretical Fourier transform of $(\chi(k)*k)$ of Ni L-edges of Ni_2P. The brackets show the range of k ($Å^{-1}$) for the FT performed on $(\chi(k)*k)$ function. The onset of the L_1-edge is just before k = $6.4Å^{-1}$.

ACKNOWLEDGEMENTS

Mr. D.W. Ackland and Mrs. K.A. Repa are thanked for their technical support. This work is supported by the US Department of Energy (DE-FG02-95ER45540).

REFERENCES

1. M. Qian, M. Sarikaya, E.A. Stern, Ultramicroscopy, **59** 137 (1995).

2. S.I. Zabinsky, J.J. Rehr, A. Ankudinov, R.C. Albers, M. Eller, J.Phys. Rev. B **52** 2995 (1995).

3. F.M. Almagir, Y. Ito, H. Jain, D.B. Williams, R.B. Schwarz, 1998 MRS Fall Meeting, Boston, MA, 1998 (these proceedings).

4. R.F. Egerton, *Electron energy-loss spectroscopy in the electron microscope*. 2nd ed. (Plenum Press, New York, 1996). p. 485.

5. C.B. Boothroyd, K. Sato, K. Yamada, in *Proc. XIIth Int. Cong. Electron Microscopy*. (San Francisco Press, San Francisco, 1990) p80.

6. T. Ressler, J. Physique IV, **7** C2-269 (1997).

7. J.K. Okamoto, C.C. Ahn, B. Fultz, in *Microbeam analysis*, edited by D.G. Howitt. (San Francisco Press, San Francisco, 1991) pp. 273.

8. P.H. Gaskell, J. Non Cryst. Solids, **32** 207 (1979).

9. R.D. Leapman, L.A. Grunes, P.L. Fejes, J. Silcox, in *EXAFS spectroscopy*, edited by B.K. Teo and D.C. joy (Plenum Press, New York, 1981). p217.

SIMULATION OF STRESS DISTRIBUTION AND MECHANICAL RESPONSE OF METALLIC GLASSES

Y. KOGURE, M. DOYAMA
Teikyo University of Science & Technology, Uenohara, Yamanashi 409-01, Japan

ABSTRACT

Molecular dynamics simulation of the metallic glasses has been done. The embedded atom method potential function for copper is used to express the atomic interaction. The stress distribution in the glassy state is evaluated from specific volume occupied by single atom and local density in divided cells. The displacements of individual atom under the shear stress are calculated and the correlation between the displacements and the atomic volumes are investigated.

INTRODUCTION

The materials of glass structure are known to show peculiar properties, which is related with the disordered structure of atoms without long range order. The metallic glasses are suitable system to study the fundamental relation between the atomic configuration and the mechanical property, because the interaction between atoms can be expressed by rather simple potential functions. As a model material for the molecular dynamics simulation the copper is chosen. Although glassy copper is not a stable structure in natural condition, it can be realized in the computer simulation by quenching the high temperature liquid state in 10^{-14} sec. The purpose of the present study is to develop the method for evaluating the internal stress and to investigate the fundamental mechanism of the stress interaction in a simple glass. These results may be applied to real materials in future.

METHOD OF SIMULATION

The EAM potentials developed by present authors [1,2] is used in the simuration. The potential energy for i-th atom is expressed as

$$E_i = F(\rho_i) + \frac{1}{2}\sum_{j\neq i}\phi(r_{ij}), \qquad (1)$$

where, ρ is the electron density function and it is a sum of the density of the neighbor atoms labeled by j.

$$\rho_i = \sum_{j\neq i} f(r_{ij}) \qquad (2)$$

37

$$F(\rho) = D\rho \ln \rho, \quad \rho = \sum_{j \neq i} f(r_{ij}). \tag{3}$$

where, $F(\rho)$ is the embedding energy for i-th atom, ρ is the electron density, and r_{ij} is the distance between i-th and j-th atom. The functions $\phi(r)$ and $f(r)$ are

$$\phi(r) = A(r_c - r)^2 \exp(-c_1 r), \tag{4}$$

$$f(r) = B(r_c - r)^2 \exp(-c_2 r), \tag{5}$$

where, r_c is a cut off distance of the potential. These functions contain five parameters A, B, C_1, C_2, and D. They are determined by fitting the potential to experimental values of physical properties for crystals. The determined values are $A = 7052.194$ eV, $B = 0.02288028$ eV, $c_1 = 11.10854r_0$, $c_2 = 0.4670802r_0$, and $D = 13.35956$ eV, where r_0 is the nearest neighbor distance.

As an initial condition 1372 Cu atoms are arranged in the fcc crystal structure. The shape of the crystal is a cube surrounded by six [100] surfaces. The periodic boundary condition is adopted. The side lengths L of the periodic cube are chosen to be $1.00L_0$, $1.03L_0$, $1.06L_0$ and $1.09L_0$ to make the glassy states of four different densities, where $L_0 = 25.3$Å is the side length for perfect crystal. Here, we define specific density as $\rho_s = (L_0/L)^3$, which correspond to the relative density to the crystal. The specific densities correspond to above four side lengths are 1.00, 0.92, 0.84, and 0.77, respectively. The time interval Δt for the molecular dynamics simulation is chosen to be 1×10^{-15}sec. Initially a particle velocity corresponds to the temperature $T = 7000$K is given to each atoms. After the simulation is started the temperature drop to about 3500 K, because half of the thermal energy is partitioned to the potential energy. The liquid state is maintained for 5000 MD steps. Then the molten state is quenched and relaxed for 10000 MD steps to obtain a metastable glassy state.

RESULTS AND DISCUSSION

Atomic Configuration

Results of the atomic configuration and the radial distribution function (RDF) in the liquid and the glass state for the case of $\rho_s = 1.0$ are shown in Fig. 1. The potential energy of individual atom, E_i, is calculated by Eq.(1), and solid circles in the figure are atoms with higher potential energy ($E_i > -3.3$ eV, value of E_i for an atom in crystal $E_0 = 3.436$ eV). The potential energy is seen to be decreased in glass state. The second peak in RDF split to two subpeaks in glassy state, which is characteristic for glass structure, and it is observed in $Ni_{76}P_{24}$ system, experimentally [3]. The atomic configuration for the cases of $\rho_s = 0.92$ and 0.77 are shown in Fig.2. In the case of very low density $\rho_s = 0.77$, the system becomes porous under the condition of fixed total volume.

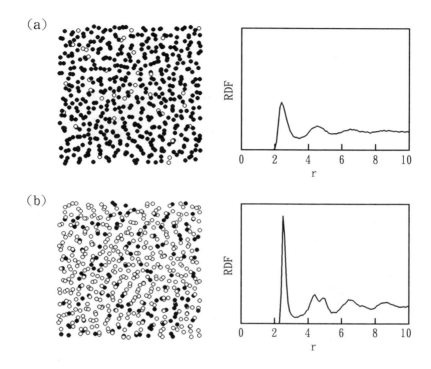

Fig.1 Atomic configuration and the radial distribution function ($\rho_s = 1.00$).
(a) liquid state and (b) glass state.

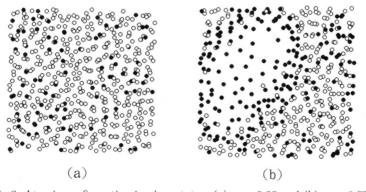

Fig.2 Atomic configuration in glass states. (a) $\rho_s = 0.93$ and (b) $\rho_s = 0.77$.

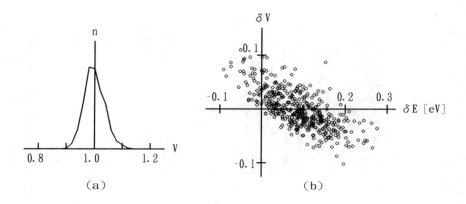

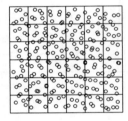

Fig.3 Distribution of atomic volume (a) and Correlation between specific volume
and potential energy.

Fig.4 6 × 6 × 6 cells to calculate local
density.

Distribution of Atomic Volume

The space occupied by an atom can be expressed as a polyhedron, and the volume of
the polyhedron for each atom is calculated numerically, and normalized value by the atomic
volume in crystal is denoted as v_i. The distribution of v_i may be a measure of internal
stress distribution. The distribution of v_i for $\rho_s = 1.0$ is shown in Fig.3 (a). The correlation
between the deviation of the specific volume from the crystal value, $\delta v = v_i - 1$, and the
deviation of the potential energy from the perfect crystal $\delta E = E_i - E_0$ is also shown in
Fig.3 (b). It is seen that δv decreases with δE and the potential energy is larger for atoms
in contracted region.

Local Density

There are no systematic method to evaluate the stress and the strain in the glass structure.
As a first step distribution of local density is evaluated. The model system is divided into
$6 × 6 × 6 = 216$ cells.(Fig. 4) The local density P_j in j-th cell is calculated from the number

40

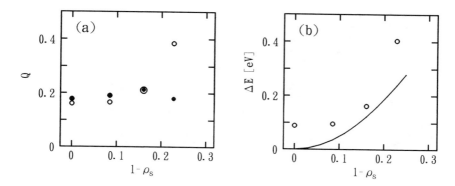

Fig.5 (a) Root mean square deviation of local density. Open circles shows glass state and solid circles show liquid state. (b) Mean potential energy, $< E_i >_{\text{mean}} - E_0$ of single atom in glass. Solid line shows the elastic energy of perfect crystal by uniform dilatation.

of atoms in the cell, which is normalized by the crystal value. The root mean square deviation of the P_j,

$$Q = \sqrt{\sum_j (\frac{P_j - P_0}{P_0})^2 \frac{1}{216}} \qquad (6)$$

is calculated and the results are shown as a function of $1 - \rho_s$ in Fig.5(a), where open circles show glass and solid circles show liquid state values. The density fluctuation Q for liquid is mostly constant. In glasses the value is slightly smaller than liquid for higher specific density samples, but suddenly increases for lowest density sample due to the inhomogeneity (see Fig.2 (b)). The Q value is a measure of shear strain. The deviation of the mean potential energy from the perfect crystal,

$$\Delta E = \sum_{i=1}^{n} \frac{E_i - E_0}{n}, \qquad (7)$$

is shown in Fig.5 (b). The solid curve in the figure is the energy change for one atom of the crystal when the crystal is uniformly expanded, namely, the elastic energy. The difference between ΔE and the elastic energy may be due to the internal strain energy or due to the energy originated from disorder structure.

Response to Shear Stress

To investigate the stress response, 10 % shear strain is applied in the glass state ($\rho_s = 0.92$) sample, and the displacements of atoms are calculated. The method of stress application is described elsewhere [4]. Atoms displaced more than 0.3 Åare shown by solid circle in Fig.6(a). Largely displaced atoms are distributed randomly through the model system. The correlation between the atomic volume and the displacement is shown in Fig.6(b). It is seen that the displacements are larger in the expanded region.

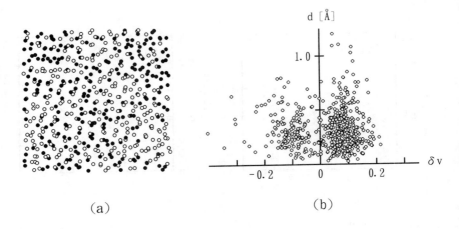

Fig.6 (a) Displacement of atoms under 10 % shear stress. Solid sircles show atoms displaced more than 0.3 Å. (b) Correlation of displacement and the specific volume.

SUMMARY

The glassy state of the metallic element, Cu, is produced at various density in this study. The stress distribution in the glassy state is investigated through the distributions of the specific volume and the potential energy of individual atom. Fluctuation of local density is also examined by calculating the number of atoms in the subcells.

The internal stress in the glass structure is important to understand the mechanical relaxation mechanism in the system, The widely distributed relaxation time is one of the characteristic properties of glass, which is also related with the internal stress. More detailed investigation on the subject is now in progress.

REFERENCES

1. M.S. M. Doyama and Y. Kogure; Radiation Effects and Defects in Solid, **142**, (1997) pp. 107-114.

2. Y. Kogure and M. Doyama; Tight-Binding approach to Computational Materials Science, MRS Symposium Proceedings **491** (1998)pp 359-364.

3. R. Zallen; *The Physics of Amorphous Solids* (John Wiley and Sons, New York, 1983) pp. 73-85.

4. Y. Kogure, H. Masuyama and M. Doyama; Proc. Int. Conf. on Mocrostructure and Functions of Materials (Science University of Tokyo, 1996) pp. 113-116.

CALCULATION OF MECHANICAL, THERMODYNAMIC AND TRANSPORT PROPERTIES OF METALLIC GLASS FORMERS

Tahir Çağın,[a,*] Yoshitaka Kimura,[a] Yue Qi,[a] Hao Li,[a] Hideyuki Ikeda,[a,b]
William L. Johnson[b] and William A. Goddard, III[a,*]

a) Materials and Process Simulation Center, 139-74
California Institute of Technology, Pasadena, CA 91125, U.S.A.

b) Materials Science Department,
California Institute of Technology, Pasadena, CA 91125, U.S.A.

Abstract

Recently, we have parametrized Sutton-Chen type empirical many body force fields for FCC transition metals to study the thermodynamic, mechanical, transport and phase behavior of metals and their alloys. We have utilized these potentials in lattice dynamics calculations and molecular dynamics simulations to describe the structure, thermodynamic, mechanical and transport properties of pure metals and binary alloys in solid, liquid and glass phases. Here, we will describe these applications: mechanical properties of binary alloys $(Pt - Rh)$ and viscosity of a binary alloy, $(Au - Cu)$, as a function of composition, temperature, and shear rate, crystal-liquid, liquid-crystal phase transformation in $(Ni - Cu)$, liquid to glass transformation in a model glass former, $(Ag - Cu)$.

1 Introduction

The development of advanced high performance materials in industrial world is increasingly coupled with the theoretical and computational modeling. In this process focus is on research areas having direct impact on innovative development of such materials. The high performance metalic alloys find use in various segments of materials and chemical industry as catalysts, low weight and high strength structural materials. The theory and computational efforts require and strive for i) *a priori* determination of the ultimate properties of metals, metallic alloys, ii) simulation and modeling of the processing conditions, iii) investigating the performance characteristics of these metals and alloys. All these are extremely important for timely, cost efficient and environmentally compliant development of such advanced materials. With the advances in computational speed and the emerging new computational algorithms, the theory and computer simulations are positioned in the midst of this innovative process.

Understanding the kinetics and thermodynamics of supercooled bulk metallic glass forming liquids is of critical importance in developing light weight high-performance amorphous metallic glasses [1]. Especially, determination of viscosity as a function of temperature and concentration and microscopic level studies on the kinetics of crystallization and glass formation are amenable through computer simulations. Here, we applied these new FF parameters in MD simulation of metals and alloys to

1. determine the mechanical properties (eg. elastic constants of Pt-Rh alloy as a function of concetration and temperature)

2. determine the shear viscosity of Au:Cu binary alloys as a function of temperature and concentration from nonequilibrium molecular dynamics (NEMD).

3. study the role of atomic size in crystalization and glass formation processes in metallic alloys from equilibrium molecular dynamics (EMD).

In the next section we present the parameters used in metal and alloy simulations. Subsequently, we will describe the three applications.

Table 1: Parameters sets for the Quantum Sutton-Chen many-body potential for fcc transition metals.

	n	m	$D(eV)$	c	$\alpha(\mathring{A})$
Ni	10	5	7.3767E-3	84.745	3.5157
Cu	10	5	5.7921E-3	84.843	3.6030
Rh	13	5	2.4612E-3	305.499	3.7984
Pd	12	6	3.2864E-3	148.205	3.8813
Ag	11	6	3.9450E-3	96.524	4.0691
Ir	13	6	3.7674E-3	224.815	3.8344
Pt	11	7	9.7894E-3	71.336	3.9163
Au	11	8	7.8052E-3	53.581	4.0651

2 Parameters for FCC transition metals

Computer simulations on various model systems usually use simple pair potentials. On many occasions to account for the directionality of bonding three body interactions were also employed. But, the interactions in metals and metal alloys can not be represented by simple pairwise interactions. In these systems the electron density plays a dominant role in interactions and resulting physical properties. Therefore interactions in metals and metal alloys are dominated by the many-body interactions. In simple sp-bonded metals this effect may be represented by the interaction potentials derived from model pseudopotentials using the second order perturbation theory [2]. However,for d-band metal and metal alloys the model pseudopotential approach gives way to newer techniques evolved over the past ten years to account for the many body effects. Among these approaches we can list the empirical many body potentials based on Norskov's Effective Medium Theory [3], Daw and Baskes' Embedded Atom Method [4], Finnis and Sinclair's [5] empirical many body potentials, and more recently the many body potentials developed by Sutton and co-workers [6,7] within the context of tight binding approach [8].

We adopted the Sutton-Chen form, where the total potential energy of the metal is given by Eq. (1)

$$U_{tot} = \sum_i U_i = \sum_i \epsilon \left[\sum_{j \neq i} \frac{1}{2} V(r_{ij}) - c\rho_i^{1/2} \right] \tag{1}$$

Here $V(r_{ij})$ is a pair potential defined by Eq. (2);

$$V(r_{ij}) = \left(\frac{a}{r_{ij}} \right)^n \tag{2}$$

accounting for the repulsion (Pauli orthogonality) between the i and j atomic cores and ρ_i is a local density accounting for cohesion associated with atom i defined by Eq. (3);

$$\rho_i = \sum_{j \neq i} \phi(r_{ij}) = \sum_{j \neq i} \left(\frac{a}{r_{ij}} \right)^m \tag{3}$$

In Eqs. (1)-(3), r_{ij} is the distance between atoms i and j, a is a length parameter scaling all spacings (leading to dimensionless V and ρ), c is a dimensionless parameter scaling the attractive terms, ϵ sets the overall energy scale, and n, m are integer parameters such that $n > m$. [The interaction cut-off range in (1) and (3) is chosen to be twice the lattice parameter of the fundamental cubic unit cell.] Given the exponents (n,m), c is determined by the equilibrium lattice parameter and ϵ is determined by the total cohesive energy (E_{coh}). The parameters for transition metals, Ni, Cu,Rh, Pd, Ag, Ir, Au, Pt are given in Table 1.

To simulate alloys, we use the following combination rules to represent the interaction between A-B pairs

$$D_{ij} = \sqrt{D_i D_j} \quad m_{ij} = \frac{(m_i + m_j)}{2} \quad n_{ij} = \frac{(n_i + n_j)}{2} \quad \alpha_{ij} = \frac{(\alpha_i + \alpha_j)}{2} \tag{4}$$

3 Applications

3.1 Elastic constants of Pt-Rh alloy at elevated temperatures

We used MD simulations to calculate the elastic properties of metals and alloys. As an example here we present the results for Pt-Rh alloy simulated at various temperatures by varying the concentration.

In our computations at each concentration and at each temperature, first the zero strain state, h_o, of the system is determined by performing constant temperature and constant stress simulations (NPT) at zero stress. This yields the reference shape and size matrix, h_o in Parrinello-Rahman formalism [9]. In determining elastic constants this reference state is used in constant temperature constant volume simulations (NVE) of 50000 steps for each state point. The elastic constants are evaluated using the following statistical fluctuation formulas

$$
\begin{aligned}
C_{ijkl}^T &= -\frac{\Omega_o}{k_B T}(< P_{ij}P_{kl} > - < P_{ij} >< P_{kl} >) \\
&+ \frac{2Nk_B T(\delta_{ik}\delta_{jl} + \delta_{il}\delta_{jk})}{\Omega_o} + < \chi_{ijkl} >
\end{aligned}
\tag{5}
$$

where $<>$ denotes the averaging over time and $\Omega_o = det h_o$ is the reference volume for the model system. The first term represents the contribution from the fluctuation of the microscopic stress tensor, P_{ij}, the second term represents the kinetic energy contribution, and the third term is the Born term.

After equilibration of 20,000 steps (20 ps with a time step of 1 fs) we calculated the elastic constants at 6 different concentrations, 0, 20, 40 ,60, and 100 % at 300 K after collecting statistics over 50,000 steps for the fluctuation terms. As reported earlier the botn terms converges within first few 1000 steps. The results are given in Table 2.

Table:2 Elastic constants and bulk modulus of Pt-Rh binary alloy calculated at 300^0K as obtained from NVE MD Simulation after 50000 steps.

Percent Rh @ Pt	C_{11} (GPa)	C_{12} (GPa)	C_{44} (GPa)	B (GPa)
0	253.4 ±1.3	208.1±0.5	60.1 ±0.2	223.4 ±0.8
20	268.4 ±0.4	213.8 ±1.0	73.4 ±0.9	232.3 ±0.8
40	282.2 ±0.3	216.2 ±0.3	88.1 ±0.5	238.0 ±0.3
60	289.1 ±1.2	212.6 ±0.8	101.8 ±1.0	237.3 ±0.9
80	298.1 ±0.4	211.9 ±0.3	114.0 ±0.6	240.5 ±0.4
100	303.5 ±0.4	208.5 ±0.4	125.1 ±0.2	240.4 ±0.4

3.2 Shear viscosity of metals and alloys

Using Nonequilibrium molecular dynamics (NEMD) methods, we calculated the shear viscosity of Cu-Au alloys subjected to a planar Couette shear flow. We used the parameters given in Table 1 above. These parameters are shown to give accurate values for surface energies, vacancies energies, and stacking faults.

We considered temperatures of 1500K, 1750K, and 2000K and alloys Au_x-Cu_{1-x} with $x = 0$, 0.25, 0.5, 0.75, and 1.0. The NEMD simulation used periodic boundary conditions with (500 atoms per cubic unit cell). The unit cell length was based on extroplating the experimental density at the corresponding temperature. For alloys we assumed that the density is given by the linear combination rule

$$\rho_T (Au_x Cu_{1-x}) = \rho_T (Au) \cdot x + \rho_T (Cu) \cdot (1 - x)$$

We applied shear rates of $\dot{\gamma} = 2$, 1, 0.5, and 0.25 ps^{-1}. For each shear rate and temperature, we first equilibrated (with NEMD) for N_{EQ} steps followed by $N_{meas} = 9N_{EQ}$ steps which were used to calculate properties. As the shear rate decreases we observed larger fluctuations and slower convergence. Thus, we increased the number of steps. The total number of steps ranged from 20,000 ($\dot{\gamma} = 2$) to 100,000 ($\dot{\gamma} = 0.25$). The calculated viscosity scaled as $\dot{\gamma}$ as shown in Figure 1. This relation was used to extrapolate the $\dot{\gamma} = 0$ for comparison to experiment. The results are summarized in Table 3. The calculations are within 1 to 3 %

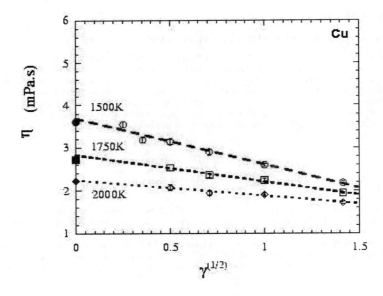

Figure 1: Viscosity as a function of shear rate.

for Cu but high by 12 to 30 % for Au. These exploration calculations will be extended next year by using much slower rates and by considering NPT MD rather than NVT MD.

Table:3 Viscosity at zero shear rate, η_0 (mPas).

T (K)	Cu Sim.	Cu Exper.	Cu_3Au_1 Sim.	Cu_1Au_1 Sim.	Cu_1Au_3 Sim.	Au Sim.	Au Exper.
1500	3.68	3.6	2.64	2.88	3.97	5.86	4.5
1750	2.82	2.72	2.29	2.47	3.21	4.56	3.7
2000	2.24	2.22	1.87	2.04	2.70	3.67	3.2

3.3 Glass formation and Crystallization in metals and alloys

In this section, we use molecular dynamics to examine melting and quenching of CuNi and CuAg alloys. These two model systems were particularly chosen, since Cu and Ag have very different sizes, making them good candidates for forming a metal glass, while Cu and Ni have similar sizes thus making them good candidates for forming a crystal even at high quenching rates.

The simulations in this section performed in constant temperature, constant thermodynamic tension (TtN) MD conditions [10]. The TtN MD simulations started from a cubic box with 500 atoms subject to periodic boundary conditions. To obtain the stress free reference size and shape of the unit cell, we performed 25 ps of simulation with constant enthalpy, constant thermodynamic tension (HtN) at zero pressure. The TtN MD simulations were carried out in a series of increasing temperatures from 300 to 1500K in 100K increments. The final temperature of 1500K is a few hundred degrees above the melting temperature. At every temperature the MD time step was taken as 1 fs and the simulation time for determining the properties was 25 ps.

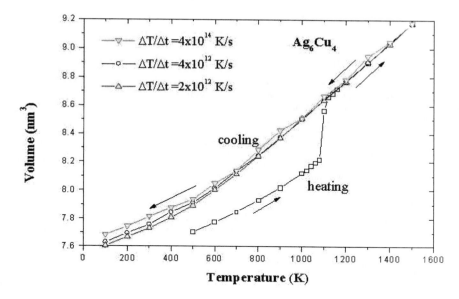

Figure 2: Ag-Cu alloy at the eutectic composition.

After equilibrating the structure in the liquid phase at 1500K, we cooled the system using different quenching rates from 1500K down to 300K in 100K decrements in the TtN ensemble. To achieve the fast, intermediate, and slow cooling rates, we kept the model system at the same temperature for times of 50 ps, 25 ps, and 0.25 ps. This leads to cooling rates of 2×10^{12}, 4×10^{12}, and 4×10^{14} K/s, respectively. To ensure convergence of the results for the fastest cooling rate (0.25 ps per 100K), using the conditions at the end of each 0.25 ps interval (for each 100 K drop) we performed additional 25 ps long TtN simulations for thermodynamic averaging.

Figure 2 shows the variation of the volume as Ag_6Cu_4 is heated and cooled. The large jump in volume in the temperature range of 1000K to 1200K for the heating process is due to the melting of the Ag-Cu alloy. In order to obtain a more refined estimate of T_{melt}, we used a smaller increment in temperature, namely 20K, from 1000K to 1200K. This leads to a theoretical melting temperature, $T_{melt} = 1090K$, in reasonable agreement with experimental melting temperature of 1053K. One reason for the melting temperature being a bit high is that our system is homogeneous without a free surface. In addition, we started with a perfect crystal; given the rapid rate of heating, the system might not have had time to generate an equilibrium distribution of defects, thus leading to a slightly higher T_{melt}.

Using cooling rates in the range of 2×10^{12} to 4×10^{14} K/s, we find that CuNi and pure Cu always form a face-centered cubic (fcc) crystal while Cu_4Ag_6 always forms a glass (with T_g decreasing as the quench rate increases). The crystal formers have radius ratios of 1.025 (CuNi) and 1.00 (Cu) while the glass former (CuAg) has a ratio of 1.13, confirming the role of size mismatch in biasing toward glass formation.

We have considered several cooling rates to investigate its effect on the glass transition temperature T_g. Each rate leads to a slightly different value for the temperature at which the slope changes. A parameter often used to define the glass transition temperature is the Wendt-Abraham parameter [11] defined by $R = g_{min}/g_{max}$. Here $g_{min}(g_{max})$ is the value of $g(r)$ at the first minimum (maximum) in the RDF. The Wendt-Abraham parameter stresses the local character of $g(r)$, permitting a direct comparison between structures and leading to a better estimate of glass transition temperatures. The Wendt-Abraham transition temperature, T_g^{WA}, are $T_g^{WA} \approx 500K$ at $\Delta T/\Delta t = 2 \cdot 10^{12}$ K/s, $T_g^{WA} \approx 550$ K at $\Delta T/dt = 4 \times 10^{12}$ K/s, and $T_g^{WA} \approx 700$ K at $\Delta T/\Delta t \approx 4 \cdot 10^{14}$ K/s. Thus, the glass transition temperature increases with increased cooling rate. The fastest cooling rates result in shorter times for the atoms to relax, thus leading to formation of the glass at a higher temperature than at lower cooling rate.

4 Concluding Remarks

We presented a many body potential parameterized to simulate fcc transition metals and their alloys. We have used these potentails and described combination rules to study elastic properties of binary alloy Pt-Rh, to study the shear viscosity of Au-Cu binary melt and finally to study the glass formation from melt at the eutectic concentration for Ag-Cu.

Acknowldegements This research was supported partially by a grant (ARO-DAAH 95-1-0233) from the Army Research Office and by grants from the NSF (ASC 92-17368 and CHE 95-12279). In addition, support for the Materials Simulation Center (MSC) facilities came from DOE-ASCI, ARO-DURIP (DAAG55-97-1-0140), and MSC industrial associates.

References

[1] P. Duwez, R. H. Willens, and W. Klement Jr. *J. Appl. Phys.* **31**, 1137 (1960). A. Peker and W. L. Johnson *Appl. Phys. Lett.* **63** 2342 (1993); X. H. Liu and W. L. Johnson *J. Appl. Phys.* **78** 6514 (1995).

[2] W. A. Harrison, **Solid State Theory** (Dover, NY, 1979)
W. A. Harrison, **Electronic Structure and Properties of Solids** (Dover, NY, 1980)

[3] J. K. Norskov, Phys. Rev. B 26, 2875 (1982)

[4] M. S. Daw, M. L. Baskes, Phys. Rev. B 29, 6443 (1984)

[5] M. W. Finnis, J. F. Sinclair, Phil. Mag. A, 45 (1984)

[6] A. P. Sutton, J. Chen, Phil. Mag. Lett. 61, 139 (1990)

[7] H. Rafii-Tabar, A. P. Sutton, Phil. Mag. Lett. 63, 217 (1991)

[8] D. D. Koleske, S. J. Sibener, Surf. Sci. 290, 179 (1993)

[9] S. Nosè *Mol. Phys.* **52** , 255 (1984); S. Nosè *J. Chem. Phys.* **81**, 511 (1984). M. Parrinello and A. Rahman *Phys. Rev. Lett.* **45** 1196 (1980).

[10] J. R. Ray and A. Rahman *J. Chem. Phys.* **82** 4243 (1985).

[11] H. R. Wendt and F. F. Abraham *Phys. Rev. Lett.* **41** 1244 (1978).

Part II

Atomic and Electronic Structure II

THERMODYNAMICS AND GLASS FORMING ABILITY
FROM THE LIQUID STATE

P.J. DESRE
Laboratoire de Thermodynamique et Physicochimie Métallurgiques (CNRS UMR 5614), Institut National Polytechnique de Grenoble, University Joseph Fourier
BP 75 , 38402 Saint Martin d'Heres France

ABSTRACT

This study is mainly devoted to the establisment of relations between easy glass forming ability of some multicomponent liquid alloys and strong heteroatomic bonding in connection with nucleation .The chemical short range order in typical liquid bulk glass formers as Zr-Ni-Al is evaluated from a statistical model and presented as Cowley's order parameters versus concentration. The effect of the nature and of the number of components on the chemical contribution to the liquid-crystal interface energy and on the Gibbs energy of crystallisation is analysed and discussed. A specific mechanism of nucleation based on a distribution of concentration fluctuations in the undercooled liquid is proposed . This homophase fluctuation mechanism, which is thermodynamically and kinetically justified, leads to a lower preexponential factor in the expression of the frequency of nucleation as a function of the number of components . Furthermore, this frequency of nucleation can be strongly lowered by an augmentation of the energy barrier of nucleation depending on the nature and the number of components in the liquid phase.

INTRODUCTION

The possibility of obtaining bulk metallic glasses[1-4] by moderate quenching rate from the liquid state demonstrates the high stability of some multicomponent liquid alloys against crystalline nucleation .The particular dense random packing of these liquid bulk glass formers (LBGF) , which results both from atomic size differences between the components and strong hetero-atomic bonding confers to these liquids a particular high viscosity contributing to lower the frequency of nucleation in the highly undercooled range.

This work is mainly devoted to the study of the stability of these liquids and its consequence on crystalline nucleation with emphasis on the effect of the number of components. The first part concerns the evaluation of the chemical short range order in a typical liquid glass formers : Zr-Al-Ni . The second part is focussed on the effect of strong heteroatomic bonding on liquid-crystal interface energy and driving force of crystallisation and their consequence on nucleation when this transition is constrained to large partitioning .

1 - Chemical short range order (CSRO) in liquid bulk glass formers Zr-Al-Ni

In our knowledge, no accurate thermodynamic data on the Gibbs energy of mixing of the liquid alloys Zr-Ni-Al are available in the literature. However, the highly negative enthalpies of mixing of the three constitutive binary alloys [5] leads to expect strong heteroatomic bonding and CSRO in the ternary liquid .

An evaluation of the CSRO as a function of concentration can be performed by applying the so called quasi-chemical model [6] . This model is based on the quasi-lattice an the nearest atomic neighbour approximations including the concept of nearest neighbour pair energy .

The configuration partition function Q is expressed versus the pair energies (which in the present application are designated by : ε_{AlNi}, ε_{NiZr}, ε_{AlZr}, ε_{AlAl}, ε_{NiNi}, ε_{ZrZr}) the number of heroatomic pairs noted zm, zn and zp for Ni-Zr, Al-Zr and Al-Ni respectively and the mole fractions of Al and Ni noted x and y respectively . z is the coordination number .

The maximisation of the partition function Q versus m , n and p leads to equilibrium conditions. The corresponding values of m* n* and p* are given by the following relations:

$$\frac{p^{*2}}{(x-p^*-n^*)(y-m^*-p^*)} = \exp-\frac{2\lambda}{zRT} \quad ; \quad \frac{m^{*2}}{(y-m^*-p^*)(1-x-y-n^*-m^*)} = \exp-\frac{2\mu}{zRT}$$

$$\frac{n^{*2}}{(x-p^*-n^*)(1-x-y-n^*-m^*)} = \exp-\frac{2\upsilon}{zRT} \tag{1}$$

The interaction parameters λ, μ, ν which appear in eqn (1) are such :

$$\lambda = z\left(\varepsilon_{AlNi} - \frac{\varepsilon_{AlAl} + \varepsilon_{NiNi}}{2}\right) ; \mu = z\left(\varepsilon_{NiZr} - \frac{\varepsilon_{NiNi} + \varepsilon_{ZrZr}}{2}\right) ; \nu = z\left(\varepsilon_{AlZr} - \frac{\varepsilon_{AlAl} + \varepsilon_{ZrZr}}{2}\right)$$

A numerical resolution of eqns (1) leads to the values of m*, n* and p* versus composition for given λ, μ, ν, z and temperature T . The model has been applied for T = 1000 K , z = 10 and the following values of the interaction parameters deduced from the heat of mixing of the constitutive binary alloys drawn from Miedema's tables (4) : $\lambda = -88.5$ KJ, $\mu = -207$ KJ , $\nu = -182$ KJ .

From the calculated p*, m* and n* in the whole range of concentration of the ternary alloy it is possible to deduce the corresponding Cowley's order parameters η_{ij} which measure the CSRO and are such : $\eta_{ij} = 1 - \frac{z_{ij}}{z\, x_j}$. Where z_{ij} is the partial coordination number of j atoms around an atom i located at the center of the first sphere of coordination. x_j is the average mole fraction of component j in the liquid. In the present application the η_{ij} are related to the average number of pairs (per atom) zp*, zm* and zn* through the simple relations :

$$\eta_{AlNi} = 1 - \frac{p^*_{AlNi}}{x\, y} \quad ; \quad \eta_{NiZr} = 1 - \frac{m^*_{NiZr}}{y(1-x-y)} \quad ; \quad \eta_{AlZr} = 1 - \frac{n^*_{AlZr}}{x\, (1-x-y)} \tag{2}$$

Isovalues of the order parameters η_{NiZr} and η_{AlZr} versus concentrations are given in figures (1) and(2) .

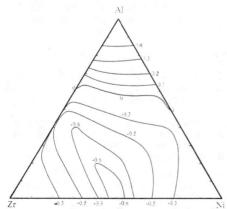

Figure 1 : Isovalues of the order parameter η_{NiZr} at T = 1000K

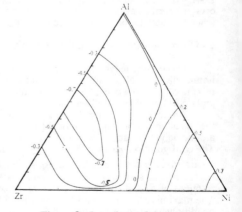

Figure 2 : Isovalues of the order parameter η_{AlZr} at T =1000K

Negative values of the order parameter are obtained over an extended composition range ; this would mean that CSRO dominates in these liquid . More specifically there is a region localised around $x_{Al} = 0.17$, $x_{Ni} = 0.33$ where both η_{AlZr} and η_{NiZr} are stongly negative ($\eta_{AlZr} = -0.7$; $\eta_{NiZr} = -0.8$). Note that this domain is not far from the composition $Zr_{60}Al_{15}Ni_{25}$ where the maximum difference between Tx (onset temperature of crytallisation) and Tg (the glass transition temperature) has been found from DSC measurements [7].

The concentration region where both η_{AlZr} and η_{NiZr} are negative covers practically the domain where ductile metallic glasses have been obtained [8] .

In addition to the previous evaluation, it is possible to calculate the thermodynamic fluctuations of the order parameters around their average value .

The probability of a fluctuation ω around equilibrium is proportional to $\exp(-\Delta F/kT)$. Where ΔF represents the free energy increase associated with fluctuations of the number of heteroatomic pairs ij .
By developing ΔF versus the number of pairs ij around their average values and after integration the following expression of the maximum probability ω_{max} is obtained .

$$\omega_{max} = \frac{|F(c_v^2)|^{1/2}}{(2\pi\, NRT)^{v(v-1)/4}} \tag{3}$$

where v is the number of components . N is the number of atoms involved in the fluctuation .

$|F(c_v^2)|$ is the determinant of order c_v^2 whose elements are such $\partial^2 F / \partial p_{ij}\, \partial\, p_{kl} = F_{ijkl}$

(with $i \neq j$ and $k \neq l$). p_{ij} and p_{kl} are the probabilities of pairs ij and kl respectively.
ω_{max} which is the probability of no-fluctuation event around the average value can be taken as a measure of the liquid stability against fluctuations of the order parameters (or of the number of heteroatomic pairs) .

By applying the quasi chemical model to ZrAlNi liquid at T = 1000 K the determinant $|F(c_3^2)|$ can be calculated for different compositions .

For the alloy $Zr_{60}Al_{15}Ni_{25}$ the value of $|F(c_3^2)|$ is found to be $2.2\ 10^8$ Joules^3 . The value of $|F(c_3^2)|$ which would be obtained in the alloy completely disordered is $1.8\ 10^5$ Joules^3 .

This gives a ratio $\omega_{max}^{QC} / \omega_{max}^{disorder}$ equal to 30 which means a significant lowering of the fluctuations of the order parameter due to hetero-atomic bonding .
Such a lowering of the fluctuations of the order parameter is consistent with small fluctuations in time t of the force F(t) exerted on an atom by its neighbours which is measured by the autocorrelation function $<F(t) . F(t + \tau) >$ ($<\ >$ means an ensemble average). The classical fluctuation- dissipation theorem yields a simple relationship between the friction coefficient ξ_s and the autocorrelation function [9] :

$$\xi_s = \frac{1}{3kT} \int_0^\infty < F(t)\, F(t+\tau) > d\tau. \tag{4}$$

This relation means that a strong force exerted on any atom together with small fluctuations in time of this force give large ξ_s and consequently large shear viscosity . Experiments have already proven such correlation between strong heteroatomic bonding and excess viscosity in liquid alloys [10][11].

2 - Influence of strong heteroatomic bonding in LBGF on liquid- crystal interface energy and Gibbs energy of crystallisation .

The addition of components to a basic liquid glass former which are insoluble (or partially soluble) in the expected nucleating compound, can change both the liquid-crystal interface energy and the Gibbs energy of crystallisation . Note that these energies are determinant on the values of the thermodynamic energy barrier of nucleation . An approach of these effects is presented in the the following paragraphs.

a) Effect of the number of components on the liquid-crystal interface energy .

The liquid-crystal interface energy in alloys is classically decomposed in structural and chemical contributions . The following analysis concerns the chemical contribution brought from partitioning associated with crystallisation on the liquid-crystal interface energy . More specifically, the study is focussed on the effect of components in the liquid which are insoluble in the potentially nucleating compound :

The interface energy between a binary compound chosen as $A_{0.5} B_{0.5}$ and a liquid of same composition is designated by σ_o . Components i insoluble in the compound are called " added components " ; the liquid-crystal interface energy in the presence of added components is noted σ . The interface energy difference $\Delta\sigma = \sigma - \sigma_o$ is evaluated by applying the broken bonds method with the nearest atomic neighbour interaction and the quasi-lattice approximations with a sharp representation of a non-chemically relaxed nterface.

As the liquid alloy, which contains v components, exhibits strong heteroatomic bonding, the classical Bragg and Williams approximation for the evaluation of the probability of heteroatomic pairs whose energy is noted ε_{ij} ($i \neq j$) can not be used in this case . Cowley's order parameters η_{ij} in the liquid must be introduced in order to take account of the CSRO in the evaluation of $\Delta\sigma$. The mole fractions of components A and B in the liquid are such $c_A = c_B = c_o$ .The mole fractions of the added components are taken identical and are such $c_a = (1 - 2c_o) / (v - 2)$. After algebra[12], the following expression of $\Delta\sigma$ is obtained :

$$\Delta\sigma = \frac{1}{8\Omega} \left[\{ <\lambda_{Aj}\eta_{Aj}> + <\lambda_{Bj}\eta_{Bj}> - \lambda_{AB}\eta_{AB} \} (1 - 2 c_o) + \cdots\cdots \right.$$

$$\left. \cdots + \{ <\lambda_{Aj}(1-\eta_{Aj})> + <\lambda_{Bj}(1-\eta_{Bj})> - \frac{\lambda_{AB}}{2}(1-\eta_{AB}) - \frac{v-3}{v-2}<\lambda_{ij} (1-\eta_{ij})>\} (1- 2 c_o)^2 \right] \qquad (5)$$

Ω is the average area occupied by one mole at the interface . $\lambda_{AB}, \lambda_{Aj}, \lambda_{Bj}, \lambda_{ij}$ are the binary interaction parameters. $v - 2$ is the number of added components .The brackets < > mean the average of the products as $\lambda_{ij} (1-\eta_{ij})$ taken over the added components .

As the interface is considered as chemically unrelaxed, the evaluation of $\Delta\sigma$, derived from eqn (5), would be slightly overestimated. However, as we are concerned here with the effect of the liquid-crystal energy on nucleation, the approximation of a nonequilibrium interface appears to be plausible .

Due to the lack of data on the order parameters, only qualitative arguments can be derived from eqn (5) in favor of an increase of the interface energy .

Strong hetero-atomic bonding between the added components ($v > 3$)will contribute to increase $\Delta\sigma$. Note that this effect is enhanced by an increase of the number of components and this for a double reason : an increase of v , at given c_a , increases $(v-2)c_a$, which appears to the square in eqn (5), but also enlarges the ratio $(v-3)/(v-2)$.But inversely, strong hetero-atomic bonding between the added components and compounds components in the liquid contributes to lower $\Delta\sigma$.

Owing to the presence in eqn (5) of the term $\lambda_{ij}(1-\eta_{ij})$ (where i and j are added components) it must be pointed out that a significant increase of σ can be expected when more than one component is added which means at least a quaternary liquid when the crystallisation of a binary compound is concerned .

The conclusion which can be drawn from this approach is that strong heteroatomic bonding between the added components together with weak interactions between each of the added components and the compound components favor an increase of the interface energy . On a physical point of view, a positive (chemical) excess interface energy can be understood as resulting from an increase of the liquid cohesive energy brought by suitable added components .

b) Effect of the number of components on the Gibbs energy of crystallisation .

The absolute stability of a liquid alloy measured by its Gibbs energy of mixing, is expected to increase with the number of components provided that the constitutive binary alloys have negative enthalpy of mixing of the same order of magnitude . This can be easily demonstrated by taking an expression of the molar enthalpy of mixing of a liquid containing v components given by :

$$\Delta H/RT = \sum_{i \neq j} \lambda_{ij}\, c_i c_j \qquad (6)$$

Where c_i and c_j are the mole fractions of i and j respectively .

The interaction parameters λ_{ij} are characteristic of the heteroatomic pairs interactions in each constitutive binary alloy. Let us consider the situation where heteroatomic bonding dominates in each binary liquid, in such a case the parameters λ_{ij} are all negative .

Furthermore, in order to eliminate a possible phase separation in the liquid, the λ_{ij} are taken identical $\lambda_{ij} = \lambda$ with $\lambda < 0$. As the application concerns liquids exhibiting strong hetero-atomic bonding, the entropy of mixing can be neglected (in front of the enthalpy of mixing) leading to the following expression of the Gibbs energy of mixing :

$$\Delta G(v)/RT = \Delta H/RT = \lambda \sum_{i \neq j} c_i c_j \qquad (7)$$

The minimum value of ΔG is obtained when $\quad c_1 = c_2 = \cdots = c_i \cdots = c_v = \dfrac{1}{v}$

This yields : $\Delta G_{min}(v) = \dfrac{\lambda}{2}\dfrac{v-1}{v}$

The relative variation of $\Delta G_{min}(v)$ when v varies from v to $v+1$ is independant of λ and is such :

$$\frac{\Delta G_{min}(v+1) - \Delta G_{min}(v)}{\Delta G_{min}(v)} = \frac{1}{v^2 - 1} \qquad (8)$$

As $\nu > 2$, the maximum relative increase of the absolute stability is obtained when going from a binary to a ternary liquid where the gain is of 33%. When going from a ternary to a quaternary liquid, the gain becomes only of 12.5%.

Under the previous conditions (negative and nearly equal interaction parameters) such an increase of the absolute stability with the number of components is accompanied with an augmentation of the Gibbs stability wich is measured by the determinant $\Delta = |\,G_{\nu-1}\,|$ whose elements are such : $G_{ij} = \partial^2 \Delta G\,(\,c_1....c_{\nu-1})\,/\,\partial\,c_i\,c_j$. ΔG is the molar Gibbs energy of mixing of the liquid alloy .

The Gibbs energy of crystallisation of compounds in multicomponent liquids can be expressed in a relatively simple way as a function of the Gibbs stability of the liquid. Designating by $C_o(c_{1o}\quad c_{io}\quad c_{\nu-1\,o})$ and by $C^*\,(c^*_1\quad c^*_i\quad c^*_{\nu-1})$ the "vectors" defining the composition of the alloy and of the compound respectively and after expending the Gibbs energy of mixing around C_o, to the second order in $C^* - C_o$, the following approximated expression of the molar Gibbs energy of crystallisation, given in a matrix form, is obtained[13][14] :

$$\Delta G_c = \Delta G_{pc} + \frac{1}{2}\Delta C_t\,(G_{\nu-1})\Delta C \tag{9}$$

ΔG_{pc} is the polymorphous Gibbs energy of crystallisation of a liquid of composition C^*; this is a negative quantity . ΔC_t is the line matrix $\Delta C_t = (\,c_{1o} - c^*_1,\quad c_{io} - c^*_i\quad,c_{\nu-1,o} - c^*_{\nu-1})$ and ΔC the corresponding column matrix . $(G_{\nu-1})$ is the square matrix constructed with the elements G_{ij} previously defined . The second term of the RHS of eqn (9) is a positive quadratic quantity which represents the Gibbs energy associated with the partition of the liquid .This quantity is the partitioning term of the Gibbs energy of crystallisation; it will be noted ΔG_f in the following. It is of interest to discuss the case where one component is added to the liquid including initially ν components , such a component being insoluble in the compound (which is supposed to contain also ν components). Under this condition, ΔG_{pc} is unchanged by the added component . When the added component increases the liquid Gibbs stability , i.e. the partitioning term $\Delta G_f = 1/2\ \Delta C_t\,(G_{\nu-1})\Delta C$, the crystallisation driving force is lowered. In order to illustrate this situation, we have reported in figure (3) a typical isotherm molar Gibbs energy diagram ($-\Delta G$) of a ternary system ABC where the Gibbs energy of mixing of the liquid and the Gibbs energy of formation of a binary line compound A B are represented versus concentration .

The liquids whose compositions are represented by the points E and Q' in the isothermal diagram are both in equilibrium with the compound AB . For a supersaturated binary liquid of concentration c_A , the driving force leading to the formation of the compound AB is obtained by application of the classical tangent rule; this driving force is represented by the segment MN in the diagram. Owing to the convexity of the Gibbs energy surface,which corresponds to an increase of the Gibbs stability in the ternary liquid, any tangent plane drawn to this surface along the quasi-binary section (corresponding to c_A) between Q and Q' cut the Gibbs energy of the line compound AB between points M and N which indicates a lowering of the crystallisation Gibbs energy when the component C is added to the initial binary liquid.

As a conclusion, when the constitutive binary alloys have negative interaction parameters of the same order of magnitude an increase of the number of added components, which preserve the previous property, contributes to lower the crystallisation driving force.

When the interaction parameters of the binaries remain all negative but with significant differences between their enthalpy of mixing, phase separation may occur in a certain concentration range of the liquid . In a such composition range the determinant Δ becomes negative .

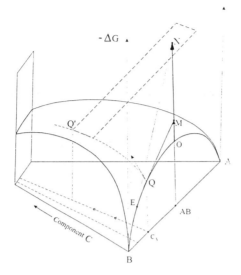

Figure 3:Typical isothermal Gibbs energy diagram of a ternary liquid A-B-C (see the text)

3) Homophase concentration fluctuations mechanism of nucleation in multicomponent liquid alloys .

In strongly interacting multicomponent liquids, the expected nucleating phases are intermetallic compounds whose structure is generally complex with a significant number of atoms per unit cell. Due to this complexity, the first stage of formation of an intermetallic embryo in the liquid would require a specific local composition and CSRO, which could be generated by thermal homophase concentration fluctuations. The second stage would be the genesis of the fitting crystalline symetry where the metallic ions are driven towards their crystalline sites through indirect ion-ion interactions via bonding electrons. Independently of the previous comments, it has been established that in the undercooled range of multicomponent liquid alloys, the energy barrier associated with the formation of a critical nucleus becomes higher than the energy of formation of a suitable homophase fluctuation of concentration of same size and concentration as the critical embryo [13] .

The previous considerations leads to the idea of a nucleation mechanism where the critical nucleus is formed by a polymorphous nucleation process inside thermal excited homophase concentration fluctuations .

On a kinetic point of view, such a process is significant if the relaxation time τ_{relax} of a as formed thermal concentration fluctuations is at least of the same order of magnitude than the transient time τ_t of polymorphous nucleation inside the homophase fluctuation . If it is so, a significant time would be allowed for the development of intra-fluctuation nucleation confirming the assistance of nucleation via concentration fluctuations .

The Kashchiev approach [15] of the characteristic transiant time of polymorphous nucleation ,which has been found to be in good agreement with numerical simulations [16], can be used for the evaluation of τ_t .

In theframe of the present application, the Kashchiev expression of τ_t is of the form :

$$\tau_t = \frac{24 \; RT \; n_c}{\pi^2 \; v_c \; \gamma \; (- \Delta G_{pc})} \tag{10}$$

n_c is the intra-fluctuation critical number of atoms . ΔG_{pc} is the polymorphous Gibbs energy of crystallisation previously defined . v_c is the number of atomic sites at the critical embryo interface which is taken as $v_c = 4 \; n_c^{2/3}$ [16] . γ is the atomic jump rate at the interface which is taken as $\gamma = 6 \; D/\lambda_d^2$ where D is the diffusion coefficient and λ_d the average jump distance .

Taking $\lambda = (V_m / N_a)^{2/3}$ (V_m is the molar volume, N_a the Avogadro number) , eqn (10) becomes:

$$\tau_t = \frac{RT \; (V_m/N_a)^{2/3}}{\pi^2 \; D(- \Delta G_{pc})} \; n_c^{1/3} \tag{11}$$

For the sake of simplicity, let us consider the case where the nucleating crystal is in its pure state.

Applying the interface energy expression proposed by Turnbull and Spaepen for the liquid-crystal interface energy [17] $\sigma = \alpha \; \Delta H_m / N_a^{1/3} \; V_m^{2/3}$ (ΔH_m is the enthalpy of melting and α is a constant) , ΔG_{pc} can be expressed as :

$$\Delta G_{pc} = - 2 \; \alpha \; \Delta H_m \; (V_m/N_a)^{1/3} \; \frac{1}{r_c} \tag{12}$$

r_c is here the polymorphous critical radius of the embryo formed inside the fluctuation ; r_c is given by the relation :

$$r_c = 2 \; \sigma_{poly} V_m / \; |\Delta G_{pc}| \tag{13}$$

where σ_{poly} is the liquid-crystal interface energy for the polymorphous nucleation mechanism . Taking account of eqn (12) the expression of τ_t becomes :

$$\tau_t = (\frac{4\pi}{3})^{1/3} \; \frac{R}{2 \; S_m \; \pi^2 \; D \; \alpha} \; [\; 1 - \frac{2 \; \alpha \; (V_m/N_a)^{1/3}}{r_c}] \; r_c^2 \tag{14}$$

where S_m is the entropy of melting .

Furthermore, the relaxation time of a spherical fluctuation (which must extends up to the classical critical size R_c as it will be discussed in the following paragraph) is controlled by heterodiffusion and can be approximated for a 50% concentration relaxation [18] by the relation : $\tau_{relax} = 0.17 \; R_c^2 / D_h$ and for a 95% relaxation by : $\tau_{relax} = 0.4 \; R_c^2/ D_h$

D_h is the heterodiffusion coefficient and R_c the classical critical nucleus in the alloy given by the relation : $R_c = 2 \; \sigma \; V_m / (\; |\Delta G_{pc}| - \frac{1}{2} \Delta C_t \; (G_{v-1}) \; \Delta C \;) .$ \tag{15}

σ is the liquid-crystal interface energy between the classical critical embryo and the liquid of composition C_o.

From the previous expressions of τ_{relax} and eqn (13) and taking $\alpha = 0.6$, $S_m = 10$ J /K, $V_m = 10^{-5}$ m^3/ mole and $r_c = 1$ nm, the ratio of the characteristic times τ_t / τ_{relax} is given by :

$$\frac{\tau_t}{\tau_{relax}} = \kappa \frac{D_h}{D} \left(\frac{r_c}{R_c}\right)^2 \tag{16}$$

with $\kappa = 0.47$ for a 50% concentration relaxation and $\kappa = 0.19$ for a 95% relaxation of the concentration fluctuation .

It appears that these characteristic times are of the same order of magnitude . But owing to the numerical coefficient κ together with the fact that the intra-fluctuation critical radius r_c is always smaller than the classical critical radius R_c it results $\tau_t < \tau_{relax}$

From eqn (13) and(15), and taking $\sigma = \sigma_{poly}$ the ratio of the critical radii is given with a good approximation by the following relation :

$$\frac{r_c}{R_c} = 1 - \frac{\frac{1}{2}\Delta C_t (G_{v-1})\Delta C}{|\Delta G_{pc}|} \tag{17}$$

As previously mentioned, strong heteroatomic bonding increases the partitioning Gibbs energy $\Delta G_f = (1/2) \Delta C_t (G_{v-1})\Delta C$, this contributes to lower r_c /R_c and consequently the ratio τ_t / τ_{relax} .

The ratio D_h / D is more difficult to evaluate specifically in the case of liquids including more than two components .The heterodiffusion coefficient D_h is a function of a thermodynamic factor which is expected to increase with the partitioning Gibbs energy term , this could yield a ratio D_h / D greater than unity . But inversely, as mentioned before, such an increase of the partitioning term would lower the ratio r_c /R_c which, appearing to the power two in (16), would maintain the ratio τ_t / τ_{relax} lower than unity . The following development is based on the validity of this latter property .

Note that any initial fluctuation which would only extend up to the intra-fluctuation critical radius r_c generates clusters of subcritical size which would dissolve in the mother phase; this means that the initial suitable fluctuation must extend beyond the size r_c (in fact up to R_c) in order that the relaxing fluctuations might lead these subcritical clusters to their critical size R_c.

At all events, the Homophase Fluctuation Mechanism (HFM) of nucleation remains strongly dependent of the probability of creating suitable fluctuations in the liquid which are able to yield a critical nucleus by relaxation.

It has been established [12] that the distribution of the probability of concentration fluctuations in a liquid alloy is given by the following relation :

$$\omega (n) = \frac{\Phi}{(2 \pi n)^{(v-1)/2}} \exp\left[-\frac{n}{RT}\Delta G_f (C^*, C_o)\right] \tag{18}$$

with $\quad \Phi = \frac{|G_{v-1}|^{1/2}}{(R T)^{(v-1)/2}}$

n is the number of atoms in the fluctuation of concentration between C^* and C_o in a liquid containing v components. The size independant dimensionless function Φ can be decomposed as follows : $\Phi = \Phi^{id} + \Phi^{xs}$ where Φ^{id} is the value Φ for an ideal solution which is of the form :

$$\Phi^{id} = \{ c_1 ... c_i ... c_v \}^{-1/2} \qquad (19)$$

Φ^{xs} measures the deviation from ideality .
Accounting for the fact that the bulk metallic glasses are concentrated alloys, and for simplicity, the concentrations are taken such that : $c_1 = c_2 = ----- = c_i = ------ = c_v = 1/v$

From eqn (18) and (19), the logarithm of $\omega(n)$ can be expressed as follows :

$$\ln \omega(n) = - \frac{(v-1)}{2} \ln (2\pi n) + \ln [v^{v/2} + \Phi^{xs}] \qquad (20)$$

which for an ideal solution reduces to :

$$\ln \omega(n) = - \frac{(v-1)}{2} \ln (2\pi n) + \frac{v}{2} \ln v \qquad (21)$$

As the number of components v is obviously much smaller than the number of atoms n constituting the concentration fluctuation , the first term of the RHS of eqn (20) is preponderant unless Φ^{xs} approaches a critical value corresponding to a phase separation which would yield an infinite value for the second term of eqn (20). Thus, provided the liquid alloy remains far from critical conditions, eqn (21) gives a fairly good approximation of $\omega(n)$ which for strongly interacting liquids will be slightly underestimated . It must also be added that the shortage which has consisted in taking identical concentrations is not restrictive at all as the second term of eqn (20) remains generally much smaller than the first one.
The previous arguments lead to the following expression of the frequency of nucleation J_{HFM} via the HFM mechanism versus the polymorphous nucleation frequency J_{Poly} :

$$J_{HFM} = \frac{v^{v/2}}{(2\pi n)^{(v-1)/2}} \exp [- \frac{n}{RT} \Delta G_f (C^*, C_o)] \frac{1}{\tau_{relax}} \int_0^{\tau_{relax}} J_{poly}(t) \, dt \qquad (22)$$

Considering that the transient time of polymorphous nucleation τ_t is smaller than the life time of the fluctuation τ_{relax} and approximating the average value of $J_{poly}(t)$ between $t = 0$ and $t = \tau_t$ by $1/2 \, J_{Poly}^{st}$, where $J_{Poly}^{st} = K_v \exp - \Delta G^* (r_c)/kT$ is the expression of the frequency of polymorphous nucleation in the stationary state, eqn (22) becomes :

$$J_{HFM} = K_v \frac{v^{v/2}}{(2\pi n)^{(v-1)/2}} (1 - \frac{1}{2} \frac{\tau_t}{\tau_{relax}}) \exp [- \frac{n}{N_a kT} \Delta G_f(C^*, C_o)] \exp[- \Delta G^* (r_c) /kT] \qquad (23)$$

K_v is the preexponential factor of the classical polymorphous nucleation process and

ΔG^* (r_c) is the corresponding Gibbs energy barrier at $r = r_c$ (the intra-fluctuation critical radius).

$\frac{n}{Na} \Delta G_f(C^*,Co)$ is the Gibbs energy required to form a suitable fluctuation including n atoms .

It must be emphasized that the initial homophase fluctuation requires an energy which allows the embryo to attain its critical radius R_c . In fact the minimum energy required to form such a fluctuation, noted $\frac{N_c}{N_a} \Delta G_f^{min}$ (where N_c is the number of atoms in the critical embryo) must be such :

$$\frac{N_c}{N_a} \Delta G_f^{min} = \Delta G^* (R_c) - \Delta G^* (r_c) \tag{24}$$

Relation (24) results merely from the fact that the minimum total energy required to form a critical nucleus by the HFM process is necessarily equal to the classical energy barrier of nucleation as the two processes have the same initial and final states .
Combining eqns (21) and (22) the expression of J_{HFM} can be condensed to :

$$J_{HFM} = K_v \frac{v^{v/2}}{(2\pi N_c)^{(v-1)/2}} (1 - \frac{1}{2} \frac{\tau_t}{\tau_{relax}}) \exp [- \Delta G^*(R_c)] /kT] \tag{25}$$

Where $\quad \Delta G^*(R_c) = \frac{16}{3} \frac{\pi \sigma^3 V_m^2}{[\Delta G_{pc}+(1/2) \Delta C_t (G_{v-1})\Delta C]^2}$

σ is the liquid-crystal interface energy .

It appears that the homophase fluctuation mechanism introduces a specific preexponential factor in the expression of the frequency of nucleation. Note that as the term $(1 - \tau_t / 2\tau_{relax})$ in eqn (25) varies with τ_t / τ_{relax} only between 0.5 and 1 it does not modify appreciably the preexponential factor . Inversely, this latter can be significantly lowered by an increase of the number of components. For a critical nucleus including 250 atoms (R_c = 1nm ; $V_m = 10^{-5}$ m^3) formed in a liquid which contains five components, the prefactor would be reduced by approximatively five powers of ten .
More generally, and by application of eqn (25) it is found that each component added to a binary liquid, lowers the prefactor of a factor of ten .
As it has been previously discussed, the classical energy barrier of nucleation can be increased in multicomponent liquid where strong heteroatomic bonding dominates and this specifically when several liquid components are insoluble (or partially soluble) in the potentially nucleating compounds . As a consequence, the number of atoms in the critical nucleus N_c would increase and would enhance the lowering of the preexponential factor as it appears in eqn (25) .

CONCLUSION

This study has been an attempt to understand the particular stability of some multicomponent liquids against crystalline nucleation. The application of the quasi chemical model to the basic liquid bulk glass former Zr-Ni-Al has shown that strong chemical short range order is expected in these liquids; the strongest CSRO has been found in the composition range where the easiest amorphisability has been obtained from experiments.

It has been demonstrated that added components in the liquid, insoluble in the nucleant compounds, act in order to lower the driving force of crystallisation . When these added components are strongly interacting from one to each other but exhibit weaker heteroatomic bonding with the compound components in the liquid phase, an increase of the liquid-crystal interface energy can be expected . The previous effects act in favor of an increase of the energy barrier of nucleation.

Furthemore, a homophase fluctuation mechanism of nucleation applied to multicomponent liquid alloys has been proposed. This model considers that a prexisting distribution of concentration fluctuations in the liquid is able to assist nucleation via suitable fluctuations in which polymorphous nucleation can develop . This model yields the same total energy barrier as for the classical mechanism but introduces a preexponential factor as a function of the number of components which is significantly lowered as the number of components is increased .

However, the low frequency of nucleation which is experimented in liquid bulk glass formers remains mainly due to the strong heteroatomic bonding in the liquid state which lowers the driving force of crystallisation and increases the liquid viscosity.

References

[1] A.Inoue,T.Zhang and T.Masumoto, Mater. Trans. JIM , 31,425 (1990)
[2] A.Peker, W.L.Johnson , Appl.Phys.Lett. , 63, 2342 (1993)
[3] A.Inoue, T.Zhang and T.Masumoto, Mater.Sc. Forum 79,181,497 (1995)
[4] K.Busch, Y.J.Kimrand and W.L.Johnson , J. Appl. Phys. 77,4039,(1995)
[5] A.R.Miedema, F.R.de Boer, R.Boom,J .F.W. Dorleijn , Calphad, 1, 33, 359 (1987)
[6] E.A.Guggenheim, Mixtures , Oxford Clarendon Press (1952)
[7] A.Inoue,T.Zhang and T.Masumoto, Mater. Trans. JIM ,31,177 (1990)
[8] A.Inoue,T.Zhang and T.Masumoto, Mater. Trans. JIM ,31,3 (1990)
[9] J.Ross J.Chem.Phys. 24,375 (1956)
[10] M.Kitajima, T.Itami, M.Shimoji, Phil.Mag. 30;285 (1974)
[11] E.A. Moelwyn Hughes , Physical Chemistry Oxford , Pergamon (1964)
[12] To be published .
[13] P.J. Desre , T.Itami, I.Ansara Z. Metallkd,3, 84 (1993)
[14] P.J. Desre Mater. Trans. JIM 38,7,583 (1997)
[15] D.Kashchiev , Surf. Sci., 4, 205 (1966)
[16] K.F. Kelton, A.L.Greer, C.V. Thomson , J. Chem.Phys. 79, 12 (1983)
[17] F. Spaepen and D.Turnbull RQM 2 Eds M.J.Grant, B.C.Giessen MIT Press Cambridge MA, 205 (1976) .
[18] J.Crank, Clarendon Press , p 92 Oxford (1992).

VISCOSITY MEASUREMENTS FOR La-Al-Ni LIQUID ALLOYS BY AN OSCILLATING CRUCIBLE METHOD

Tohru YAMASAKI*, Tomohiro TATIBANA**, Yoshikiyo OGINO* and Akihisa INOUE***
* Department of Materials Science & Engineering, Faculty of Engineering, Himeji Institute of Technology, 2167 Shosha, Himeji, Hyogo 671-2201, JAPAN.
** Graduate School, Himeji Institute of Technology, Himeji, JAPAN.
*** Institute for Materials Research, Tohoku University, Sendai 980-8577, JAPAN.

ABSTRACT

The viscosity of liquid lanthanum-based and aluminum-based La-Al-Ni alloys has been measured by an oscillating crucible method of the inverse suspending type in the temperature range from melting temperature (Tm) up to about 1400 K. In the case of $La_{55}Al_{45-x}Ni_x$ (x = 10 ~ 40 at. %) alloys, the viscosity increased with increasing Ni content up to about 20 at. % Ni and then decreased with increasing the Ni content, while the activation energy for viscous flow decreased to a minimum value at about 20 at. % Ni. This composition is well consistent with that of the La-Al-Ni alloy having largest glass-forming ability.

INTRODUCTION

Lanthanum-based and aluminum-based La-Al-Ni amorphous alloys exhibit a large glass-forming ability in the compositional regions of 25 to 78 at. % La and 5 to 50 at. % Ni for lanthanum-rich alloys and 5 to 9 at. % La and 8 to 12 at. % Ni for aluminum-rich alloys [1]. Especially, $La_{55}Al_{25}Ni_{20}$ alloy can be amorphized in the bulk form by casting of the molten alloy into a copper mold [2]. In the glass-forming process from the liquid, the most important factor is the temperature dependence of the viscosity of the supercooled liquid, and the viscosity is sensitive to the liquid structure at the molecular level. In the present study, compositional dependence of the viscosity of lanthanum-based and aluminum-based La-Al-Ni liquid alloys and La-Al-Ni-Co-Cu alloy has been measured by an oscillating crucible method of the inverse suspending type in the temperature range from melting temperature, T_m, up to about 1400 K. The viscosity below T_m was estimated from the Fulcher relation.

On the other hand, very different values of the viscosity of pure aluminum and aluminum-based alloys have been reported [3 - 6]. This discrepancy may be due to the effects of oxide, causing a skin effect which increases the value of the viscosity. Therefore, in the present study, special care has been taken to avoid the skin effect of oxide.

EXPERIMENTAL PROCEDURES

Pure Al, $Al_{90-x}Ni_{10}La_x$ (x = 0, 1, 2, 5, 7 and 10 at.%), $La_{55}Al_{45-x}Ni_x$ (x =10, 20, 30 and 40 at.%) and $La_{55}Al_{25}Ni_{10}Co_5Cu_5$ alloys were treated in the present study. The ingots of the alloys were prepared by arc-melting a mixture of Al (99.99 mass%), La (99.9mass%), Ni (99.9mass%), Co (99.9mass%) and Cu (99.9mass%) metals in a purified argon atmosphere.

The oscillating crucible method of the inverse suspending type was used for measuring the viscosity under a purified helium atmosphere [7]. The logarithmic decrement was found by measuring successive time intervals as the reflected beam the distance between two phototransistors. The period of oscillation and the logarithmic decrement were 6.2 s and 0.015, respectively, and the moment of inertia of the oscillating system was about 4200 g/cm^2. In order to obtain the viscosity coefficients from the measured logarithmic decrements, the following equations proposed by Roscoe and coworker [8, 9] have been used in the present study.

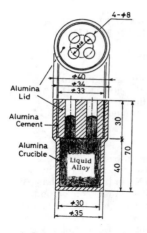

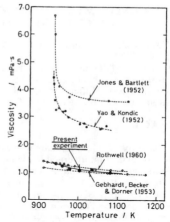

Figure 1 Schematic explanation of the vessel for the liquid.

Figure 2 The changes in viscosity of liquid Al as a function of temperature.

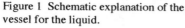

$$\lambda/\rho = A(\eta/\rho)^{1/2} - B(\eta/\rho) + C(\eta/\rho)^{3/2}, \quad (1)$$
$$A = (\pi^{3/2}/I)\{1 + R/(4H)\}HR^3\tau^{1/2}, \quad (2)$$
$$B = (\pi/I)\{3/2 + 4R/(\pi H)\}HR^2\tau, \quad (3)$$
$$C = (\pi^{1/2}/I)\{3/8 + 9R/(4H)\}HR\tau^{3/2}, \quad (4)$$

where λ, η and ρ are the logarithmic decrement, the viscosity and the density of the liquid, respectively, and R, H, I and τ are the inner radius of a cylindrical crucible, the depth of the liquid, the moment of inertia are the period of oscillation, respectively. As the vessel for the liquid, a cylindrical alumina crucible sealed with an alumina lid was used in order to remove the skin effect of the oxide.   A schematic presentation of the vessel is shown in Figure 1. When the vessel is used, twice the value of H is substituted in eqs. (2)-(4) for correcting the sealing effect of the lid [10]. Because of the irregularity in the shape of the alumina crucible, the constants A, B and C should be different in every crucible. Therefore, as a standard liquid mercury was chosen, and the constants were determined for each crucible. The numerical equation obtained is

$$\lambda/\rho = 0.0058(\eta/\rho)^{1/2} - 0.0616(\eta/\rho) + 0.0383(\eta/\rho)^{3/2}. \quad (5)$$

The density of lanthanum-based and aluminum-based La-Al-Ni and La-Al-Ni-Co-Cu liquid alloys was estimated as follows: the density of the alloy ingots at room temperature was measured by the Archimedes method, and assuming that the thermal expansion coefficients of these alloys are the same of as those of pure Al or pure La in the solid and liquid states [11], the density of the liquid alloys could be estimated.

RESULTS

Viscosity of Pure Liquid Aluminum

Figure 2 shows the change of viscosity of liquid Al as a function of temperature. The solid line indicates the values of the present study and the dotted lines indicate the literature values [3-6]. As shown in this figure, the present data are in good agreement with the data reported by Rothwell [5] and Gebhardt et al. [6]. However, very different data have been

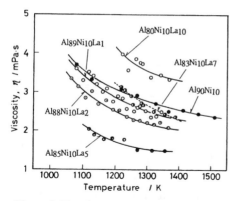

Figure 3 The change in viscosity of liquid Al$_{90-x}$Ni$_{10}$La$_x$ (x =0 ~ 10 at.%) alloys as a function of temperature.

also reported by others [3,4]: they are very much larger than those of the present study. This discrepancy may be due to the skin effect of oxide which increases the value of the viscosity.

Viscosity of the Aluminum-based Al-Ni-La Liquid Alloys

Figure 3 shows the changes of viscosity, η, of aluminum-based Al$_{90-x}$Ni$_{10}$La$_x$ (x =0, 1, 2, 5, 7 and 10 at.%) liquid alloys as a function of temperature. Figure 4 also shows their relationships between ln η and the inverse absolute temperature. The viscosity, η, of the Al-Ni-La alloys is higher than that of pure Al and can be represented by the Arrhenius relation.

Figure 5 shows the variation of the η at 1123 K, 1223 K and 1323 K, and of the activation energy for viscous flow with the La

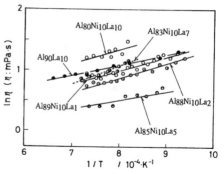

Figure 4 Relationships between ln η and 1/T of liquid Al$_{90-x}$Ni$_{10}$La$_x$ (x = 0 ~ 10 at.%) alloys.

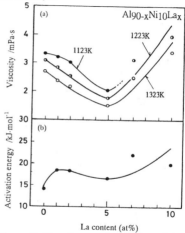

Figure 5 The changes in (a) viscosity and (b) activation energy for the liquid Al$_{90-x}$NI$_{10}$La$_x$ (x =0 ~ 10 at.%) alloys.

content in a liquid Al$_{90-x}$Ni$_{10}$La$_x$ alloys. The η decreases to minimum values at the La content of about 5 at. % and then increases with increasing the La content. The activation energy for the viscous flow initially increases to 18.5 kJ/mol by the addition of 1 at. % La. With increasing the La content of more than 1 at. %, the activation energy tends to decrease to a minimum value of 16.5 kJ/mol at the La content of 5 at. %.

Viscosity of the Lanthanum-based La-Al-Ni Liquid Alloys

Figure 6 shows the changes of viscosity, η, of lanthanum-based La$_{55}$Al$_{45-x}$Ni$_x$ (x = 10, 20, 30 and 40 at.%) and La$_{55}$Al$_{25}$Ni$_{10}$Co$_5$Cu$_5$ liquid alloys as a function of temperature. Figure 7 shows their relationships between ln η and the inverse absolute

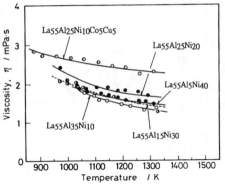

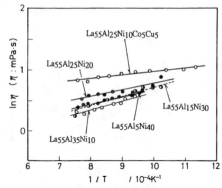

Figure 6 The change in viscosity of liquid La$_{55}$Al$_{45-x}$Ni$_x$ (x = 10 ~ 40 at.%) and La$_{55}$Al$_{25}$Ni$_{10}$Co$_5$Cu$_5$ alloys as a

Figure 7 Relationships between ln η and 1/T of liquid La$_{55}$Al$_{45-x}$Ni$_x$ (x =10~40 at.%) and La$_{55}$Al$_{25}$Ni$_{10}$Co$_5$Cu$_5$ alloys.

temperature. The viscosity, η, of the La$_{55}$Al$_{25}$Ni$_{10}$Co$_5$Cu$_5$ alloy is much higher than those of the La-Al-Ni alloys and the η of the La-Al-Ni and La-Al-Ni-Co-Cu alloys can be represented by the Arrhenius relation.

Figure 8 shows the variation of the η at 1123 K, 1223 K and 1323 K, and of the activation energy for viscous flow with the Ni content in a liquid La$_{55}$Al$_{45-x}$Ni$_x$ alloys. The η increases to maximum values at the Ni content of about 20 at. % and then decreases with increasing the Ni content. The activation energy for viscous flow, however, decreases to a minimum value of about 9 kJ/mol at the Ni content of 20 at. %. This composition is well consistent with that of the La-Al-Ni alloy having largest glass-forming ability [1].

DISCUSSION

It is well known that the viscosity decreases as the solute concentration approaches the eutectic value, and reaches a

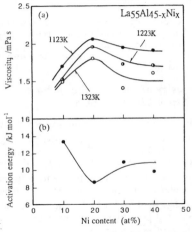

Figure 8 The change in (a) viscosity and (b) activation energy of liquid La$_{55}$-Al$_{45-x}$Ni$_x$ (x =10 ~ 40 at.%) alloys as a function of Ni content.

minimum at the eutectic composition [3, 12]. In the case of the Al$_{90-x}$Ni$_{10}$La$_x$ alloys shown in Fig. 5, the viscosity decreases largely to the minimum values of 2.1~1.5 mPa·s (T= 1123-1323 K) at the La content of about 5 at. %.

The activation energy for viscous flow also decreases to the minimum value of 16.5 kJ/mol at the same composition. Judging from the Al-Ni, Al-La and Ni-La phase diagrams [13], this decrease may be due to the alloy composition which approaches the eutectic values. As the reason for this decrease, it may be assumed that the decrease of amounts of some particles with sizes larger than La atom which are suspended in the liquid. As these particles, clusters having similar structures to those of the primary crystallites which are

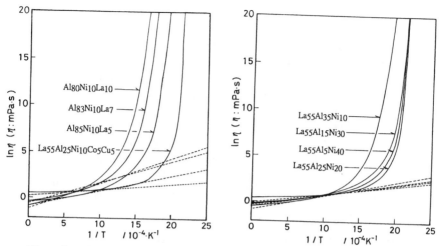

Figure 9 Relationships between ln η and 1/T of lanthanum-based and aluminum-
based La-Al-Ni liquid alloys.
-------the Fulcher relation,   - - - - the Arrhenius relation.

formed during solidification may be assumed and they influence the strong force acting
between particles of the liquid. The amounts of these particles may decrease when the
composition of the liquid approaches the eutectic.

In the case of the $La_{55}Al_{45-x}Ni_x$ (x =10 - 40 at.%) alloys shown in Fig. 8, however, the
viscosity increases to the maximum values of 2.0~1.4 mPa-s (T= 1123-1323 K) at the Ni
content of 20 at. %, and the activation energy for viscous flow decreases to the minimum
value of 9 kJ/mol. This composition is well consistent with that of the La-Al-Ni alloy having
largest glass-forming ability.   Gilman [12] has speculated that the formation of small
particles having strong chemical interactions between constituent atoms tends to decrease the
liquid viscosity because these particles interact relatively weak.   For larger particles the
dependence of viscosity on particles size reverses. According to the theory by Frenkel[14]
and Glasstone et al. [15], the activation energy for viscous flow is equivalent to the energy
necessary to form the hole in the liquid. Therefore, the formation of the particles that interact
relatively weak might decrease the activation energy. In the case of the lanthanum-based La-
Al-Ni liquid alloys, it is considered that La atoms have strong interactions with Al and Ni
atoms, resulting in the formation of some kinds of particles that decrease the activation
energy. When Co and Cu are added in La-Al-Ni alloys, the liquid viscosity increased while
the activation energy for viscous flow further decreased. Since the particle structure may be
expected to be best developed at the particular composition, this can account for the glass
forming tendencies of these alloys.

In order to obtain information about the temperature dependence of the viscosity below
Tm of lanthanum-based and aluminum-based La-Al-Ni liquid alloys was estimated from a
Fulcher relation [16].

$$\ln \eta = A + B/(T - T_0) \qquad (6)$$

where η, T and T_0 are the viscosity, the absolute temperature and the temperature to reach an
infinite value of viscosity, respectively. In order to estimate the viscosity over the entire
temperature range, two sets of data are required, one for the high- and one for the low-
viscosity region. Therefore, assuming that the viscosity at the glass temperature, Tg, is
equal to 10^8 mPa-s, the viscosity over the entire temperature range could be estimated.

The numerical equations obtained are as follows:

$$\ln \eta = -0.475 + 513.9/(T-492.8) \qquad \text{for Al}_{85}\text{Ni}_{10}\text{La}_5 \text{ alloy,}$$
$$\ln \eta = -0.329 + 1018.1/(T-515.7) \qquad \text{for Al}_{83}\text{Ni}_{10}\text{La}_7 \text{ alloy,}$$
$$\ln \eta = -0.787 + 1573.2/(T-528.1) \qquad \text{for Al}_{80}\text{Ni}_{10}\text{La}_{10} \text{ alloy,}$$
$$\ln \eta = -0.318 + 565.9/(T-486.8) \qquad \text{for La}_{55}\text{Al}_{35}\text{Ni}_{10} \text{ alloy,}$$
$$\ln \eta = 0.432 + 194.3/(T-466.2) \qquad \text{for La}_{55}\text{Al}_{25}\text{Ni}_{20} \text{ alloy,}$$
$$\ln \eta = -0.155 + 533.1/(T-436.3) \qquad \text{for La}_{55}\text{Al}_{15}\text{Ni}_{30} \text{ alloy,}$$
$$\ln \eta = 0.0087 + 416.0/(T-444.4) \qquad \text{for La}_{55}\text{Al}_5\text{Ni}_{40} \text{ alloy,}$$
$$\ln \eta = 0.554 + 280.5/(T-454.3) \qquad \text{for La}_{55}\text{Al}_{25}\text{Ni}_{10}\text{Co}_5\text{Cu}_5 \text{ alloy.} \quad (7)$$

They are plotted as solid curves in Figure 9. Compared with the dotted lines which represent the Arrhenius relation, the viscosity increases rapidly with falling temperature.

CONCLUSIONS

In the case of the $\text{Al}_{90-X}\text{Ni}_{10}\text{La}_X$ alloys, the viscosity decreases largely to the minimum values at the La content of about 5 at. %. The activation energy for viscous flow also decreases to the minimum value at the same composition. Judging from the Al-Ni, Al-La and Ni-La phase diagrams, this decrease may be due to the alloy composition which approaches the eutectic values.

In the case of $\text{La}_{55}\text{Al}_{45-X}\text{Ni}_X$ ($x = 10 \sim 40$ at. %) alloys, the viscosity increased to the maximum values at the Ni content of 20 at. %, while the activation energy for viscous flow decreased to the minimum value. This composition is well consistent with that of the La-Al-Ni alloy having largest glass-forming ability. This may be due to the formation of the particles in the liquid having strong chemical interactions between La, Al and Ni atoms, and these particles interact relatively weak resulting the decrease of the activation energy.

REFERENCES

1. A. Inoue, T. Zhang and T. Masumoto, Mater. Trans. JIM, **30**, 965 (1989).
2. A. Inoue, T. Zhang and T. Masumoto, Mater. Trans. JIM, **31**, 425 (1989).
3. W. R. D. Jones and W. L. Bartlett, J. Inst. Met. **81**, 145 (1952-53).
4. T. P. Yao and V. Kondic, J. Inst. Met. **81**, 17 (1952-53).
5. E. Rothwell, J. Inst. Met.**90**, 384 (1960).
6. E. Gebhardt, M. Becker and S. Dorner, Z. Metalkd. **44**, 510 (1953).
7. Y. Ogino, F. O. Borgmann and M. G. Frohberg, Trans. ISIJ, **14**, 82 (1974).
8. R. Roscoe, Proc. Phys. Sco. **72**, 576 (1958).
9. R. Roscoe and W. Bainbridge, Proc. Phys. Soc. **73**, 585 (1959).
10. H. Schneck, M.G. Frohberg and K. Hoffmann, Archiv. Eisenhutt. **2**, 1633 (1963).
11. *Metals Data Book* (Japan Inst. of Metals, Sendai, 1974), p. 14.
12. J. J. Gilman, Philo. Mag. **37**, 577 (1978).
13. M. Hansen, *Constitution of Binary Alloys* (McGraw-Hill, New York, 1958).
14. J. Frenkel, Kinetic Theory of Liquids (Dover, New York, 1955), p. 200.
15. S. Glasstone, K. J. Laidler and H. Eyring, *The Theory of Rate Processes* (McGraw-Hill, New York, 1941), p. 477.
16. D. E. Polk and D. Turnbull, Acta Metall. **20**, 493 (1972).

IN-SITU ELECTRON DIFFRACTION STUDY OF STRUCTURAL CHANGE IN THE SUPER-COOLED LIQUID REGION IN AMORPHOUS La-Al-Ni ALLOY

T. Ohkubo*, T. Hiroshima*, Y. Hirotsu*, S. Ochiai**, A. Inoue***

* Institute of Science and Industrial Research, Osaka University, 567-0047, Japan
** Teikyo University of Science and Technology, Yamanashi, 409-0193, Japan
*** Institute of Materials Research, Tohoku University, Sendai, 980-8577, Japan

ABSTRACT

Atomic structures of an amorphous $La_{55}Al_{25}Ni_{20}$ alloy in the supercooled liquid state have been investigated by *in-situ* electron diffraction using a specimen-heating stage in TEM and the imaging-plate intensity recording. From the analysis of atomic pair distribution functions, changes in interatomic distances and coordination numbers were clearly observed at temperatures in the supercooled liquid state. From the reverse Monte Carlo simulations, structural units (icosahedral, Archimedean anti-prism and trigonal prism atomic clusters) typical of the metallic glass structure were found and increased in the supercooled liquid region. In addition, the deformation behavior was investigated using tensile test. The superplastic elongation was confirmed at optimum strain rates in the supercooled liquid region. From the TEM observation of tensile tested specimens with superplastic elongation, β-La (fcc) nano precipitates in the amorphous matrix were confirmed. The superplasticity in this alloy is thought to originate in viscous flow due to the glassy structure formation but is closely related to an additional flow mode with the microcrystalline precipitation from the amorphous state during the deformation.

INTRODUCTION

It is known that the amorphous $La_{55}Al_{25}Ni_{20}$ alloy has an extremely high thermal stability [1] and an excellent superplasticity in the stable supercooled liquid state between temperatures from about 470 to 515K [2]. These temperatures, 470 and 515K, correspond to those of the glass transition (Tg) and the crystallization (Tx), respectively, and ΔT (=Tx-Tg) the supercooled liquid temperature range. The glass transition and the crystallization of amorphous alloys depend on the heating rate [3] and the holding time [4], since the structural change in supercooled liquid is a kinetic phenomenon. Two-stage glass-transitions have been observed in the ΔT range by measuring a change of the dynamic elastic modulus [3].

In order to investigate the origin of the superplastic behavior of the amorphous alloy, it is important to know a temperature dependence of the superplastic behavior and its relation to a microstructural change. However, no investigation has been made about atomic structural change in the ΔT range and also at temperatures before and after the glass transition. In the present study, atomic structures at temperatures before and after Tg and in the ΔT range on annealing have been investigated by *in-situ* electron diffraction and high-resolution electron microscopy (HREM) using a specimen-heating stage with imaging-plate (IP) intensity recording. From electron diffraction intensity analysis, atomic partial pair distribution functions (PDF) were obtained, followed by structure modeling with spatial atomic configurations by reverse Monte Carlo (RMC) simulation [5]. In this study, we also made tensile tests at various temperatures in the supercooled liquid region and the tested specimens were observed by TEM. Relationship between the structure change and the deformation behavior is discussed.

EXPERIMENTAL

TTT Diagram

In the *in-situ* TEM observation in the Δ T range, time-temperature-transformation (TTT) diagram is important to know the temperature dependence of the time for crystallization. The TTT diagram of the amorphous $La_{55}Al_{25}Ni_{20}$ specimen was made using differential scanning calorimetry (DSC: Rigaku 8230B). Referring to this TTT diagram, the *in-situ* TEM observation in the supercooled liquid region was made.

PDF Analysis

TEM specimens were prepared from amorphous $La_{55}Al_{25}Ni_{20}$ ribbons by ion-milling. A liquid nitrogen cooling stage was used in the milling. Structural changes in and near the supercooled liquid region on annealing have been investigated by *in-situ* selected area electron diffraction (SAED) using a specimen heating stage. Diffraction intensity was recorded on IP in a 200kV electron microscope (JEM-2010). The pixel size of IP was 25×25 μm^2. The intensities were digitized using IP reader (FDL-5000) on 16,384 gray levels. The camera length was set at 50cm but the length was corrected by using a reference diffraction pattern from gold particles. The intensities were normalized at a common scattering angle position. An averaged intensity curve for each IP data ($I(Q)$: $Q=4\pi\sin\theta/\lambda$, θ is the half scattering angle and λ, the electron wavelength) along the radial direction in reciprocal space was obtained up to the scattering vector $Q=150$ nm^{-1}. The reduced interference function $i(Q)$ was obtained by subtracting a smooth cubic spline curve that penetrates through the halo intensity ripples of $I(Q)$ along Q. The reduced radial distribution function $4\pi r\rho_0(g(r)-1)$ (ρ_0 is the number density and $g(r)$ the pair distribution function) were calculated by Fourier transform of the product of $Qi(Q)$ and a window function. After correcting the spline curve so that reduced radial distribution function from r=0 to the rising edge of the first peak became straight, a conclusive $g(r)$ was determined.

RMC Simulation

RMC simulation has now become a general method [5] in modeling the structure of disordered materials based on experimental diffraction data. In the present structure modeling, the structural composition was fixed as $La_{55}Al_{25}Ni_{20}$ and the density was chosen as $5.83Mg/m^3$ which correspond to the experimental composition and density of the as-quenched sample [6]. All of 2,346 atoms were distributed randomly in a structure cell (4.115 nm ×4.115 nm ×4.115 nm). The sizes of heated structure cells were extended in accordance with the first halo peak position obtained from the *in-situ* observation. After that, for each structure all, total potential energy of the system was minimized using a static structure relaxation where Lennard-Jones interatomic potentials were assumed between atoms under the periodic boundary condition. The relaxed structure was used as an initial structural model for each RMC simulation.

In the present RMC structural simulation, total PDF: $g_{total}(r)$ for each of the diffraction data was used as a reference data for the corresponding structure. The partial PDFs between constituent atoms, $g_{La-La}(r)$, $g_{La-Al}(r)$, $g_{La-Ni}(r)$, $g_{Al-Al}(r)$, $g_{Al-Ni}(r)$ and $g_{Ni-Ni}(r)$, were determined by the Gaussian distribution on the basis of experimental interatomic distances. A 'goodness-of-fit' parameter, x^2, for the total and partial PDFs was obtained from the experimental $g(r)$ and the calculated $g^{cal}(r)$ as follows

$$x^2 = [g(r)-g^{cal}(r)]^2/g(r). \tag{1}$$

Then deviation parameter D which is composed of the product of each x^2 and the corresponding weight parameter W was calculated as

$$D = W_{total} \, x^2_{total} + W_{La-La} \, x^2_{La-La} + W_{La-Al} \, x^2_{La-Al} + W_{La-Ni} \, x^2_{La-Ni} +$$
$$W_{Al-Al} \, x^2_{Al-Al} + W_{Al-Ni} \, x^2_{Al-Ni} + W_{Ni-Ni} \, x^2_{Ni-Ni}. \tag{2}$$

In this simulation, it was necessary to set W_{total} larger than the partial weight parameters for minimizing x^2_{total} values ($W_{total} = 6W_{m-n}$ was used, m,n: La, Al, Ni). One iteration in the simulation consists of choosing an atom at random and moving it in an arbitrary direction with a arbitrary displacement. When the move reduces the value of D, it is accepted. This process is continued until the D value becomes less than a certain amount. Here, the maximum displacement is confined to be ±0.05nm. The obtained structures were then examined by the Voronoi polyhedra analysis.

Tensile Test

Tensile tests of the melt-spun amorphous $La_{55}Al_{25}Ni_{20}$ alloy were made at strain rates between 8.3×10^{-4} and 8.3×10^{-1} s^{-1} at temperatures between 478 and 513 K. Ribbon specimens with a width of 4 mm and with a thickness of 60 μm were prepared. The gauge length of specimens was 10 mm. Tensile tests were performed in a silicon oil bath. Before starting each tensile test a time interval of 300 s was kept after immersing the specimen into oil. Microstructural observations of the tensile tested specimens were made by HREM combined with nanobeam diffraction. For the observations, different parts of the tested specimens were cut along in the elongation direction, and were thinned using ion milling.

RESULTS

TTT Diagram

Figure 1 shows the TTT diagram for the start and finish of crystallization obtained from isothermal annealing in the DSC furnace. By fitting the experimental values of the start of crystallization to the Arrhenius equation, activation energy for crystallization was obtained as 2.77 eV. It is smaller than the activation energies obtained from radio tracer diffusion experiment of Ni (3.76 eV) in $Zr_{55}Al_{10}Ni_{10}Cu_{25}$ alloy [7] and Be (4.47 eV) in $Zr_{42.1}Ti_{13.8}Cu_{12.5}Ni_{10}Be_{22.5}$ alloy [8].

In-situ Observation

Figure 2 shows the *in-situ* HREM image and SAED obtained from the specimen heated at 493K. No local structural change was observed by HREM before and after Tg. From the Gaussian fitting of atomic PDFs, increases in interatomic distances of La-La, La-Al and La-Ni by an amount of 2 to 4% were clearly observed at temperatures higher than Tg as shown in Fig. 3 (filled circles), which exhibits one of the characteristics of the supercooled liquid state. In the present *in-situ* TEM observation, Tg was estimated to be about 430 K and Tx about 505

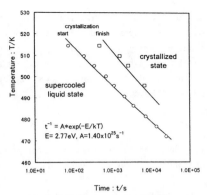

Fig. 1. TTT diagram for crystallization.

K. Just before the crystallization, the interatomic distance of La-La increased further. The atomic coordination number for La-La changes inversely in accordance with the change of La-La distance. In the supercooled liquid state, the increase of interatomic distance between La-La atoms was observed in two stages, while atomic distances between La-Al and La-Ni were observed in three stages. The multi stage structural change may correspond to the two stage glass-transitions [3].

RMC Simulation

In Fig. 3 also shown is the change of interatomic distances obtained from partial PDFs by the RMC simulations on heating (marked with triangles). Since a similar tendency was obtained in the experimental and the calculated atomic distances, the present structural models by the RMC simulations are judged to be preferable ones. The distribution of the Voronoi polyhedra in a simulated model for the structure at 430K is shown in Fig. 4. Some polyhedral clusters known as typical structural units of metallic glasses [9,10] were observed in the analysis shown in the

Fig. 2. In situ HREM image and SAED at 493K.

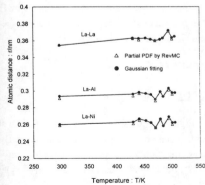

Fig. 3. Changes in interatomic distances during heating up.

figure. Namely, icosahedral, Archimedean anti-prism and trigonal prism atomic clusters were frequently found as polyhedral clusters. Many of other polyhedra were judged to be approximants of these polyhedra. Figure 5 shows fractional changes of some of the Voronoi polyhedra on annealing. The fraction of polyhedra tend to increase at temperatures higher than Tg. Since such polyhedral clusters are the well-known structural components of ideal amorphous alloys, it is understood that the increase of these clusters near Tg largely contributes to the stabilization of the supercooled liquid state. Although fcc clusters are found at temperatures near Tg, the volume fractions is small. On annealing in the Δ T range, the fraction once decreases, but starts to increase near the Tx. The tendency is consistent with the HREM observation of β-La (fcc) nanocrystals on the tensile tests.

Deformation Behavior

From the tensile test, the superplastic elongation that exceeded 400% was confirmed at some optimum strain rates in each temperature as shown in Fig. 6. The maximum elongation of 960 % was obtained at 498 K under a strain rate of $8.3 \times 10^{-2} s^{-1}$. Figure 6 shows the relationship between the elongation and the time to fracture in the tensile tests at various temperatures. The times for crystallization at these temperatures were evaluated from the TTT curve as 13600 s (478 K), 700 s (498 K) and 87 s (513 K). The tendency of the elongation decrease under the small strain rates was observed when the crystallization time becomes short. In the case of 513 K, the crystallization is judged to have already started during the period for 300 s after immersing the specimens into the oil bath before the tensile tests. When the crystallization time becomes long, the elongation increases largely even under the low strain rate (478 K). It is, therefore, understood that the maximum elongation can be obtained under the optimum strain rate and temperature in the $La_{55}Al_{25}Ni_{20}$ alloy.

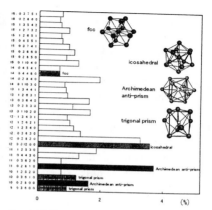

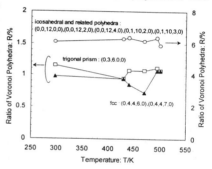

Fig. 4. Distibution of Voronoi polyhedra at 430K.

Fig. 5. Fractional changes of Voronoi polyhedra on annealing.

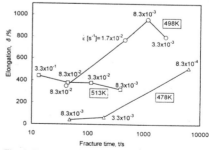

Fig. 6. The relationship between elongation and the time to fracture of the tensile tests.

73

Figure 7 shows crystalline precipitates observed in the amorphous matrix of the specimens deformed at 498 K with 3.3×10^{-3} s^{-1}. From the analysis of nanobeam diffraction patterns from the precipitates the structure was confirmed as that of β-La (fcc). In the present study, it has been confirmed that the superplastic flow originates in the viscous flow since the amorphous matrix region always remained after all the tensile tests in the temperature range of the supercooled liquid. The superplasticity suffers from the microcrystalline precipitation during the deformation.

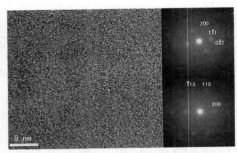

Fig. 7. HREM image and nanodifraction patterns of specimens tested at 498 K with 3.3×10^{-3}s^{-1}. Diffraction spots are indexed after β-La(fcc) structure. Beam incidences are along [031] and [011] of the fcc nanoprecipitate.

CONCLUSIONS

From the tensile test in oil bath, the superplastic elongation was confirmed at some optimum strain rates in supercooled liquid region. From the TEM observation of tensile tested specimen with superplastic elongation, a precipitation of β-La nano crystals was confirmed in the amorphous matrix. The *in-situ* observation using IP is effective to analyze the supercooled liquid structure, since precise diffraction intensity recording can be possible in several seconds. From the RMC simulations, typical structural units of amorphous alloy were found and they were increased near Tg. Therefore, superplasticity in this alloy is thought to originate in viscous flow by the glassy structure formation but is closely related to an additional flow mode with the microcrystalline precipitation from amorphous state during the deformation.

ACKNOWLEDGMENTS

This work was supported by Grant-in-Aid for Scientific Research from the Ministry of Education, Science, Sports and Culture, Japan.

REFERENCES

1   A. Inoue, T. Zhang and T. Masumoto, Mater., Trans. JIM, **30**, p. 965 (1989).
2   H. Okumura, A. Inoue and T. Masumoto, Mater., Trans. JIM, **32**, p. 593 (1991).
3   H. Okumura, A. Inoue and T. Masumoto, Acta metall. mater., **41**, p. 915 (1993).
4   Y.J. Kim, R. Busch, W.L. Johnson, A.J. Rulison, W.K. Rhim, Appl. Phys. Lett., **68**, p.1057 (1996).
5   R.L. McGreevy and L. Pusztai, Mol Simul. **1**, p. 359 (1988).
6   E. Matsubara, T. Tamura, Y. Waseda, T. Zhang, A. Inue and T. Masumoto, J. Non-Cryst. Solids, **150**, p. 380 (1992).
7   K. Nonaka, Y. Kimura, K, Yamauchi, H. Nakajima, T. Zhang, A. Inoue and T. Masumoto, Defect and Diffusion Forum, **143-147** p. 837 (1997).
8   U. Geyer, S. Schneider and W.L. Johnson, Phys. Rev. Lett., **75**, p. 2364 (1995).
9   D.J. Bernal, Proc. Roy. Soc. (Lond.), **A280**, p. 299 (1964).
10   A.L. Mackey, Physica, **A114**, p. 609 (1982)

FREE VOLUME CHANGES IN BULK AMORPHOUS ALLOYS DURING STRUCTURAL RELAXATION AND IN THE SUPERCOOLED LIQUID STATE

C. NAGEL, K. RÄTZKE, E. SCHMIDTKE, F. FAUPEL
Universität Kiel, Technische Fakultät, Lehrstuhl für Materialverbunde, Kaiserstr. 2, D-24143
Kiel, Germany

ABSTRACT

Volume changes in $Zr_{46.7}Ti_{8.3}Cu_{7.5}Ni_{10}Be_{27.5}$ and $Zr_{65}Al_{7.5}Ni_{10}Cu_{17.5}$ bulk metallic glasses have been observed by positron annihilation and density measurements. At low cooling rates excess volume of the order of 0.1 % is quenched in both glasses. Isothermal relaxation kinetics below the glass transition temperature obey a Kohlrausch law with exponents of $\beta \approx (0.3 \pm 0.1)$. Structural relaxation is not accompanied by embrittlement, as indicated by simple mechanical tests. The outer surface plays a crucial role in annealing of excess volume, which can be restored by annealing above T_g. The observed free volume changes are at variance with the behavior of a perfectly strong glass. The temperature dependence of the positron lifetime is discussed in terms of thermal detrapping from shallow traps.

INTRODUCTION

Metallic glasses, also termed amorphous alloys, are non-equilibrium structures with respect to the crystalline state. Therefore, while moderately heated, they undergo structural changes from the as-prepared state to the metastable structurally relaxed state and finally to the crystalline state. Due to the high quenching rates for conventional metallic alloys only thin samples can be prepared. Recently Inoue and coworkers [1] and Johnson and Peker [2] established Zr-based bulk metallic glasses, which can be prepared using low cooling rates and which are stable above the caloric glass transition on experimentally accessible time scales. Besides technical applications [3] these alloys allow detailed investigations of the glass transition and the supercooled liquid state.

Due to the high cooling rates, rapidly quenched metallic glasses are prone to irreversible structural relaxation that affects almost all properties [4]. This relaxation is irreversible and generally related to changes of topological short range order and annealing of excess volume. However, for the new, slowly cooled bulk glasses one might be tempted, not to expect a significant excess volume due to the low cooling rates sufficient for their preparation. On the other hand, the question arises whether free volume can be restored in the bulk glasses by annealing above T_g and slow cooling, and how this can be determined experimentally [5].

After injection and thermalization positrons in condensed matter reside in regions of reduced atomic density and annihilate with electrons. The positron lifetime τ is very sensitive to changes in electronic density. In metallic glasses one single lifetime is generally observed and interpreted in terms of trapping of positrons into the high number of cavities of different size on the atomic scale representing irregular arrays of potential wells with different binding strengths. Around a cavity, the electron density and especially the core electron density will be reduced resulting in a prolonged positron lifetime compared to the positron lifetime for the surroundings. Additionally, the reduced core electron density causes a deficit of annihilations with electrons having large momenta. This results in a narrowing of the electron-positron momentum density spectrum, which is actually measured as the Doppler broadened 511keV annihilation line (for a detailed description see e.g. [6]). In amorphous metals, the annihilation characteristics are regarded as statistically averaged quantities over the annihilation sites.

Positron annihilation measurements on free volume changes in bulk metallic glasses should therefore help to clarify the following questions: Is there any structural relaxation in these new bulk metallic glasses? Can the free volume be restored by annealing above T_g and what are the sinks for free volume? Does the free volume change around the caloric glass transition or are these alloys perfectly „strong glasses" [7]? What is the nature of the trapping centers for positrons in metallic glasses? Are the recent observations for a „Johnson" alloy [8] generally valid?

EXPERIMENT

Rapidly quenched glasses were prepared by splat quenching (10^8 K/s, $Zr_{46.7}Ti_{8.3}Cu_{7.5}Ni_{10}Be_{27.5}$) and melt spinning ($10^6$ K/s, $Zr_{65}Al_{7.5}Ni_{10}Cu_{17.5}$), respectively. Bulk samples were produced by slow cooling (10 - 100 K/s). A standard positron annihilation setup and a sample preparation procedure as described elsewhere [8,9] were used for lifetime and Doppler broadening experiments. Annealing of the sample was either performed isochronally in high vacuum (10^{-7} mbar) in a water-cooled sample holder allowing cooling rates of 1 - 2K/s and measurements were carried out at room temperature, or measuring was performed in-situ in a liquid-nitrogen-cooled sample holder, thus exploring the temperature dependence of annihilation characteristics. Lifetime spectra were evaluated using PATFIT88 [10] taking into account usual background and source corrections and one single lifetime, hereafter denoted as average positron lifetime τ_{av}. Doppler broadening spectra were evaluated in terms of S and W parameters after background corrections. The S or W parameters were defined by simply relating either the central area or the outer left and right areas to the total area of the 511keV peak, respectively. Density measurements were carried out using the Archimedian method before and after the various annealing treatments. X-ray and electron beam microprobe analysis were performed to check for amorphousness and composition, respectively.

RESULTS AND DISCUSSION

Isochronal annealing up to T_g leads to a decrease of τ_{av} in all samples, as can be seen from Fig. 1. A second run shows the irreversible nature of the changes, which were also detected by means of Doppler broadening measurements (not shown). The irreversible decrease in the average positron lifetime τ_{av} upon annealing is a well-known phenomenon for rapidly quenched conventional metallic glasses and was also observed in amorphous Fe-Si-B ribbon during our investigations [8]. It can be attributed to annealing of excess volume quenched in from the liquid state. In particular, for the $Zr_{46.7}Ti_{8.3}Cu_{7.5}Ni_{10}Be_{27.5}$ „Johnson" alloy this interpretation is strongly supported by the higher absolute τ_{av} value in the splat quenched glass and the larger relative change in τ_{av} in the splat quenched vs. the slowly cooled glass. For the Inoue glass this difference is not obvious, mainly because the cooling rate ratio between melt spinning for the ribbon and casting for the „bulk sample" is smaller than for the $Zr_{46.7}Ti_{8.3}Cu_{7.5}Ni_{10}Be_{27.5}$ alloy and excess free volume was apparently able to annihilate at the surface of the thin Inoue ribbon during cooling. Annihilation of excess volume is directly reflected in the density measurements revealing an increase in density of about 0.1 % during irreversible structural relaxation and approx. 0.6 to 0.8 % during crystallization. However, in contrast to conventional metallic glasses, e.g., the Fe-Si-B ribbon under investigation, where drastical embrittlement occurs with increasing degree of relaxation, no change in mechanical properties, e.g., elongation at break, was observed within error bars.

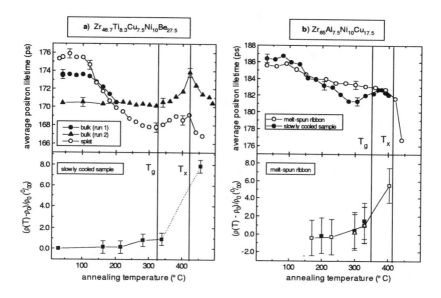

Fig. 1 Room temperature average positron lifetime and relative change of density for cumulative annealing as indicated in the plots of a) $Zr_{46.7}Ti_{8.3}Cu_{7.5}Ni_{10}Be_{27.5}$ and b) $Zr_{65}Al_{7.5}Ni_{10}Cu_{17.5}$. Lines between data points are to guide the eye. Straight lines denote approximate values for caloric glass transition temperatures and crystallization temperatures.

After heating above T_g and cooling at 1 - 2 K/s, the average positron lifetime, measured at room temperature, shows a distinct increase in the bulk samples, but not in the ribbons. This increase for the bulk samples shows that free volume, generated above T_g, can be quenched in, at least partially, even at cooling rates as low as 1 - 2 K/s. Further annealing leads to partial crystallization, which was also detected by X-ray diffraction. The decrease of the average lifetime may be a net effect because the reduced lifetime due to increasing crystal density obviously overcompensates the longer lifetime due to amorphous-crystalline interphases [11].

From the comparison of bulk samples and thin ribbons (Fig. 1, upper plots) it can be seen, that restoration of free volume by annealing above T_g and slow cooling (1 - 2 K/s) is only possible for bulk samples. Therefore, the surface must play an important role in annealing and restoration of free volume. This geometry effect rules out major contributions from annihilation mechanisms that are based on the recombination of regions of higher and lower density, so-called n and p defects on a microscopic scale. Here, the net reduction in free volume arises from the anharmonicity of the interatomic potentials [12].

Judging from well-established empirical relations between free volume and diffusivity or viscosity [13] the observed increase in free volume above T_g implies significant changes in the effective activation energies for diffusion and viscosity at the glass transition. This is clearly in contrast with the characteristics of almost perfectly strong glasses like covalently bound amorphous oxides or semiconductors [7]. As mentioned above, changes in the effective activation energy have indeed been reported for viscous flow in the „Inoue" glass under consideration here [14] and for some diffusants in the „Johnson" glass [15]. However, as seen by isotope-effect measurements, no change in the diffusion mechanism takes place at the glass transition [16].

Isothermal annealing measurements in both alloys below the caloric glass transition temperature revealed the well-known Kohlrausch behavior [17]

$$\Phi(t) = \Phi_0 \exp\left(-\left[t / t_0\right]^\beta\right) \qquad (1)$$

for structural relaxation with the relaxation function $\Phi(t) = \tau(t) - \tau_{relaxed}$ and $\Phi_0 = \tau_{as-quenched} - \tau_{relaxed}$ (examples are shown in Fig. 2). For both alloys the small value of $\beta \approx$ 0.3 suggests a broad distribution of activation energies for structural relaxation far below T_g. The effective activation energy E_a for structural relaxation can be estimated from

$$t = t_0 \exp(E_a / k_B T) \qquad (2)$$

to $E_a = (0.7 \pm 0.2)$ eV for $Zr_{46.7}Ti_{8.3}Cu_{7.5}Ni_{10}Be_{27.5}$ and (1.2 ± 0.2) eV for $Zr_{65}Al_{7.5}Ni_{10}Cu_{17.5}$. These low values suggest that either annealing of excess volume requires no long-range mass transport, which involves much higher activation energies [e.g. 15], or that mass transport is strongly facilitated through the presence of the excess volume.

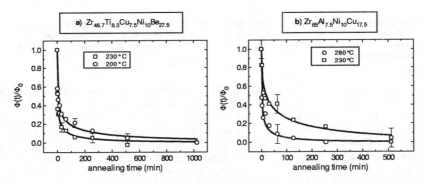

Fig. 2 Normalized changes in average positron lifetime, measured at room temperature, during annealing at different temperatures. Solid lines are fits according to the Kohlrausch law (see text).

Positron lifetime and Doppler broadening experiments have also been performed in-situ at different temperatures with both alloys using a specially designed liquid-nitrogen-cooled sample holder. Representative data for the Inoue alloy are shown in Fig. 3. For the fully crystallized sample the usual reversible increase of the average lifetime with increasing temperature is observed, which is generally attributed to the reduced average electron density due to thermal expansion. For the amorphous samples, the behavior is rather complicated. Starting at low temperature, the lifetime first increases up to a maximum at approx. - 50 °C and then decreases to a minimum at about 300 °C. This behavior is reversible. The slightly lower τ_{av} values for the relaxed sample are expected because of annealing of excess volume. The S parameter of the Doppler measurements and the corresponding experiments in the Johnson alloy show the same unusual behavior, which is not observed for crystalline materials (see Fig. 3 right).

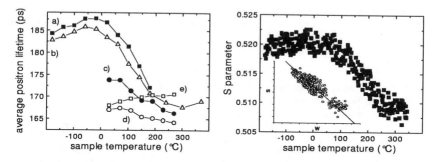

Fig. 3 Left: temperature dependence of average positron lifetime parameters for $Zr_{65}Al_{7.5}Ni_{10}Cu_{17.5}$ bulk metallic glass a) as-prepared sample b) structurally relaxed sample c,d) partially crystallized sample e) fully crystallized sample. Right: temperature dependence of the S parameter for relaxed Inoue alloy. Inset: S vs. W parameter, explanation see text.

Lifetime experiments showing a decrease of τ_{av} have already been performed by Dittmar et al. [18] in $Zr_{65}Al_{7.5}Ni_{10}Cu_{17.5}$ above room temperature. Their explanation by thermal detrapping of positrons from free volume traps can be adapted to explain the present data, assuming that the trapping centers for positrons and their binding energies are distributed around a mean value. At very low temperatures the positrons are captured in shallow traps with a distribution of binding energies. Here the positrons are not able to find the deepest traps and the average lifetime is below its maximum value. With increasing temperature, detrapping of positrons from the shallow traps is possible, resulting in an increase in lifetime. With further increase in temperature also detrapping from deep traps occurs, and annihilation mainly takes place between the atoms, which results in a decrease of lifetime. Using some simplifying assumptions about the binding energy distribution, one can estimate the binding energy of the deepest traps to be of the order of some tenth of eV, which is in good accordance with values in literature [6]. On the other hand, reversible changes in chemical short range order, which could lead to different chemical surroundings of the positrons and therefore to different lifetimes depending on temperature would also be able to explain this behavior. However, the variation of the S parameter with temperature shows the same behavior as the mean lifetime which implies that changes in free volume dominate and thus chemical effects are negligible. Additionally, we observed a linear correlation between S and W parameters (see inset in Fig. 3 right), which is indicative of a single defect type [19].

In conclusion, it has been shown that positron annihilation is a suitable tool to detect free volume changes in two representative bulk metallic glasses. In particular, structural relaxation and restoration of free volume via annealing above T_g could be clearly identified. Therefore, bulk metallic glasses are not perfectly strong glasses. By comparison of thin ribbons and bulk samples, it could be shown that the outer surface plays a crucial role as sink of free volume. Additionally, chemical short range order seems to be of minor importance for the positron annihilation characteristics.

ACKNOWLEDGEMENTS

The authors would like to thank W. Ulfert, Max-Planck-Institut Stuttgart, and U. Geyer, University of Göttingen, for sample preparation and helpful discussions. We are also indebted to R. Gerling, Forschungszentrum Geesthacht, for providing the microbalance and various hints.

REFERENCES

1.   T. Zhang, A. Inoue, T. Masumoto, Mater. Trans. JIM **32**, 1005 (1991).
2.   A. Peker and W. L. Johnson, Appl. Phys. Lett. **63**, 2342 (1993).
3.   http://www.liquidmetalgolf.com/.
4.   R.W. Cahn, in *Materials Science and Technology* Vol. 9, edited by P. Haasen, P. Kramer, R.W. Cahn (VCH Verlag, Weinheim, 1991), chap. 9.
5.   R. Gerling, F. P. Schimansky, R. Wagner, Mater. Sci. Engng. **97**, 515 (1988).
6.   W. Triftshäuser and G. Kögel, in: NATO ASI Series E **118**, edited by E. Lüscher and G. Fritsch, 218 (1987).
7.   C. A. Angell, Science **267**, 1924 (1995).
8.   C. Nagel, K. Rätzke, E. Schmidtke, J. Wolff, U. Geyer, F. Faupel, Phys. Rev. B. **57**, 10224, (1998).
9.   C. Nagel, K. Rätzke, E. Schmidtke, F. Faupel, W. Ulfert, submitted to Phys. Rev. B. Brief Reports (1998).
10.   P. Kirkegaard, N.J. Pedersen, M. Eldrup, Report of Risø Nat. Lab. (Risø-M-2740) (1989).
11.   H. S. Chen and S. Y. Chuang, Phys. Stat. Sol. (a), **25**, 581 (1974).
12.   T. Egami, Ann. N. Y. Acad. Sci. **371**, 238 (1981).
13.   F. Spaepen, *Physics of Defects*, Les Houches Lectures XXXV, edited by R. Balian, M. Kléman, J.P. Poirer (North Holland, Amsterdam, 1981), p.133.
14.   W. Ulfert and H. Kronmüller, J. Phys. C **8**, 617 (1996).
15.   P. Fielitz, M.P. Macht, V. Naundorf, G. Frohberg, J. non-cryst. sol. (1999).
16.   H. Ehmler, A. Heesemann, K. Rätzke, F. Faupel, U. Geyer, Phys. Rev. Lett. **80**, 4919 (1998).
17.   J. Jäckle, Rep. Prog. Phys. **49**, 171 (1986).
18.   R. Dittmar, R. Würschum, W. Ulfert, H. Kronmüller, H.-E. Schaefer, Solid State Communications **105**, 221 (1998).
19.   R. N. West, in *Positrons in Solids,* edited by P. Hautojärvi, (Springer Verlag, Berlin, 1979), pp. 89-144.

HEAT TREATMENT OF MOLTEN RAPIDLY QUENCHED PRECURSOR AS A METHOD TO IMPROVE THE GLASS FORMING ABILITY OF ALLOYS

V.Manov*, E.Brook-Levinson*, V.V.Molokanov**, M.I.Petrzhik**, T.N.Mikhailova**
*Advanced Metal Technologies Ltd., Even Yehuda 40500, Israel

**A.A.Baikov Institute of Metallurgy and Materials Science, Leninsky Pr. 49, Moscow, 117911, Russia

ABSTRACT

Improvement of glass forming ability (GFA) of two soft magnetic amorphous alloys ($Fe_{75.5}Ni_{1.3}Si_{8.6}B_{13.5}$ and $Co_{69.6}Fe_{1.3}Mn_{4.5}Si_{14.3}B_{9.3}Mo_1$) by heat treatment of melts prepared from different precursors (bulk ingot, rapidly quenched ribbons and granules) was studied. An assumption that the maximum undercooling ability corresponds to the maximum GFA was used to optimize the heat treatment mode. A temperature range was found by DTA for each alloy melt, favoring its undercooling (so called "undercoolable melt"). Usage of rapidly quenched precursor expands the range towards lower temperatures. GFA of the alloys was estimated by several melt quench techniques (casting, spinning and INROWASP). Fully amorphous samples with the thickness of 0.06-0.5 mm were prepared.

INTRODUCTION

The structure and composition of amorphous alloys provide their high magnetic, mechanical, corrosion resistant, and other beneficial features. At present, amorphous state in commercially important alloys is obtained in thin layers (20-40 μm) because of their poor GFA. The metallic glass forms usually at high cooling rates (10^6–10^4 K/s) which is one of the main factors retarding the novel applications of commercial alloys.

To increase the dimensions of metallic glasses, the search of new multicomponent compositions based on deep eutectics with extremely high GFA is conducted [1-4]. Another possibility to improve GFA without changing composition of known commercial alloys relates to control of structural changes along the sequence (initial components)→(master-alloy)→(melt)→(metallic glass). In addition to the general requirement to use high purity initial components, the critical role of oxygen was found recently. The oxygen concentration above 300 ppm decreases dramatically GFA in Zr-based alloys [5, 6]. Furthermore, we show in the present paper that the structures of master alloy before melting and of its molten liquid state before injection are very important for improving of GFA.

The main requirement to master alloy structure is absence of rough precipitation of primary phases. Clusters of primary refractory phases are retained in melt above the melting point for some time. They may serve as crystallization centers during subsequent solidification thus preventing melt undercooling and glass formation. The melt structure can be improved by dissolution of the clusters at heat treatment of molten master alloy.

It is known that structures with homogeneous concentration and free of any primary phases can be obtained by melt rapid quench. It was shown recently [7] that using rapidly quenched precursor in the form of amorphous granules one can produce a thick (2 mm) amorphous layer of soft magnetic Fe-based alloy using plasma sputtering deposition technique.

The relation between the state of the melt and its ability to deep undercooling was widely studied [8-10], including bulk metallic glass (BMG) forming alloy $Zr_{52.5}Cu_{17.9}Ni_{14.6}Ti_5Al_{10}$ [5]. Overheating of the melt to 1800 K was shown to result in increase of the amorphous layer thickness (up to 20 mm) in bulk metallic glass forming $Zr_{57}Ti_5Ni_8Al_{10}Cu_{20}$ alloy [11]. Furthermore a special thermal treatment of melt allows obtaining the same structure of amorphous ribbons quenched from lower temperature [12].

The effect of heat treatment on melt, undercooling and glass formation seem to be connected with existence of ranges of undercoolable melt (UCM). Using melt quench from UCM range we have shown the possibility to improve GFA for Zr-based bulk metallic glass forming alloys [4].

The aim of the work is to determine the effect of precursor structure and heat treatment of its melt on the undercooling ability and GFA using some techniques of melt quench for commercial soft magnetic $Fe_{75.5}Ni_{1.3}Si_{8.6}B_{13.5}$ and $Co_{69.6}Fe_{1.3}Mn_{4.5}Si_{14.3}B_{9.3}Mo_1$ alloys.

EXPERIMENT

The ingots with the above compositions 80 g each were prepared from the components (purity ≥99.5%) by vacuum induction melting under argon atmosphere. A part of the ingots was used to prepare rapidly quenched samples in the form of crystalline 0.3-mm granules and amorphous ribbons (thickness of 0.03 mm); they were precursors of molten alloys. The granules were prepared by quench of melt heated to 1550°C in the rotating water (INROWASP) with cooling rate $V \approx 10^4$ K/s. The ribbons were made by melt spinning, $V \approx 10^5$ K/s. The 3-mm granules were prepared by heating to 1500°C (Fe-alloy) and 1560°C (Co alloy) and subsequent cooling in DTA chamber, $V \approx 10^1$ K/s. The temperatures and cooling rates (10^1-10^5 K/s) were selected to secure the solidification from undercooled melt.

The heat treatment of the samples was carried out in DTA unit under He atmosphere using ceramic (Al_2O_3) crucibles. The samples were heated at the rate of 1.67 K/s and then cooled at the rate of 7-9 K/s. The following modes were used. 1) Thermal cycling (heating-cooling cycles) with multiple crossing of melting-crystallization range. The melt temperature was progressively increased while the annealing time was kept about 60 s. 2) Isochronal annealing (60 s) at various melt temperature with the only sample used for each temperature. 3) Isothermal annealing of some samples in the melt at various temperatures and holding times.

The undercooling below melting temperature were monitored by sharp exothermic peak on DTA curves and reduction of the structure element scale on micrographs. The undercooling was registered with high reliability for the given cooling conditions by sharp fall of the crystallization temperature by 50-100°C. The crystallization occurred in one stage and was accompanied by weak self-heating of the sample.

GFA was estimated by technique of the melt jet casting into in massive wedge shape copper mold. The castings were cut along long axis, polished and etched to estimate visually the critical thickness, d_c, as maximum width of light colored non-etchable zone corresponding the amorphous phase. The amorphous state of the "thick" rapid quenched ribbons, granules and wires was also characterized by DSC and X-ray.

RESULTS

The microstructure of precursors

The microstructure analysis revealed a difference in the precursors structure. The initial ingot of the Fe-based alloy consists of large primary crystals located on the background of coarse eutectic, Fig. 1a. The main distinction in the structures of 3-mm and 0.3-mm granules prepared by crystallization of undercooled melt, Fig. 1 b,c, is absence of the primary crystals and substantial reduction in the eutectic components dimensions with increasing the cooling rate.

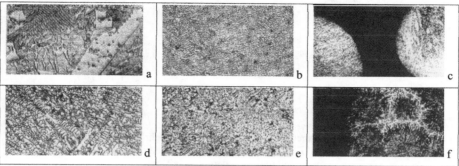

Fig.1. Optical microscopy micrographs of crystalline precursors of $Fe_{75.5}Ni_{1.3}Si_{8.6}B_{13.5}$ and $Co_{69.6}Fe_{1.3}Mn_{4.5}Si_{14.3}B_{9.3}Mo_1$ amorphous alloys prepared by different modes of melt cooling. Fe-alloy: a) an ingot (above 30 mm in cross-section); b) granules $\varnothing$ 3 mm, T_{quench}=1500°C, dT/dt $\approx$ 8 K/s; c) 0.3-mm granules, T_{quench}=1550°C, dT/dt$\approx$10^3-10^4 K/s; Co-alloy: d) an ingot (above 10 mm in cross-section); e) 3-mm granules, T_{quench}=1560°C, dT/dt$\approx$8 K/s; f) 0.3-mm granules, T_{quench}= 1550°C, dT/dt$\approx$10^3-10^4 K/s. The (a–e) figures are × 200, only the (f) one is × 500.

The structure of the initial Co ingot contains a significant amount of the primary phase in the form of dendrites. Rather dispersed eutectic components are located between the dendrites, Fig. 1d. The primary phase is preserved in the structure of 3-mm granules in form of isolated plates and drop precipitates of 5-10 μm size on the background of fine dispersed uniformly distributed eutectic, Fig. 1e. The primary phase precipitates were not found in the structure of rapid quenched 0.3-mm granules. The increase of cooling rate results in formation of unusual cellular structure in Co-based alloy granules, Fig. 1f. It consists of 10-40 μm polyhedrons; their boundaries are hardly etchable. Unlike Fe-based alloy, these granules have neither coarse dendrites, nor eutectic in the structure.

Those rapidly quenched precursors, which crystallize at the melt undercooling without primary phase formation, i.e. 3-mm and 0.3-mm granules of the Fe-based alloy and 0.3-mm granules of the Co-based alloy were selected for further investigations.

Determination of the temperature range of undercoolable melt

All melts studied could be undercooled below the melting temperature, T_m. We estimated the undercooling ability as $\Delta T = T_m - T_{cr}$, T_{cr} is the crystallization temperature). Existence was found of a specific temperature range within which so-called "undercoolable" melt (UCM) shows easy undercooling. The effect of various heat treatment and precursor type on the location of the UCM temperature range for Fe-alloy is shown in Fig. 2. Undercooling ability of the molten pieces of massive ingot at thermal cycling grows after overheating of melt to 1430°C and above, Fig. 2a. Using 3-mm granules crystallized prior to melt undercooling as melt precursors, results in significant reduction to 1320-1340°C in the lower temperature limit of UCM. The undercooling ability grew when using 0.3-mm granules prepared at higher cooling rate.

Further reduction of the low temperature limit of easy undercooling to 1290°C was obtained after isochronal annealing of the molten 0.3-mm granules, Fig. 2b. The influence of

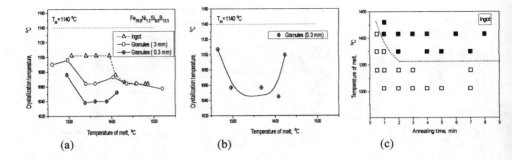

(a)   (b)   (c)

Fig. 2. The effect of heat treatment by thermal cycling (a), isochronal (b) and isothermal (c) annealing on undercooling of $Fe_{76.6}Ni_{1.3}Si_{8.6}B_{13.5}$ alloy. Closed squares at bottom (c) plot correspond to the "undercoolable" melt.

isothermal annealing of the melt on the location of UCM is shown in Fig. 2c. The minimal temperature of easy undercooling could be reduced from 1450 to 1370^0C if the melt holding time exceeded 2 min however further increasing did not expand UCM towards low temperatures.

Molten ingot of Co-alloy also keeps an undercooling ability after overheating above T_m only, as high as 1560^0C and more. Molten 0.3-mm granules show two UCM ranges: at $1200-1250^0C$ and then above 1520^0C, Fig. 3a. Isochronal annealing of melt prepared from $\varnothing 0.3mm$ granules does not change the UCM range (Fig. 3b) as compared to results of thermal cycling, Fig. 3a. Annealing of the melt prepared from amorphous ribbon also does not change the type of the dependence obtained, however it extends the low temperature UCM range, Fig. 3b. The experiments on isothermal treatment of molten rapidly quenched $\varnothing 0.3mm$ granules show that for holding times exceeding 4 min, two UCM ranges merge into one with lower temperature limit of about 1200^0C, Fig. 3c.

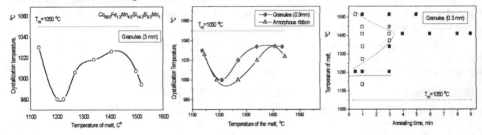

Fig.3. Effect of heat treatment by (a) thermal cycling, (b) isochronal and (c) isothermal annealing on Co-base alloy undercooling. Closed squares at the bottom of (c) correspond to UCM.

Estimation of GFA of Fe and Co alloys at quenching from UCM range.

Correct estimation of GFA value requires analysis of the melt and glass energies or physical parameters, which is problematic. The thickness of the amorphous layer usually used for GFA estimation is a technology rather than physical parameter as it strongly depends on the cooling conditions and quench method. Every method for preparation of an amorphous article must provide continuity and stability of the cooling process with certain values of the melt temperature and viscosity. We think it is impossible to determine GFA of the alloys investigated in all UCM range with one melt quenching techniques only. Each technique requires its own melt overheating above T_m: the casting processes are stable only at $T>1.2T_m$; amorphous wire at $T<1.2T_m$; and amorphous granules formation and plasma coating at $T>1.5T_m$. Therefore it seems reasonable to use several melt quench methods for GFA estimation.

The experimental results on estimation of GFA of molten precursors of Fe and Co alloys at jet casting and spinning methods, are given in Figs. 4, 5.

One can note that both undercooling and glass formation are stimulated mainly by melt quench from the same temperature range. Choosing a proper precursor and melt heat treatment (melt isothermal annealing modes are chosen according to Figs. 2, 3) enables to obtain the thickness of amorphous layer of up to 0.5 mm in cast wedge-like samples of both alloys.

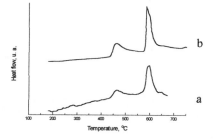 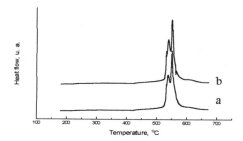

Fig. 4. DSC curves of RQ "thick" samples of Co-base alloy: a) 0.5-mm amorphous edge of wedge-shape casting; b) ribbon of 60 μm in thickness, $T_{quench}= 1270^0C$

Fig. 5. DSC curves of RQ "thick" (70 μm) Fe-base ribbons prepared from: a) 0.3-mm granules, b) ingot. $T_{quench}= 1380^0C$.

Exothermic effect on the DSC curves of the 0.5 mm light colored non etchable edge of wedge-shape casting of Co-alloy confirms its amorphous structure, Fig. 4a.

Using modes providing melt easy undercooling, "thick" amorphous ribbons were prepared also by melt spinning: 70μm in thickness for Fe-based alloy, Fig. 5a,b and 60μm for Co-based alloy, Fig. 4b. Using granules as a precursor of molten alloys improves both the thickness and quality of melt spun ribbons compared to those prepared by industrial technology for alloys with similar composition (25-30 μm or less). The lower UCM range (1200-1250⁰C) corresponds to high values of the melt viscosity for Co-based alloy 0.3-mm molten granules. It provides a stable INROWASP process to prepare ductile amorphous wire (0.14-0.18 mm).

Attempts to prepare thick amorphous layers of Fe and Co-based alloys by jet casting method at strong melt overheating (about 1600⁰C) failed. However INROWASP process for preparation of rapidly quenched granules, is stable for both alloys if the overheating temperature of the melt of massive ingots is above 1500⁰C. This temperature falls into upper UCM range and, amorphous granules with diameter of up to 0.2 mm were prepared easily for both alloys.

DISCUSSION

Melt transforms into glass if being undercooled only. GFA is determined by the chemical composition of the alloy and controlled by the viscosity value of the undercooled melt rather than by the degree of undercooling. Therefore, the value of undercooling can not be used as a glass formation criterion for alloys with different compositions. However, all the factors that favor melt undercooling like melt temperature before cooling, cooling rate, absence of impurities, etc., favor also glass formation. Obviously if UCM range is known then quenching of the melt to prepare articles with amorphous structure should be done from this range.

Crystalline structure of the initial massive ingot is non-equilibrium since it contains refractory primary phases, impurities, and inclusions. To obtain equilibrium melt, such an ingot must be hold at high temperature which results in additional melt contamination and does not favor undercooling and glass forming effects. Near eutectic alloys with certain cooling conditions provided, expose melt easy undercooling for 150-300^{0}C, which allows crystallization with formation of fine dispersed equiaxial structure. Repeat melting reduces the temperature necessary for easy undercooling and glass formation. Anticipated melt cooling range for which the undercooling effects are reliably observed, ranges 10^5 to 10^1 K/s. The utmost melt undercooling is achieved at quench with formation of amorphous phase, usually at the cooling rate close to 10^5 K/s. According to DTA data, usage of the cooling rate below 10^1 K/s can sharply reduce the melt ability to undercooling. The ability to undercooling and glass formation is increasing if there are no refractory primary phases in the precursor which is met for Fe-based and Co-based precursors prepared at the cooling rate of >10 K/s and >10^3 K/s, respectively. We can conclude that such precursors worthwhile to prepare by the method of granular metallurgy.

CONCLUSIONS

The results of the present study show that heat treatment of melts prepared from rapidly quenched precursors (ribbons or granules) enables to improve its undercooling and GF ability. For near eutectic $Co_{69.6}Fe_{1.3}Mn_{4.5}Si_{14.3}B_{9.3}Mo_1$ and $Fe_{75.5}Ni_{1.3}Si_{8.6}B_{13.5}$ alloys the approach used allows to reduce the UCM temperature to T= T_m + (130÷200)^{0}C. GFA depending on melt quench temperature was estimated by spinning, casting, and INROWASP. It was shown that the thickness of amorphous layer grows several times compared to known values due to proper choice of precursor and heat treatment of its melt.

ACKNOWLEDGMENTS

This work is partially supported by RFBR No. 97-02-17753 and INTAS No. 96-2126 Grants.

REFERENCES

1. V.V. Molokanov and V.N.Chebotnikov, Key Eng. Mater. **40-41**, p. 319 (1990).

2. A.Inoue, T.Zhang and T.Masumoto, J. Non-Cryst. Solids, **156-158**, p. 473 (1993).

3. M.Petrzhik, V.Molokanov, T.Mikhailova, Metally (Russian Metallurgy), **4**, p. 152. (1996)

4. V.V.Molokanov, M.Petrzhik, T.Mikhailova, LAM-10, 1998, Dortmund, Germany, submitted.

5. X.H.Lin, W.L.Johnson and W.K.Rhim, Mat. Trans. JIM, **38**, p. 473 (1997).

6. J.Eckert, N.Mattern, M.Zinkevitch and M.Seidel, Mat.Trans., JIM, **39**, p. 623 (1998).

7. V.Kalita, D.Komlev, V.Molokanov, et al., Metally (Russian Metallurgy), **4**, p. 132 (1996).

8. D.M.Herlach, Mat. Sci. Eng., **A226-228**, p. 348 (1997).

9. M. Schwarz, A.Karma, K.Eckler and D.M.Herlach, Phys. Rev. Lett., **73**, p. 1380 (1994).

10. R.Willnecker, D.M.Herlach and B.Feurbacher, Appl. Phys. Lett., **56**, p. 324 (1990).

11. L.Q.Xing and P.Ochin, J. Mat. Sci. Lett., **16**, p. 1277 (1997).

12. P.Manov, S.I.Popel, P.I.Buler, et al., Mat. Sci. Eng., **A133**, p. 535 (1991).

PROBING SLOW ATOMIC MOTIONS IN METALLIC GLASSES USING NMR

X.-P. Tang[1], Ralf Busch[2], William L. Johnson[2], and Yue Wu[1] †

[1]Department of Physics & Astronomy, University of North Carolina, Chapel Hill, NC 27599-3255; [2]Keck Laboratory of Engineering Materials, California Institute of Technology, Pasadena, CA 91125

ABSTRACT

We report a nuclear magnetic resonance (NMR) study of slow atomic motions in Zr-Ti-Cu-Ni-Be bulk metallic glasses. The employed ^{9}Be spin alignment echo technique is able to probe Be motions with jump rate below 0.1 Hz. It was found that the Be motion is spatially homogeneous. The jump rate follows a perfect Arrhenius temperature dependence. The measured activation enthalpy of 1.2 eV is nearly identical to that obtained by Be diffusivity measurement using elastic backscattering (EBS); this indicates that energy barriers are the same for short and long range Be motions. The present work provides direct experimental evidences that exclude vacancy-assisted and interstitial diffusion mechanisms for Be motions in these systems. The result is interpreted in terms of the spread-out free volume fluctuation mechanism.

INTRODUCTION

In the investigation of atomic motions in metallic glasses, most experiments were conducted for measuring viscosity [1] and diffusivity [2-5]. Since viscosity and diffusivity detect the long range effect of atomic motions, these measurements do not probe directly the characteristics of a single jump of atomic motion and are not sensitive to distribution of energy barriers and spatial inhomogeneity of atomic motions in amorphous systems. To reveal the microscopic nature of atomic motions in metallic glasses, characterization of short range atomic motions is necessary.

NMR is well known for its merit of directly probing local atomic motions. NMR has made significant contributions to the understanding of atomic motions in numerous systems. However, the potential of NMR has not been realized for studying motions in metallic glasses. In this work, we report an NMR study of slow atomic motions in bulk metallic glasses $Zr_{41.2}Ti_{13.8}Cu_{12.5}Ni_{10}Be_{22.5}$ (vit1) and $Zr_{46.75}Ti_{8.25}Cu_{7.5}Ni_{10}Be_{27.5}$ (vit4) [6]. The ^{9}Be spin alignment echo technique [7, 8] was used to directly probe slow atomic motions with Be jump frequency as low as 1 jump per 10 seconds. The Be motion is found to be spatially homogeneous. The Be motion detected in this work and in the EBS studies [2] follow Arrhenius behavior with identical activation enthalpy. This indicates that the energy barriers for short range Be motion is the same as that for long range motion. This work provides direct experimental evidences that are inconsistent with vacancy-assisted and interstitial diffusion mechanisms. It suggests that in vit1 and vit4 the Be motion is through a direct diffusion process involving thermal fluctuation of spread-out free volume.

EXPERIMENT

For the NMR measurement the glass ingots were cut into thin slices of 200 μm thick and vacuum-sealed in glass tubes. A home-built high temperature probe was used to conduct NMR experiments on a pulsed spectrometer at 9.4 and 4.7 Tesla. In this work the 90° rf pulse is about 4 μs which ensures non-selective excitations of ^{9}Be nuclear magnetic resonance.

The features of the ^{9}Be spectra, spin-lattice relaxation time T_1 and spin-spin relaxation time T_2 are nearly the same in vit1 and vit4 glasses [7, 8]. The spectra consist of two components. The narrow central peak is mainly broadened by a distribution of Zeeman interactions; the broad line is associated with the satellite transitions broadened by the first-order quadrupole interactions which depend on both the principal values of the electric field gradient (EFG) tensor at Be and the relative orientation of the EFG tensor with respect to the external magnetic field. Because of the randomness of the local environment around Be the EFG tensor is different from site to site in metallic glasses. This randomness is reflected by the fact that the characteristics of quadrupole powder pattern is smeared out. In the entire temperature range of the current investigation, no evidence of change of the ^{9}Be spectra and T_2 (about 1.5 ms) is observed. This indicate that fast motions with timescale shorter than 1.5 ms do not occur.

T_1 is obtained by the saturation recovery method. Figure 1 shows the nuclear magnetization M(t) as a function of the recovery time t at 300 K. It can be fitted perfectly with a single exponential function $M^*(t) = \exp(-t/T_1)$ ($M^*(t) \equiv [M(\infty) - M(t)]/M(\infty)$) with $T_1 = 3.55$ s. The inset of Fig. 1 shows $1/T_1$ vs T. In both glasses, $1/T_1$ is approximately proportional to T with $(T \cdot T_1)^{-1} = 0.00105$ $K^{-1}s^{-1}$. Thus, the ^{9}Be T_1 is dominated by the coupling with the conduction electron spins.

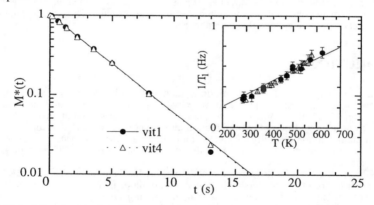

FIG.1. The saturation recovery curves of $M^*(t)$ vs t at 300 K for vit1 and vit4. The decays are perfectly exponential. The inset shows $1/T_1$ vs T and the curves follow $1/T_1 \propto T$.

The ^{9}Be spin alignment echo technique is described in detail elsewhere [7, 8]. The employed pulse sequence is the Jeener-Broekaert sequence $90°_x - \tau_1 - 45°_y - \tau - 45° - \tau_2$. In this experiment, the evolution time τ_1 (about 6 μs to 15 μs at 9.4 Tesla) is fixed; a spin alignment echo is formed at $\tau_2 = \tau_1$. The amplitude of the alignment echo is measured as a function of the diffusion time τ. The spin alignment echo is contributed by those ^{9}Be nuclear spins which experience the same quadrupole interactions during τ_1 and during τ_2. If a ^{9}Be spin experiences different quadrupole interactions during τ_2 and τ_1, it does not contribute to the alignment echo. Specifically, the amplitude of the alignment echo, $S_{echo}(\tau)$, is proportional to the single-particle correlation function $f(\tau) = <\sin(\omega\tau_1)\sin(\omega'\tau_1)>$ [7, 9], where ω and ω' are the quadrupole interactions during τ_1 and τ_2, respectively. The brackets represent ensemble average over all spins. $f(\tau)$ decays if and only if ω' and ω are different. Since the quadrupole interaction is determined by the local environment at Be which is different from site to site in glasses, ω' and ω can thus be used to label the local environments of Be during τ_1 and τ_2, respectively. Atomic motions during the long time interval τ are expected to induce changes of the local environment. This will lead to the change of the quadrupole interaction and the $f(\tau)$ decay. $f(\tau)$ thus represents the percentage of the Be atoms which remain at the unchanged local environment during τ. Assuming that the motion-induced ω change in metallic glasses can be described by a Markovian process, $f(\tau)$ is then proportional to $\exp(-\Omega\tau)$ where Ω is the jump frequency of atomic motions [7, 9]. Technically, the local environment of Be is memorized by creating pure [9]Be quadrupole order during τ under the condition $\tau > T_2$. Therefore, slow atomic motions with jump frequency lower than $1/T_2$ can be probed by measuring the alignment echo decay. The lower limit of detectable Ω is determined by the decay rate of the created quadrupole order. The quadrupole order is not affected by the aforementioned slow motion. The [9]Be spin-lattice relaxation mechanism, which is electronic in origin as shown above, causes the decay of the quadrupole order with decay rate $3/T_1$. Thus, the total decay rate of the alignment echo, $1/T_{QE}$, is given by $1/T_{QE} = 3/T_1 + \Omega$. The function $S_{echo}(\tau)$ is analogous to other types of relaxation functions used for probing the dynamics of atomic motions such as the polarization in dielectric measurements.

In order to achieve rapid data acquisition, which is crucial for conducting experiments on metallic glasses, the pulsed spin-lock technique [10] was used in this work to improve the signal-to-noise ratio. Here a train of 90° pulses is applied following the Jeener-Broekaert sequence. These 90° pulses separated by $2\tau_1$ have the same phase as the third pulse of the Jeener-Broekaert sequence. A train of echoes with the same origin and feature as the first alignment echo are detected and all the echoes are added together. With this technique, each $1/T_{QE}$ can be obtained in a few minutes. This is especially important for the structural relaxation study in which $1/T_{QE}$ is measured versus the isothermal annealing time. The present study shows that the effect of isothermal annealing on Be motion is not observable in vit1 and vit4 in agreement with the EBS results [2].

RESULTS

The alignment echo decay curves of vit4 are shown in Fig. 2. The decay curve can be fitted perfectly with a single exponential function $S_{echo}(\tau) = S_{echo}(0)\exp(-\tau/T_{QE})$. The inset of Fig. 2 displays explicitly $1/T_{QE}$ versus T from 300 K to T_g for vit4. In the low temperature range, $1/T_{QE}$ is linearly temperature dependent; moreover, $1/T_{QE}$ is exactly $3/T_1$. This shows that only the electronic mechanism contributes to the alignment echo decay in the low temperature range. The rapid increase of $1/T_{QE}$ at high temperatures indicates the drastic increase of the contribution of slow motion. Similar results were obtained for vit1. Figure 3 shows Arrhenius plot of Ω (= $1/T_{QE}$ - $3/T_1$) for both glasses. The Arrhenius behavior $\Omega(T) = \Omega_0 \exp(-E_a/k_BT)$ is followed by both glasses with the same activation enthalpy $E_a = (1.2 \pm 0.15)$ eV. Whereas the fittings give $2\times10^9 < \Omega_0 < 4\times10^{11}$ Hz for vit1 and 1×10^{10} Hz $< \Omega_0 < 2\times10^{12}$ Hz for vit4. Since Be is the smallest component in vit1 and vit4, Be motion is expected to dominate the observed motion. The measured Ω is thus attributed to the Be jump frequency. As discussed above, the [9]Be alignment echo technique detects single jump of Be motion and thus probes short-range Be motion. It is different from the previous EBS study [2] which investigates the long-range atomic motion.

Previous study suggests that the long-range and short-range atomic displacements are sensitive to different parts of the energy barriers [11]. For long-range displacement the high energy barriers provide the main hindrance to diffusion whereas for short-range displacement, atoms would most likely jump over the low energy barriers [11]. Thus, it is somewhat surprising that the observed activation enthalpies in both works are nearly identical [2]; the ratio of $\Omega(T)$ between vit1 and vit4 is also consistent with the ratio of the Be diffusivities between vit1 and vit4 measured by EBS [2]. Thus, the activation enthalpies for both local Be motion and long range motion are the same and represent an averaged value such as in the collective diffusion mechanism.

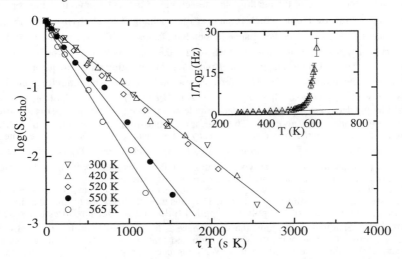

FIG.2 The normalized alignment echo intensity S_{echo} is plotted vs the diffusion time τ (scaled by T) in vit4 at various temperatures. The corresponding curves are exponential fits. The inset shows the alignment echo decay rate T_{QE} vs T for vit4.

As aforementioned, $S_{echo}(\tau) = f(\tau) \exp(-3\tau/T_1)$. The ^9Be quadrupole order decay due to the electronic origin follows $\exp(-3\tau/T_1)$. $f(\tau)$ represents the percentage of the Be atoms which do not execute a jump during the diffusion time τ. Since $S_{echo}(\tau)$ follows a single exponential function (Fig. 2), $f(\tau)$ is also a single exponential function. Thus, the mean jump probability w averaged over the Be atoms which did not execute a jump during the time interval τ should be independent of τ. Since the mean jump probability averaged over all Be atoms is generally independent of τ, this implies that the mean jump probability averaged over those Be atoms that have already executed a jump during τ is also independent of τ and is equal to w. Therefore, at each moment of τ the availability of the diffusion vehicle (vacancies, interstitials, free volume, etc.) is the same for the remaining atoms as for those that have already executed a jump during τ. This is true if the diffusion vehicles in vit1 and vit4 is spatially homogeneous relative to all Be atoms such as interstitials or homogeneous free volume. For vacancy mechanism, however, at the moment of τ the spatial distribution of the vacancies relative to the remaining Be atoms depends on the previous jumps during the time interval τ. Thus, the mean jump probability for the remaining Be atoms (which did not execute a jump during the time interval τ) is expected to be dependent on τ. Then, $f(\tau)$ would not be a single exponential function. For the vacancy-assisted diffusion, the non-exponential decay of $f(\tau)$ should be evident at short time intervals of τ. The vacancy concentration in vit1 and vit4 is expected to be low. The measured free volume in vit1 is only about 0.1% [13]. Thus, there exists a distribution of distances between Be atoms and vacancies. Given a time interval τ, there should be a distribution of Be jump probability under vacancy-assisted diffusion; a Be atom close to a vacancy at the beginning of τ will have a higher probability to execute a jump than those initially far from vacancies. In this case, $f(\tau)$ is described by: $f(\tau) = \sum_i A(w_i)\exp(-w_i\tau)$, where w_i is one of the jump rate of Be and $A(w_i)$ is the distribution of w_i. This is inconsistent with the observed single exponential function of the [9]Be alignment echo decay.

Diffusion of small atoms is frequently attributed to direct interstitial diffusion of a single atom jump through interstitial-like sites. For such a mechanism Be jump rate can be written as [3]: $\Omega = Z_i v_i \exp(S_{i,m}/k_B) \exp(-H_{i,m}/k_BT)$. $H_{i,m}$ and $S_{i,m}$ are the migration enthalpy and entropy, respectively. v_i is the attempt frequency which is comparable to the Debye frequency of about 10^{13} Hz. Z_i is the number of nearest interstitial sites. Based on the measured prefactor values of $\Omega_0 \equiv Z_i v_i \exp(S_{i,m}/k_B)$, this implies that $-8.5 < S_{i,m}/k_B < -3.2$ for vit1 and $-6.9 < S_{i,m}/k_B < -1.6$ for vit4. This excludes the conventional interstitial mechanism for which $S_{i,m}/k_B = 1$ is expected. Since vit1 and vit4 consist of small (Be), medium (Cu and Ni), and large (Zr and Ti) atoms, it favors dense atomic packing. This might be effective in reducing interstitial diffusion.

The observed significant negative effective entropy is also inconsistent with diffusion assisted by thermally generated vacancies. If Be atoms diffuse through thermally generated vacancies, the Be jump frequency can be written as [3]:

$$\Omega = Z_v v_{v,0} \exp\{(S_{v,f}+S_{v,m})/k_B\} \exp\{-(H_{v,f}+H_{v,m})/k_BT\} \qquad (1)$$

where $H_{v,f}$ and $H_{v,m}$ are the formation and migration enthalpy, respectively and $S_{v,f}$ and $S_{v,m}$ are the corresponding entropy. $v_{v,0}$ is about 10^{13} Hz and Z_v is the number of possible jump paths of a vacancy. In a similar discussion as for the interstitial diffusion mechanism, the measured Ω_0 is thus also too small to be compatible with vacancy-assisted diffusion mechanism for which $S_{v,f}/k_B = 1$ and $S_{v,m}/k_B = 1$ are expected.

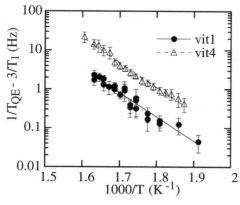

FIG. 3. $\Omega = 1/T_{QE} - 3/T_1$ is plotted versus $1/T$ for vit1 and vit4. The lines are Arrhenius fits with an activation enthalpy of 1.2 eV.

Therefore, the observed negative effective migration entropy and the homogeneous nature favor the spread-out free volume mechanism [12]. The mobility of such free volume is too low to have a direct effect on Be diffusion below T_g. However, thermal fluctuations of spread-out free volume could lead to the temporary formation of a volume larger than a critical volume; Be hopping into such temporarily formed volume leads to Be diffusion. The observed negative effective migration entropy can thus be explained by the small probability of forming such critical volume [3]. The faster atomic motions in vit4 could be due to a larger free volume concentration in vit4 than in vit1. Since such free volume fluctuation is the same near all Be atoms [12], this mechanism is also consistent with the observed pure exponential decay of $S_{echo}(\tau)$.

CONCLUSION

The present NMR study provides direct information on the characteristics of Be diffusion in the novel bulk metallic glasses vit1 and vit4. No distribution of Be jump probability is visible through alignment echo decay. The result is consistent with direct diffusion process for Be motion involving thermal fluctuations of spread-out free volume. The activation enthalpy detected by NMR is essentially the same as that determined by EBS diffusion measurement. This implies that energy barriers for long range and short range atomic motions are the same. Again, this is consistent with spread-out free volume fluctuations which involve collective motions of many atoms.

ACKNOWLEDGMENTS

This work was supported by the U.S. Army Research Office under the contract DAAH04-96-1-0185, the National Science Foundation under the contract DMR-9520477, and the Department of Energy under the contract DEFG-03-86ER45242.

REFERENCES

†E-mail: yuewu@physics.unc.edu
1. R. Busch, A. Masuhr, E. Bakke, and W. L. Johnson, Mater. Sci. Forum **269-2**, 547 (1998).
2. U. Geyer, S. Schneider, W. L. Johnson, Y. Qiu, T. A. Tombrello, and M.-P. Macht, Phys. Rev. Lett. **75**, 2364 (1995); U. Geyer, W. L. Johnson, S. Schneider, Y. Qiu, T. A. Tombrello, and M.-P. Macht, Appl. Phys. Lett. **69**, 2492 (1996).
3. W. Frank, A. Hörner, P. Scharwaechter, and H. Kronmüller, Mater. Sci. Eng. **A179**, 36 (1994); W. Frank, J. Horvath, and H. Kronmüller, ibid. **97**, 415 (1988).
4. F. Faupel, Phys. Stat. Sol. (a) **134**, 9 (1992); P. Klughist, K. Rätzke, S. Rehders, P. Troche, and F. Faupel, Phys. Rev. Lett. **80**, 3288 (1998).
5. A. Grandjean and Y. Limoge, Acta Mater. **45**, 1585 (1996).
6. A. Peker and W. L. Johnson, Appl. Phys. Lett. **63**, 2342 (1993).
7. X.-P. Tang and Y. Wu, J. Magn. Reson. **133**, 155 (1998).
8. X.-P. Tang, R. Busch, W. L. Johnson and Y. Wu, Phys. Rev. Lett., in press.
9. H. W. Spiess, J. Chem. Phys. **72**, 6755 (1980).
10. J. R. C. van der Maarel, J. Chem. Phys. **94**, 4746 (1990).
11. H. Kronmüller, W. Frank, and A. Hörner, Mater. Sci. Eng. **A133**, 410 (1991).
12. F. Spaepen, Acta Metall. **25**, 407 (1977).
13. A. Masuhr, T.A. Waniuk, R. Busch, and W. L. Johnson, to be published.

Part III

General Topics in Bulk Metallic Glasses (Poster Session)

NMR INVESTIGATIONS OF THE BULK METALLIC GLASS
$Zr_{55}Cu_{30}Al_{10}Ni_5$

W. HOFFMANN* **, M. BAENITZ* **, K. LÜDERS*, A. GEBERT***, J. ECKERT***,
L. SCHULTZ***
* Fachbereich Physik, Freie Universität Berlin, Arnimallee 14, D-14195 Berlin,
Germany
** Current Address: Max-Planck-Institute of Chemical Physics of Solids, Bayreuther
Str. 40, D-01187 Dresden, Germany
*** IFW Dresden, Institut für Metallische Werkstoffe, Helmholtzstr. 20, D-01069
Dresden, Germany

ABSTRACT

Nuclear magnetic resonance (NMR) was applied for structural investigations of the bulk metallic glass system $Zr_{55}Cu_{30}Al_{10}Ni_5$. The ^{63}Cu as well as ^{27}Al resonance was used. For both nuclei, two different spin-lattice relaxation rates were found which can be explained by different local environments of the nuclei.

INTRODUCTION

Multicomponent metallic glasses have recently attracted attention since they allow to prepare bulk samples. A prominent group are Zr-based alloys [1-3] which are known to reveal outstanding mechanical properties like high strength, low Young's modulus, some microplasticity and high wear resistance. In this contribution, nuclear magnetic resonance (NMR) investigations on the bulk metallic glass system $Zr_{55}Cu_{30}Al_{10}Ni_5$ are reported.

The method of NMR is a good tool for getting structural information of metallic glasses [4]. Its parameters depend on things like phase separation, short or long range order or cluster formation. Two of the most important NMR parameters are the resonance frequency ω and the spin-lattice relaxation time T_1. In metallic systems, a typical frequency shift occurs, first discovered by Knight. This Knight shift expresses the enhancement of the local magnetic field at the metallic nucleus relative to the field which would exist in a non-metallic non-magnetic compound of the same nucleus:

$$\omega_M = \omega_0 + \Delta\omega \quad . \tag{1}$$

95

The spin-lattice relaxation time T_1 characterizes the energy exchange of the spin system with the environment. At room temperature, T_1 values for metals range from 10^{-1} to 10^{-3} s. Mainly, the relaxation rate is proportional to the density of states at the Fermi surface, $N(E_F)$, the square of the electron wave function at the nucleus, $\rho^2(E_F)$, and the temperature T:

$$1/T_1 \propto N(E_F)\, \rho^2(E_F)\, T \quad . \tag{2}$$

Both qualities are connected and fulfill in most cases for metallic systems the Korringa relation

$$\Delta\omega^2\, T_1\, T\, S_K = \text{const.} \quad , \tag{3}$$

where S_K is a scaling factor which takes electron-electron correlations into account. For a simple Fermi gas its value is $S_K = 1$.

The application of this method to the metallic glass system $Zr_{55}Cu_{30}Al_{10}Ni_5$ allows to take advantage of two suitable nuclei: ^{63}Cu and ^{27}Al. Such experiments are described in the present contribution. The NMR parameters were determined in a wide temperature range and discussed in view of different local environments of the nuclei.

EXPERIMENTAL

Sample preparation

Bulk amorphous samples of $Zr_{55}Cu_{30}Al_{10}Ni_5$ were prepared and characterized by several methods [1]. Prealloyed ingots were obtained by arc-melting the pure elements under a Zr-gettered Ar atmosphere. To ensure complete homogeneity, cylindrical rods with 3 mm diameter were manufactured by die casting into a Cu mold under Ar atmosphere. The structure and composition was characterized by X-ray diffraction, SEM, TEM, and DSC.

Sample characterization by resistivity and magnetization measurements

A commercial multi measurement system (PPMS, Quantum Design), allowing resistivity as well as susceptibility measurements, was applied to determine the temperature dependence of these two quantities [5]. The resistivity $\rho(T)$ is plotted in Fig. 1. Its value is about 250 $\mu\Omega$cm slightly increasing with decreasing temperature in rather good agreement with the Mooij correlation typical for metallic glasses. The magnetic characterization (Fig. 2) shows a weak and temperature independent paramagnetic behaviour indicating no contributions from para- or ferromagnetic inclusions. Therefore, high field NMR measurements could be performed.

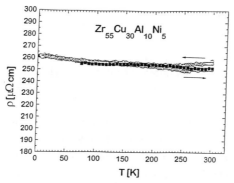

Fig. 1.

Resistivity as a function of temperature. The arrows indicate increasing
or decreasing temperature. (O: four probe method, ■: contact-free method.)

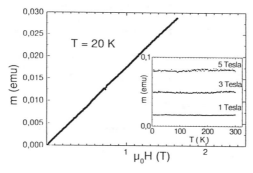

Fig. 2

Magnetic moment as a function of the magnetic field at T = 20 K
and as a funtion of temperature in different fields.

NMR experiments

^{63}Cu and ^{27}Al nuclei were used to determine the NMR parameters Knight shift $\Delta\omega$ and
spin-lattice relaxation time T_1 in the temperature range of 10 to 300 K. Using a 14 T high-field
spectrometer, it was possible to obtain sufficient signal-to-noise ratios. The spectra were
measured using a point-by-point Hahn spin-echo technique at a magnetic field of $B_0 = 13.466$ T.
As reference frequency for the Cu shift the signal of a saturated $CuSO_4$ solution was used. In
case of Al, the reference value was calculated using the gyromagnetic ratios of ^{63}Cu and ^{27}Al [6].

RESULTS AND DISCUSSION

Knight shift and line width

Fig. 3 shows NMR spectra for different temperatures. The line widths (approximately 2 MHz for ^{27}Al and 550 kHz for ^{63}Cu), reflecting the local symmetry of the nuclei environments, are comparable to those of the pure metals [7], indicating a high symmetry also in the amorphous system. The Knight shift values of both nuclei decrease smoothly and monotonically with decreasing temperature. The values for 10 and 200 K are given in Table I in comparison with metallic values. With the exception of the low temperature values for Cu, the Knight shifts in the amorphous system are slightly higher compared to those of the pure metals which might be attributed to the different kind of nearest neighbours and to local distortions in the structural environment of the nuclei.

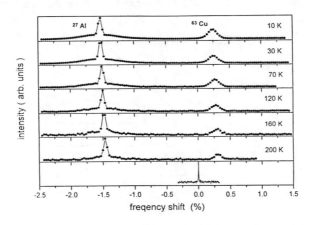

Fig. 3

^{27}Al and ^{63}Cu NMR spectra for different temperatures. For comparison, the ^{63}Cu signal of a saturated solution of $CuSO_4$ is plotted.

Tab. I Knight shifts

metallic glass		metal	
Cu 200 K	0.30 (0.05)	300 K	0.239 %
10 K	0.22 %	4 K	0.238 %
Al 200 K	0.25 %	300 K	0.164 %
10 K	0.24 %	4 K	0.161 %

Spin-lattice relaxation time

More influence of the amorphous structure is found for the spin-lattice relaxation time T_1. Fig. 4 shows a typical time decay of the signal intensity after magnetizing the ^{63}Cu spin system. Usually, this decay is monoexponential and can be described by only one relaxation time T_1:

$$I(t) = I_0\left(1 - e^{-t/T_1}\right) \ . \tag{3}$$

For $Zr_{55}Cu_{30}Al_{10}Ni_5$ such a behaviour was not found. However, the $I(t)$ curves can be fitted by biexponential functions:

$$I(t) = I_{01}\left(1 - e^{-t/T_1}\right) + I_{02}\left(1 - e^{-t/T_1'}\right) \ . \tag{4}$$

The resulting relaxation rates are plotted in Table II and one example is shown in Fig. 5. The occurrence of two relaxation rates means that at least two relaxation mechanisms are present. In the case of Al one of the relaxation rates is very similar to that of the pure metal value. This might be indicative for Al rich inclusions or small clusters in the amorphous alloy. The other relaxation rates seem to be determined by additional relaxation mechanisms caused by disordered atomic environments.

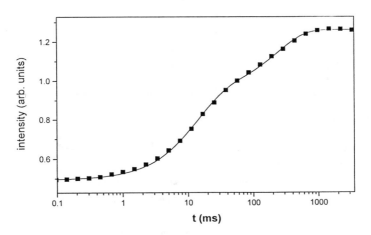

Fig. 4
Time decay of the ^{27}Al NMR signal intensity.

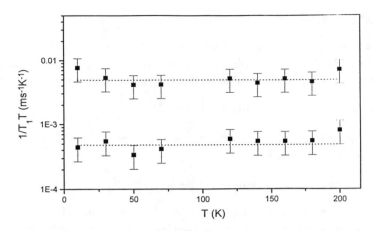

Fig. 5

Temperature dependence of $1/T_1T$ for ^{27}Al.

Tab. II Relaxation rates (300 K)

	metallic glass		metal	
Cu	$1/T_1T$ $(s^{-1}K^{-1})$	S_K	$1/T_1T$ $(s^{-1}K^{-1})$	S_K
T_1	5.5	2.30	0.79	0.52
T_1'	20.9	8.75		
Al				
T_1	0.48	0.30	0.55	0.80
T_1'	4.9	3.06		

REFERENCES

1. A. Leonhard, L.Q. Xing, M. Heilmaier, A. Gebert, J. Eckert, and L. Schultz, Nano Structured Materials **10**, p. 805 (1998).

2. A. Gebert, J. Eckert, and L. Schultz, Acta mater. **46**, p. 5475 (1998).

3. J. Eckert, N. Mattern, M. Zinkevitch, and M. Seidel, Materials Transactions, JIM **39**, p. 623 (1998).

4. H.R. Khan and K. Lüders, phys. stat. sol. (b) **108**, p. 9 (1981).

5. L. Brinker and M. Baenitz, Quantum Design Newsletter **7**, p. 2-7 (1998).

6. NMR Frequency Table, Bruker Almanac, 1992.

7. G.C. Carter, L.H. Bennett, and D.J. Kahan, "Metallic Shifts in NMR", Pergamon Press, Oxford 1977.

Fast X-ray Measurement System for Structural Study in $Zr_{60}Al_{15}Ni_{25}$ Supercooled Liquid

S. Sato*, E. Matsubara**, Y. Waseda***, T. Zhang**** and A. Inoue****
* Inoue Superliquid Glass Project, ERATO, 2-1-1 Yagiyama-minami, Sendai 982-0807, JAPAN
** Department of Materials Science and Engineering, Graduate School, Kyoto University, Kyoto 606-8317, JAPAN
*** Institute for Advanced Materials Processing, Tohoku University, Sendai 980-8577, JAPAN
**** Institute for Materials Research, Tohoku University, Sendai 980-8577, JAPAN

ABSTRACT

A fast x-ray measurement system adopting the geometry of the Debye-Scherrer camera in combination with an imaging plate has been developed for the structural study of supercooled liquid. We carried out the anomalous x-ray scattering (AXS) measurement as well as the ordinary x-ray diffraction measurement with this system in $Zr_{60}Al_{15}Ni_{25}$ supercooled liquid at 720K above the glass transition temperature (693K). A whole diffraction profile of very good counting statistics that even fits to the AXS analyses is obtained for a very short time ($\approx 600s$). The analyses of scattering data observed in the $Zr_{60}Al_{15}Ni_{25}$ supercooled liquid for various annealing times in this system provide us information on a change of local atomic structures in the liquid state.

INTRODUCTION

The conventional amorphous alloys are produced as a thin film and fine powder by quenching melts at an extremely high cooling rate of about 10^5K/s. Recently, a new type of amorphous alloy exhibiting excellent glass forming ability and good thermal stability has been reported in La-Al-TM [1, 2], Zr-Al-TM [3], and Pd-Ni-Cu-P [4] systems. They show a wide supercooled liquid region and a large ratio of a glass transition temperature to a melting temperature. This enables us to produce a bulk amorphous alloy of several cm thick.

The origin of their thermal stability has been discussed from several viewpoints, such as their crystallization behavior in supercooled liquid [5, 6], packing densities [7], heat of mixing [8] and local atomic structures [9]. The environmental structural studies in $Zr_{60}Al_{15}Ni_{25}$ [10], $La_{55}Al_{25}Ni_{20}$ [11], $Zr_{33}Y_{27}Al_{15}Ni_{25}$ [12] and $Zr_{70}Ga_{10}Ni_{20}$ [13] amorphous alloys by the anomalous x-ray scattering (AXS) method suggest that the difference of local atomic structures between amorphous and crystalline state retards nucleation and growth in these amorphous alloys and provide thermal stability [9]. In these studies, the atomic structure of supercooled liquid has been determined in a sample that is quenched to room temperature once heated up in the supercooled liquid region. It is obvious that the structure of this quenched sample is different from that of the supercooled liquid itself. Thus, for a further structural study in these amorphous alloys, a fast x-ray measurement system has been developed to observe scattering intensities of supercooled liquid at an elevated temperature between glass transition and crystallization. In the present study, this new x-ray diffraction apparatus will be introduced and the experimental results of atomic structures in $Zr_{60}Al_{15}Ni_{25}$ supercooled liquid will be discussed.

EXPERIMENTAL

An ingot of the ternary alloy with a nominal composition of 60 at% Zr, 15 at% Al and 25 at% Ni was prepared by arc-melting a mixture of pure zircon (99.9 mass%), aluminum (99.99 mass%) and nickel (99.9 mass%) in a purified argon atmosphere. From the master ingot, a $Zr_{60}Al_{15}Ni_{25}$ amorphous wire of about 0.2mm diameter was produced. For X-ray measurements, it was cut into small pieces of about 8

Mat. Res. Soc. Symp. Proc. Vol. 554 © 1999 Materials Research Society

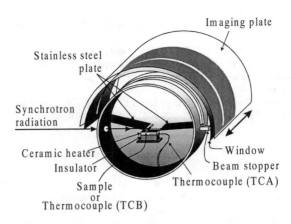

Fig. 1 The schematic diagram of the fast X-ray measurement system with an imaging plate in the geometry of the Debye-Scherrer camera.

mm long and sealed under a purified argon atmosphere in a silica capillary of 0.3mm diameter and 0.01mm thick wall. Measurements of supercooled liquid were carried out at 720K just above the glass transition temperature (693K) [3].

Figure 1 shows the newly developed X-ray diffraction system using the geometery of the Debye-Scherrer camera. The radius of the camera is 150mm. By keeping the diameter of the wire sample less than 0.2mm, the angular resolution of 1.3mrad is achieved. This angular resolution is good enough to measure a broad diffraction profile from an amorphous sample. Two stainless steel plates of 0.5mm thick placed to cover incident and transmitted beams exclude scattering from substances except for the sample. Air scattering is extremely reduced by replacing air with helium gas in the chamber. In the present measurements, the incident energy is selected below Zr K absorption edge (17.9989keV). In this way, the fluorescent radiation of Zr is eliminated. The strong fluorescent radiation of Ni is, however, emitted from the sample. An iron foil of 0.03mm thick is used to eliminate the nickel fluorescence. Fe fluorescence emitted from the iron foil as a result of absorption of nickel fluorescence is absorbed by an aluminum foil of 0.3mm thick. Incidentally, aluminum fluorescence is completely absorbed by air. With these combined filters, the nickel fluorescence is diminished less than 10^{-5} while the elastic radiation is reduced about 20%. By translating an imaging plate vertical to the diffraction plane, four different intensity profiles are recorded continuously without change of the plate. The ceramic heater (195 watts) is placed 2mm away from a sample to obtain a fast heating rate and to avoid scattering from the heater itself. Temperature of the heater was controlled with a thermocouple (TCA in Fig.1) placed under the heater. The temperature of the sample was monitored by another thermocouple (TCB in Fig.1) placed at an end of the sealed capillary tube. With this method, the sample temperature was controlled within ±5K.

X-ray diffraction measurements were carried out at the beam line of BL-3A in the Photon Factory of the National Laboratory for High-Energy Accelerator Research Organization, Tsukuba, Japan. Two incident energies of 17.699 and 17.949keV that correspond to the lower energy side of Zr K absorption edge, were used. A whole intensity profile was observed for 300s in the present system although it takes several hours in the conventional x-ray diffraction with a double-axis diffractometer. The reduction of measurement time is extremely crucial in the present study since the supercooled liquid is thermally unstable and it lasts about 3600s at most [3]. After corrections of absorption and scattering from the capillary, the observed intensity was converted to electron units per atom with the generalized Krogh-Moe-Norman method [14] using the x-ray atomic scattering factors [15], including the anomalous dispersion terms [16]. The Compton scattering was corrected by the theoretical value [17]. Then, the interference function and radial distribution function (RDF) were computed. The details of data processing are described elsewhere [18]-[20].

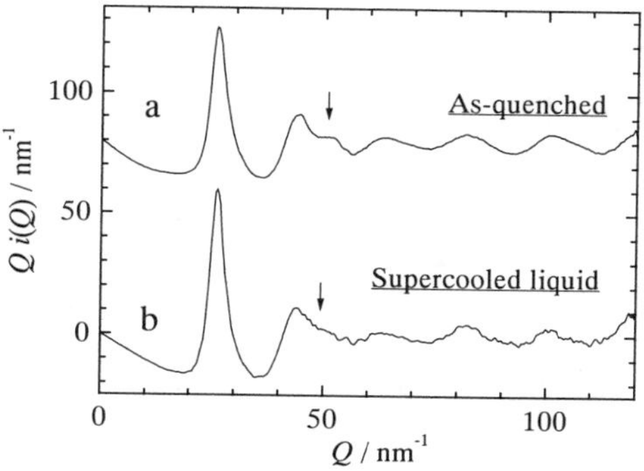

RESULTS AND DISCUSSION

The ordinary interference functions $Qi(Q)$ in the as-quenched $Zr_{60}Al_{15}Ni_{25}$ amorphous alloy and the $Zr_{60}Al_{15}Ni_{25}$ supercooled liquid at 720K observed at 17.699keV are compared in Fig. 2.

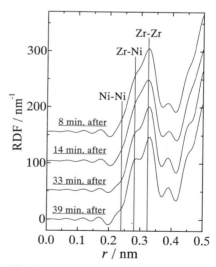

Fig. 2 Interference functions in (a) as-quenched $Zr_{60}Al_{15}Ni_{25}$ amorphous alloy and (b) $Zr_{60}Al_{15}Ni_{25}$ supercooled liquid at 720K.

In the as-quenched sample, a clear shoulder at about 50nm^{-1} is observed at the higher r-side of the second peak. This characterizes the intensity profile of metallic amorphous alloys [21] and disappears in the profile of the supercooled liquid. This indicates some local structural change in the heated sample.

Figure 3 shows the interference functions of the $Zr_{60}Al_{15}Ni_{25}$ supercooled liquid measured at 720 K for various annealing times. It should be stressed that these experimental data were obtained only with the

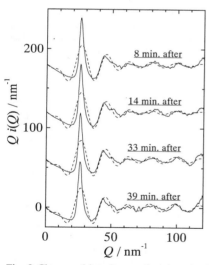

Fig. 3 Change of interference functions in the $Zr_{60}Al_{15}Ni_{25}$ supercooled liquids at 720 K. The solid and dotted curves correspond to the experimental and calculated profiles.

Fig. 4 Change of radial distribution functions (RDFs) of $Zr_{60}Al_{15}Ni_{25}$ supercooled liquids at 720K.

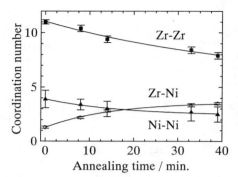

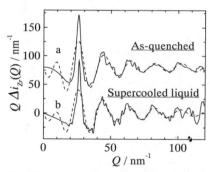

Fig. 5 The variation of the coordination numbers for Zr-Zr, Zr-Ni and Ni-Ni pairs in the $Zr_{60}Al_{15}Ni_{25}$ supercooled liquid during annealing at 720K.

present diffraction system. Oscillations in the functions are gradually magnified with annealing time, which indicates that a local short-range order is progressively developed in the $Zr_{60}Al_{15}Ni_{25}$ liquid at 720K. RDFs which have been evaluated by Fourier transform of the interference functions in Fig. 3 are shown in Fig. 4. The first peak consisting of three peaks sequentially sharpens and their separations become more distinct with the annealing time. Such variation in the RDFs is consistent with that in the interference functions in Fig. 3.

Referring to the atomic radius of each element (Zr: 0.160nm, Al: 0.143nm and Ni: 0.124nm) and the weighting factor of each atomic pair in the $Zr_{60}Al_{15}Ni_{25}$ alloy, we used three Zr-Zr, Zr-Ni and Ni-Ni pairs to fit the RDFs in Fig. 4. The structural parameters of these pairs were determined by the well-established analytical method using the non-linear least squares to fit the interference functions [19]. The resultant coordination numbers are summarized in Fig. 5. The coordination numbers of Zr-Zr and Ni-Ni pairs decrease and that of Zr-Ni pair increases with annealing time. This indicates that the total number of unlike atomic pair increases and that of like pair decreases with increase in annealing time. This change in the coordination numbers is qualitatively understood by approaching the local atomic structure of the supercooled liquid to that of the crystalline phase obtained in the crystallized sample by annealing for a long time in the supercooled liquid region [10].

The anomalous x-ray scattering (AXS) measurements were also tried in the present system. This will verify the present analytical process as well as the present measurement in the new system with an imaging plate in combination with some appropriate filters and a wire sample sealed in a capillary. The environmental interference functions $Q\Delta i_{Zr}(Q)$ for Zr in the as-quenched amorphous alloy and the

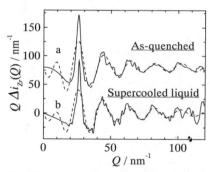

Fig. 6 Environmental interference functions for Zr (a) in as-quenched $Zr_{60}Al_{15}Ni_{25}$ amorphous alloy and (b) in $Zr_{60}Al_{15}Ni_{25}$ supercooled liquid at 720K. The solid and dotted curves correspond to the experimental and calculated results, respectively.

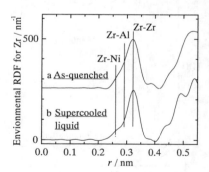

Fig. 7 The environmental RDFs for Zr of as-quenched and supercooled liquid $Zr_{60}Al_{15}Ni_{25}$ alloys.

Table 1 Coordination numbers and atomic distances calculated from the environmental interference functions for Zr in the $Zr_{60}Al_{15}Ni_{25}$ amorphous alloys.

	Zr-Zr		Zr-Ni		Zr-Al	
	r / nm	N	r / nm	N	r / nm	N
as-quenched	0.316±0.001	11.0±0.4	0.267±0.002	1.8±0.2	0.303	0.2±0.7
as-quenched (ref.[10])	0.317±0.002	10.3±0.7	0.267±0.002	2.1±0.5	0.303	-0.1±0.9
supercooled liquid	0.321±0.001	10.2±0.4	0.268±0.003	2.1±0.3	0.303	0.7±1.2

supercooled liquid that were calculated from a difference between scattering intensities observed at two incident energies of 17.699 and 17.949keV below Zr K absorption edge are shown in Fig. 6 (a) and (b), respectively. The ordinary interference functions in a ternary Zr-Al-Ni alloy is a sum of six partial structure factors of Zr-Zr, Zr-Al, Zr-Ni, Al-Al, Al-Ni and Ni-Ni pairs. On the other hand, the environmental RDF for Zr is only a sum of the three partial RDFs of Zr-Zr, Zr-Ni and Zr-Al pairs. This makes ease to evaluate the details of local atomic structure in the present alloys. The environmental RDFs for Zr calculated from Fourier transformation of the energy derivative intereference functions in Fig.6 are also shown in Fig.7 (a) and (b), respectively. The first peak located in the region between 0.22 and 0.35nm in the environmental RDFs is attributed to these three pairs. Referring to the atomic radii of these three components, we deduced that the shoulder of the first peak at about r =0.27 nm is mainly ascribed to Zr-Ni pair. This shoulder is rather distinct in the supercooled liquid state than in the amorphous state, which indicates some change in the local structure in the supercooled liquid. The dotted lines in Fig. 6 are the results obtained by fitting the experimental data by the least-squares variation method originally proposed by Narten [19]. The resultant structural parameters are summarized in Table. 1. Since the peak position of Zr-Al pair in the RDF is not resolved, the atomic distance of Zr-Al pair was fixed at 0.303 nm estimated from the Goldschmidt radii in the fitting procedure. Table 1 includes the coordination numbers and atomic distances obtained for the as-quenched $Zr_{60}Al_{15}Ni_{25}$ amorphous alloy by the conventional AXS measurement [10]. The structural parameters determined from the experimental data using the new system using an imaging plate agree well with those of the previous study. Consequently, it is safely said that the present system also works well even for the AXS measurement. Furthermore, the measurement time is cut one twentieth of the conventional one.

CONCLUDING REMARKS

A new X-ray diffraction system has been developed by introducing the Debye-Scherrer camera geometry coupled with an imaging plate. The new system makes possible a very fast X-ray measurement with a good quality of data. The usefulness and capability of this new system have been demonstrated in *in-situ* measurements in the $Zr_{60}Al_{15}Ni_{25}$ supercooled liquid at 720 K above the glass transition temperature. With this system, the AXS measurements at Zr K absorption edge were also successfully carried out. This verifies the analysis adopted in the present study as well as the present experimental method itself. This will extend the potential of the present system to apply to the local structural study in the multicomponent system.

A small shoulder of the second peak in the interference function that is known to be a typical characteristic feature of metallic amorphous alloys is not well-recognized in the profile of the supercooled liquid. This indicates some structural change occurs in the supercooled liquid. The structural observation of the supercooled liquid at 720K above the glass transition temperature reveals that such structural change gradually progresses with annealing time.

ACKNOWLEDGMENTS

The authors want to thank Dr. M. Kimura, Mr. T. Suzuki, Nippon Steel Corporation and Mr. S. Tanaka, Kyoto University for their kind help on the experiments. This work was partly supported by a Grant-in-Aid for Scientific Research from the Ministry of Education, Science and Culture, Japan (No.09242106). The authors (SS and EM) also thank the staff in the Photon Factory of the National Laboratory for High-Energy Accelerator Research Organization for their assistance (No.97G167).

REFERENCES

1. A. Inoue, T. Zhang and T. Masumoto, Mater. Trans. JIM, **30**, 965-972 (1989).
2. A. Inoue, H. Yamaguchi, T. Zhang and T. Masumoto, Mater. Trans. JIM, **31**, 104-109 (1990).
3. A. Inoue, T. Zhang and T. Masumoto, Mater. Trans. JIM, **31**, 177-183 (1990).
4. N. Nishiyama and A. Inoue, Mater. Trans. JIM, **37**, 1531-1539 (1996).
5. W. D. Bruton, T. O. Callaway, R. H. Langley, B.H. Zhang and D. G. Naugle, Mater. Sci. Eng. A, **133**, 482-485 (1991).
6. N. Mattern, J. Eckert, M. Seidel, U. Kühn, S. Doyle, I. Bächer, Mater. Sci. Eng. A, **226-228**, 468-473 (1997).
7. A. Inoue, T. Negishi, H. M. Kimura, T. Zhang and A. R. Yavari, Mater. Trans. JIM, **39**, 318-321 (1998).
8. A. Inoue, T. Zhang, and T. Masumoto, J. Non-Cryst. Solids, **156-158**, 473-480 (1993).
9. E. Matsubara and Y. Waseda, Mater. Trans. JIM, **36**, 883-889 (1995).
10. E. Matsubara, T. Tamura, Y. Waseda, A. Inoue, T. Zhang and T. Masumoto, Mater. Trans. JIM, **33**, 873-878 (1992).
11. E. Matsubara, T. Tamura, Y. Waseda, T. Zhang, A. Inoue and T. Masumoto, J. Non-Cryst. Solids, **150**, 380-385 (1992).
12. E. Matsubara, K. Sugiyama, A. H. Shinohara and Y. Waseda, Mater. Sci. Eng. A, **179-180**, 444-447 (1994).
13. T. Ikeda, E. Matsubara, Y. Waseda, A. Inoue, T. Zhang and T. Masumoto, Mater. Trans. JIM, **36**, 1093-1096 (1995).
14. C. N. J. Wagner, H. Ocken and M. L. Joshi, Zeit. Naturforsch., **20a**, 325-335 (19965).
15. *International Tables for X-ray Crystallography*, Vol. IV (Kynoch Publishers, Birmingham, 1974) p.94,99.
16. Public database *SCM-AXS* : http://www.iamp.tohoku.ac.jp/.
17. D. T. Cromer and J. B. Mann, J. Chem. Phys., **47**, 1892-1893 (1967).
18. B. E. Warren, *X-ray Diffraction*, (Dover Publisher, New York, 1990) pp.116-142.
19. A. H. Narten, J. Chem. Phys., **56**, 1905-1909 (1972).
20. K. Sugiyama, A. H. Shinohara, Y. Waseda, A. Inoue, J. Non-Cryst. Solids, **192/193**, 376-379 (1995).
21. Y. Waseda, *The Structure of Non-Crystalline Materials*, (McGraw-Hill Inc., USA, 1980). pp.90-92.

PRODUCTION OF ZIRCONIUM-BASED GLASSY ALLOY WIRES BY MELT EXTRACTION METHOD AND THEIR MECHANICAL PROPERTIES

A. KATSUYA *, A. INOUE **
*Graduate School, Tohoku University, Sendai 980-8577, Japan (NHK SPRING CO., LTD.),
a.katsuya@nhkspg.co.jp
**Institute for Materials Research, Tohoku University, Sendai 980-8577, Japan

ABSTRACT

Glassy $Zr_{65}Al_{10}Ni_{10}Cu_{15}$ wires were produced in the diameter range up to about 310 μm by a melt extraction method using a copper wheel in an argon atomospher. The grass transition temperature (T_g) and crystallization temperature (T_x) are 652K and 757K, respectively, and the temperature interval of the supercooled liquid, $\Delta T_x(=T_x-T_g)$ is 105K, in agreement with those for the corresponding melt-spun ribbon. The wires have a nearly real circular cross section and good mechanical properties.

INTRODUCTION

Recently, we have succeeded in producing continuous amorphous alloy wires in Fe- [1,2,3], Co- [1], Ni- [4], Al- [5] and Ti-based [5] systems by the melt extraction method [6,7]. It has been realized that this method can produce directly the wires with a fine diameter and new compositions which cannot be produced by an in-rotating-water melt spinning method [8]. On the other hand, our group succeeded in finding a multicomponent Zr-Al-Ni-Cu system with a wide supercooled liquid region before crystallization exceeding 100K and reported that this alloy has a high glass-forming ability [9]. As the maximum wire diameter depends on the glass-forming ability, the use of the alloy system is expected to produce thick glass alloy wires. The purpose of this paper is to investigate the production of Zr-based glassy alloy wires which have not been obtained up to date and to examine their morphology, thermal stability and mechanical properties.

EXPERIMENT

A quaternary $Zr_{65}Al_{10}Ni_{10}Cu_{15}$ alloy was examined in the present study. The ingots were prepared by arc melting a mixture of pure Zr, Al ,Ni and Cu metals in a purified argon atmosphere. The composition is given in nominal atomic percent. The master alloy ingot was remelted in a ceramic crucible in an argon atmosphere and the wires with diameters ranging from 20 to 350 μm and the length of about 10 meters were prepared by melt extraction using a copper wheel with a steep edge in an argon atmosphere. The diameter of the copper wheel used for the melt extraction was 200 mm and the angle of the edge was fixed 60 degrees. The circumferential speed of the wheel was changed in the range of 2.6 to 63 ms^{-1}. A glassy ribbon with a width of 0.4 mm and a thickness of about 15 μm was also prepared for comparison by single roller melt spinning in an argon atmosphere. The amorphicity of the melt-extracted wires and the crystallization behavior of the glassy phase were examined by X-ray diffraction using Cu Kα radiation, optical microscopy (OM) and differential scanning calorimetry (DSC). The OM observation was made after etching for 10 s at 298 K in a solution of 1% fluoric acid aqueous solution. The heating rate in the DSC measurement was fixed at 0.67 Ks^{-1}. Ductility was evaluated by means of a simple bend test in which bending through 180 degrees without fracture was taken as the criterion of good bend ductility. Hardness and tensile strength of the wire specimens were measured by a Vickers microhardness tester with a 0.49N load and an Instron testing machine at a strain rate of $8.3 \times 10^{-4}s^{-1}$, respectively.

RESULTS

Figure 1 shows the transverse cross sectional shape of the wires with diameters 60 to 305 μm which were prepared in different extraction conditions. The cross section of the wires tends to exhibit concavity in the cross section with increasing wire diameter. This concavity is thought to correspond to the trace of extraction by the copper wheel. However, the wires have a nearly circular cross section with diameters less than about 200 μm and no contrast corresponding to the precipitation of a crystalline phase is seen. Figure 2 shows the outer surface appearance of the melt-extracted wires with diameters 53 and 300 μm. These wires have a smooth outer surface and no distinct ruggedness associated with precipitation of any crystalline phase is seen. The X-ray diffraction patterns of the wires with diameters 50 to 320 μm are shown in Fig. 3, in comparison with that of their melt-spun glassy ribbon of 15 μm in thickness. The X-ray diffraction patterns consist only of a broad peak for the wire samples with diameters less than 300 μm and the ribbon sample, indicating that these samples are composed of a glassy phase. As indexed in the figure, some diffraction peaks are seen for the wire sample with a diameter of 320 μm. We also examined the thermal stability and crystallization behavior of the wires with diameters of 50 and 300 μm, for comparison with the result for the melt-spun ribbon. As shown for the wires in Fig. 4, the grass transition temperature (T_g) and the onset temperature of crystallization (T_x) values are measured to be about 652K and 757K, respectively, and the temperature interval of the supercooled liquid, $\Delta T_x (=T_x-T_g)$ is 105K for the wire samples, in agreement with those for the melt-spun ribbon. From the consistent results for OM observation, outer surface appearance, X-ray diffraction patterns and DSC curves, we can conclude that these alloy wires prepared by the melt extraction method are mostly composed of a glassy phase in the diameter range up to about 310 μm. It is noticed that the critical sample diameter is about 170 μm larger than that for the Fe-Si-B wire sample [1]. The significant increase indicates clearly that this alloy system has high glass-forming ability.

In order to investigate the difference in the quenched structures due to different quenching conditions between the wires prepared by melt extraction and the ribbons prepared by melt spinning, the temperature dependence of the apparent specific heat (C_p) of the glassy $Zr_{65}Al_{10}Ni_{10}Cu_{15}$ alloy in the shapes of wire and ribbon was examined with a DSC. Figure 5 shows the thermograms of the

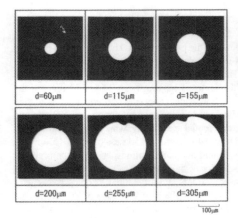

Fig. 1 Optical micrographs of the transverse cross sectional structure in an etched state of the glassy $Zr_{65}Al_{10}Ni_{10}Cu_{15}$ wires.

Fig. 2 Scanning electron micrographs of the outer surface appearance of the glassy $Zr_{65}Al_{10}Ni_{10}Cu_{15}$ wires.

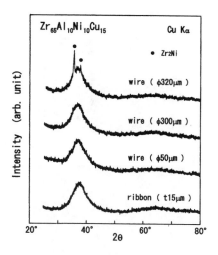

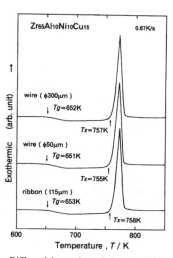

Fig. 3 X-ray diffraction patterns of the glassy $Zr_{65}Al_{10}Ni_{10}Cu_{15}$ wires. The result of the melt-spun amorphous ribbon of 15 μm in thickness is also shown for comparison.

Fig. 4 Differential scanning calorimetric (DSC) curves of the glassy $Zr_{65}Al_{10}Ni_{10}Cu_{15}$ wires. The result of the melt-spun amorphous ribbon of 15 μm in thickness is also shown for comparison.

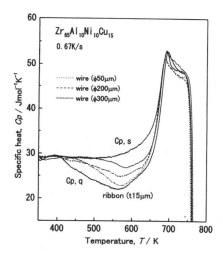

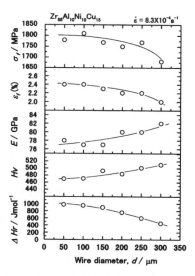

Fig. 5 Change in the apparent specific heat (C_p) of the glassy $Zr_{65}Al_{10}Ni_{10}Cu_{15}$ alloy in wire and ribbon shapes. The $C_{p,q}$ and $C_{p,s}$ represent the specific heats of the samples in as-quenched state and annealed state for 60s at 680K, respectively.

Fig. 8 Changes in ΔH_r, H_v, E, ε_f and σ_f as a function of sample diameter for the glassy $Zr_{65}Al_{10}Ni_{10}Cu_{15}$ wires.

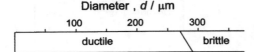

Diameter , d / μm

Fig. 6   Change in the bending ductility with diameter (d) for the glassy $Zr_{65}Al_{10}Ni_{10}Cu_{15}$ wires. "Ductile" and "brittle" represent that the wire can be bent through 180 degrees without fracture and fractures during bending, respectively.

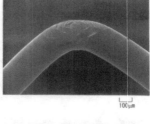

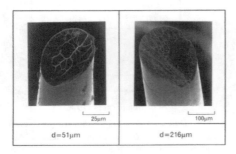

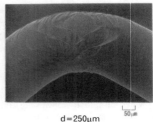

Fig. 9   Scanning electron micrographs revealing the tensile fracture surface appearance of the glassy $Zr_{65}Al_{10}Ni_{10}Cu_{15}$ wires with diameters of 51 and 216 μm.

Fig. 7   Scanning electron micrographs revealing the deformation structure of the glassy $Zr_{65}Al_{10}Ni_{10}Cu_{15}$ wire which were bent through 180 degrees.

glassy $Zr_{65}Al_{10}Ni_{10}Cu_{15}$ wires with diameters of 50, 200 and 300μm, along with the data on the melt-spun ribbon. The heat of structural relaxation (ΔH_r) defined by $\int \Delta C_p (=C_{p,s} - C_{p,q})dT$ where $C_{p,s}$ and $C_{p,q}$ are the heat capacities of the relaxed and as-quenched glassy phases, respectively, is measured to be 975J/mol for the 50μm wire, 755J/mol for the 200μm wire, 455J/mol for the 300μm wire and 1480J/mol for the ribbon and decreases in the order of ribbon > 50μm wire > 200μm wire > 300μm wire. It can therefore be said that the cooling rate in the temperature range below T_g decreases with an increase in the wire diameter.

Figure 6 shows the wire diameter range in which the glassy wires with good bending ductility are prepared for the $Zr_{65}Al_{10}Ni_{10}Cu_{15}$ alloy. The ductile-brittle transition takes place in the vicinity of about 280 μm in diameter. .Figure 7 shows the outer surface appearance of the glassy $Zr_{65}Al_{10}Ni_{10}Cu_{15}$ wire with a diameter of 250 μm which was bent through 180 degrees. Although a number of deformation markings are observed in the severely bent region, no appreciable cracks are seen. The feature of the deformation markings in the bent region is the same as that for Pd-, Pt-, Fe-, Co- and Ni- based amorphous alloy wires produced by the in-rotating-water melt spinning method.

Table 1 summarizes the mechanical properties of Young's modulus (E), tensile fracture strength (σ_f), fracture elongation (ε_f) and Vickers hardness (H_v) for the glassy wire with a diameter of 50 μm. The data of the glassy ribbon prepared by melt spinning are also shown for comparison. In respect of the mechanical properties, although the E, ε_f and H_v of the wire are nearly the same as those for the ribbon, the σ_f value is considerably higher for the wire. The tendency is analogous for the amorphous alloy wires prepared by the in-rotating-water melt spinning method and the reason for the difference is thought to result from the decrease in the degree of stress concentration caused by the increase in smoothness on the outer surface for the wire sample. In any event, it is notable that the glassy

Table 1   Mechanical properties of the glassy $Zr_{65}Al_{10}Ni_{10}Cu_{15}$ wire of 50 μm in diameter. The data of the melt-spun amorphous ribbon of 15 μm in thickness are also shown for comparison.

Form	E (GPa)	σf (MPa)	εf (%)	Hv
wire (φ50μm)	78	1770	2.4	470
ribbon (t15μm × w0.4mm)	77	1440	2.0	470

$Zr_{65}Al_{10}Ni_{10}Cu_{15}$ wire exhibits high σf value reaching 1.8 GPa.

In figure 8 are plotted ΔH_r, H_v, E, εf and σf as a function of sample diameter for the glassy $Zr_{65}Al_{10}Ni_{10}Cu_{15}$ wires. With increasing wire diameter, the H_v and E values increase monotonously. This result corresponds to a decrease of ΔH_r. On the other hand, the σf and εf values remain almost constant excluding the data of the ductile wire with a diameter of 300μm. The reason why the σf value does not increase with an increase of H_v value can be thought to be due to the decrease of wire roundness with increasing wire diameter.

Figure 9 shows the tensile fracture surface appearance of the glassy $Zr_{65}Al_{10}Ni_{10}Cu_{15}$ wires with diameters of 51 and 216 μm. The fracture surfaces consist of smooth and veined regions. This fracture surface appearance is independent of wire diameter and just the same as that for other glassy alloys with good bending ductility [10].

CONCLUSIONS

Glassy $Zr_{65}Al_{10}Ni_{10}Cu_{15}$ wires are produced in the diameter range up to about 310 μm by a melt extraction method using a copper wheel in an argon atomospher. The grass transition temperature (T_g) and crystallization temperature (T_x) are 652K and 757K, respectively, and the temperature interval of the supercooled liquid, $\Delta T_x (=T_x-T_g)$ is 105K, in agreement with those for the melt-spun ribbon. The wires have a nearly real circular cross section and good mechanical properties.

REFERENCES

[1] A. Inoue, A. Katsuya, K. Amiya and T. Masumoto, Mater. Trans., JIM, **36**, 802 (1995).
[2] A. Katsuya, A. Inoue, T. Masumoto, Mater. Sci. Eng., **A226-228**, 104 (1997).
[3] A. Katsuya and A. Inoue, Proceeding of the Special Symposium on Advanced Materials, Nagoya, 87 (1998).
[4] A. Katsuya, A. Inoue and K. Amiya, Inter. J. Rapid Solidification, **9**, 137 (1996).
[5] A. Inoue, K. Amiya, A. Katsuya and T. Masumoto, Mater. Trans., JIM, **36**, 858 (1995).
[6] R. E. Maringer and C. E. Mobley, Vac. Sci. Technol., **11**, 1067 (1974).
[7] J. Strom-Olsen, Mater. Sci. Eng., **A178**, 239 (1994).
[8] I. Ohnaka, T. Fukusako and T. Matsui, J. Japan Inst. Metals, **45**, 751 (1981).
[9] T. Chang, A. Inoue and T. Masumoto, Mater. Trans., JIM, **32**, 1005 (1991).
[10] H. S. Chen, Rep. Prog. Phys., **43**, 353 (1980).

BULK METALLIC GLASSES PRODUCED BY DETONATION GUN SPRAYING

T. P. SHMYREVA

Metallurgical Academy, 4 Gagarin Ave.Dnipropetrovsk 320635, Ukraine, shmyreva@tafa.com

ABSTRACT

Bulk metallic glasses in the form of amorphous coatings were produced by detonation gun spraying of eutectic alloy containing iron, chromium, phosphorus and carbon. The thickness of the resulting homogeneous amorphous deposit may be up to 2-3 millimeters.

The heat capacity, the coefficient of thermal expansion, the corrosion resistance, the hardness and the short range order structure parameters were studied for the bulk metallic glasses and thin ribbon. The investigation revealed that time (for 3 year) and temperature (300-670K) related structural relaxation in amorphous alloys is a nonmonotonic wavelike process having an irreversible nature.

The amorphous detonation sprayed coatings have been successfully employed for improving service properties of medical instruments.

INTRODUCTION

Protection and hardening of surfaces is an important and actual problem for a lot of industrial fields. When high resistance to wear and corrosion is a desired, good candidates are alloy coatings having an amorphous structure. For the first time, bulk amorphous glasses in the form of amorphous detonation spray coatings a few millimeters thick were developed [1,2].

Studies of structure relaxation behavior in amorphous alloys give insight into their nature and allow prediction of how stable amorphous alloy properties will be with changing time and temperature. The latter consideration is of particular importance for protective amorphous coatings. With this in mind, a comprehensive study into relaxation behavior of Fe-Cr-P-C amorphous alloys was carried out.

EXPERIMENT

Amorphous coatings were produced by detonation spraying of Fe-Cr-P-C eutectic alloy. The feed stock powders were made by combination of cast alloy and had crystalline structure. The powder was fed into the detonation gun tube at its closed end. The detonation mixture consisted of oxygen and acetylene. The gas detonation products are hot enough to melt the powder particles. Their velocity and consequently the velocity of particles blasted out from the gun open-end ranges into ultrasonic region. As the molten particles strike the substrate, they spread out and splatter to a thickness of about 10 micrometers. This result in melt solidification rates exceeding 10^6 to 10^7 K/s. It is known from the literature [3] that cooling rates above 10^6K/s are needed to produce an amorphous structure in iron-base eutectic alloys. Thus a single amorphous layer is formed. Another layer and so forth follow it, until a required coating thickness is attained. The final thickness of amorphous deposit may run into 2-3 mm. This is 100 to 300 times more the thickness of amorphous ribbons produced from iron base alloys. An amorphous deposit thus produced may either be used as a coating or constitute a part in itself. For example, an iron-base layer may be deposited on an aluminum substrate. After the aluminum is dissolved in an alkali, the sprayed layer replicating the target shape of a sheet, a hollow cylinder etc.may serve as finished part.

Amorphous plates 2 mm thick deposited by detonation spraying and separated from the substrate as in an earlier work [2] and splat ribbons about 30 µm thick were characterized. Some specimens were studied as made and after 1 year and 3 year holding. Other specimens were heat treated by heating at 10 K/min to 370, 420, 520, 610, 670 K and water quenching. These specimens were characterized as-heat-treated and after 1 and 3 years holding as well. X-ray diffraction method was used for G(r) building and short-range structure calculation. The heat capacity (C_p) was determined using differential scanning microcalorimetry. Potentiodynamic electrochemical measurements were carried out for corrosion property characterization. The standard cell with Ag/AgCl reference electrode and 10% HCl solution were used. The micro hardness was determined by DPH_{50} technique. The thermal expansion behavior was measured in heating at 10 K/min using well-known JWT-102 technique [4].

RESULTS

Fig.1 shows curves of C_p change for bulky amorphous specimen and ribbon during heating at 10 K/min from 300K to 670K. Fig.1 also contains curves of the passivation current density (i_p), the corrosion potential (E), the microhardness (DPH_{50}), the radius of the second coordination sphere (r_2), the structure disordering parameter (r_s) and the coordination number (CN). The samples after water quenching from temperatures 370, 470, 520, 610 and 670 K were investigated for taking all these characteristics.

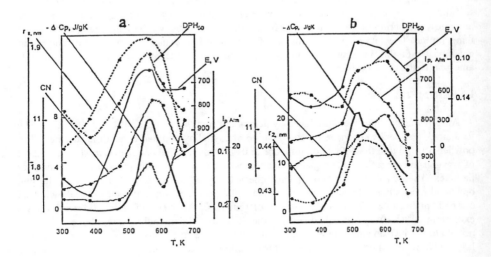

Fig.1. Temperature dependence of the amorphous Fe-Cr-P-C alloy structure and properties:
a-bulky sample; b-thin ribbon

Fig.1 shows the curves of all structure and property characteristics have the profile similar C_p curve for the both bulk (coating) and think (ribbon) amorphous samples. For the bulk sample maximum change in C_p were observed at 520 K, for ribbon maximum change C_p were observed at 570 K. The bulky sample displayed more pronounced absolute relaxation thermal effect than the ribbon. The maximum changes for i_p, E, DPH_{50}, r_2, r_s, "C N" corresponded to the temperatures of relaxation maximum.

Fig.2 depicts behavior of some short-range order parameters and passivation current density in relation to the heat treatment temperature and the holding time.

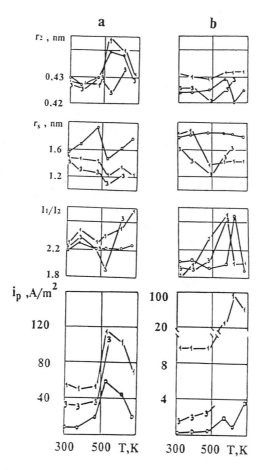

Fig.2. Structural characteristics and passivation current of bulky sample (a) and ribbon (b) of amorphous Fe-Cr-P-C alloy subsequent to quenching(0); 1 yr. holding (1) and 3 yr. holding (3)

The nearest-neighbor atom distance remained constant. By contrast, the radius r_2 of the second coordination sphere (Fig. 2a), the I_1/I_2-the ratio of the first to second G(r) maxima (Fig. 2b) and the structure disordering parameter r_s (Fig.2c) displayed significant changes. The values of I_1/I_2 reached their maxima after 1 yr. and became close to the initial level following 3 yr. The value r_2 showed a rise after 1 yr. but was lower than initial after 3 yr. The r_s declined with time and increasing temperature; this may indicate a stage of further disordering prior to crystallization. The maximum changes in I_1/I_2, r_2 and r_s coincided to the temperatures of relaxation maximum.

Fig.2.d illustrates change in the passivation current i_p, which indicates corrosion resistance of material. Comparison with Fig.2a-c shows that the corrosion resistance behavior with temperature and time is similar to that of the alloy's short-range parameters. The change in passivation current is the greatest on1 yr. holding; after 3 yr. holding i_p nearly regains its initial values. In ribbon heated above the relaxation maximum, depassivation was observed subsequent to 3-yr. holding.

All quantities show maximum change when the heat of relaxation is at its maximum.

The quantitative differences in short-range order variation and in corrosion resistance between the bulky deposit and the ribbon are due to the specific segregation network in the sprayed coatings. The above research suggested a three-tier model of the segregation network in bulky amorphous detonation spray deposits [5]. The network includes: (I) solidified molten particles up to 10 μm in diameter varying composition, (ii) the network that holds them and is mainly contains oxygen but also rich in other light elements, and (iii) intarparticle segregations in the form of spherical amorphous inclusions up to 0.1μm in diameter embedded in the amorphous matrix. The relative temporal change in the corrosion resistance is less significant in detonation-sprayed amorphous deposits than in ribbon. Therefore, they offer possibilities as corrosion coatings capable of maintaining their properties fairly stable at temperatures ranging to 520 K.

It is important that a coating should retain its dimensions in heating. With this in mind, thermal expansion behavior of amorphous coating was investigated. The date is plotted on Fig.3.

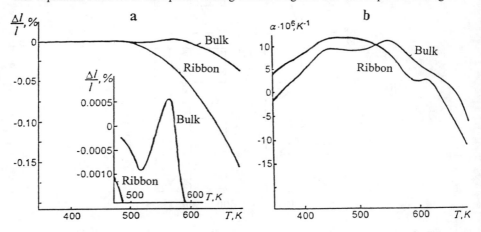

Fig.3. Temperature dependence of the relative expansion/contraction Δl/l parameter (a) and of the thermal expansion coefficient (b) during heating of the bulky amorphous sample and the thin ribbon

The relative expansion/contraction was measured in heating at 10 K/min on specimen of coating stripped from the substrate and of an amorphous ribbon. It is seen from Fig.3a that the both specimens showed practically no dimensional change in the low temperature region up to 470 K. As the temperature was raised from 470 to 520 K, the coating exhibited a slight decrease in $\Delta l/l$, than heating up to 570 K lead to its dimensions expansion. Further heating gave way to increasing contraction again. The ribbon displayed monotonic contraction when heated above 470 K. Fig.3b shows the of coefficient of thermal expansion α vs. T plots both for ribbon and coating. In the range from 420 to 470 K the respective values of α were equal to 11.3×10^{-6} and 9.4×10^{-6} K^{-1}.

According to a well-known theory, the amorphous alloy contract in heating due to decrease of inside free volume [6]. Bulky amorphous sample displayed a lower decrease of the volume in heating than the ribbon and responsible for coating's relatively low value of $\Delta l/l$ between 420 and 470 K. It can be explained by its special bield up described previously. The existence of segregations within any of the solidified molten particles making the coating, their chemical differences, the resultant complexity of rearrangements within any such microregion on any level, and their mutual isolation play an important role. This implies that expansion or contraction may simultaneously occur within an individual microregion and in any of its neighbors.

The detonation sprayed amorphous coatings have been employed to improve service properties of medical instruments made from stainless steel and tools from carbon steel [7]. This resulted in a more than twofold increase in surface hardness and a five- to tenfold improvement in corrosion resistance. For example: stainless steel has passivation current near 60 A/m^2 but amorphous coating has $i_p \approx 3.6$ A/m^2 in 10% HCl solution. It is well known that corrosion resistance is inversely proportional to passivation current [8]. From this point the amorphous coating shows corrosion resistance more than 10 times hier than stainless steel. Stainless steel has hardness only near 450-550 DPH_{50}, but amorphous coatings from FeCrPC alloy may have hardness from 800 to 1100 DPH_{50} depend of detonation spraying parameters and hight treatment. The amorphous coatings prolong service term for the same types of instruments from two to five times.

CONCLUSIONS

The investigation thus revealed that time- and temperature-related structural relaxation and properties changes in an amorphous Fe-Cr-P-C alloy are the nonmonotonic wavelike processes having an irreversible nature. This can be regarded, as an experimental validation that the amorphous state in metals is labile.

The maximum changes for passivation current, microhardness, r_2, r_s, and over amorphous structure characteristic corresponded to the temperatures of relaxation maximum.

Bulky amorphous sample displayed a lower decrease of the volume in heating than the ribbon and responsible for coating's relatively low value of $\Delta l/l$ between 420 and 470 K.

The quantitative differences in short-range order variation and in properties between the bulky deposit and the ribbon are due to the specific segregation network in the detonation sprayed amorphous coatings.

The relative temporal (during 3 years) change in the corrosion resistance is less significant in detonation-sprayed amorphous deposits than in ribbon. Therefore, they offer possibilities as corrosion coatings capable of maintaining their properties fairly stable at temperatures ranging to 520 K. Amorphous coatings have been successfully employed to improve service properties of medical instruments.

REFERENCES

1. T. Shmyreva, in *Chemical Physics of Combustion and Explosion Processes*, (Nauka, Moscow, Russia, 1980), pp.119-121.
2. T. Shmyreva, G.Vorob'ev, Izv. AN SSSR, Metally 6, 204-206 (1983).
3. W.Heineman in *Proceedings Conf. on Rapidly Quenched Metals*, (Wurtzburg, Germany, Sept.1984), p.12-19
4. B.Kantor, in *Bystrozakalenye Metally (Rapidly Quenched Metals)*, (Metallurgiya, Moscow, 1983), pp.52-56.
5. T. Shmyreva, E. Bereza, *Rapidly Quenched Eutectic Alloys*, (Teknika, Kiev, Ukraine, 1990), 144 p.
6. H.S. Chen, in *Amorphous Metal Alloys, Transl. from English, F.E. Lubarskii*, (Metallurgiya, Moscow, 1987), pp.164-183.
7. T. Shmyreva, in *Proceeding of the 7th Thermal Spray Conference*, edited by C. Berndt Boston, MA, 20-24 June, 1994), pp.201-204.
8. Standard ASTM G-89 (Reapproved 1994) Standard Practice for Convention Applicable to Electrochemical Measurements in Corrosion Testing.

ON THE EFFECT OF AL ON THE
FORMATION OF AMORPHOUS Mg-Al-Cu-Y ALLOYS

M. OHNUMA*, S. LINDEROTH, N. PRYDS, M. ELDRUP, AND A.S. PEDERSEN
Materials Research Department, Risø National Laboratory, DK-4000 Roskilde, Denmark,
soren.linderoth@risoe.dk
*Permanent address: National Research Institute for Metals, 305 Tsukuba, Japan

ABSTRACT

The bulk amorphous alloy $(Mg_{1-x}Al_x)_{60}Cu_{30}Y_{10}$ has been studied, in particular the influence of Al concentration on the glass forming ability and on the various transition temperatures. The amorphous single phase has been obtained for x up to 0.07 by casting into a wedge-shaped copper mold. The amorphous alloys were investigated by differential scannning calorimetry (DSC). All the specimens with x = 0 - 0.07 show a clear glass transition. The crystallization temperature decreases with increasing Al concentration, while the temperatures of the glass transition, melting and solidification change only slightly. The DSC measurements show that for Al contents below 0.05 the first exothermic peak which corresponds to crystallization, consists of two overlapping peaks. To clarify the origin of the splitting of the first exothermic peak, the crystallization process of $Mg_{60}Cu_{30}Y_{10}$ alloy has also been studied by x-ray diffraction.

INTRODUCTION

A number of new alloys have recently been discovered that can be made amorphous in bulk form, e.g. Zr-Al-TM [1, 2] and Mg-TM-Ln [3, 4] alloys (TM is a transition metal and Ln is a lanthanide metal). The general features of these new bulk alloys are that they are multicomponent (at least three components) and they exhibit large values of $\Delta T = T_X - T_g$, where T_X and T_g are the crystallization and glass transition temperatures of the amorphous alloy, respectively.

The Mg-based alloys are particularly interesting due to the possibility of achieving a high strength to weight ratio. A tensile strength of more than 800 MPa has been obtained for amorphous Mg-Cu-Y alloys which is about twice the tensile strength of conventional Mg-based crystalline alloys [5]. The glass forming ability (GFA) of $Mg_{90-x}Cu_xY_{10}$ alloys has been investigated by Inoue and Masumoto [5]. They found that at least for Cu contents varying between 10 and 30 at% bulk amorphous alloys were obtained.

The present work forms a part of a study which aims at modifying the mechanical properties and the thermal stability of amorphous $Mg_{60}Cu_{30}Y_{10}$ by the addition of Al (i.e. $(Mg_{1-x}Al_x)_{60}Cu_{30}Y_{10}$). According to the empirical rules suggested by Inoue, increasing the number of elements may increase the GFA of the alloy. With respect to the mechanical properties, the addition of Al is expected to be beneficial – still maintaining a light weight alloy. In this paper we report the influence of addition of Al to the Mg-Cu-Y system on the GFA. Furthermore, the dependence of transition temperatures on the Al content has been investigated in order to clarify the thermal stability of this system.

EXPERIMENTAL

Alloys with the composition $(Mg_{1-x}Al_x)_{60}Cu_{30}Y_{10}$ (x = 0, 0.01, 0.02, 0.04, 0.05, 0.07, 0.17) have been prepared in the following way. A $Cu_{75}Y_{25}$ master alloy (Cu: 99.99%, Y: 99.9% purity) was prepared by induction heating in an alumina crucible. An appropriate mixture of this alloy with Mg (99.99% purity) and Al (99.999% purity) was then melted, at about 1100 K, and

119

quenched into a copper mold. This was done in order to obtain a homogeneous distribution of the elements. The alloy was remelted at about 735 K, which is just above the melting point of the alloy and then quenched into a wedge-shaped copper mold. All preparations were done in an Ar atmosphere.

Differential sacanning calorimetry (DSC) measurements were carried out by the use of a high sensitivity type equipment (SII-DSC120). The sample, typically with a weight of 3 - 4 mg , was placed in an Al crucible. Pure Al was used as reference. DSC scans were made between room temperature and 765 K with heating rates varied between 0.0033 Ks^{-1} and 0.33 Ks^{-1} in order to determine T_g, T_X, melting temperature (T_m) and other transition temperatures associated with a crystallization process. Controlled cooling rates were also employed in order to determine the solidification temperature (T_s). In this paper all the transition temperatures are defined as their onset points.

X-ray diffraction (XRD) studies were carried out by using Cu-K_α in a STOE & CIE diffractometer equipped with a solid state Ge detector. The samples were investigated either in the as-quenched state or after heating in vacuum to 473 K or to 523 K. A constant heating rate of 0.033 Ks^{-1} was applied without any holding time at the maximum temperature. The samples were subsequently cooled rapidly to room temperature.

RESULTS AND DISCUSSION

In order to evaluate the GFA of the various alloys, the wedge-shaped mold was chosen because it gives rise to different cooling rates [6]. At the lower part of the wedge-formed samples, characterized by relatively high cooling rates, the alloy consists of an amorphous phase as evidenced by XRD. The XRD patterns in Fig. 1 show no indication of crystalline phases. Figure 2 shows the maximum thickness (t) of the amorphous single phase as a function of Al content. The thickness is seen to decrease with increasing Al content, reaching zero for x ≈ 0.1. In fact, a $(Mg_{0.83}Al_{0.17})_{60}Cu_{30}Y_{10}$ alloy solidified in the wedge mold exhibited no amorphous region.

Measured DSC curves for amorphous specimens with different Al contents are shown in Fig. 3. Three main features are observed: (1) At 400 – 420 K all specimens exhibit a glass transition (onset temperature T_g),

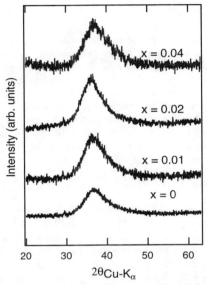

Fig. 1 XRD profiles for as-quenched $(Mg_{1-x}Al_x)_{60}Cu_{30}Y_{10}$ alloys with various aluminum contents.

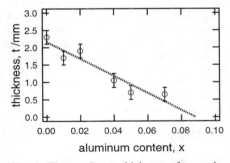

Fig. 2 The maximum thickness of amorphous $(Mg_{1-x}Al_x)_{60}Cu_{30}Y_{10}$ single phase obtained by casting in a wedge-shaped copper mold.

(2) at 450 – 475 K a first stage of crystallization takes place and (3) at 550 – 570 K a second stage of crystallization (onset temperature T_{X3}) takes place. The first crystallization stage seems to consists of two exothermic peaks. The onset temperatures of the first one we call T_{X1} and of the second one T_{X2} (see Fig. 3). Figure 4 shows the variation of the various temperatures as functions of Al content.

T_g changes only little, while T_{X1} decreases by adding Al. Accordingly, the temperature difference between T_g and T_{X1} decreases by addition of Al. The temperatures of melting, T_m and solidification, T_s during cooling (not shown in Fig. 3) depend only little on the Al contents as shown in Fig. 4.

The exothermic peak at T_{X3} is thought to correspond to recrystallization or transformation from a metastable phase to a stable phase. T_{X3} shifts to higher temperature with increasing Al content. This exothermic peak is not visible in x = 0.04, but exists in the curves of x = 0.05 and 0.07. However, it becomes extremely small and diffuse.

With increasing aluminum content, the highest exothermic peak of the first stage of crystallization does not only shift to the lower temperature but also becomes diffuse. The total heat of crystallization of the first stage can roughly be estimated from the area under the double-peak. The heats of crystallization for the specimens with x = 0 - 0.02 are the same within the error and those for the specimens with x = 0.04 - 0.07 are about 80% of the one for x = 0.

In the first stage of crystallization for x = 0, 0.01, and 0.02, a low peak at T_{X1} appears before the higher peak at T_{X2}. The difference between the onset temperatures, $T_{X2} - T_{X1}$, increases with increasing aluminum content up to x = 0.02. For x = 0.04 the peak on the higher temperature side is the smaller

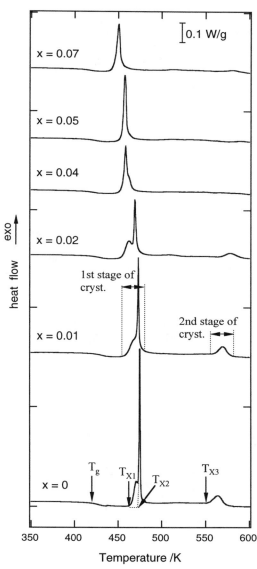

Fig. 3 DSC profiles for as-quenched $(Mg_{1-x}Al_x)_{60}Cu_{30}Y_{10}$ alloys with various aluminum contents. All profiles were measured with a heating rate of 0.033 Ks^{-1}. The different transition temperatures are indicated by arrows.

121

one. However, the splitting of the peak is not visible in the specimen with x = 0.05 and 0.07. A small pre-peak accompanied with a main crystallization peak seems also to be present in the DSC measurements by Busch et al. [7] on $Mg_{65}Cu_{25}Y_{10}$, but they did not discuss this point.

We have also performed DSC measurements using different heating rates (0.0033 Ks^{-1} up to 0.33 Ks^{-1}). In Fig. 5 DSC results are shown for $Mg_{60}Cu_{30}Y_{10}$ for heating rates of 0.01 Ks^{-1} and 0.33 Ks^{-1}. It can be noticed that the small pre-peak becomes most evident for lower cooling rates. The use of rather high heating rates in other works [5, 7] may be the reason that the appearance of the small peak has not been discussed previously.

To clarify the origin of the low peak, T_{X1} (see Fig. 3), XRD profiles were measured for the $Mg_{60}Cu_{30}Y_{10}$ alloy after heating to 473 K (~ T_{X2}) and 523 K (Fig. 6). The latter temperature is between T_{X2} and T_{X3}. The heating rate is the same as for the DSC measurement shown in Fig. 3. Since only the T_{X2}-peak of the first crystallization stage is visible between T_{X2} and T_{X3}, the difference in XRD profile must be attributed to crystallization that takes place at the T_{X2}-peak. The XRD profile for the specimen heated to 473 K exhibits a few sharp peaks and diffuse peaks which are lying in the same range as the amorphous halo. The set of sharp peaks are identified as Mg_2Cu phase by comparison with its simulated profile, which is also plotted in Fig. 6. Because there are some differences between simulated and observed profiles, the lattice parameters and/or atomic sites of the phase seem to deviate a little from those for the stoichiometric and equilibrium Mg_2Cu phase. The remaining diffuse peaks may indicate the existence of a certain amount of other crystalline phases with small grain size and/or an amorphous phase. On the other hand the XRD profile of the specimen heated to 523 K includes many sharp peaks including

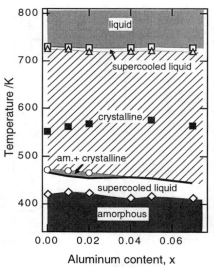

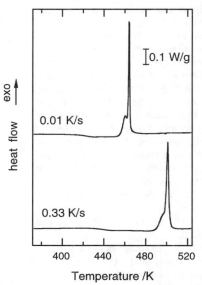

Fig. 4 The dependence of the various transition temperatures, T_g($\diamond$), T_{X1} (thick line), T_{X2} ($\bigcirc$), T_{X3} ($\blacksquare$), T_m ($\square$), T_s ($\triangle$) on Al concentration. All temperatures are defined by arrows in Fig. 3.

Fig. 5 DSC profiles of $Mg_{60}Cu_{30}Y_{10}$ measured with heating rates of 0.33 and 0.01 Ks^{-1}.

those from the Mg_2Cu phase, but does not show any diffuse ones. A hcp phase which has nearly the same lattice constants as pure Y probably exists. However, other peaks can hardly be identified by any phases which are listed in Pearson's Handbook [8]. Most of the unidentified peaks are thought to come from a crystal with large lattice constants because of the existence of a peak in the 2θ range below 20 degree. From the change of relative intensity marked by arrows in Fig. 6, we take it that at least two other phases exist in the specimen heated to 523 K. Thus we may conclude that the T_{X1}-peak in the DSC signal originates from the crystallization of the Mg_2Cu phase, while the T_{X2}-peak corresponds to the transformation of the remaining amorphous phase to the unidentified complex compounds.

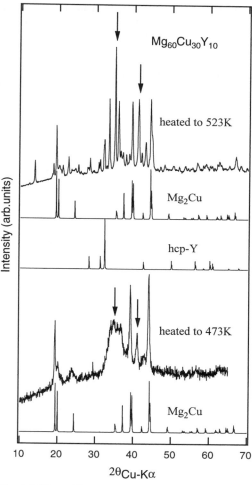

Fig. 6 XRD profiles of alloy heated to 473 and 523 K. The simulated profiles of Mg_2Cu and hcp-Y phases are also plotted

CONCLUSIONS

The effect of an increasing Al concentration on $(Mg_{1-x}Al_x)_{60}Cu_{30}Y_{10}$ bulk amorphous alloys has been studied. Adding Al causes a decrease of GFA. The amorphous single phase in bulk form can be obtained up to an Al concentration of x = 0.07. All the alloys (x = 0 - 0.07) show a glass transition. The crystallization temperature, T_{X1} decreases with increasing aluminum content, while T_g, T_m and T_s change only little. A splitting of the crystallization peak is observed in the DSC curves for x = 0 – 0.04. The XRD measurements indicates that this splitting in the first crystallization stage of the $Mg_{60}Cu_{30}Y_{10}$ alloy can be attributed to the difference between the onset temperatures for the precipitation of Mg_2Cu and of unidentified complex compounds.

REFERENCES

1.   A. Inoue, T. Zhang, and T. Masumoto, Mater. Trans., JIM **31**, 177 (1990).
2.   A. Inoue, T. Zhang, and T. Masumoto, J. Non-Cryst. Solids **156-158**, 473 (1993).

3. A. Inoue, K. Ohtera, K. Kita, and T. Masumoto, Jpn. J. Appl. Phys. **27**, L2248 (1988).
4. S.G. Kim, A. Inoue, and T. Masumoto, Mater. Trans., JIM **31**, 929 (1990).
5. A. Inoue and T. Masumoto, Mater. Sci. Eng. **A173**, 1 (1993).
6. N. H. Pryds and X. Huang, Scripta Materalia **36**, 1219(1997).
7. R. Busch, W. Liu, and W.L. Johnson, J. Appl. Phys. **83**, 4134 (1998).
8. P. Villars, *Pearson's Handbook Desk Edition*, ASM International, 1997, p. 1541 (for Cu-Mg), p. 1602 (for Cu-Y), p. 2350 – 2351 (for Mg-Y).

EXPERIMENTAL MEASUREMENTS OF NUCLEATION FREQUENCY AND CRYSTAL GROWTH RATE IN Pd-Cu-Ni-P METALLIC GLASS

Nobuyuki Nishiyama* and Akihisa Inoue*,**
*Inoue Superliquid Glass Project, ERATO, Japan Science and Technology Corporation (JST), Sendai 982-0807, Japan, nisiyama@sendai.jst.go.jp
**Institute for Materials Research, Tohoku University, Sendai 980-8577, Japan

ABSTRACT

Crystallization mechanism and kinetics of a $Pd_{40}Cu_{30}Ni_{10}P_{20}$ glass was investigated in a wide temperature range from 603 (near the glass transition temperature) to 764 K (near the equilibrium melting temperature) by using an isothermal annealing treatment for nucleation and growth. The nucleus density (n_v) is about 5×10^{13} nuclei/m^3 and is independent of annealing temperature. Therefore, it is assumed that the crystallization of the alloy was dominated by heterogeneous nucleation due to "quenched-in nuclei". On the other hand, the crystal growth rate (U_c) increases from 1.07×10^{-11} to 5.68×10^{-5} m/s with rising annealing temperature from 603 to 764 K. These values of U_c are 2-3 orders of magnitude larger than the calculated U_c on the basis of Classical Nucleation and Growth Theory (CNT). Furthermore, the glass-forming ability of the alloy will be discussed in the framework of the present results.

INTRODUCTION

Some molten alloys with particular alloy compositions can be frozen into a glass. The glass-forming ability (GFA) of molten alloy is the competition between the cooling constraint and the crystallization kinetics. Therefore, it is important to understand the crystallization kinetics of the glass-forming alloy. Great effort has been devoted to clarify the crystallization kinetics with theoretical and experimental approach, though the information was rather limited in a narrow temperature range near the glass transition temperature (T_g) because that the thermal analysis of the glass could be done with heating condition until crystallizing. If crystallization kinetics can be investigated in a wide temperature range from T_g to equilibrium melting temperature (T_m), the obtained information will be much useful for understanding the GFA of the alloy.

Recently, a number of metallic glass with large GFA have been found in malticomponent alloy systems such as Mg-TM-Ln[1], Ln-Al-TM[2], Zr-Al-TM[3] and Zr-Ti-TM-Be[4] (Ln = lanthanide metal, TM = transition metal). Particularly, as the results of examination of improving the GFA of Pd-Ni-P[5-7] alloys, it has been found that a $Pd_{40}Cu_{30}Ni_{10}P_{20}$ alloy[8-11] exhibits an outstanding GFA and shows the lowest critical cooling rate for glass formation (R_c) of 0.10 K/s. Therefore, it is assumed that the $Pd_{40}Cu_{30}Ni_{10}P_{20}$ alloy has the highest resistance against crystallization in the metallic glasses reported up to date. Using this alloy, it may be possible to analyze the crystallization mechanism and kinetics in a wide temperature range from T_g to T_m and to discuss the GFA. This paper intends to determine the crystallization mechanism of the $Pd_{40}Cu_{30}Ni_{10}P_{20}$ glass on the basis of the Avrami kinetics and to clarify the temperature dependence of nucleus density (n_v) and crystal growth rate (U_c) by use of the experimental method[12-17] of estimation of nucleation and growth for the oxide glasses. Finally, the GFA of the alloy is discussed by comparing the present results with the results calculated from the CNT.

THEORY

Avrami kinetics

In the kinetic analysis of crystallization of metallic glasses, the following Avrami kinetics[18] has been usually used[19-23]:

$$X(t) = 1 - \exp\left(-\pi/3 \cdot I_v U_c^3 t^4\right) \qquad (1)$$

where I_v and U_c are nucleation frequency and crystal growth rate, respectively. However, it is

Mat. Res. Soc. Symp. Proc. Vol. 554 © 1999 Materials Research Society

necessary to consider that Eq. (1) involved some assumptions. That is, (1) nucleation frequency will be constant, (2) growth length will be proportional to the time and (3) transformed particles will be spherical. In fact, it seems to be rare that above-mentioned assumptions are satisfied. Thus, the kinetic analysis of crystallization must be analyzed by taking nucleation mechanism into account. If crystallization behavior is dominated only by quenched-in nuclei, $X(t)$ can be given as:

$$X(t) = 1 - \exp\left(-4\pi/3 \cdot n_v U_c^3 t^3\right) \qquad (2)$$

where n_v is the nucleus density .

I_v and U_c

Turnbull analyzed the homogeneous nucleation frequency to use the thermodynamic parameters such as reduced glass transition temperature $(T_{r,g})$ defined as T_g/T_m, interfacial tension and entropy of fusion, and successfully presented the overall features of their effect[24]. Then based on the Turnbull's approach, several forms of equation were introduced to express the homogeneous nucleation frequency. Uhlmann was the first who introduced formal transformation theory into the kinetic analyses for glass formation[25]. According to the Onorato – Uhlmann's expression[26], the homogeneous nucleation frequency (I_v^{hom}) is shown as:

$$I_v^{hom} = N_v^0 v \exp\left(-1.229/T_r^3 \Delta T_r^2\right) \qquad (3)$$

where Nv^0 is the number of single molecules per unit volume, T_r is reduced temperature defined as T/T_m and ΔT_r is reduced undercooling defined as $(T_m-T)/T_m$. The exponential factor is based on the standard treatment of homogeneous nucleation with an energy barrier for nucleation of $60kT$ at $\Delta T_r=0.2$. The frequency factor v will be approximated by Stokes-Einstein relation as:

$$v = kT/\left(3\pi a_0^2 \eta\right) \qquad (4)$$

where k, a_0 and η are the Boltzmann constant, the mean atomic or ionic diameter for the diffusive jump and the viscosity, respectively.

Using the Turnbull's approximation, the crystal growth rate can be expressed as:

$$U_c = \frac{fD_g}{a_0}\left[1 - \exp\left(-\Delta T_r \Delta H_m^f/RT\right)\right] \qquad (5)$$

where D_g is a diffusion coefficient for atomic motion required for crystal growth, f is the fraction of sites at the interface where growth occurs and ΔH_m^f is the molar heat of fusion. If all sites are considered to be available for growth, the fraction of sites becomes $f=1$.

EXPERIMENTAL

A quaternary $Pd_{40}Cu_{30}Ni_{10}P_{20}$ alloy was used in the present study because the alloy exhibited an outstanding GFA and showed the lowest R_c of 0.10 K/s. The master ingot was prepared by arc-melting mixture of pre-alloyed $Pd_{60}P_{40}$, pure Cu, Ni, and Pd metals (99.99% pure) in a purified argon atmosphere. From the ingot, a glassy $Pd_{40}Cu_{30}Ni_{10}P_{20}$ ribbon was prepared by a single roller melt spinning method. The ribbon sample (10 ± 0.5mg) was charged in an Al_2O_3 crucible together with B_2O_3 flux, and remelted and fluxed at 1173 K for 300s in DTA equipment (RIGAKU thermoflex TG-DTA with IR furnace). Then, the spherical glassy sample (about 1mm in diameter) was obtained by quenching at a cooling rate of about 3.3 K/s. The annealing for nucleation and crystal growth was carried out in the following condition. The spherical samples, covered by the flux medium in the Al_2O_3 pan, were heated at 603 to 764 K ($0.75T_m$ to 0.95 T_m) for 10 s to 200 ks in a muffle furnace. The transverse cross-sections of annealed samples were polished, and the number of the crystals was counted with an optical microscope (OM). The average radius (d_{av}), inter particle distance and crystal growth rate (U_c) were also determined. Using these results, nucleus density (n_v) was determined, and then crystallization mechanism was estimated by use of the Avrami kinetics

The heats of crystallization of annealed samples were measured by DSC at a heating rate of 0.67 K/s to determine the volume fraction (v_f) of crystal. Based on the crystallization mechanism and v_f of the alloy, apparent crystal growth rate was calculated.

RESULTS and DISCUSSION

Figure 1(a) shows typical optical micrographs of the partially crystallized $Pd_{40}Cu_{30}Ni_{10}P_{20}$ glass by annealing at 684 K for 110 s. One can notice that crystalline precipitates are dispersed homogeneously. Neither surface nor foreign substrate-induced crystallization is detected. Therefore oxidation- or interface -induced nucleation effect seems to be negligible in the present study. Figure 1(b) to (f) shows optical micrographs of the partially crystallized $Pd_{40}Cu_{30}Ni_{10}P_{20}$ glasses annealed at various temperatures (T) and different times (t). At 764 K, the crystallization was completed within a short time interval between 40 to 42 s and hence it was impossible to obtain the partially crystallized sample. These micrographs reveal that the crystal diameter increases at higher T in spite of shorter t. Thus, it is concluded that the U_c increases with increasing T.

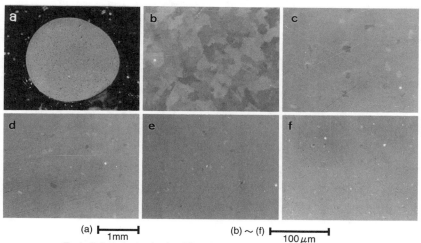

(a) ⊢————⊣ 1mm   (b) ~ (f) ⊢————⊣ 100 μm

Fig. 1. Optical micrographs of partially crystallized $Pd_{40}Cu_{30}Ni_{10}P_{20}$ glasses by annealing.
(a) 683 K for 110 s at low magnification, (b) 764 K for 42 s, (c) 724K for 70 s,
(d) 683 K for 110 s, (e) 643 K for 500 s, (f) 603 K for 50 ks.

Subsequently, n_v is calculated by counting crystalline particles grown during the annealing. Figure 2 shows the n_v as a function of reduced annealing time $(t/t_f$, where t_f is the time of crystallization completed). It is seen that the values of n_v lie around 5×10^{13} nuclei/m^3, and these are independent of the progress of crystallization and T.

If the nuclei generate during the annealing, apparent I_v can be obtained since n_v is divided by t. Figure 3 shows the apparent I_v calculated by using the obtained n_v and t, together with the I_v calculated from Eq. (2) for comparison. The temperature dependence of viscosity of the alloy used in the Eq. (2) was described elsewhere[27]. This figure points out that the crystallization of the alloy dose not occurred by homogeneous nucleation because the values of apparent I_v are much larger than calculated I_v. Therefore, it is assumed that the crystallization of the alloy is dominated by the heterogeneous nucleation mechanism due to quenched-in nuclei. Eq. (2) is used in the kinetic analyses hereafter.

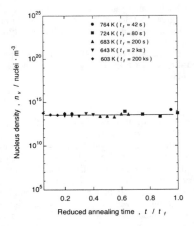

Fig. 2. Nucleus density as a function of reduced annealing time for the Pd $_{40}$Cu $_{30}$Ni $_{10}$P $_{20}$ glass.

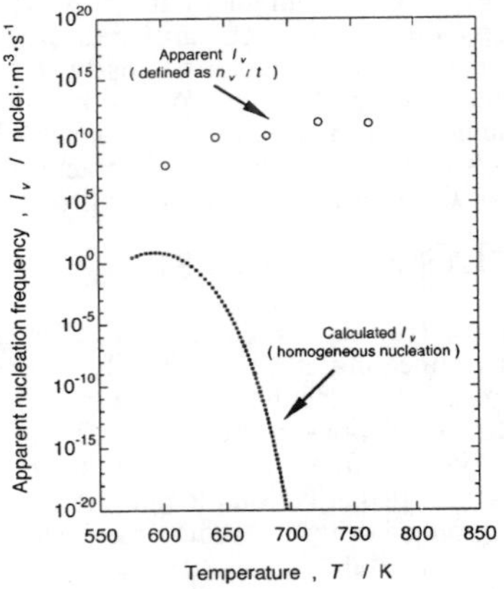

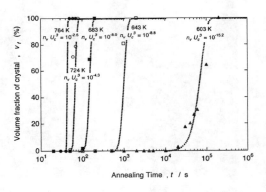

Fig. 3. Apparent I_v as a function of annealing temperature, together with I_v calculated from the CNT.

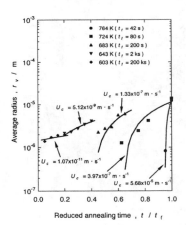

Fig. 4. Average radius of precipitated crystals as a function of annealing time for the Pd $_{40}$Cu $_{30}$Ni $_{10}$P $_{20}$ glass.

Fig.5. Relationship between volume fraction of precipitates and annealing time. The fitting curves calculated by using Eq. (2) are also shown.

Then U_c can be calculated from the average radius (r_{av}) of precipitate crystals. r_{av} at different T and t is plotted in Fig. 4 as a function of t/t_f. The slope of each line gives U_c. As shown in Fig.4, the relations between r_{av} and t/t_f are almost linear in the case of annealing at 764K, the crystallization progress is so fast so that only 2 plots are obtained. Therefore the value of U_c at 764 K obtained by tracing the line between 2 plots is doubtful. At any events, the U_c increases from 1.07×10^{-11} to 3.97×10^{-7} m/s with rising T from 603 to 724 K.

In order to confirm the transient behavior in the thermal analysis, the crystallization kinetics of the annealed samples was measured by DSC. The volume fraction (v_f) of crystals precipitated by annealing was obtained from the heat of crystallization as the following equation;

$$v_f = \Delta H_{annealed} / \Delta H_{as-quenched} \qquad (6)$$

where $\Delta H_{annealed}$ and $\Delta H_{as-quenched}$ are heat of crystallization of annealed sample and that of as-quenched sample, respectively.

Figure 5 summarizes the relationship between v_f and t at different annealing conditions, together with the fitting curves calculated from Eq. (2). The fitting curves indicate that the multiple of n_v by U_c^3 increases from $10^{-15.2}$ to $10^{-2.5}$ nuclei/s^3 with rising T from 603 to 764 K. Therefore, the apparent U_c can be obtained since the n_v is already known. The apparent U_c, the calculated U_c by using Eq. (5) and experimental U_c are shown in Fig.6. It is seen that the apparent U_c is nearly the same as the experimental U_c while the calculated U_c is underestimated by 2-3 orders of magnitude[28].

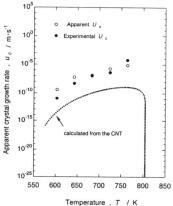

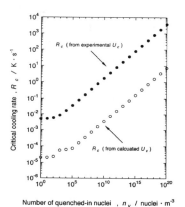

Fig.6. Apparent U_c calculated from n_v as a function of temperature. The data of the U_c calculated from the CNT are shown for comparison.

Fig. 7. Change in the calculated R_c as a function of n_v for the Pd$_{40}$Cu$_{30}$Ni$_{10}$P$_{20}$ alloy.

This underestimating of the calculated U_c may suggest that there maybe an error of the parameter in Eq. (5). Let us remind the used parameters f was chosen as $0.2 \Delta T$, along with the suggestion by Onorato-Uhlmann[26] and the ΔH_m^f used the experimental value[11]. It is assumed that calculated U_c may not be underestimated by 2-3 orders of magnitude if there is an error in either parameter. Therefore there remains an uncertainty on D_g. It is known that D_g has been used with the assumption of $D=D_l=D_g$ in order to simplify, where D is the bulk diffusion coefficient and D_l is diffusion coefficient for atomic transport across the nucleus matrix interface. Turnbull already pointed out[29] that D_g was larger than the corresponding D in the Fe-Ni alloys. Therefore the underestimating of the calculated U_c seems to be caused by the underestimating of D_g.

Finally, the GFA of the alloy was evaluated by constructing the continuous-cooling-transformation (C-C-T) diagrams at different n_v ranging from 0 to 10^{20} nuclei/m^3. For constructing the diagrams, we used the approximately values of calculated U_c times 500 as actual U_c. In order to simplify, approximated Eq. (1) and (2) were used because X (t) was set the small value of 10^{-6}. R_c obtained from C-C-T diagrams are plotted in Fig.7 as a function of n_v. It is clear that higher n_v requires larger R_c for glass formation. In brief, if n_v increases 3 orders of magnitude, R_c increases 1 order. The small bump around the 10^3 nuclei/m^3 in the open dots of R_c based on the CNT is due to the effect of both growths of quenched-in nuclei and homogeneous nucleation. The underestimating of the calculated U_c causes the underestimating R_c by almost 3 orders of magnitude. Therefore the nature of nucleation of the glass seems to be concealed by the growth of quenched-in nuclei. In other words, decreasing n_v is required to discuss the nucleation frequency in detail. At all events, the value of n_v strongly affects the R_c of the alloy.

CONCLUSIONS

The crystallization mechanism and kinetics of the $Pd_{40}Cu_{30}Ni_{10}P_{20}$ glass were investigated in a wide temperature range from 603 to 764 K by using an isothermal annealing treatment for nucleation and growth. The results obtained are summarized as follows.

(1) The nucleus density (n_v) is about 5×10^{13} nuclei/m³ and these remain constant at different annealing temperatures and times. The apparent I_v obtained from n_v divided by t are much larger than the calculated I_v. Therefore, it is assumed that the crystallization of the alloy is dominated by the heterogeneous nucleation due to "quenched-in nuclei".

(2) The crystal growth rate (U_c) increases from 1.07×10^{-11} to 5.68×10^{-5} m/s with rising annealing temperature from 603 to 764 K. These values of U_c are 2-3 orders of magnitude larger than the calculated U_c on the basis of the CNT. This underestimating of the calculated U_c seems to be caused by the underestimating of D_g.

(3) The GFA was evaluated by constructing the C-C-T diagrams at different n_v based on the present results. The R_c obtained from C-C-T diagrams indicates that higher n_v requires larger R_c for glass formation. If n_v increases by 3 orders of magnitude, R_c increases by 1 order. The nature of nucleation of the alloy seems to be concealed by the growth of quenched-in nuclei. At all events, the value of n_v strongly affects the GFA of the alloy.

ACKNOWLEDGMENTS

The authors wish to thank Professor R. Ota (Kyoto Inst. of Tech, JAPAN) and Dr. O. Haruyama (Sci. Univ. of Tokyo, JAPAN) for their very helpful information and discussion.

REFERENCES

1. A. Inoue, K. Ohtera, K. Kita and T. Masumoto, Jpn. J. Appl. Phys., **27**, L2248 (1988).
2. A. Inoue, T. Zhang and T. Masumoto, Mater. Trans., JIM, **30**, p.965 (1989).
3. A. Inoue, T. Zhang and T. Masumoto, Mater. Trans., JIM, **31**, p.177 (1990).
4. A. Peker and W.L.Johnson, Appl. Phys. Lett., **63**, p.2342 (1993).
5. H. S. Chen, Acta Metall., **22**, p.1505 (1974).
6 .A. J. Drehman, A. L. Greer and D.Turnbull, Appl. Phys. Lett., **41** p.716 (1982).
7. H. W. Kui, A. L. Greer and D.Turnbull, Appl. Phys. Lett., **45** p.615 (1984).
8. A. Inoue, N. Nishiyama and T. Matsuda, Mater. Trans., JIM, **37**, p.181 (1996).
9. N. Nishiyama and A.Inoue, Mater. Trans., JIM, **37**, p.1531 (1996).
10. A. Inoue and N. Nishiyama, Mater. Trans., JIM, **38**, p.181 (1997).
11. N. Nishiyama and A.Inoue, Mater. Trans., JIM, **38**, p.464 (1997).
12. Z. Strand and R. W. Douglas, Phys. Chem. Glasses, **11**, p.125 (1970).
13. D. G. Burnett and R. W. Douglas, Phys. Chem. Glasses, **11**, p.125 (1970).
14. Z. Strand and R. W. Douglas, Phys. Chem. Glasses, **14**, p.33 (1973).
15. K. Matusita and M. Tashiro, Phys. Chem. Glasses, **14**, p.77 (1973).
16. N. Mishima, R. Ota, T. Wakasugi and J. Fukunaga, J. Non-Cryst. Solids, **197**, p.19 (1996).
17. R. Ota, N. Mishima, T.Wakasugi and J.Fukumaga, J. Non-Cryst. Solids, **219**, p.70 (1997).
18. M. Avrami, J. Chem. Phys., **7**, p.1103 (1939).
19. B. G. Bagley and E. M. Vogel, J. Non-Cryst. Solids, **18**, p.29 (1975).
20. C. –P. Peter Chou and D. Turnbull, J. Non-Cryst. Solids, **17**, p.169 (1975).
21. P. G. Bosewell, J. Mater. Sci., **15**, p.1926 (1980).
22. S. Ranganathan and M. von Heimendahl, J. Mater. Sci., **16**, p.2401 (1981).
23. K. Matusita, T. Komatsu and R. Yokota, J. Mater. Sci., **19**, p.291 (1984).
24. D. Turnbull, Contemp. Phys., **10**, p.473 (1969).
25. D. R. Uhlmann, J. Non-Cryst. Solids, **7**, p.337 (1972).
26. P. I. K. Onorato and D. R. Uhlmann, J. Non-Cryst. Solids, **22**, p.367 (1976).
27. N. Nishiyama and A. Inoue, Mater. Trans., JIM, to be published
28. N. Nishiyama and A. Inoue, Acta Mater., to be submitted.
29. D. Turnbull, Trans. Metall. Soc. AIME, **221**, p.422 (1961).

PHASE EQUILIBRIA IN THE ZR-CU-AL SYSTEM: IMPLICATIONS FOR BULK METALLIC GLASS FORMATION

S. A. SYED, D. SWENSON*
Department of Metallurgical and Materials Engineering, Michigan Technological University
1400 Townsend Drive, Houghton, MI 49931
*dswenson@mtu.edu

ABSTRACT

Preliminary phase equilibrium relationships have been established in the Zr-Cu-Al system at 800 °C, using a combination of X-ray diffraction and electron probe microanalysis. These results are similar to previous investigations that have been reported in the literature. Several ternary phases are found to exist in this system, many of which lie within the gross compositional vicinity of interest to bulk amorphous alloy formation. The equilibrium phases present in the alloy $Zr_{65}Cu_{27.5}Al_{7.5}$, which exhibits a particularly high T_x-T_g in the amorphous state, are Zr_2Cu and minor amounts of two additional phases: Zr_3Al and what may be a ternary phase with a composition near Zr_6CuAl_3. When the 800 °C phase diagram isotherm is correlated with the known glass forming composition range of the Zr-Cu-Al system, it is found that the best glass forming behavior is confined to those regions of the diagram in which all equilibria include Zr-Cu constituent binary phases and Al-poor ternary phases. This may suggest that difficulties in the nucleation of these binary phases plays a role in the glass forming ability of Zr-Cu-Al and related higher order alloys.

INTRODUCTION

Recently, there has been much interest in Zr-Cu-Al alloys, owing to their relative ease of glass forming ability as well as their tremendous resistance to crystallization when held in the supercooled liquid state [1-2]. Moreover, through further alloying additions, especially Ni, it has proven possible to enhance this resistance to crystallization, and form bulk amorphous alloys of very large dimensions directly from the melt at relatively low cooling rates[3-5]. Accordingly, the properties of both ternary and higher order alloys comprising Zr, Cu and Al have been studied by many different researchers (see, for example, [6-10]).

There has, however, been very little study of phase equilibria in the Zr-Cu-Al and related systems. Such information is potentially very useful, as it indicates for a given overall composition of alloy what phases one would expect to find at thermodynamic equilibrium. Since the propensity for amorphous alloy formation is undoubtedly related to the ease of nucleation and growth of crystalline, equilibrium phases, a correlation between phase equilibria and glass forming ability could prove provide significant insight into the ease or difficulty of glass formation in a given alloy system.

Current knowledge of phase equilibria in the Zr-Cu-Al system has been summarized by [11]. Except for a study by Panseri and Leoni[12] of the Al-rich corner of the diagram, all that is known about phase equilibria in the Zr-Cu-Al system has been reported by Markiv and Burnashova[13], who established a majority of a ternary isotherm at 800 °C using X-ray diffraction analysis. They reported this system to be complex, containing 8 ternary phases. Owing to the experimental technique employed by Markiv and Burnashova, the compositions of these ternary phases are known only approximately, and many of the tie-lines reported by these researchers were drawn as dashed, reflecting uncertainty in their data.

The present investigation reports preliminary data from a complete redetermination of phase equilibria in the Zr-Cu-Al system at 800 °C, employing both X-ray diffraction (XRD) for phase identification and electron probe microanalysis (EPMA) for compositional analysis. These results are then compared with those of Markiv and Burnashova, and are also correlated with the known glass forming ability of Zr-Cu-Al alloys.

Mat. Res. Soc. Symp. Proc. Vol. 554 © 1999 Materials Research Society

EXPERIMENTAL PROCEDURE

Elemental Zr (99.5%), Cu (99.999%), and Al(99.999%) were utilized to prepare 9 phase equilibrium samples, each with a nominal weight of 10 g. After weighing appropriate amounts of each element, the samples were arc-melted under high purity Ar in a system that had been purged several times using a vacuum pump. Each alloy button was arc-melted at least 4 times to ensure good mixing. All samples were weighed again after melting as a precaution against significant loss of Al by volatilization. Subsequent to arc melting, each sample was encapsulated in a quartz tube under a vacuum of approximately 10^{-6} Torr and annealed at 800 °C for at least 40 days. After annealing, the samples were quenched in water. Upon entering the quenching water, each quartz tube was broken in order to allow the samples to come directly into contact with the water so that the quenching rate would be increased.

Quenched samples were sectioned using a diamond saw and mounted in epoxy. They were then polished using a Leco autopolisher down to 0.05 μm alumina. The polished samples were analyzed by XRD for phase identification using a Scintag XDS 2000 diffractometer, operating at 45 kV and 35 mA and employing Cu K_α radiation. Most scans were performed over a range of about 20-90 ° 2θ, using a step size of 0.03° and a scan rate of 0.8-1.8 °/min.

EPMA was accomplished using a JEOL JXA 8600 Superprobe, operating at 15kV with a 20 nA beam current. Wavelength-dispersive spectroscopy (WDS) was used for compositional analysis, employing the ZAF correction routine and utilizing high purity elements as standards. For each phase, at least 5 EPMA data points were taken. These multiple data points were then averaged in order to obtain the compositions of the phases In a few cases, grains of individual phases were too small (1-2 μm) for effective compositional analysis using this technique.

RESULTS AND DISCUSSION

Phase Equilibria in the Zr-Cu-Al System at 800 °C

Phase equilibria in the Zr-Cu-Al system at 800 °C, as determined in the present investigation, are shown in Fig. 1. In the figure, gross sample compositions are depicted by the symbol "X". These gross sample compositions are also given in Table 1, along with the phases present in each sample and compositions of phases as determined by EPMA. For comparison, the isotherm determined by Markiv and Burnashova is shown in Fig. 2.

By comparing Fig. 1 with Fig. 2, it may be seen that the results of the present investigation are largely consistent with those of Markiv and Burnashova. The major differences are that in the present investigation, no evidence of the ternary phase τ_2 (crystal structure unknown; composition $Zr_{73}Cu_{14}Al_{13}$) was found. Additionally, whereas Markiv and Burnashova reported a significant ternary solubility of Cu in the phase Zr_4Al_3, the XRD pattern obtained in the present investigation does not conclusively show this phase to be Zr_4Al_3; rather, it appears that it may be a ternary phase with an approximate stoichiometry of Zr_6CuAl_3. Nevertheless, in the present paper, the Zr_4Al_3 designation will be maintained.

Additionally, according to the results of the present investigation, for alloys of the composition $Zr_{65}Cu_{27.5}Al_{7.5}$, which have been reported to exhibit the greatest resistance to crystallization of any Zr-Cu-Al alloy[2], one would expect to find Zr_2Cu with minor amounts of Zr_3Al and Zr_4Al_3 under conditions of thermodynamic equilibrium. Fig. 3 shows an SEM micrograph of sample ?, which depicts the coexistence of these three phases. This is partly inconsistent with the results of Inoue *et al.*[2], who reported that for amorphous alloys of the composition $Zr_{65}Cu_{35-x}Al_x$ ($0 \leq x \leq$ 20), amorphous alloys crystallized into single phase Zr_2Cu for $0 \leq x \leq 7.5$, whereas they formed Zr_2Cu and Zr_2Al for $10 \leq x \leq 20$. Because Inoue *et al.*'s samples were annealed at 600 °C rather than 800 °C, it is possible that the phase equilibria are slightly different at this lower temperature. However, it should be pointed out that the diffraction pattern of Zr_6Cu_3Al, while not indexed successfully, contains strong peaks that are similar to those of Zr_2Al. It would therefore be

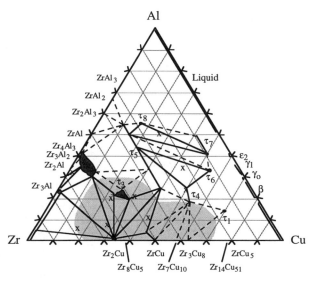

Fig. 1. Zr-rich phase equilibria in the Zr-Cu-Al system at 800 °C. Gross sample compositions are denoted by the symbol "X". Tie-lines that have been determined experimentally are solid, whereas additional tie-lines inferred from the phase rule and considered likely for thermodynamic reasons are shown as dashed. Bold lines and medium shaded regions represent solution phases. The lightly shaded area near the Zr-Cu binary system corresponds to the compositional range of glass forming alloys in the Zr-Cu-Al system as reported by Inoue et al.[2].

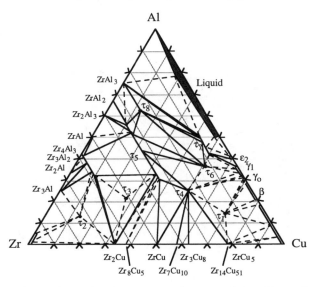

Fig. 2 Phase Equilibria in the Zr-Cu-Al system according to Markiv and Burnashova[13].

Sample Number	Gross Composition (at.%)			Phase	Composition of Phase (at.%)		
	Zr	Cu	Al		Zr	Cu	Al
1	77.0	17.0	6.0	Zr_2Cu	67.7	31.3	1.6
				Zr_3Al	77.1	2.0	20.9
				$Zr(\alpha)$	94.1	1.4	4.5
2	67.0	1.0	32.0	Zr_3Al	78.9	0.5	20.7
				Zr_2Al	70.8	0.3	28.9
				Zr_4Al_3	62.5	4.5	33.0
3	65.0	27.5	7.5	Zr_2Cu	67.7	30.9	1.4
				Zr_3Al	76.2	2.3	21.5
				Zr_4Al_3	59.9	10.4	29.6
4	57.0	23.0	20.0	Zr_2Cu	69.2	29.8	1.0
				Zr_4Al_3	60.3	10.4	28.5
				τ_3	54.7	22.6	22.7
5	47.0	42.0	11.0	Zr_2Cu	68.0	31.5	0.5
				$ZrCu$	52.9	42.8	4.4
				τ_3*	52.8	28.8	18.4
				τ_5	39.4	36.8	23.8
6	47.0	32.0	21.0	Zr_2Cu	69.2	28.9	1.3
				τ_3	50.3	29.2	20.5
				τ_5	36.0	39.5	24.5
7	26.0	50.0	24.0	τ_4	27.2	53.0	19.8
				τ_5	36.9	29.6	33.6
8	23.0	27.0	50.0	τ_5	34.7	15.2	50.0
				τ_7	9.3	50.3	40.4
				τ_8	28.5	17.0	54.5
9	22.0	44.0	34.0	τ_4	25.9	50.4	23.7
				τ_5	35.2	20.6	44.1
				τ_6	12.1	54.8	33.1

Table 1. Gross Sample Compositions, Phases Identified by X-Ray Diffraction, and the Compositions of Phases in the Zr-Cu-Al System as Determined by EPMA

*Nonequilibrium phase; See text for details

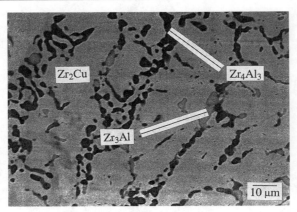

Fig. 3. SEM micrograph of Sample 3, showing the coexistence of Zr_2Cu, Zr_3Al and Zr_4Al_3 in an alloy with the gross composition $Zr_{65}Cu_{27.5}Al_{7.5}$.

possible to mistake the former phase for the latter phase on the basis of XRD experiments.

It should be noted that owing to significant absorption/fluorescence effects between Zr and Al, even when employing the ZAF correction routine for EPMA, the amount of Al present in a phase is under-reported, whereas the amount of Zr in a phase is over-reported. This is apparent in Table 1 by examining the compositions of constituent binary phases that are known to be line compounds. Fig. 1 has been drawn to be consistent with the EPMA data; however, the reader should take note that the actual composition of each Al-bearing phase is likely to be a few atomic percent richer in Al than Fig. 1 would indicate.

Additionally, it should be noted that Sample 5 contained four phases, which is unlikely in a ternary system. This suggests that it did not reach thermodynamic equilibrium. Considering the phases present in other samples in the compositional vicinity of Sample 5, it is most likely that τ_3 is the nonequilibrium phase.

Phase Equilibria in the Zr-Cu-Al System and Glass Forming Ability

In Figure 1, the compositional regions of glass forming ability in the Zr-Cu-Al system as reported by Inoue et al.[2] have been superimposed on the 800 °C ternary phase diagram isotherm as determined in the present investigation. It may be seen that the region of good glass forming ability corresponds almost exclusively to regions where Zr-Cu intermetallics and ternary Zr-Cu-Al phases coexist under equilibrium conditions. While one would expect the glass forming region to lie in close proximity to the Zr-Cu binary[13], this nevertheless suggests that even in ternary alloys, the thermodynamic properties of the Zr-Cu binary system play a role in glass forming ability.

CONCLUSIONS

Preliminary phase equilibrium relationships were established in a majority of the Zr-Cu-Al system at 800 °C, using X-ray diffraction and electron probe microanalysis. The results of the phase diagram study were similar to previous studies reported in the literature that were conducted using X-ray diffraction only. Six ternary were confirmed to exist in the system, and what was reported previously to be the phase Zr_4Al_3 with significant ternary solubility may in fact be an additional ternary phase with the approximate composition Zr_6CuAl_3. Four of the six ternary phases, as well as Zr_4Al_3, lie in compositional regions of relevance to metallic glass formation. For an alloy of the gross composition $Zr_{65}Cu_{27.5}Al_{7.5}$, which exhibits the greatest resistance to crystallization of any ternary Zr-Cu-Al alloy, the phases present at thermodynamic equilibrium were determined to be Zr_2Cu, plus minor amounts of Zr_3Al and Zr_4Al_3. Compositional analysis by EPMA proved difficult, owing to absorption/fluorescence interactions between Zr and Al. The compositional data obtained in the present investigation over-reports Zr and under-reports Al in each phase by up to a few atomic percent.

When the 800 °C Zr-Cu-Al isotherm was compared with previous reports of metallic glass formation among Zr-Cu-Al alloys, it was found that those alloys exhibiting good glass forming ability lie exclusively in compositional regions whose phase equilibria include Zr-Cu constituent binary phases and Al-poor ternary phases. This suggests that perhaps the properties of the Zr-Cu binary phases, especially as they pertain to nucleation, play an important role in the glass forming ability of Zr-Cu-Al alloys.

ACKNOWLEDGMENTS

The present investigators wish to thank the National Science Foundation for its financial support of this work through Grant Number DMR-97-02263.

REFERENCES

1. A. Inoue, K. Kita, T. Zhang and T. Masumoto, Mater. Trans. JIM **30**, 722 (1989).
2. A. Inoue, D. Kawase, A. P. Tsai, T. Zhang and T. Masumoto, Mater. Sci. Engin. A **178** 255 (1994).
3. T. Zhang, A. Inoue and T. Masumoto, Mater. Trans. JIM **32**, 1005 (1991).
4. A. Inoue, T. Zhang, N. Nishiyama, K. Ohba and T. Masumoto, Mater. Trans. JIM **34**, 1234 (1993).
5. A. Inoue and T. Zhang, Mater. Trans. JIM **37**, 185 (1996).
6. A. Inoue, T. Zhang, K. Ohba and T. Shibata, Mater. Trans. JIM **36**, 876 (1995).
7. A. Inoue, T. Zhang and T. Masumoto, Mater. Sci. Engin. A **134**, 1125 (1991).
8. H. Kimura, M. Kishida, T. Kaneko, A. Inoue and T. Masumoto, Mater. Trans. JIM **36**, 890 (1995).
9. A. Inoue, Y. Kawamura, T. Shibata and K. Sasamori, Mater. Trans. JIM **37**,1337 (1996).
10. H. Schumacher, U. Herr, D. Oelgeschlaeger, A. Traverse and K. Samwer, J. Appl. Phys. **82**, 155 (1997).
11. G. Petzow and G. Effenberg, eds., *Ternary Alloys- A Comprehensive Compendium of Evaluated Constitutional Data and Phase Diagrams* (VCH Publishers, Mannheim FRG, 1988).
12. C. Panseri and M. Leoni, Aluminio **33**, 63 (1964).
13. V. Ya. Markiv and V. V. Burnashova, Poroshk. Metall. No. 12, 53 (1970).
14. A. J. Kerns, D. E. Polk, R. Ray nad B. C. Giessen, Mater. Sci. Engin. **38**, 49 (1982).

DEFORMATION BEHAVIOR OF BULK AMORPHOUS Zr-BASE ALLOYS

A. LEONHARD, M. HEILMAIER, J. ECKERT and L. SCHULTZ
IFW Dresden, Institute of Metallic Materials, D-01069 Dresden, Germany

ABSTRACT

Bulk Zr-Al-Cu-Ni alloys were produced by die casting into a copper mold under Ar-atmosphere. The microstructure of fully amorphous as well as partially crystalline samples was analyzed by X-ray diffraction (XRD), scanning (SEM) and transmission electron microscopy (TEM), and chemical analysis with special emphasis on the size and composition of the crystallites. The mechanical behavior of the different samples was investigated by constant compression rate tests. At room temperature the samples show inhomogeneous deformation and, independent of the chosen composition, relatively low Young's moduli of about 70 GPa, flow stresses around 2 GPa and elastic strains of up to 3 %. Fully amorphous samples show microplasticity of up to 2 % strain without significant work hardening while specimens with a fairly high volume fraction of crystalline phases are extremely brittle. In contrast, at high temperatures around the glass transition temperature T_g both amorphous and partially crystalline specimens exhibit at low strain rates homogeneous deformation with an initial stress overshoot followed by an extended region of plastic flow. As compared to room temperature, the peak stresses are much lower and are hardly influenced by the presence of small volume fractions of crystalline phases. The observed thermal stability against crystallization provides a promising possibility for easy shaping of complex parts at temperatures around T_g.

INTRODUCTION

To produce large-scale bulk glassy samples metallic alloy systems with a large glass-forming ability are necessary. Zr-base alloys exhibit low cooling rates to vitrify without crystallization and are characterized by a wide supercooled liquid region and a high thermal stability against crystallization [1, 2]. These characteristics allow manufacturing of cylindrical samples with diameters up to 30 mm [3] by die casting, thus, enabling the measurement of their bulk mechanical properties.

At room temperature fully amorphous Zr-base alloys show high strength, relatively low Young's modulus, some microplasticity and high wear resistance, while crystalline specimens are brittle [4, 5]. Currently, there are only a few investigations dealing with the mechanical properties at elevated temperatures [1, 6]. However, a detailed knowledge of these properties is a mandatory prerequisite for evaluating the perspectives of shaping of complex parts from as-cast specimens. Additionally, the effect of different volume fractions of crystallites on the mechanical properties at temperatures around the glass transition temperature T_g of partially crystalline bulk samples has not been investigated.

Amorphous samples containing crystalline precipitates can be obtained by two different ways. On one hand, crystalline phases can already form during casting due to nucleation caused by impurities stemming from the starting materials, reaction with the crucible or due to insufficient cooling rate upon quenching [7, 8]. On the other hand, partial devitrification of glassy alloys upon annealing or hot working above the glass transition temperature can lead to formation of nanocrystalline precipitates in the glassy matrix [8, 9]. Consequently, this work

Mat. Res. Soc. Symp. Proc. Vol. 554 © 1999 Materials Research Society

focuses on the effect of crystalline precipitates on the mechanical behavior of bulk glass forming Zr-base alloys at both room temperature and elevated temperatures around T_g.

EXPERIMENTAL

Prealloyed $Zr_{55}Cu_{30}Al_{10}Ni_5$ ingots were prepared by arc-melting the pure elements under a zirconium-gettered argon atmosphere. Bulk samples with 3 mm diameter and 50 mm length were obtained by die casting the prealloy into a copper mold under argon atmosphere. In order to produce fully amorphous as well as partially crystalline samples, the oxygen content of the samples was varied by adjusting the oxygen partial pressure in the inert gas atmosphere upon casting. In addition, partially crystalline samples were produced by subsequent isothermal annealing of initially amorphous rods in a differential scanning calorimeter (DSC).

The structure and composition of the samples was characterized by X-ray diffraction (XRD) (Philips PW 1050 diffractometer using $CoK\alpha$ radiation), scanning electron microscopy (SEM) (JEOL JSM 6400) and transmission electron microscopy (TEM) (JEOL 2000 FX). TEM samples were prepared by cutting thin slices from the cylindrical bulk samples. After polishing and dimpling these slices were ion milled using a Gatan Precision Ion Polishing System (PIPS). The thermal stability of the specimens was examined by DSC (Perkin-Elmer DSC 7) at a heating rate of 20 $Kmin^{-1}$ under argon atmosphere.

For mechanical testing cylindrical samples of 3 mm in diameter and 6 mm in height were cut from the cast cylinders and polished on both sides. The compression tests at room temperature were conducted at a strain rate of $\dot{\varepsilon} = 1 \times 10^{-3}$ s^{-1} using an electromechanical Instron 8562 device. Tests at the onset of the glass transition (as determined from the DSC scan, see Fig. 1) were carried out at constant true strain rates ranging from $\dot{\varepsilon} = 1 \times 10^{-3}$ - 3×10^{-5} s^{-1} using a high temperature furnace including a laser extensometer, which provide testing under protective (argon) atmosphere as well as continuous monitoring of the creep strain by measuring the displacement of the compression punches inside the furnace.

RESULTS AND DISCUSSION

Microstructure and thermal stability of the samples

Figure 1 shows the DSC plots of the $Zr_{55}Cu_{30}Al_{10}Ni_5$ alloy in the fully amorphous (curve (a)) and the partially nanocrystallized condition (curve (b)) at a heating rate of 20 $Kmin^{-1}$. The amorphous sample exhibits a supercooled liquid region of $\Delta T_x = 90$ K between the glass transition temperature T_g (onset = 685 K) and the crystallization temperature T_x (onset = 765 K). The partially nanocrystallized sample shows nearly the same T_g, while the crystallization peak is markedly broader and T_x is shifted to a lower temperature (onset = 750 K). The difference in the area under both crystallization peaks was evaluated to estimate the volume fraction of nano-crystallites giving a volume fraction of 25 % nanocrystals for the annealed sample.

Before and after the compression tests the samples were analyzed by XRD (Figs. 2(a) and (b)). Curve (I) in Figure 2(a) reveals the common broad amorphous maxima for the as-cast specimens. In contrast, curves (II) and (III) in Figure 2(a) exhibit additional superimposed diffraction peaks due to crystalline particles embedded in the amorphous matrix. Additionally, the location of the diffraction peaks in curves (II) and (III) show that the crystalline phases formed after casting and annealing are different. Figure 2(b) reveals XRD patterns of the same specimens after the compression tests. The fully amorphous sample shows nearly the same broad amorphous maxima (curve (I)), while the peak height of the as-cast partially crystalline sample is slightly increased (curve (II)). The peak broadening after deformation of the partially nanocrystallized samples (curve (III)) indicates the formation of supplementary precipitates during testing at elevated temperature.

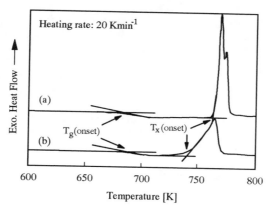

Figure 1: DSC traces of the amorphous (curve (a)) and the partially nanocrystallized (curve (b)) $Zr_{55}Cu_{30}Al_{10}Ni_5$ alloy.

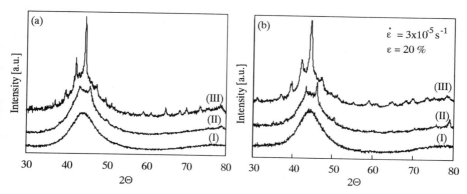

Figure 2: X-ray diffraction patterns of (I) the amorphous, (II) the as-cast partially crystalline (volume fraction of 8 % crystallites) and (III) the partially nanocrystallized samples (volume fraction of 25 % crystallites): (a) before and (b) after high temperature deformation.

In Figure 3(a) a SEM image of a partially crystalline $Zr_{55}Cu_{30}Al_{10}Ni_5$ sample with a microstructure corresponding to curve (II) in Fig. 2(a) is displayed. Dendritic crystallites (dark spots in the SEM image) with an average size of up to 5 μm and an areal fraction of about 8 % are homogeneously distributed in the amorphous matrix. In contrast, the bright field TEM image in Figure 3(b) shows the microstructure of the annealed partially crystalline sample (Fig. 2(a), curve (III)). In this sample elliptical crystallites with a size up to 200 nm in length (volume fraction: 25 %) are distributed in the amorphous matrix.

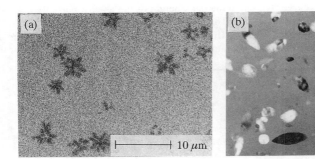

Figure 3: (a) Backscattering SEM image of as-cast $Zr_{55}Cu_{30}Al_{10}Ni_5$ showing dendritic precipitates (dark spots) (cf. curve (II)), (b) Bright field TEM image of the annealed $Zr_{55}Cu_{30}Al_{10}Ni_5$ sample showing elliptical, nanosize precipitates (cf. curve (III))

Deformation behavior at room temperature

Figure 4 shows compressive stress-strain curves of as-cast fully amorphous and partially crystalline samples (curves (a) and (b), respectively). The stress-strain curve of the fully amorphous sample (curve (a)) shows an elastic strain ε_{el} of 2.7 % and a Young's modulus E of 70 GPa. Beyond the elastic limit, serrated flow is observed without work hardening. Consequently, the yield strength and the flow stress are equal and amount to about 1800 MPa. The fully amorphous state exhibits microplasticity of $\varepsilon_{pl} \approx 1$ %. The partially crystalline as-cast $Zr_{55}Cu_{30}Al_{10}Ni_5$ specimens (curve (b)) show nearly the same values for Young's modulus, elastic strain and flow stress but are extremely brittle.

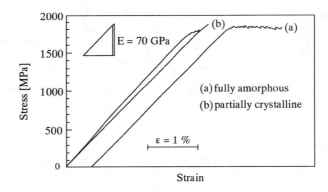

Figure 4: Mechanical behavior at room temperature for (a) fully amorphous and (b) partially crystalline as-cast $Zr_{55}Cu_{30}Al_{10}Ni_5$ samples.

The observed deformation behavior can be ascribed to inhomogeneous flow of the material [10], *i.e.* the strain is localized in very thin shear bands. In agreement with results on tensile tests of Zr-based amorphous alloys [11] we also observe an angle of the slip plane with the compression axis of $\approx 45°$.

Deformation behavior at the onset of T_g

To investigate the deformation characteristics of the supercooled liquid a heating rate of 20 Kmin^{-1} was applied for all creep specimens because of the dependence of T_g on the heating rate. The test temperature was set to the corresponding T_g-value (see Fig. 1) to ensure the test to be done in the supercooled liquid state.

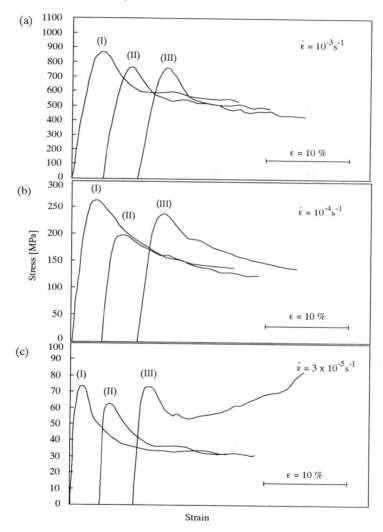

Figure 5: Compressive true stress – true strain curves of (I) amorphous, (II) as-cast partially crystalline and (III) the annealed partially nanocrystallized samples at different $\dot{\varepsilon}$ as indicated.

Figures 5(a) to (c) show compressive true stress-true strain curves of fully amorphous and partially crystalline samples at strain rates of $\dot{\varepsilon} = 1\times10^{-3}$, 1×10^{-4} and 3×10^{-5} s^{-1}, respectively. In contrast to the room temperature measurements the stress level is much lower and predominantly homogeneous flow is observed resulting in extended plasticity without failure. It is obvious that plastic deformation starts with a stress maximum (stress overshoot) and then levels off to a stress plateau at plastic strains of about 20%. This stress overshoot becomes less pronounced with decreasing strain rate. Since we can exclude structural relaxation effects due to the transition from the amorphous into the supercooled liquid state, we attribute the generation of slip planes over the whole sample, which results in atomic reordering processes and destroys the short-range order of the atoms, to cause the observed stress overshoot. A particular behavior is observed for the partially nanocrystalline sample at the lowest strain rate of $\dot{\varepsilon} = 3\times10^{-5}$ s^{-1} (Fig. 5(c)): it reveals a significant stress increase after the stress plateau even topping its initial peak stress value. This increase could be attributed to the formation of new crystallites during long-term thermal exposure during the compression test (Fig. 2 (a)) and due to a connected change of the viscosity of the remaining amorphous matrix.

CONCLUSIONS

The deformation behavior of the bulk metallic glass forming alloy $Zr_{55}Cu_{30}Al_{10}Ni_5$ was investigated by constant compression rate tests at room temperature and at the onset of the glass transition temperature T_g, respectively. At room temperature as-cast crystalline precipitates embrittle the material. In contrast, they have a negligible influence on the flow stresses at T_g, while the increase in the volume fraction of nanosized crystallites causes an increasing flow stress during deformation at low strain rates and, thus, long-term exposure at elevated temperatures. However, the observed extended plasticity without failure demonstrates the potential of hot working of Zr-based amorphous and partially crystalline alloys.

ACKNOWLEDGMENTS

Financial support by the German Science Foundation through the DFG Schwerpunktprogramm "Unterkühlte Metallschmelzen" under Grant He 1872/6-1 is gratefully acknowledged.

REFERENCES

1. A. Inoue et al., Mater. Trans. JIM **32**, 609 (1991).
2. A. Peker, and W.L. Johnson, Appl. Phys. Lett. **63**, 2342 (1993)
3. A. Inoue, and T. Zhang, Mater. Trans. JIM **37**, 185 (1996)
4. W.L. Johnson, Mater. Sci. Forum **225-227**, 35 (1996)
5. A. Inoue, T. Zhang, and T. Masumoto, Mater. Trans. JIM **36**, 391 (1995)
6. Y. Kawamura et al., Appl. Phys. Lett. **69**, 1208 (1996)
7. X.H. Lin, and W.L. Johnson, Mater. Trans. JIM **38**, 473 (1997)
8. A. Gebert, J. Eckert, and L. Schultz, Acta Mater. **46**, 5475 (1998)
9. J. Eckert et al., Mater. Trans. JIM **39**, 623 (1998)
10. F. Spaepen, Acta Metall. **25**, 407 (1977)
11. C. Gilbert, R.O. Ritchie, and W.L. Johnson, Appl. Phys. Lett. **71**, 476 (1997)

HIGH-STRENGTH BULK NANOSTRUCTURE ALLOYS CONSISTING OF COMPOUND AND AMORPHOUS PHASES

Akihisa Inoue*, Cang Fan**
*Institute for Materials Research, Tohoku University, Sendai 980-8577, Japan, ainoue@imr.tohoku.ac.jp
**Inoue superliquid glass project, ERATO, Japan Science and Technology Corporation (JST), Sendai 982-0807, Japan

ABSTRACT

Bulk nanocrystalline alloys with good ductility and high tensile strength (σ_f) in Zr-Al-Cu-Pd and Zr-Al-Cu-Ni-Ti systems were formed by partial crystallization of cast bulk amorphous alloys. The nanostructure alloys consist of $Zr_2(Cu, Pd)$ or $(Zr, Ti)_2Al$ surrounded by the remaining amorphous phase. The particle size and interparticle spacing of their compounds are less than 10 and 2 nm, respectively. The crystallization of a $Zr_{60}Al_{10}Cu_{30}$ amorphous alloy occurs by the simultaneous precipitation of Zr_2Al and Zr_2Cu with a large particle size of 500 nm and hence the addition of Pd or Ti is effective for formation of the nanostructure. The Pd or Ti has much larger negative heats of mixing against Zr or Al, respectively, and the Zr-Pd or Ti-Al atomic pair seems to act as preferential nucleation sites leading to the primary precipitation of $Zr_2(Cu, Pd)$ or $(Zr, Ti)_2Al$. The nanostructure alloy cylinders of 2 to 3 mm in diameter keep good ductility in the volume fraction (V_f) range of the compounds below 40 %. The σ_f and Young's modulus (E) increase from 1760 MPa and 81.5 GPa, respectively, at V_f=0 % to 1880 MPa and 89.5 Gpa, respectively, at V_f =40 % for the $Zr_{60}Al_{10}Cu_{20}Pd_{10}$ alloy and from 1830 MPa and 89.0 GPa, respectively, at V_f =0 % to 1940 MPa and 95.2 GPa, respectively, at V_f =28 % for the $Zr_{53}Al_{12}Cu_{20}Ni_{10}Ti_5$ alloy. The formation of the bulk nanostructure alloys with high σ_f is presumably due to the reentrance of free volumes into the remaining amorphous phase caused by quenching from the supercooled liquid region.

INTRODUCTION

Nanocrystalline alloys with useful characteristics have been synthesized by partial crystallization of an amorphous phase in a number of melt-spun alloys such as Fe-Nd-B, Fe-Si-B-Nb-Cu, Fe-(Zr, Nb)-B, Al-Ln-TM (Ln=lanthanide metal, TM=transition metal) and Mg-Ln-TM etc [1]. The nanocrystallization process has a great advantage of forming a nanocrystalline alloy in a ribbon form. The material form has enabled the application of their nanocrystalline alloys as engineering materials [1]. As engineering characteristics which have not been obtained for conventional crystalline alloys, one can list up hard magnetism, soft magnetism, high tensile strength and high sensitive magnetostriction. However, the shape and dimension of the nanocrystalline alloys have been limited to the ribbon form with a thickness less than about 30 μ m and the powder form with a diameter less than about 40 μ m. Thus, no bulk nanocrystalline alloys containing a residual amorphous phase have been synthesized by any kinds of preparation techniques reported hitherto. The synthesis of a bulk nanocrystalline alloy with similar useful characteristics is expected to cause a further extension of application fields of nanocrystalline alloys. Recently, a number of bulk amorphous alloys have been formed by copper mold casting processes in Mg-Ln-TM, Ln-Al-TM, Zr-Al-TM, Zr-(Ti, Nb, Pd)-Al-TM, Zr-Ti-TM-Be, Fe-(Al, Ga)-(P, C, B, Si), Pd-Cu-Ni-P, Pd-Ni-Fe-P, (Fe, Co, Ni)-(Zr, Nb, Ta)-B, Fe-Co-(Zr, Nb)-(Mo, W)-B and Co-Fe-(Zr, Nb, Ta)-B systems [2]-[6]. Consequently, if a bulk amorphous alloy with nanostructure in crystallization process is found, we can develop a simple preparation process of a bulk nanocrystalline alloy consisting of the formation of a bulk amorphous alloy, followed by partial crystallization. More recently, it has been found [7] that the addition of Pd, Au, Pt or Ti element into melt-spun Zr-Al-Cu amorphous alloys induces the formation of a nanostructure consisting of $Zr_2(Cu, M)$ (M=Pd, Au or Pt) or $(Zr, Ti)_2Al$ and remaining amorphous phases and the nanostructure alloys can keep better bending ductility and higher tensile fracture strength as compared with those for the corresponding amorphous single

143

phase alloys. These novel results allow us to expect that bulk nanocrystalline alloys with high tensile fracture strength and good ductility are synthesized in the Zr-Al-Cu alloys containing an appropriate amount of Pd, Au, Pt or Ti element. This paper is intended to present the changes in the structure and mechanical properties by nanocrystallization for the Zr-Al-Cu-Pd and Zr-Al-Cu-Ni-Ti amorphous alloys in melt-spun ribbon and cast bulk forms and to investigate the mechanisms for the formation of the nanostructure and for the achievement of high tensile strength and good ductility even for the bulk alloys.

EXPERIMENTAL PROCEDURE

Multicomponent alloys with compositions of $Zr_{60}Al_{10}Cu_{30-x}Pd_x$ (x=0 and 10 at%) and $Zr_{53}Al_{12}Cu_{20}Ni_{10}Ti_5$ were examined in the present study because the $Zr_{60}Al_{10}Cu_{30}$ amorphous alloy has a wide supercooled liquid region before crystallization and a high glass-forming ability leading to the formation of a bulk amorphous alloy with diameters up to about 10 mm. The alloy ingots were prepared by arc melting the mixtures of pure metals in an argon atmosphere. Rapidly solidified ribbons with a thickness of 30 μm were prepared by melt spinning. Bulk cylindrical alloys with diameters of 4 to 6 mm were prepared by the copper mold casting method. Annealing was made for different times in the temperature range between T_g and T_x. The structure was examined by X-ray diffractometry and transmission electron microscopy (TEM). The distribution of alloy components was examined by high-resolution TEM and nanobeam energy dispersive X-ray (EDX) spectroscopy. Thermal stability was examined by differential scanning calorimetry (DSC). Tensile strength, Young's modulus and elongation were measured with an Instron testing machine. Tensile fracture surface appearance was examined by scanning electron microscopy (SEM).

RESULTS

Nanocrystallized Structure and Mechanical Properties of the Melt-spun Alloys

Figure 1 shows DSC curves of the melt-spun amorphous $Zr_{60}Al_{10}Cu_{30-x}Pd_x$ (x=0 and 10 at%) and $Zr_{53}Al_{12}Cu_{20}Ni_{10}Ti_5$ alloys. Although the Zr-Al-Cu amorphous alloy crystallizes by a single exothermic reaction in the supercooled liquid region, the Pd- or Ti-containing alloys show two exothermic peaks. It is seen that the extra exothermic peak appears at the higher temperature side. The X-ray diffraction patterns of the $Zr_{60}Al_{10}Cu_{20}Pd_{10}$ and $Zr_{53}Al_{12}Cu_{20}Ni_{10}Ti_5$ alloys annealed in the supercooled liquid region are identified to consist of amorphous and Zr_2Cu or Zr_2Al phases, respectively. The intensity decreases and the half width increases by the addition of Pd or Ti, suggesting that the Zr_2Cu or Zr_2Al phase has a much smaller grain size. In order to confirm the formation of the nanoscale mixed structure, the TEM images and selected-area electron diffraction patterns of the $Zr_{60}Al_{10}Cu_{20}Pd_{10}$ and $Zr_{53}Al_{12}Cu_{20}Ni_{10}Ti_5$

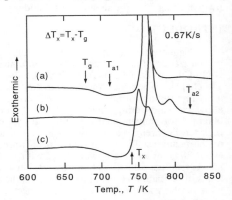

Fig. 1 Change in the DSC curves of the melt-spun amorphous $Zr_{60}Al_{10}Cu_{30-x}Pd_x$ ((a) x=0 and (b) 10 at%) and (c) $Zr_{53}Al_{12}Cu_{20}Ni_{10}Ti_5$ alloys.

alloys annealed for 1.8 ks at 726 K and 705K, respectively, are shown in Fig. 2. The annealed alloys consist mainly of very fine Zr_2Cu or Zr_2Al grains with sizes of less than 10 nm. Besides, the diffraction patterns contain distinct halo rings, in addition to the spotty reflection rings resulting from Zr_2Cu or Zr_2Al phase. The V_f of the Zr_2Cu or Zr_2Al phase is estimated to be 75 % or 55%, respectively, from the change in the heat of exothermic reaction due to the precipitation of their compounds.

Fig. 2 Bright-field electron micrographs and selected-area electron diffraction patterns of the melt-spun amorphous (a), (b) $Zr_{60}Al_{10}Cu_{20}Pd_{10}$ and (c), (d) $Zr_{53}Al_{12}Cu_{20}Ni_{10}Ti_5$ alloys annealed at 726 K and 705 K, respectively, for 1.8 ks.

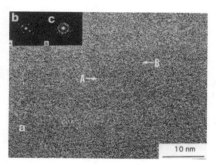

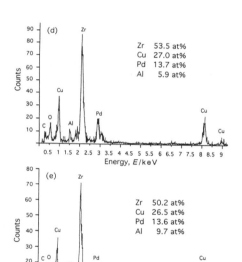

Figure 3 shows the high-resolution TEM image and EDX profiles taken from the Zr_2Cu compound and amorphous phase regions in the $Zr_{60}Al_{10}Cu_{20}Pd_{10}$ alloy. It is recognized that Pd is enriched into the Zr_2Cu phase while Al is segregated into the remaining amorphous phase. Consequently, the chemical formula of the compound is expressed as $Zr_2(Cu, Pd)$. Besides, the enrichment of Al into the remaining amorphous phase may be the origin for the formation of the mixed structure by the increase in the thermal stability of the remaining amorphous phase. On the other hand, the Zr_2Al phase in the Zr-Al-Cu-Ni-Ti alloy was identified to contain an enriched Al content, while the remaining amorphous phase was enriched by Cu and Ni. The enrichment of Cu and Ni elements is not effective for the increase in the thermal stability of the remaining amorphous phase, in agr-

Fig. 3 (a) High-resolution TEM image, (b and c) nanobeam electron diffraction pattern and (d and e) nanobeam EDX spectroscopy profiles taken from the amorphous and compound phases in the melt-spun amorphous $Zr_{60}Al_{10}Cu_{20}Pd_{10}$ alloy annealed at 726 K for 1.2 ks.

eement with the result that the crystallization temperature of the Zr-Al-Cu-Ni-Ti amorphous alloys decreases slightly by the addition of Ti.

It is important to point out that the mixed phase alloys in the ribbon form exhibit good bending ductility and can be bent through 180 degrees without fracture. By using the ductile ribbon samples, the mechanical properties were examined as a function of V_f of Zr_2Cu phase. Figure 4 shows the changes in the tensile fracture strength (σ_f), Vickers hardness (Hv) and Young's modulus (E) with volume fraction (V_f) of the Zr_2Cu phase for the melt-spun $Zr_{60}Al_{10}Cu_{20}Pd_{10}$ amorphous ribbons, respectively. The σ_f, Hv and E increase almost linearly with increasing V_f in the wide V_f range up to about 75 % and the increasing ratio is about 18 % for σ_f, 20 % for Hv and 19 % for E. With further increasing V_f, the ribbon sample becomes brittle, leading to a rapid decrease in σ_f. The highest values of σ_f and Hv are 1960 MPa and 605, respectively, being nearly the same as those for the Pd-containing alloy.

Preparation and Mechanical Properties of Bulk Nanocrystalline Alloys

We prepared cast bulk amorphous Zr_{60}-$Al_{10}Cu_{20}Pd_{10}$ and $Zr_{53}Al_{12}Cu_{20}Ni_{10}Ti_5$ cylinders

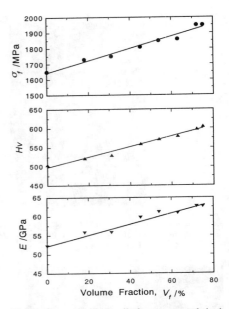

Fig. 4   Changes in the tensile fracture strength (σ_f), Vickers hardness (Hv) and Young's modulus (E) with V_f of Zr_2Cu phase for the melt-spun amorphous $Zr_{60}Al_{10}Cu_{20}Pd_{10}$ alloy.

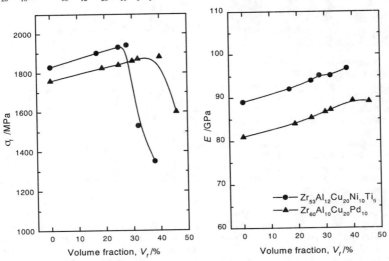

Fig. 5   Changes in the σ_f and E with V_f of the Zr_2Cu and Zr_2Al for the cast bulk amorphous $Zr_{60}Al_{10}Cu_{20}Pd_{10}$ and $Zr_{53}Al_{12}Cu_{20}Ni_{10}Ti_5$ alloys, respectively.

with diameters of 4 and 6 mm, respectively. These cylindrical samples had smooth surface and metallic luster. Neither ruggedness nor cavity was seen over the whole surface, indicating that the cast alloy is composed of an amorphous phase. Figures 5 show the changes in σ_f and E with V_f for the cast bulk amorphous $Zr_{60}Al_{10}Cu_{20}Pd_{10}$ and $Zr_{53}Al_{12}Cu_{20}Ni_{10}Ti_5$ cylinders with diameters of 2.2 mm. The σ_f and E of the Pd-containing cast cylinder are 1760 MPa and 81.5 GPa, respectively, at V_f =0 % and increase to 1880 MPa and 89.5 GPa, respectively, at V_f =40 %. The similar increase in mechanical properties by precipitation of the nanoscale $(Zr, Ti)_2Al$ phase is recognized for the Ti-containing bulk alloy. The σ_f and E increase from 1830 MPa and 89.0 GPa, respectively, at V_f =0 % to 1940 MPa and 95.2 GPa, respectively, at V_f =28 %. Furthermore, as exemplified in Fig. 6, we have confirmed that the tensile fracture mode consisting of a shear sliding, followed by an adiabatic failure remains unchanged in the V_f range up to 40 % where the highest σ_f value is obtained and the further increase in V_f induces the change in the fracture mode to a brittle type characterized by the generation of a perpendicular-type fracture surface against the direction of tensile load.

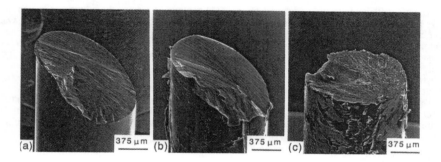

Fig. 6   Tensile fracture surface appearance of the cast bulk amorphous $Zr_{60}Al_{10}Cu_{20}Pd_{10}$ alloy
(a) V_f=0%, (b) V_f=40%, and (c) V_f=46%.

DISCUSSION

Formation of the Nanocrystalline Structure

It is generally known that the formation of a nanostructure from an amorphous phase requires the following three factors, i.e., (1) a multistage crystallization mode leading to the precipitation of a primary crystalline phase, (2) the ease of homogeneous nucleation of the primary phase, (3) the difficulty of subsequent crystal growth reaction, and (4) a high thermal stability of the remaining amorphous phase [1][8]. The addition of Pd or Ti element with much larger negative heats of mixing against Zr or Al, respectively, causes the change of the crystallization mode from the single stage to the two stages as well as the homogeneous generation of Zr-Pd or Ti-Al rich clusters which can act as a nucleation site of $Zr_2(Cu, Pd)$ or $(Zr, Ti)_2Al$ phase. The subsequent growth of the Zr-Pd or Ti-Al rich cluster is also difficult because of the enrichment of Al in the remaining amorphous or the compound Ti_2Al phase near the interface between amorphous and compound phases resulting from the segregation of Al from the constituent phases. The difficulty of the growth reaction seems to result in the high activation energy for the precipitation of their compound phases. Furthermore, the redistribution of the Al element plays a dominant role in the thermal stability of the remaining amorphous phase against crystallization.

147

High Tensile Strength Combined with Good Ductility for the Nanostructure Alloys

We briefly consider the reason why the nanoscale compound base alloys can have good ductility and high tensile strength. In particular, the nanostructure alloys in a melt-spun ribbon form are also regarded as a new type of mixture consisting of nanoscale compound grains (with a size of 5 to 15 nm) and amorphous intergranular phase (with a V_f of about 30 %). The deformation of the nanostructure alloys by bending or tensile stress is thought to occur by sliding of the amorphous intergranular phase. The present result is believed to be the first synthesis of high-strength and high-ductility alloys via amorphous intergranular sliding among intermetallic compounds at room temperature. The maintenance of good bending ductility even at the high V_f of 75 % is in contrast to the previous data [9][10] that the formation of the ductile nanostructure alloys is limited to the small V_f range below about 30 %. The above-described anomalous ductile behavior has been recognized only for the amorphous alloy exhibiting the glass transition phenomenon. Consequently, the reason why the nanostructure alloys consisting of $Zr_2(Cu, Pd)$ or $(Zr, Ti)_2Al$ phase surrounded by the amorphous grain boundary phase exhibit good bending ductility and high tensile strength is presumably due to the combination of the following four factors, i.e., (1) the residual amorphous phase can contain a large amount of free volumes by annealing in the supercooled liquid region, followed by water quenching, (2) the amorphous/ Zr_2Cu or amorphous/ Zr_2Al interface has a high degree of packing fraction because of a much lower interface energy at the liquid/solid interface[11], (3) their compounds disperse homogeneously and isolately in coexistent with an amorphous intergranular phase, and (4) the nanostructure has high thermal stability resulting from the redistribution of solute elements. The optimum V_f leading to the maximum values of σ_f is about 20 to 40 % for the cast bulk alloys and considerably lower than that (about 70 %) for the melt-spun ribbons. The difference seems to originate from the lower cooling rates during water quenching from the annealing temperature for the bulk samples. The amount of free volumes introduced by water quenching is considerably lower for the bulk samples, leading to the lower values of the optimum V_f for the bulk samples.

CONCLUSIONS

The bulk nanocrystalline alloys have the following features: (I) The structure can be regarded as a new type of mixture consisting of nanoscale compound grains and amorphous intergranular phase, and (II) The deformation of the mixed phase alloys by bending or tensile stress occurs by sliding in the amorphous intergranular phase. Consequently, this is the expected synthesis of high-strength and high-ductility alloys via amorphous intergranular sliding among intermetallic compounds at room temperature. The subsequent development of bulk nano-crystalline alloys along the present alloy design concept is expected to cause the appearance of a new type of engineering bulk materials exhibiting useful properties which cannot be obtained for bulk amorphous and crystalline alloys.

REFERENCES

1. A. Inoue, Bulletin Japan Inst. Metals, 36, 926 (1997).
2. A. Inoue, Mater. Trans., JIM, 36, 866 (1995).
3. A. Inoue, Sci. Rep. Res. Inst. Tohoku Univ., A42, 1 (1996).
4. A. Inoue, Proc. Japan Academy, 73B, 19 (1997).
5. A. Inoue, Mater. Sci. Eng., A226-228, 357 (1997).
6. W. L. Johnson, Mater. Sci. Forum, 225-227, 35 (1996).
7. C. Fan and A. Inoue, Mater. Trans., JIM, 38, 1040 (1997).
8. A. Inoue, Nanostruct. Mater., 6, 53 (1995).
9. Y. H. Kim, A. Inoue and T. Masumoto, Mater. Trans., JIM, 31, 747 (1990).
10. H. Chen, Y. He, G.J. Shiflet and S.J. Poon, Scripta Met., 25, 1421 (1991).
11. E.S. Machlin, Thermodynamics and Kinetics, Giro Press, New York (1991), p.125.

MECHANICAL PROPERTIES OF Zr-Al-Cu-Ni ALLOY GLASS AT DIFFERENT QUENCHING CONDITION

H. YABE, R. TOYOSHIMA, Y. HIRAIWA, K. OGURI, Y. MIYAZAWA & Y. NISHI
Department of Materials Science, Tokai University, 1117 Kita-kaname, Hiratsuka, Kanagawa, 259-1292 JAPAN, am026429@keyaki.cc.u-tokai.ac.jp

ABSTRACT

The samples of $Zr_{56}Al_9Cu_5Ni_{30}$ alloy were prepared by liquid-quenching. The mechanical properties depend on the cooling condition. The Hv value (25g load) is 580 for slow cooled rate sample, whereas the Hv value is over 800 for fast cooled sample. Since the fast cooling enhances the sample hardness, these samples become brittle. The brittleness is evaluated by the critical deformation energy (E_c) of collapse.

INTRODUCTION

The mechanical, electric and magnetic properties of liquid-quenched glassy alloys have been investigated. It is important to know the effect of cooling condition on the properties. Zr-based amorphous alloys exhibit a large glass-forming ability [1-2]. It is useful to know the various properties for Zr-based amorphous alloys. Rapidly cooled metallic glasses are, in general, ductile, whereas slowly cooled or aged glasses are brittle. From the engineering point of view, it is important to get the ductile sample. On the other hand, the stress intensity factor (K_{IC}) has been used as a standard to evaluate the brittleness [3-5]. However, it is difficult to prepare test specimens to measure the brittleness of materials by a conventional method. Thus, we have suggested a convenient method to evaluate ductility by means of a standard hardness tester for small brittle samples [6-9]. When collapse forms around a Vickers indentation under a certain diagonal, a critical deformation energy (E_c) of collapse generation can be defined. It is easy to measure on small samples such as metallic the brittleness glasses. The purpose of the present work is to investigate the effects of cooling condition on the ductility of liquid-quenched Zr-Al-Cu-Ni alloy.

EXPERIMENTAL PROCEDURE

The $Zr_{56}Al_9Ni_5Cu_{30}$ alloy was prepared by melting pure Zr (99.9%), Al (99.9%), Ni (99.9%) and Cu (99.9%) in an infrared furnace under a protective atmosphere of Ar-5vol%H_2. The liquid-quenching was performed by a piston-anvil apparatus [10]. The quenched samples were from 64 to 186 μ m in thickness. The temperature of molten alloy was just above the melting point. The cooling rate (R; K/s) was changed by controlling sample thickness. Since it could control the cooling rate, it was easy to control the glass randomness [11]. The cooling rate (R; K/s) was varied by controlling the thickness (D; m) of as-quenched sample [12]. The R value was determined by the D value as a following equation.

$$\log R = - \log D + 0.9 \quad \cdots (1)$$

The composition was not deviated on the solidification, because the maximum temperature of molten alloy was just above the melting point. The chemical compositions analyzed by

EDX (JSM-6301F, JEOL Ltd. TOKYO) were shown in Table I.   The structure of the sample was examined by means of X-ray diffraction (see Fig. 1).   The crystalline diffraction peaks were not observed for the glassy samples.   The resistance to plastic deformation and a precise calculation of ductility was performed with a standard Vickers microhardness tester [6].   The glass transition temperature (676K) was determined by using DSC.   The DSC heating rate was 40 K/min.   The size of the specimen was 12.5 mm in length and 2.2 mm in width.

Table I EDX chemical analysis of master and liquid-quenched alloys.

	Zr (at%)	Al (at%)	Ni (at%)	Cu (at%)
Specimen	55.97	9.13	5.09	29.81

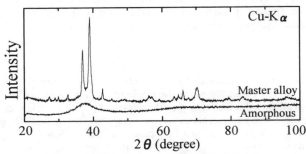

Figure. 1 XRD patterns of $Zr_{56}Al_9Cu_5Ni_{30}$ alloy.

RESULTS AND DISCUSSIONS

Hardness

The Vickers hardness value (Hv) is one of indicators to evaluate the micro-resistance to plastic deformation.   Fig. 2 shows change in the Hv value against the cooling rate. The Hv value (25g load) is 580 for the slow cooled glass.   On the other hand, the Hv value is over 800 for the fast cooled glass. The fast cooling enhances the hardness of the $Zr_{56}Al_9Ni_5Cu_{30}$ metallic glass.

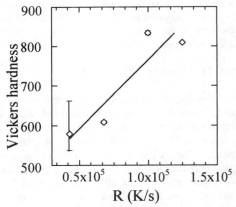

Figure 2 Change in resistance to plastic deformation Hv with cooling rates R (K/s) in $Zr_{56}Al_9Cu_5Ni_{30}$ alloy glasses.

Evaluation of brittleness

Figure. 3 shows an optical photograph and schematic profile of a Vickers indentation with an accompanying collapse with the brittle $Zr_{56}Al_9Ni_5Cu_{30}$ glass

The collapse exists at the edge of the hill. The shape of the collapse is apparently different from those of other samples previously reported [6-8]. When a collapse is found outside the indentation, the critical values of P and d are defined. The critical load (P_c^{mid}) and critical diagonal (d_c^{mid}) are defined as the midpoint between the minimum and maximum values [8], i.e.

$$P_c^{mid} = (P_c^{min} + P_c^{max})/2 \quad \cdots(2)$$
$$d_c^{mid} = (d_c^{min} + d_c^{max})/2 \quad \cdots(3)$$

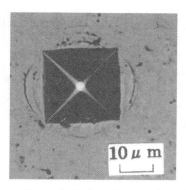
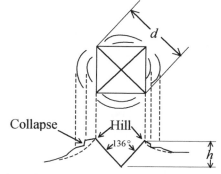

Figure. 3 Optical photograph and schematic profile of a Vickers indentation with an accompanying collapse of brittle Zr-Al-Cu-Ni alloy glass.

When one of three indentations exhibits a collapse at a certain load, the minimum load (P_c^{min}) and the minimum diagonal (d_c^{min}) are defined. P_c^{max} & d_c^{max} are the maximum load and maximum diagonal, respectively, for which one of three indentations is collapse-free. Therefore, critical deformation energy (E_c^{mid}) of collapse is suggested here as a convenient parameter to evaluate brittleness. The deformation energy can be expressed by the following equation [8].

$$E_c^{mid} = 0.0477 \, P_c^{mid} \, d_c^{mid} \quad \cdots(4)$$

The critical values, P_c^{mid}, d_c^{mid} and E_c^{mid} of the liquid-quenched Zr-10at%Al-30at%Cu-5at%Ni alloy glasses are summarized in TABLE II. Figs. 4, 5 & 6 show changes in P_c^{mid}, d_c^{mid} & E_c^{mid} against the cooling rate R. The critical values for the slow cooled glass are larger than that for the fast cooled glass. Namely, the $Zr_{56}Al_9Ni_5Cu_{30}$ bulk metallic glass shows ductility.

TABLE II Critical values of liquid-quenched $Zr_{56}Al_9Ni_5Cu_{30}$ alloy glasses

Cooling rate R (K/s)	P_c^{mid} (gf)	d_c^{mid} (μ m)	E_c^{mid} (erg)
4.27×10^4	175.0	22.5	18.393
6.77×10^4	137.5	20.0	12.846
4.98×10^4	125.0	17.8	10.394
1.25×10^5	112.5	17.2	9.039

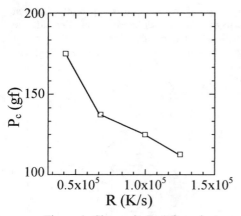

Figure 4. Change in P_c (gf) against the cooling rate R (K/s).

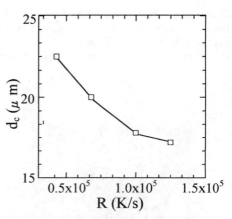

Figure 5. Change in d_c (μ m) against the cooling rate R (K/s).

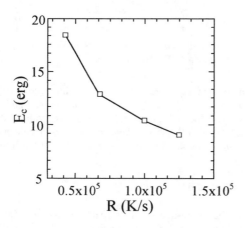

Figure 6. Change in E_c (erg) against cooling rate R (K/s).

If the critical deformation energy (E_c) of collapse is assumed to be equal to that (E_f) of crack [8], the value of K_{IC} can be estimated [8]. Figs. 7 & 8 show changes in K_{IC} and G_{IC} against the cooling rate R. The values of K_{IC} and G_{IC} can be determined through the following equations.

$$\log_{10} K_{IC} = 0.44\log_{10}E_f\text{-}0.63 \quad \cdots(5)$$
$$\log_{10} G_{IC} = 0.52\log_{10}E_f\text{-}0.32 \quad \cdots(6)$$

Although the K_{IC} and G_{IC} values are dominated by the growth mechanism of the crack, the E_f & E_c values are determined by the nucleation and growth of crack.

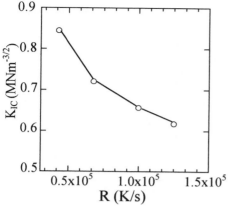

Figure 7. Change in K_{IC} ($MNm^{-3/2}$) against cooling rate R (K/s).

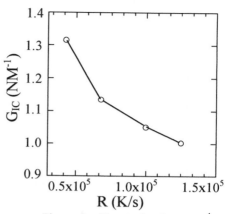

Figure 8. Change in G_{IC} (NM^{-1}) against cooling rate R (K/s).

Discussion based on rate process

Based on the rate process [13-15], the volume-fraction change (X) is generally expressed by the following equation in relation to the cooling rate (R).

$$X = \exp(-kR^{-1}) \quad \cdots(7)$$

Here, k and n are constants.

$$X = [(E_c^{mid} - {}^0E_c^{mid})/({}^mE_c^{mid} - {}^0E_c^{mid})] \quad \cdots(8)$$

${}^mE_c^{mid}$ and ${}^0E_c^{mid}$ are the E_c^{mid} of the glass randomized for extremely fast cooling rate and of the glass randomized for extremely slow cooling rate, respectively. E_c^{mid} of the randomized glass approaches the estimated ${}^mE_c^{mid}$ value in Fig. 9. When the correlation coefficient (F) of Equation (8) is maximum, as shown in FIG.10 [16], the estimated ${}^mE_c^{mid}$ value is 8.22 erg. From the results, X is expressed by the following equation for $Zr_{56}Al_9Ni_5Cu_{30}$ alloy glasses (see FIG.11).

$$\log_{10}[-\ln(1-X)] = -\log_{10}R + \log_{10}k \quad \cdots(9)$$

Equation (9) is plotted as the solid line in FIG. 10. The linear plot of Equation (9) confirms the assumption of the rate process (see Equation (7)). Therefore, E_c^{mid} can be obtained by controlling the cooling rate (R) as expected.

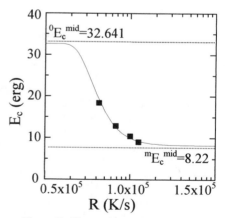

Figure 9. Change in E_c with cooling rate, of liquid-quenched $Zr_{56}Al_9Ni_5Cu_{30}$ alloy glass. ${}^mE_c^{mid}$ is the E_c at extremely fast cooling rate, ${}^0E_c^{mid}$ is E_c at extremely slow cooling rate.

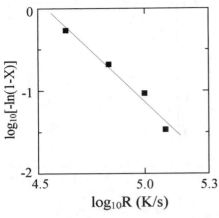

Figure.10 Linear plot of $\log_{10}[-\ln(1-X)]$ against R.

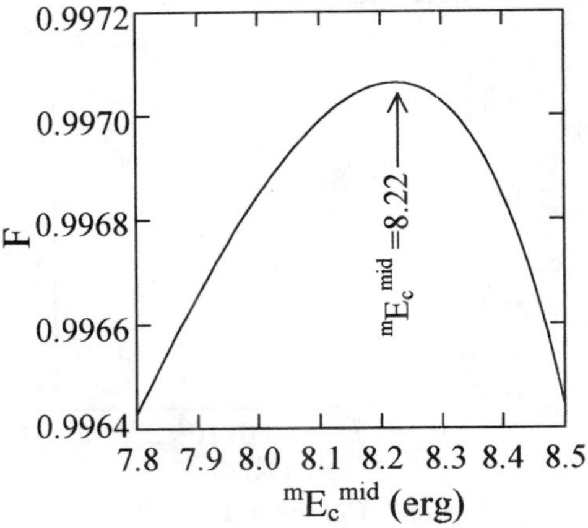

Figure. 11 Change in correlation coefficient (F) with $^{m}E_{c}{}^{mid}$ for $Zr_{56}Al_{9}Ni_{5}Cu_{30}$ alloy glass.

CONCLUSION

The ductility (brittleness) for the liquid-quenched $Zr_{56}Al_{9}Ni_{5}Cu_{30}$ alloy glasses has been evaluated by the use of a critical deformation energy of collapse. The slow cooling increases the ductility of the metallic glass. The critical deformation energy can be controlled by the cooling condition.

REFFERENCES

1. A. Inoue, T. Zhang and T. Masumoto. J. Non-Cryst. Sol. 156-158 (1993) 473-480.
2. A. Inoue, T. Zhang and T. Masumoto. Mater. Trans., JIM 31 (1990) 929.
3. A. Brenner, D. E. Couch and E. K. Willoams. J. Res. Nat. Bur. Stand. 44 (1950) 109.
4. W. Buckel and R. Hisch. Z. Phys. 132 (1952) 420.
5. A. G. Evans and E. A. Charls. J. Am. Ceram. Soc. 59 (1976) 371.
6. Y. Nishi, N. Ninomiya, F. Kanai, S. Uchida and S. Moriya. J. Appl. Phys. 66 (1989) 2069.
7. Y. Nishi, N. Ninomiya, T. Katagiri and H. Harano. J. Mater. Sci. Lett. 7 (1988) 1126.
8. Y. Nishi, T. Katagiri, T. Yamano, F. Kanai, N. Ninomiya, S. Uchida, K. Oguri, T. Morishita, T. Endo and M. Kawakami. Appl. Phys. Lett. 58 (1991) 2084.
9. Y. Nishi, K. Suzuki and T. Masumoto, edited by T. Masumoto and K. Suzuki (Fourth International Conference on Rapidly Quenched Metals Proc. Sendai, 1981) pp. 217-220.
10. Y. Nishi, H. Harano, S. Uchida, K. Oguri. J. Materials Science 25 (1990) 4477-4482.
11. Y. Nishi, K. Suzuki and T. Masumoto, J. Jpn Inst. Met. 45 1300-1305 (1981).
12. S. Uchida, O. Shibata, K.Oguri & Y.Nishi. J. Mater. Sci. 29 (1994) 5589-5592.
13. Y. Nishi and H.Harano. J. Appl. Phys. 63 (1988) 1141.
14. Y. Nishi, H. Harano, T. Fukunaga and K. Suzuki. Phys. Rev. B 37 (1988) 2855.
15. Y. Nishi, H. Harano, H. Ishizuki, M. Kawakami & E. Yajima. Mater.Sci.Eng. 98 (1988) 505.
16. M. Kendall and A. Stuart, "The Advanced Theory of Statistics", Vol. 1, (Charles Griffin, London, 1969) pp. 383-5.

COMPOSITIONAL DEPENDENCE OF THERMAL STABILITY AND SOFT MAGNETIC PROPERTIES FOR Fe-Al-Ga-P-C-B GLASSY ALLOYS

T. Mizushima*, A. Makino*, S. Yoshida* and A. Inoue**
*Central Res. Lab., Alps Electric Co., Ltd., 1-3-5 Higashitakami, Nagaoka 940-8572, Japan,
mizushim@alps.co.jp
**A. Inoue, Inst. for Materials Res., Tohoku University, 2-1-1 Katahira, Aoba-ku, Sendai 980-77, Japan.

ABSTRACT

Structure, glass forming ability and soft magnetic properties for Fe-Al-Ga-P-C-B glassy alloy system were investigated in the compositional range of Fe from 69 to 78 at%, (Al+Ga) from 2 to 12 and (P+C+B) from 17 to 28. The saturation magnetization (σ_s) rises gradually with increase of Fe concentration. The maximum value of 70K for supercooled liquid region ($\Delta T_x = T_x - T_g$, T_x: crystallization temperature, T_g: glass transition temperature.) and the maximum thickness of 180 μm for glass formation (t_{max}) are found in the composition range around Fe=70at% and (Al+Ga)=7at%. The highest permeability (μ_e) of 20,000 at 1kHz and the lowest coercive force (H_c) of 2 A/m at the sample thickness of 30 μm can be also obtained at this composition. It was ascertained that the composition regions to yield the maximum glass forming ability and lowest magnetostriction were in agreement with that in which the most excellent soft magnetic properties were yielded. This results allow us to assume that the excellent soft magnetic properties for this glassy alloy system in the limited composition range are presumably due to high structural homogeneity resulting from significantly high glass-forming ability.

INTRODUCTION

It is well known that Fe- and Co- based amorphous alloys exhibit good soft magnetic properties[1]. The soft magnetic properties have been characterized as the achievement of high saturation magnetization for Fe-based alloys and high permeability (μ) and zero magnetostriction for Co-based alloys. However, the shape of these soft magnetic amorphous alloys is usually limited to sheet, wire, film and powder because of the necessity of high cooling rate resulting from their low glass forming ability. For instance, the maximum thickness (t_{max}) to form a single amorphous phase for $Fe_{78}Si_9B_{13}$ amorphous ribbon is below 100 μm[2]. This limitation deters the extension of its application fields. Consequently, the bulky amorphous alloys are desirable from the point of view of application.

Recently, the large glass forming ability combined with the wide supercooled liquid region (ΔT_x) above 50 K has been found in a number of multicomponent alloy systems of Mg-Ln-TM[3-4], Ln-Al-TM[5-6], Zr-Al-TM[7-8], Zr-Ti-Al-TM[9-10], Zr-Ti-TM-Be[11] and Pd-Cu-Ni-P[12] etc. (Ln=lanthanoid metal, TM= transition metal). These alloy systems consistently satisfy the empirical rule which is common with the glassy alloys having a large glass forming ability with a wide ΔT_x; i.e. multicomponents with significantly different atomic size ratios higher than approximately 12% and negative heats of mixing[13-17]. Based on the above empirical rule, Fe-based amorphous alloy with higher glass forming ability has been searched. As a result, we have already reported that a $Fe_{73}Al_5Ga_2P_{11}C_5B_4$ amorphous alloy sheet has a wide supercooled liquid region exceeding 40K before crystallization and maximum thickness of 135μm for t_{max} and exhibits good soft magnetic properties as compared with $Fe_{78}Si_9B_{13}$[18].

This paper is intended to deal with compositional dependence of the glass forming ability based on the ΔT_x and soft magnetic properties for Fe-Al-Ga-P-C-B glassy alloys prepared by melt-spinning method.

EXPERIMENTAL PROCEDURE

Multicomponent alloys with composition $Fe_{100-x-y}(Al_{0.71}Ga_{0.29})_x(P_{0.55}C_{0.25}B_{0.2})_y$ ($x=2\sim12$, $y=17\sim28$) were used because the supercooled liquid region before crystallization and good soft magnetic properties in Fe-Al-Ga-P-C-B system have been obtained for $Fe_{73}Al_5Ga_2P_{11}C_5B_4$ alloy[19]. Their alloy ingots were prepared by induction melting the mixtures of pure Fe, Al, and Ga metals, premelted Fe-P and Fe-C alloys and pure crystal boron in an argon atmosphere. Rapidly solidified alloy ribbons with the width of 1mm and the various thickness ranging from 15 to 305 μm were prepared through the control of wheel velocity, nozzle diameter and ejection pressure by a single roller melt spinning method. The amorphous nature was examined by X-ray diffraction and also optical and transmission electron microscopy (OM and TEM). Thermal stability associated with glass transition, supercooled liquid region and crystallization was examined at a heating rate of 0.67K/s by differential scanning calorimetry (DSC). Magnetic properties of saturation magnetization (σ_s), coercive force (H_c) and permeability (μ_e) at 1kHz were measured at room temperature with a vibrating sample magnetometer (VSM) under 800kA/m, a B-H loop tracer under 1.6 kA/m and an impedance analyzer under 0.8A/m, respectively. Saturated magnetostriction was also evaluated by three capacitance method under 240 kA/m.

RESULTS AND DISCUSSION

1. Structure and thermal properties of Fe-based glassy alloys

The Fe-Al-Ga-P-C-B alloy system follows the empirical rules above mentioned. We can classify Al and Ga as large size, Fe as middle size and P,C,B as small size atoms. In this study, we investigate the relation between the glass forming ability, magnetic properties and the combination of constituent elements. The ratios of Al: Ga and P:C:B are kept on constant values of 0.71:0.29 and 0.55:0.25:0.2, respectively, because the high glass forming ability and good soft magnetic properties are obtained for $Fe_{73}Al_5Ga_2P_{11}C_5B_4$ glassy alloy[19].

We first investigate the changes in the structure and thermal properties for Fe-Al-Ga-P-C-B glassy alloys as a function of Fe concentration at Al+Ga=7 at%. Figure 1 shows the X-ray diffraction patterns of the melt-spun $Fe_{100-x-y}(Al_{0.71}Ga_{0.29})_x(P_{0.55}C_{0.25}B_{0.2})_y$ ($x=7$, $y=17,18,21,22$) alloy ribbons with a thickness of 30 μm. Although crystalline peaks of a bcc phase are superimposed on the broad peak only for the sample containing 78 at% Fe, the other alloys show only of a broad peak at a wave vector ($K_p=4\pi\sin\theta/\lambda$) of about 31 nm^{-1}. Thus, the increase in Fe content above 78 at% causes a rapid decrease in the glass forming ability.

In order to examine the existence of the supercooled liquid region before crystallization, the DSC curves are shown in Figure 2. The 78 at% Fe alloy consisting of amorphous and bcc-Fe phases

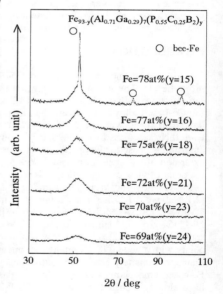

Figure 1. The X-ray diffraction patterns of the melt-spun $Fe_{100-x-y}(Al_{0.71}Ga_{0.29})_7(P_{0.55}C_{0.25}B_{0.2})_y$ ($y=15,16,18,21,23,24$) alloy ribbons with the thickness of 30 μm.

crystallizes through multiple stages; amorphous + bcc-Fe → amorphous + bcc-Fe +Fe₃B → bcc-Fe + Fe₃B + Fe₃P + Fe₂B + Fe₃C. On the other hand, the other alloys exhibit the glass transition, followed by the supercooled liquid region and then a single exothermic reaction. As marked with T_g and T_x, the glass transition temperature (T_g) and crystallization temperature (T_x) increase from 732 to 752 and 770 to 822 K, respectively, with decreasing Fe content from 75 to 69 at%. The rise is larger for T_x than for T_g, leading to the extension of the supercooled liquid region (ΔT_x).

The compositional dependence of ΔT_x and reduced glass transition temperature (T_g/T_m) for $Fe_{100-x-y}(Al_{0.71}Ga_{0.29})_x(P_{0.55}C_{0.25}B_{0.2})_y$ (x=2~12, y=17~28) alloys was studied in order to find the alloy composition with high glass forming ability. These results are shown in Figure 3. The maximum ΔT_x of 70K is found in the compositional range around Fe=70at% and (Al+Ga)=7at%. The value of ΔT_x becomes smaller with the deviation from the compositional range. The T_g/T_m increases with increase of P+C+B concentration and indicates a large value of 0.60 over P+C+B>30at%. The relatively large value of 0.58 is also found at the composition range of Fe content over 77 at% and Al+Ga content over 2 at%.

Figure 4 shows the compositional dependence of the t_{max}. The maximum t_{max} is obtained.

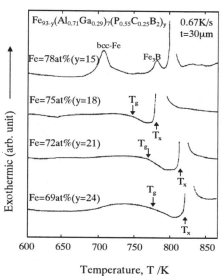

Figure 2. DSC curves of the melt-spun $Fe_{100-x-y}(Al_{0.71}Ga_{0.29})_7(P_{0.55}C_{0.25}B_{0.2})_y$ (y=15, 18,21,24) alloy ribbons with the thickness of 30 μm.

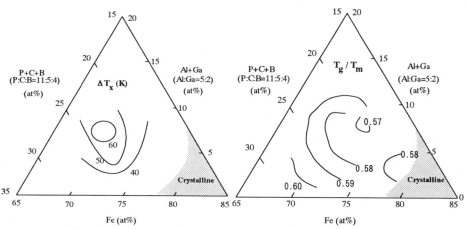

Figure 3. The compositional dependence of supercooled liquid region (ΔT_x) and reduced glass transition temperature (T_g/T_m) for $Fe_{100-x-y}(Al_{0.71}Ga_{0.29})_x(P_{0.55}C_{0.25}B_{0.2})_y$ (x=2~12, y=17~28) alloys.

157

around Fe=70at% and (Al+Ga)=7at%. The composition range in which the maximum value of t_{max} can be obtained, corresponds with that in which the largest ΔT_x is found. The t_{max} increases slightly with increasing Fe content over 75at%, presumably caused by the result that high T_g/T_m is obtained in the Fe content.

Figure 5 shows the relation between the t_{max} and the ΔT_x for $Fe_{100-x-y}(Al_{0.71}Ga_{0.29})_x(P_{0.55}C_{0.25}B_{0.2})_y$ ($x=2 \sim 12$, $y=17\sim28$) alloys. There is a tendency for t_{max} to increase with increasing ΔT_x. It is, therefore, concluded that the extension of the supercooled liquid region causes the increase in the glass-forming ability.

2. Magnetic properties of Fe-based glassy alloys

Figure 6 shows the compositional dependence of saturation magnetization (σ_s) for the Fe-based glassy alloys. The σ_s increases from 170×10^{-6} to 210×10^{-6} Wbm/kg monotonously, with increasing Fe content from 69 to 77 at%.

Figure 7 shows the compositional dependence of effective permeability (μ_e) at 1 kHz and coercive force for the Fe-based glassy alloys with the thickness of 30 μm. The maximum μ_e of 20000 and the minimum H_c of 2A/m are obtained around the composition range of Fe=70at% and (Al+Ga)=7at% and both become inferior with deviation from the composition.

Figure 8 shows the relation between the μ_e at 1kHz with sample thickness of 30μm and the t_{max} for Fe-based glassy alloys. There is a clear tendency for μ_e to increase with increasing t_{max}. This result indicates that the composition range in which the higher μ_e can be obtained is in agreement with that in which the higher glass forming ability can be obtained. The magnetostriction (λ_s) for Fe-based glassy alloy shows the value from 20×10^{-6} to 30×10^{-6} in the entire composition range in this study. These values of the λ_s are nearly as same as those of conventional Fe-based amorphous alloys.[20] It is, therefore, concluded that the excellent soft magnetic properties for the Fe-based glassy alloy are mainly due to their high structural

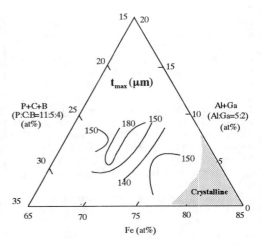

Figure 4. The compositional dependence of maximum thickness of glass formation (t_{max}) for $Fe_{100-x-y}(Al_{0.71}Ga_{0.29})_x(P_{0.55}C_{0.25}B_{0.2})_y$ ($x=2\sim12$, $y=17\sim28$) alloys.

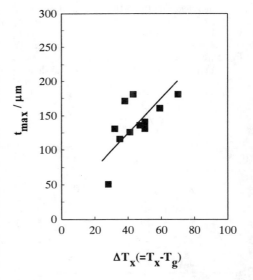

Figure 5. Relation between the t_{max} and the ΔT_x for $Fe_{100-x-y}(Al_{0.71}Ga_{0.29})_x(P_{0.55}C_{0.25}B_{0.2})_y$ ($x=2 \sim 12$, $y=17\sim28$) alloys

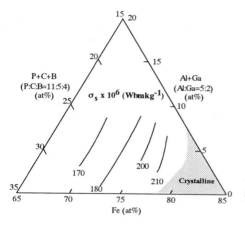

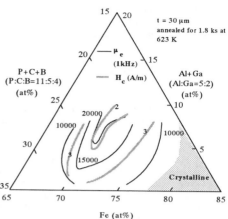

Figure 6. The compositional dependence of saturation magnetization (σ_s) for $Fe_{100-x-y}$ $(Al_{0.71}Ga_{0.29})_x(P_{0.55}C_{0.25}B_{0.2})_y$ $(x=2\sim12, y=17\sim28)$ alloys.

Figure 7. The compositional dependence of effective permeability (μ_e) at 1KHz and coercive force (H_c) for $Fe_{100-x-y}$ $(Al_{0.71}Ga_{0.29})_x$ $(P_{0.55}C_{0.25}B_{0.2})_y$ $(x=2\sim12, y=17\sim28)$ alloys.

homogeneity based on the significantly high glass-forming ability caused by the appearance of the wide supercooled liquid region before crystallization.

CONCLUSIONS

With the aim of developing a new Fe-based amorphous alloy with large glass forming ability and excellent soft magnetic properties, we examined the compositional dependence of ΔT_x, T_g/T_m, t_{max}, σ_s, μ_e, H_c and λ_s for alloy series of $Fe_{100-x-y}$ $(Al_{0.71}Ga_{0.29})_x(P_{0.55}C_{0.25}B_{0.2})_y$ $(x=2\sim12, y=17\sim28)$. The results obtained are summarized as follows.

(1) The glass transition phenomena are observed in the temperature range below T_x in the Fe concentration range of 69 to 77 at% and Al+Ga concentration range of 2 at% to 12 at%.

(2) The maximum values of ΔT_x and t_{max} are obtained around the composition range of Fe=70at% and (Al+Ga)=7at%. The increase in t_{max} is due to the extension of the supercooled liquid region.

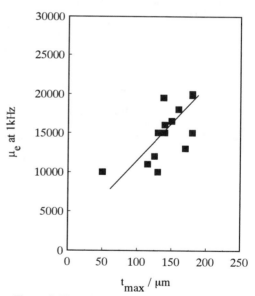

Figure 8. The relation between the μ_e at 1kHz with thickness of 30μm and the t_{max} for $Fe_{100-x-y}$ $(Al_{0.71}Ga_{0.29})_x(P_{0.55}C_{0.25}B_{0.2})_y$ $(x=2\sim12, y=17\sim28)$ alloys.

159

(3) The excellent soft magnetic properties for this Fe-based glassy alloy system are mainly due to their high structural homogeneity based on the significantly high glass-forming ability caused by the appearance of the wide supercooled liquid region before crystallization..

ACKNOWLEDGMENT

This work has been financially supported by the New Energy and Industrial Technology Development Organization (NEDO) under Super Metal Technology Project through the R&D Institute of Metals and Composites for Future Industries (RIMCOF).

REFERENCES

1. C. H. Smith, Rapidly Solidified Materials, edited by H. H. Liebermann and Marcel Dekker (New York 1993),P617.
2. M. Hagiwara, A. Inoue and T. Masumoto, Met. Trans. A13, p.373(1982).
3. A. Inoue, K. Ohtera, K. Kita and T. Masumoto, Jpn. J. Appl. Phys. 27, p. 2248(1988).
4. A. Inoue, M. Kohinata, A. P. Tsai and T. Masumoto, Mater. Trans. JIM 30, p. 378(1989).
5. A. Inoue, T. Zhang and T. Masumoto, Mater. Trans. JIM 30, p. 965(1989).
6. A. Inoue, H. Yamaguchi and T. Masumoto, Mater. Trans. JIM 31, p. 104(1990).
7. A. Inoue, T. Zhang and T. Masumoto, Mater. Trans. JIM 31, p. 177(1990).
8. A. Inoue, T. Zhang and T. Masumoto, Mater. Trans. JIM 32, p. 1005(1991).
9. A. Inoue, T. Zhang, N. Nishiyama, K. Ohba and T. Masumoto, Mater. Trans. JIM 34, p. 1234(1993).
10. A. Inoue, T. Shibata, and T. Zhang, Mater. Trans. JIM 37, p. 1,420(1995).
11. A. Peker and W. L. Johnson, App. Phys. Lett.63, p. 2,342(1993).
12. A. Inoue, N Nishiyama and T. Matsuda., Mater. Trans. JIM 37, p. 181(1996).
13. A. J. Drehman, A. L. Greer and D. Turnbull, Appl. Phys. Lett.,41, p. 716(1982).
14. A. Inoue; Advanced Materials and Processing, edited by K. S. Shin, J. K. Yoon and S. J. Kim (The Korean Inst. Metals and Materials, Seoul 1995),p1849.
15. A. Inoue; Mater. Sci. Forum 179-181, p. 691(1995).
16. A. Inoue; Sic. Rep. Res. Inst. Tohoku Univ. A42, p. 1(1996).
17. A. Inoue, T. Zhang and T. Masumoto, J. Non-Cryst. Solid.473, p. 156(1993).
18. T. Mizushima, A. Makino and A. Inoue, Mater. Sci. Eng. A226-228, p. 721(1997).
19. A. Inoue and J. S. Gook, Mater. Trans. JIM 36, p. 1,180(1995).
20. Y. Makino, K. Asoh, S. Uehira and M. Itoh, J. Mag. Soc. Jpn., 3-3, p. 43(1978).

INVESTIGATIONS ON THE ELECTROCHEMICAL BEHAVIOR OF ZR-AL-CU-NI BULK METALLIC GLASS

K. BUCHHOLZ [a], A. GEBERT [a, b], K. MUMMERT [a], J. ECKERT [a], L. SCHULTZ [a]
[a] IFW Dresden, Institut für Metallische Werkstoffe, Postfach 27 00 16,
    D-01171 Dresden, Germany
[b] Present address: École Polytechnique de Montréal, Dep. de génie physique et de génie des
    materiaux, C. P. 6079, Montréal (Québec), H3C 3A7 Canada

ABSTRACT

The passivation behavior of bulk glassy $Zr_{55}Al_{10}Cu_{30}Ni_5$ alloy samples in weakly alkaline sulphate solution (pH = 8) is investigated in comparison to the arc-melted crystalline alloy, to the main alloying component zirconium and to aluminium. Results of potentiodynamic and potentiostatic polarization measurements reveal the formation of a stable passivating surface film on the glassy alloy grown by a high-field mechanism. Auger electron spectroscopic investigations of anodized sample surfaces show that the films formed on the glassy Zr-Al-Cu-Ni alloys exhibit a composition gradient in cross-sectional direction. In 0.001M NaCl electrolytes bulk glassy Zr-Al-Cu-Ni samples are susceptible to pitting corrosion, which is due to the existence of crystalline inclusions.

INTRODUCTION

Typical for a glassy alloy is its single phase nature without imperfections such as grain boundaries or dislocations. This chemical homogeneity provides an excellent corrosion resistance. Furthermore, the composition of glassy alloys is not restricted by solubility limits allowing selective alloying of strongly passivating elements. Thus, the formation of a stable and uniform passive film is possible [1, 2]. A number of glassy alloys exhibit superior corrosion resistance in comparison to crystalline alloys of same overall chemical composition. But, up to now, glassy alloys were typically prepared as melt-spun thin ribbons or sputter-deposited films [3, 4] and no details about the corrosion properties of bulk glasses are known. Hence, it is tempting to investigate if a superior corrosion resistance can also be observed for bulk glassy Zr-Al-Cu-Ni alloys.

The aim of our work is to basically characterize the electrochemical behavior of zirconium-based bulk glassy alloys with respect to their passivating ability and to local corrosion susceptibility. In this paper we present investigations on the passive film formation of the glassy $Zr_{55}Al_{10}Cu_{30}Ni_5$ alloy in weakly alkaline electrolytes in comparison to the crystalline counterpart, as well as to elemental zirconium and aluminium.

EXPERIMENTAL METHODS

Cylindrical bulk glassy $Zr_{55}Al_{10}Cu_{30}Ni_5$ samples with a diameter of 5 mm and a length of 50 mm were prepared by die casting into a copper mold. A detailed description of the preparation conditions is given elsewhere [5]. In addition, crystalline samples of the $Zr_{55}Al_{10}Cu_{30}Ni_5$ alloy were prepared by arc-melting. The samples were characterized by X-ray diffraction (XRD), differential scanning calorimetry (DSC) and transmission electron microscopy (TEM) [5].

For electrochemical investigations the samples were embedded into epoxy resin and measurements were carried out on the mechanically polished cross-sectional areas of the samples. In the electrochemical cell the reference electrode was a saturated calomel electrode (SCE) with

Mat. Res. Soc. Symp. Proc. Vol. 554 © 1999 Materials Research Society

U(SHE) = 0.24 V. A Pt net was chosen as counter electrode. Cyclic potentiodynamic polarization measurements were performed with a scanrate (sr) of 10 mV/s in a 0.1M Na_2SO_4 electrolyte with pH = 8, air-saturated or nitrogen-purged. For further investigations of anodic passive film growth processes current transient measurements were conducted by potential step rise from the open circuit potential to anodic potentials in the passive region. Anodized sample surfaces were characterized by Auger electron spectroscopy (AES) depth profile analysis.

Pitting suscptibility of the bulk glassy $Zr_{55}Al_{10}Cu_{30}Ni_5$ samples was investigated in 0.001M NaCl electrolyte with pH = 8 (adjusted with NaOH) in comparison to the crystalline alloy samples and to zirconium. Slow anodic polarization measurements were conducted on freshly polished samples and pre-passivated samples.

RESULTS AND DISCUSSION

Structural characterization of the glassy and crystalline $Zr_{55}Al_{10}Cu_{30}Ni_5$ samples was carried out by XRD, DSC and optical microscopy. These investigations revealed that the as-cast samples are almost completely glassy except only some few crystalline inclusions which are homogeneously distributed in the samples. Despite the presence of the crystalline inclusions the samples showed the typical wide undercooled liquid region before crystallization of about 95 K in the DSC (heating rate 40 K/min). The as-cast crystalline alloy consists of several crystalline phases: $CuZr_2$, CuZr and other unidentified phases. Corresponding to its multiphase nature the sample shows a coarse-grained microstructure.

Potentiodynamic polarization measurements showed, that the glassy and the crystalline alloy form in general strongly passivating films, similar to the behavior of zirconium [6, 7]. However, a more detailed comparison of current densities in the passive region reveals small differences (Table I). The anodic current densities were determined from the second cycle since the film is formed in the first cycle and so anodic current densities in the passive region cannot determined.

Table I

Anodic current densities in the passive region of glassy and crystalline $Zr_{55}Al_{10}Cu_{30}Ni_5$ alloy samples and zirconium in air-saturated and nitrogen-purged 0.1M Na_2SO_4 electrolyte with pH = 8 determined from cyclic potentiodynamic polarization measurements recorded with 10 mV/s.

cycle number	anodic current densities [$\mu A/cm^2$]					
	air- saturated			nitrogen- purged		
	Zr	crystalline alloy	glassy alloy	Zr	crystalline alloy	glassy alloy
1			film	formation		
2	0.250	3.300	0.632	3.670	3.75	3.571
3	0.200	2.690	0.409	1.320	4.06	4.850
4	0.198	2.410	0.345	0.560	2.61	2.450
5	0.217	2.280	0.302	1.000	2.48	3.520
6	0.195	2.180	0.293	1.980	1.96	0.704

The current densities of Zr reach values of about 0.2 μA/cm² after several polarization cycles indicating the formation of a barrier-type surface film, which is typical for valve-metals. The anodic current densities of the glassy alloy are slightly higher compared of those of Zr suggesting the formation of a slightly less protective anodic film. This may result from a difference in the film formation on the surface of the multicomponent glassy alloy and on the pure element zirconium. In contrast to zirconium, which forms a relatively homogeneous film of zirconium oxides only, anodic films on Zr-Al-Cu-Ni obviously consist of a mixure of several oxides [6]. The anodic current densities of the crystalline alloy are one order of magnitude higher than the current densities determined for the glassy alloy, which is supposed to be due to an inhomogeneous film growth in lateral direction on the surface of the multiphase coarse-grained alloy. In nitrogen-purged electrolyte the anodic current densities in the passive region are in general higher and less differentiated, which is explained by inhibited oxide film growth in a medium with low oxygen concentration.

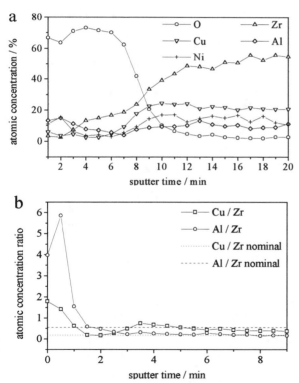

AES depth profile analysis has been applied in order to get more detailed information about the film composition on anodically passivated bulk glassy samples. The bulk glassy samples were potentio-statically polarized at U(SHE) = 0.84 V for 20 hours in air-saturated Na₂SO₄ solution (pH = 8). A typical depth profile of the glassy sample surface and the corresponding atomic concentration ratio as a function of the sputter time are shown in Figure 1. The AES depth profile analysis reveals that all alloying elements take part in the anodic film formation. From this depth profile, Cu/Zr- and Al/Zr-atomic concen-tration ratios were calculated. The Al/Zr-atomic concentration ratio decreases from the outer film region towards the film/alloy interface indicating an enrichment of Al in the film. This can be explained by the strong affinity of aluminium to oxygen. Beyond that, the Cu/Zr-atomic concentration

Figure 1:   AES depth profiles (a) and atomic concentration ratios   (b) as a function of the sputter time of an anodically passivated glassy sample (0.84 V, 20 h, O₂, 0.1M Na₂SO₄, pH = 8).

ratio has a maximum in the outer region of the film as well as a small second maximum nearby the film/alloy interface. Similar as it was derived from electrochemical investigations [5, 6], results of AES investigations confirm, that anodic surface films on Zr-Al-Cu-Ni alloys consist not only of zirconium oxides but of a mixture of several oxides of all alloying elements. Furthermore, the film reveals a gradient in composition in vertical direction.

Current transient measurements have been applied to get information about the anodic passive layer growth processes on the glassy and crystalline alloys and on Zr and Al. Current transients were recorded when the samples were polarized in one step from the open circuit potential to a potential in the passive region. Double-logarithmic current density - time plots are shown in Figure 2. Similar to the growth mechanism on Zr the anodic film formation on the glassy and crystalline alloys is determined mainly by a high field growth mechanism. This means, that the migration of film-forming ions under the effect of an electric field across an existing field is the determing process of growth [8, 9]. With increasing film thickness the field strength decreases and an inverse-logarithmic growth results. After several seconds the curves of the crystalline alloy deviates from the slope of about -1 and tends towards a higher current density level. This agree with results from potentiodynamic polarization tests, revealing that the anodic current densities in the passive region of the crystalline are higher than those of the glassy alloy. The current transient recorded for aluminium is significantly different from those measured on the Zr-Al-Cu-Ni alloys. This indicates that an aluminium oxide formation is not a determing process in the anodic film growth on Zr-Al-Cu-Ni alloys although AES investigations reveal an enrichment of aluminium in the surface film.

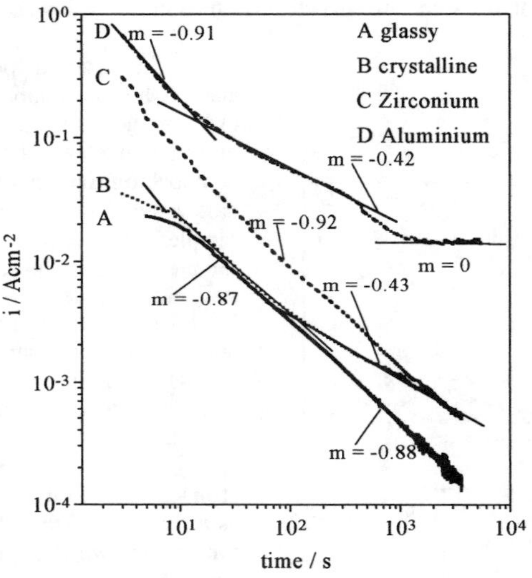

Figure 2: Double-logarithmic plot of current transients, $U_{anod.}(SHE) = 0.84$ V, for a glassy and crystalline $Zr_{55}Al_{10}Cu_{30}Ni_5$ alloy, Zr and Al (0.1M Na_2SO_4, O_2, pH = 8).

In order to investigate the susceptibility of the glassy alloy to chloride-induced pitting corrosion, freshly polished and pre-passivated samples were anodically polarized at a low scanrate. The pre-passivation was carried out at U_{anod} (SHE) = 0.84 V for t = 30 min or 20 h in 0.1M Na_2SO_4 at pH = 8, or 0.1M NaOH at pH = 13. Polarization curves of glassy alloy samples recorded in 0.001M sodium chloride electrolyte with pH = 8 are shown in Figure 3. With increasing pre-passivation time in sulphate electrolyte at pH = 8 the critical pitting potential region shifts to more positive values indicating a protective effect of anodically formed films against chloride attack. But even after long-term pre-passivation initiation of pitting cannot be completely suppressed in the low chloride concentrated electrolyte. Short-term anodization in strongly alkaline solution does not result in a significant shift of the critical pitting potential region. But in

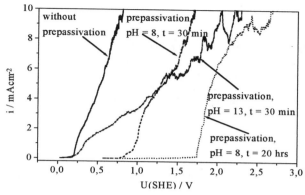

contrast to the mostly observed steep increase of current density values at potentials above the critical potential, which indicates an extreme pit growth, in this case the current density increases only gradually. This suggests a tendency to an alternating process of passive film breakdown and repassivation. The occurence of pitting processes on bulk glassy samples was found to result from the existence of single micrometer-sized crystalline inclusions, which are formed during sample preparation by slow cooling due to oxygen impurities in

Figure 3: Potentiodynamic anodic polarization of a glassy sample in 0.001 M NaCl solution (pH = 8), sr = 0.2 mV/s, pre-passivated in Na_2SO_4 solution (pH = 8) and 0.1M NaOH solution (pH = 13).

the melt [5].

From comparison of the anodic polarization behavior of pre-passivated samples of the glassy and the crystalline alloy and of zirconium in low concentrated chloride solution shown in Figure 4 it is obvious, that the corrosion resistance of our bulk glassy samples with a number of

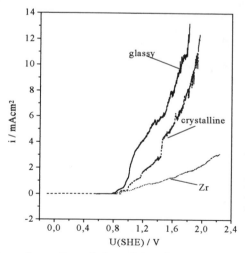

crystalline inclusions is not higher than that of the multiphase crystalline alloy. It is known that also the fully glassy $Zr_{41.2}Ti_{13.8}Cu_{12.5}Ni_{10}Be_{22.5}$ alloy samples are susceptible to chloride-induced pitting corrosion [10]. Furthermore, pit growth processes on zirconium seem to be more inhibited compared to those on Zr-Al-Cu-Ni alloy surfaces indicated by a more gradual increase of current densities after initial surface film breakdown. This may result from a higher repassivation ability of the single component corresponding to the initially discussed formation of more protective homogeneous passive films on zirconium compared to anodic films grown on the multicomponent alloy.

Figure 4: Potentiodynamic anodic polarization curves of glassy and crystalline alloy samples and Zr in 0.001M NaCl solution (pH = 8), sr = 0.2 mV/s, samples prepassivated in 0.1M Na_2SO_4 solution (pH = 8, U(SHE) = 0.84 V, t = 30 min).

CONCLUSIONS

Electrochemical investigations on bulk glassy $Zr_{55}Al_{10}Cu_{30}Ni_5$ alloy samples show the formation of strongly passivating anodic surface films grown by a high-field mechanism which is quite similar to anodic film formation on Zr. These films exhibit a lower protective ability. Anodically formed films on Zr-Al-Cu-Ni alloys consist of oxides of all alloying elements exhibiting a gradient in composition in cross-sectional direction. Slowly cooled bulk glassy Zr-Al-Cu-Ni alloy samples are susceptible to chloride induced pitting even after long-term pre-passivation. This was found to be due to the existence of crystalline inclusions. The corrosion resistance of such bulk samples is not higher than that of crystalline alloy samples.

ACKNOWLEDGMENTS

The authors would like to thank A. John for AES investigations and T. Fahr for stimulating discussions.

REFERENCES

1.   T. Masumoto, Sci. Rep. RITU **A39,** 91 (1994).
2.   U. Köster and H. Alves, Proceedings (Supplement) RQ 9 Bratislava 1997, eds. P. Duhaj, P. Mrafko, P. Svec, ELSEVIER Amsterdam-Oxford-New York-Tokyo, 1997.
3.   M. Janik-Czachor, Corrosion **49**, 763 (1993).
4.   M. Janik-Czachor, A. Wolowik, A. Szummer, K. Lublińska, S. Hofmann and K. Kraus, Electrochimica Acta **43**, 875 (1998).
5.   A. Gebert, J. Eckert, H.-D. Bauer and L. Schultz, Mater. Sci. Forum **269-272**, 797 (1998).
6.   A. Gebert, K. Buchholz, A. Leonhardt, K. Mummert, J. Eckert and L. Schultz, Proceedings of the 5th Int. Symposium on Electrochemical /Chemical Reactivity of Novel Materials, Sendai 1998, to be published in Mater. Sci. Eng. A.
7.   A. Gebert, K. Mummert, J. Eckert, L. Schultz and A. Inoue, Materials and Corrosion **48**, 293 (1997).
8.   M.M. Lohrengel, Mater. Sci. Eng. **R11**, 243 (1993).
9.   N. Cabrera and N.F. Mott, Rep. Pogr. Phys. **12**, 163 (1948-49).
10.   V. Schroeder, C.J. Gilbert and R.O. Ritchie, Scripta Materialia **38**, 1481 (1998).

OXIDE FORMATION ON THE BULK METALLIC GLASS $Zr_{46.75}Ti_{8.25}Cu_{7.5}Ni_{10}Be_{27.5}$

M. KIENE, T. STRUNSKUS, G. HASSE, AND F. FAUPEL
Lehrstuhl für Materialverbunde, Technische Fakultät der CAU Kiel, Kaiserstr. 2, 24143 Kiel,
GERMANY

ABSTRACT

In this investigation depth profiling by Ar^+ ion sputtering in combination with x-ray photoelectron spectroscopy was used to determine the thickness and the chemical composition of the oxide layers formed on the bulk metallic glass $Zr_{46.75}Ti_{8.25}Cu_{7.5}Ni_{10}Be_{27.5}$ (V4) under different preparation conditions. The thickness of the oxide layer was in the range between 4 nm and 30 nm for samples oxidized in air at room temperature and annealed in air at 573 K for 15 hours, respectively. The oxide layers showed an enrichment of Be at the outermost surface followed by a Be, Zr and Ti rich phase. Cu and Ni are depleted within the whole oxide layer. BeO, ZrO_2 and various Ti oxides were detected in the oxide layer, whereas Cu and Ni occurred only in a non-oxidized zerovalent state. Our data indicate that the oxidation is controlled by the inward and outward diffusion of the metallic components.

INTRODUCTION

The surface properties of amorphous metallic alloys are determined by the oxide layer formed under standard processing conditions. It is therefore important to study the oxide formation under real world conditions. On the other hand the oxidation behavior of alloys has also been of interest from a scientific point of view for many years [1]. For alloys containing both a relatively active element and a more noble one preferential oxidation of the active element may take place, and the back-diffusion of the noble element can become rate controlling. This behavior has been observed for many binary alloys, e.g., for amorphous ZrNi [2], and also recently for the dry oxidation of the ternary amorphous alloy $Zr_{60}Ni_{25}Al_{15}$ in the temperature range between 310 °C and 430 °C [3, 4]. Using Rutherford backscattering (RBS) and cross-sectional transmission electron microscopy (TEM) the formation of an uniform oxide layer consisting of ZrO_2 and Al_2O_3 in a similar Zr/Al ratio as in the alloy was found, i.e. no preferential oxidation of Zr or Al was detected. Ni was depleted completely in the oxide layer and enriched at the interface. The growth kinetics of the oxide seemed to be controlled by the Ni back-diffusion into the alloy [4]. The aim of this work is to study the oxidation behavior of the quinternary alloy $Zr_{46.75}Ti_{8.25}Cu_{7.5}Ni_{10}Be_{27.5}$ (Vitreloy 4, V4) [5] consisting of three reactive and two noble elements under real world conditions. This alloy is an important representative of the new bulk metallic glasses [5]. X-ray photoelectron spectroscopy (XPS) was chosen to obtain information about the oxidation state of the different metals and their quantitative distribution within the oxide.

EXPERIMENT

Samples of amorphous $Zr_{46.75}Ti_{8.25}Cu_{7.5}Ni_{10}Be_{27.5}$ were prepared as described in [6]. The samples were sputter cleaned by 3 keV Ar^+ ion sputtering until no oxygen was detectable by XPS. The sputter cleaned samples were exposed to laboratory atmosphere and annealed under different conditions. The different treatments of the samples investigated are summarized in Table I.

Tab. I: Treatments of the samples investigated

Sample #	Initial Oxidation	Heat treatment
1	several weeks in air	None
2	one day in air	None
3	one day in air	15 h at 300 °C at 10^{-6} mbar
4	one day in air	15 h at 300 °C in air

After exposure to oxygen and the heat treatment the samples were transferred into the UHV-system and kept there for the complete analysis. Depth profiles of all samples were taken using a combination of argon ion sputtering and photoelectron spectroscopy measurements. The sputter rate was calibrated gravimetrically and was assumed to be constant throughout the sampling depth, i.e., the same sputter rate (1.4 nm/min) was assumed for the oxide layer and the metallic bulk.

XPS investigations were performed in situ using a hemispherical electron analyzer (VG MKII) and a non-monochromatized Al K_α x-ray source operated at 500 W. The pressure in the vacuum system was $< 5 \times 10^{-8}$ Pa during XPS analysis. All XPS measurements were done with electron emission normal to the sample surface. The spectra were taken at 20 eV pass energy with an energy resolution of 1.2 eV measured at the $Ag3d_{5/2}$ line. The peak areas used for the quantitative analysis were determined after a Shirley-type background removal.

RESULTS

Selected XPS overview spectra of the sample annealed for 15 hours in air obtained at different sputtering depths are shown in Fig. 1. The unsputtered surface shows strong O 1s and O KLL features together with strong peaks due to Zr and weaker ones due to Ti and Be. Some carbon contamination is visible around 285 eV. The typical Ni 2p and Cu 2p peaks are essentially absent in this layer. After sputtering into the oxide layer the spectrum is dominated by the typical XPS features of oxygen and zirconium. Small features due to nickel, copper and titanium can be identified as well, together with a very weak signal originating from beryllium. Sputtering this sample further leads to the expected decrease of the oxygen signal, and the signals of the metals increase in intensity. High resolution spectra were taken for the following lines and used to determine the chemical state of the element and its abundance in the surface layer: Cu 2p, Ni 2p, O 1s, Ti 2p, Zr 3d, Be 1s. The Be 1s line shows some overlap with the Ni 3p line, therefore the intensity of the Ni 3s line was determined under the same experimental conditions and used to eliminate the contribution of the Ni 3p line to the Be 1s intensity. Note, that the XPS spectrum does not resemble the true surface composition. XPS spectra were taken at normal emission and the typical probe depth (three times the inelastic electron mean free path) varies with electron energy roughly between 4 nm for the Cu 2p line and 8 nm for Be 1s line (estimated using the values for Zr given in table 2 in [7]). We also can not exclude preferential sputtering effects, which can lead to a change of the surface composition prior to the oxidation and during the depth profiling. In this connection we would like to point out, that we observe a much higher than expected Ni 2p intensity compared to the Cu 2p intensity, which can not be explained with differences in the probing depth for the two different XPS lines.

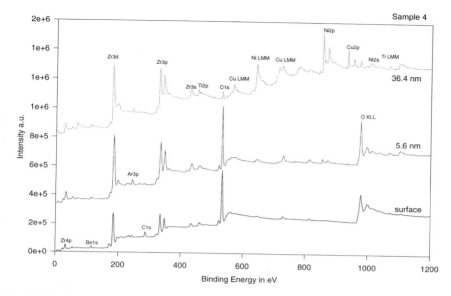

Fig. 1 Selected overview spectra of the sample annealed for 15 h in air. The sputtering depth is indicated for each plot.

Although the absolute composition determined by our XPS analysis may deviate substantially from the true composition of the analyzed surface layer, our conclusions will still remain valid, because the XPS analysis has been performed under identical conditions for all samples, and the changes measured in the XPS analysis will reflect a similar change in the true composition.

In the depth profiles shown in Fig. 2 and 3 the atomic concentration of the metals was adjusted so that they always sum up to 100%. The atomic concentration of oxygen was determined relative to the total atomic concentration of the metals assuming a homogeneous distribution within the probing depth, i.e., a value of 100% would indicate then that there is one oxygen atom for every metal atom. It is, however, obvious that oxygen and the metals are usually not distributed homogeneously within the probing depth, e.g. more oxygen is located at the surface and less deeper in the bulk. Therefore the absolute values are only a rough indication of how much oxygen is present within the probed sample region. In the depth profiles shown in Fig.2 and 3 the oxygen concentration was further scaled down by a factor of five to keep it below the highest metal concentration.

All depth profiles show the same general behavior, but there are also important differences between the unannealed and the annealed samples. At the surface all samples show a strong depletion of nickel and copper. Zirconium shows a smaller depletion, titanium is almost unchanged and beryllium is strongly enriched at the outer surface. In the unannealed samples the composition changes steadily to the XPS bulk metallic composition which is reached after about 10 nm sputtering depth. Using the attenuation of the Zr 2p and Ti 2p metal signal the oxide thickness is estimated to be about 4 nm. The larger thickness determined in the depth profile is thus due to the poorer depth resolution caused by sputtering effects.

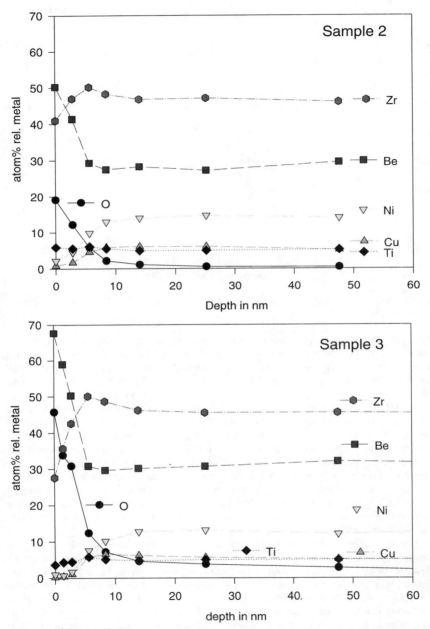

Fig. 2 Depth profiles of samples oxidized at room temperature. Sample 3 (lower panel) was subsequently annealed in high vacuum (10^{-6} mbar) for 15 hours.

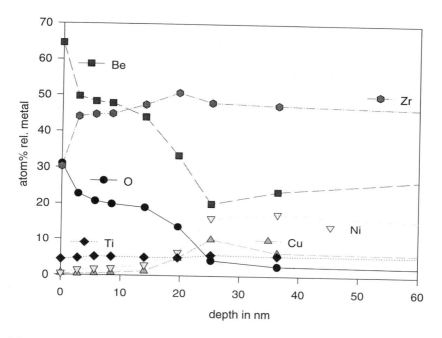

Fig. 3 Depth profile of the sample annealed in air for 15 hours.

Rather similar results were obtained for sample 1 which was exposed to air for several weeks. Thus, the prolonged exposure to air at room temperature does not lead to a significant increase in the oxide layer thickness or a change of its composition. For the vacuum annealed sample (sample 3) one notes a much higher oxygen intensity for the same sputtering depth and the oxygen intensity remains much higher going deeper into the sample. This indicates that the sample has picked up more oxygen during annealing in a 10^{-6} mbar vacuum and oxygen has diffused deeper into the bulk. The enrichment of Be at the surface is much larger, and the depletion of Cu and Ni in the surface region is also more pronounced.

The results of the XPS depth profiles for the unannealed and the vacuum annealed samples were checked by independent time of flight secondary ion mass spectroscopy TOF-SIMS depth profiles [8]. In both cases we obtained qualitatively the same results as in the XPS measurements.

An even thicker oxide layer of more than 15 nm is observed for the sample annealed in air shown in Fig. 3. Again Be is enriched strongly at the surface and Ni and Cu are depleted at the outer surface. Below this outer very thin surface layer one observes another thicker oxide layer which consists of more zirconium and less beryllium than the outer oxide layer. This layer also shows a depletion of Ni and Cu. After sputtering deeper into the bulk one notes a drop of the Be concentration and an increase in the Ni and Cu concentration. Copper appears to be enriched at the interface between the oxide and the metal bulk.

Analysis of the high resolution spectra (not shown here) revealed no indications for copper or nickel in an oxidized state. The 2p spectra always showed the typical line shape and position of the metals and for copper the Cu LMM Auger line was also at the position expected for Cu(0) excluding the formation of a Cu(I) species. Zr, Ti, and Be are in a fully oxidized state at the surface. At the surface the lineshape of the O 1s spectrum (not shown here) indicates that 30% of the oxygen is present in the form of hydroxides rather than oxides. This is supported by the quantitative analysis which shows an higher amount of oxygen at the surface than is expected for pure oxides. After sputtering only one peak typical for oxides is observed in the O 1s region. For the metals the spectra of Zr and Be could be fitted satisfactorily using the fully oxidized and the metallic state only. Therefore there seems to be a sharp transition from the oxide to the metal. For Ti we detect +4 as the highest oxidation state, but also Ti in lower oxidation states, in particular in the sputtered samples. This indicates that any oxygen deficiency in the oxide layer is compensated by lower oxidation states of Ti. It was not possible to identify any oxide peaks at either of the metals below an oxygen concentration of 10 atom%. It is possible that our measurements are not sensitive enough to identify the oxides or that the oxygen is present in a dissolved form.

CONCLUSIONS

In summary we obtain the following picture for the oxidation behavior of V4. Each sample consists of an outermost oxide/hydroxide layer which is enriched in Be(II) and contains Zr(IV) and Ti(IV). Ni and Cu are not oxidized and are therefore depleted in this layer. Below this layer is an oxide layer with less Be and more Zr, but also in their fully oxidized state. For Ti some suboxides may be present within this layer. For a thick oxide layer copper and nickel are enriched above their bulk level at the interface between the oxide and the bulk.

These results are fully consistent with the results observed previously for a ternary metallic glass. Nickel and copper diffuse from the oxide layer into the bulk and the inward diffusion of copper and nickel seems to be the rate limiting process for the oxide formation.

ACKNOWLEDGMENTS

We thank U. Geyer for the sample preparation and H. Ehmler for stimulating discussions. We are also grateful to Scott Bryan and the Physical Electronics company who enabled us to check and confirm these results with complementary TOF-SIMS measurements.

REFERENCES

1. C.E. Birchenall, *Oxidation of Alloys*, American Society for Metals, Metals Park, OH, 1970, Chap. 13.
2. K. Asami, H. M. Kimura, K. Hashimoto, and T. Masumoto, Mater. Trans. JIM **36**, 988 (1995).
3. S. Schneider, X. Sun, M.-A. Nicolet, and W. L. Johnson in *Science and Technology of Rapid Solidification and Processing*, edited by M. A. Otooni (Kluwer Academic Publishers, NL 1995), p. 317-326.
4. X. Sun, S. Schneider, U. Geyer, W. L. Johnson, and M.-A. Nicolet, J. Mater. Res. **11**, 2738 (1996).
5. A. Peker and W.L. Johnson, Appl. Phys. Lett. **63**, 2342 (1993).
6. H. Ehmler, A. Heesemann, K. Rätzke, F. Faupel, and U. Geyer, Phys. Rev. Lett. **80**, 4919 (1998).
7. S. Tanuma, C.J. Powell and D. R. Penn, Surf. Interface Anal. **17**, p. 911 (1991).
8. S. Bryan (unpublished results).

CONTINUOUS AMORPHIZATION OF Zr-BASED ALLOYS BY CONTROLLED MECHANICAL INTERMIXING

G. WILDE, H. SIEBER and J.H. PEREPEZKO,
Department of Material Science and Engineering, University of Wisconsin, Madison, WI, USA

ABSTRACT

Binary Zr-(Cu, Ni, Al) alloys were mechanically intermixed by cold-rolling stacks of elemental foils. The results indicate that solid-state amorphization is initiated if the grain size of the Zr-Cu and Zr-Ni alloys falls below a critical value. Amorphization was not observed for the Zr-Al alloy. These results are in accordance to the predictions of a model for solid-state amorphization. The comparison with the results on a quaternary Zr-Cu-Ni-Al alloy indicate the influence of multi-component alloying on the glass-forming ability of Zr-rich alloys by mechanical working.

INTRODUCTION

Different models have been developed for the solid-state amorphization of mechanically driven alloys based on chemical thermodynamics as revealed in the slopes of phase boundaries or atomic size differences [1-3]. An important prediction of these models is the composition range where amorphization by mechanical alloying is feasible. Yet, the comparison of the reliability of different models is difficult because of the large width of the composition intervals accessible for solid-state amorphization and the sensitivity of the observed amorphization composition range on the experimental procedure. Often, the experimental observation of a solid-state amorphization reaction at a given alloy composition may be in accordance with model calculations. However, the failure to observe an amorphization reaction outside of the composition range given by model calculations does not necessarily support the model. The absence of amorphization may simply be due to the use of unfavorable processing conditions or insufficient processing time, which are usually not identified in the model analysis.

Recently, Desré proposed a thermodynamic model for the nanocrystalline-to-glass transition, which treats the nanocrystal size as a pertinent variable in the description of the process [4]. As a result, a lower critical grain size of the polycrystalline material at the onset of the solid-state amorphization is obtained, which depends on the alloy system and the composition. The model considerations show that a larger critical grain size is obtained for systems with a rather small difference of the Gibbs free energy between the amorphous phase and a supersaturated solid solution, i.e. easy glass-forming alloys. Thus, the aim of the present study is to test for the existence of a critical grain size for amorphization in mechanically intermixed binary Zr-Cu, -Ni, -Al alloys and to compare the observed amorphization behavior and the relative glass-forming abilities with the results on a quaternary Zr-Cu-Ni-Al bulk glass-forming alloy [5] which has been processed under similar conditions.

THEORETICAL BACKGROUND

The nanocrystalline-to-glass transformation is described by Desré's model as a partitionless first-order transition analogous to the order/disorder transition for wetted antiphase boundaries in intermetallic compounds [6]. In this model it is considered that the ultra grain refinement obtainable by mechanical alloying generates a disordered layer at the nanograin boundaries by

173

dislocation accumulation in the grain boundary zone. On calculating the free energy per unit area of the nanograin boundaries, a size range is obtained where amorphous and nanocrystalline phases are expected to coexist as:

$$r_0 - r^* = \frac{2 \cdot \sigma \cdot S}{\xi \cdot \Delta G_V^0 (c) \cdot \left[\Delta G_V^0 (c) - S/\xi \right]}$$

(1)

with ΔG_V, the Gibbs free energy difference between the amorphous and a polycrystalline phase per unit volume, the correlation length of the amorphous layer, ξ and a so-called disorder spreading coefficient, S: $S = \sigma_{GB}^0 - 2 \cdot \sigma$, which represents the Gibbs energy difference for the interface with and without the presence of an amorphous interlayer between grains. σ is the (anisotropic) Gibbs free energy of the order/disorder sharp interface and σ_{GB}^0 denotes the Gibbs free energy of a sharp grain boundary which does not contain an amorphous layer. For radii $r > r_0$ grain boundary amorphization should not occur, while the wetted nanocrystals become unstable with respect to complete amorphization for $r < r^*$.

It is well known, that the grain size during intense cold working decreases significantly, eventually leading to nanocrystalline structures [7] and the formation of supersaturated solid solutions [8]. However, the microstructural refinement level of the crystalline state is limited by recrystallization and recombination processes that depend on the alloy system. It is obvious that amorphization, as indicated by the model, only occurs for compositions where this limiting grain size is smaller than the critical grain size. The prediction of a critical grain size thus corresponds to the prediction of a composition range feasible for solid-state amorphization and the prediction of a relative experimental time scale for the amorphization to occur as well. The second implication allows the opportunity for direct experimental verification of this amorphization model. Equation (1) also shows, that the predicted size range for amorphization is shifted to larger values for alloys with a small Gibbs free energy difference between amorphous and polycrystalline state. Therefore, easy glass formers such as bulk metallic glasses should transform into the amorphous state at rather large nanograin sizes i.e. at relatively short working times.

EXPERIMENTAL DETAILS

Recently, it has been shown that solid-state alloying and vitrification to a compositionally uniform glass can be obtained by cold-rolling stacks of elemental foils [5, 9]. This low strain rate process represents a controlled intermixing mechanism at ambient temperature and defined chemical composition. Moreover, it provides the possibility to obtain detailed microstructural information on all stages of the alloying process which is of special significance for the present study.

Foils of pure Zr, Ni, Cu and Al (Goodfellow, all $\geq$ 99.9+ %) with a thickness of 7.5 µm (Cu, Al) and 20 µm (Zr, Ni) respectively were stacked to form arrays of compositions $Zr_{50}Cu_{50}$, $Zr_{67}Cu_{33}$, $Zr_{67}Ni_{33}$, and $Zr_{25}Al_{75}$ and folded four times. The folded samples were rolled at a strain rate of approximately 0.1 s^{-1} to a thickness of about 80 µm then folded to double the thickness and rolled again to a minimum thickness of 80 µm. This procedure (one pass) was repeated until the final material was cold-rolled for up to 100- to 120 passes.

The microstructural changes during cold-rolling were investigated at different stages of the mechanical intermixing process for each alloy by X-ray diffraction (XRD, Philips) using Cu-K$_\alpha$

radiation, scanning electron microscopy (SEM, Leo 982 FESEM), transmission electron microscopy/selected area electron diffraction (TEM/SAED, JEOL 200CX and Philips CM200). Energy dispersive X-ray analyses (EDX) were carried out in the Philips CM200 TEM and the Leo 982 FESEM to verify the composition of the samples at different alloying stages.

RESULTS

During the first stages of the cold-rolling procedure a multilayered structure develops in the samples. SEM investigations in cross-section indicate that the layer thickness falls below 100 nm after 60 passes for all four binary systems. EDX measurements on samples which had been rolled for more than 80 passes indicated that the nominal composition was maintained and that the impurity pick-up amounted to less than the resolution of the method. XRD measurements at different rolling stages show that the elemental peak height decreases, the peak centers shift slightly and the peaks broaden considerably with increasing numbers of passes. This behavior is characteristic for a continuous alloying/grain size reduction process. Figure 1 shows as an example the XRD-spectra obtained on $Zr_{67}Cu_{33}$ samples at different levels of cold working. Qualitatively, these results also hold for the other binary alloys studied.

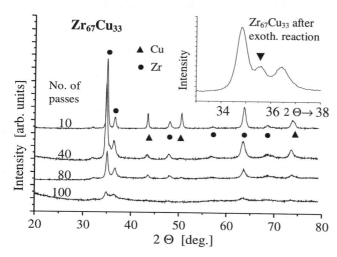

Fig. 1: XRD traces of cold-rolled $Zr_{67}Cu_{33}$ after different working stages. First the fundamental Ni - peaks vanish in accordance with the fact that Ni preferably diffuses into Zr and not vice versa. At later stages, the heights of the Zr- peaks decrease. The insert shows the presence of a new phase (▼) after a massive exothermic reaction that has occurred during cold-rolling at 113 passes.

The shape of the diffraction curves at angles of about $2\Theta = 35°\text{-}40°$ indicate the presence of an amorphous fraction in samples which have been rolled for more than 70 passes. Additionally, the XRD traces show that within the resolution of the measurement no intermediate phase has been detected. Thus, the amount of any compound phase, which might have formed during the cold-rolling process, amounts to less than about 3% of the sample volume. The XRD curve of the $Zr_{67}Cu_{33}$ sample which has been rolled for 100 times shows that only the maxima of the largest intensity remain detectable besides distinct broad maxima between 35° - 40° and 60° which are due to the amorphous product phase. However, the remaining peaks clearly reveal the presence of a crystalline fraction in the sample. Attempts to vitrify the samples further by continued cold rolling resulted in a massive exothermic solid-state reaction by which parts of the samples oxidized rapidly and intermetallic phases formed in the remaining fraction of the material. The insert in Fig. 1 shows a section of the XRD trace of a $Zr_{67}Cu_{33}$ sample after this reaction occurred. The additional maximum at 35.7° indicates the formation of an intermediate phase.

TEM investigations have been performed to obtain a quantitative measure of the degree of the amorphization and of the grain size of the residual crystalline fractions, which are in contact with the disordered regions. Figure 2-a shows a bright field image obtained on a $Zr_{67}Cu_{33}$ sample which has been cold-rolled for 80 passes.

Fig. 2-a: High-resolution TEM micrograph of a $Zr_{67}Cu_{33}$ sample that has been rolled for 80 passes. The insert at the upper left corner shows the SAED-pattern, which further indicates the presence of a mixture of nanocrystalline and amorphous phases.

Fig. 2-b: Bright-field image of a large amorphous region within a $Zr_{67}Cu_{33}$ sample that has been rolled for 110 passes. The broad rings in the SAED-pattern shown as insert in the upper left corner are characteristic for electron diffraction at a disordered structure.

The microstructure, which is typical for the Zr-Cu and Zr-Ni alloys, consists of a mixture of an amorphous phase and nanocrystalline, grains with an average diameter of 8-10 nm. Crystalline grains smaller than 5 nm in diameter have not been observed. $Zr_{67}Cu_{33}$ samples which have been rolled for 110 passes show large amorphous regions of several micrometers in size (Fig. 2-b) separated by a matrix which consists of a mixture of nanocrystalline and amorphous phases as shown in Fig. 2-a. Within the resolution of the TEM, the amorphous regions were observed to be free from nanocrystals. TEM investigations on the cold rolled $Zr_{50}Cu_{50}$ and $Zr_{67}Ni_{33}$ samples revealed similar results as found for the $Zr_{67}Cu_{33}$: the formation of a mixture of amorphous and crystalline phases on a scale of about 10 nm at about 80 rolling passes followed by the occurrence of large μm-sized amorphous regions after continued working. Table I summarizes the maximum, d_{max} and minimum, d_{min}, diameters of the grains for the different alloys which have been observed within the partially amorphous regions.

Table I: Largest and smallest grain sizes observed within the partially amorphous regions.

	$Zr_{67}Cu_{33}$	$Zr_{50}Cu_{50}$	$Zr_{67}Ni_{33}$	$Zr_{25}Al_{75}$
d_{max} [nm]	8-10	8-10	9-11	-
d_{min} [nm]	5	5	7	3

In contrast to the observations on Zr-Cu and Zr-Ni alloys, amorphization did not occur on cold rolling the $Zr_{25}Al_{75}$ alloy. After 100 folding and rolling passes TEM investigations indicate a nanocrystalline microstructure with an average grain size of about 8 nm and a fraction of smaller grains of diameter $d \geq 3$ nm present as well. Thus, only the smallest grain size of 3 nm, which has been observed for the Zr-Al alloy is listed in table I.

DISCUSSION

In order to compare the results of the present study with the model prediction according to [4], distinct values for the radii r_0 and r^* according to eq. (1) are necessary. However, the microstructures of samples, which have been alloyed by cold rolling, are intrinsically inhomogeneous as long as crystalline fractions are present and there is a local distribution of shear stresses introduced by the working procedure. Yet, when the spatial area of the nanocrystalline/amorphous regions (≥ 200 nm) exceeds the size of the nanocrystals by more than one order of magnitude, the mixed nanocrystalline and amorphous phase regions can be regarded as independent of the polycrystalline fraction of the sample volume. Thus, the experimentally determined values for d_{max} and d_{min} were taken as representative for $2r_0$ and $2r^*$, respectively. Because of the complexity of the observed microstructures, the first onset of the predicted grain boundary amorphization can not precisely be determined which leads to an underestimation of r_0. Yet, taking r_0 as the maximum grain size within the amorphous matrix should lead to a small difference between d_{max} and $2r_0$, considering that the amorphous layer thickens by replacing the nanocrystalline phase [4].

For a quantitative evaluation of the model, ΔG_V must be calculated as a function of the local composition. Application of the model for Zr-Ni using data on the Gibbs free energy of mixing at T = 600 K for the Zr-based solid solution and for the liquid alloy [10] and extrapolating the data to ambient temperature allows the calculation of a coexistence range of amorphous and nanocrystalline phases in terms of nanograin size, d=2r and concentration, c. This yields a roughly parabolic dependence of $2r_0$ and $2r^*$ on c [4]. The local composition, as required for a direct comparison of the model calculations with the experimentally obtained values for d_{max} and d_{min}, could not be determined because of the small scale of the microstructure. However, we may consider that the grain size decreases initially faster than the development of concentration variations. This assumption is in accordance with experimental observations on the mechanical alloying behavior of immiscible Fe-Cu alloys [8] and is further supported by investigations on the metastable solubility of Cu-Cd alloys prepared by cold rolling [11]. Thus, because of the topology of the calculated phase diagram boundaries, experimental pathways of d(c) are likely to follow a course which intersects with the curves for $2r_0$ and $2r^*$ at values of about $2r_0$=10 nm and $2r^*$=7 nm for the Zr-Ni system [4]. These values as well as the predicted grain size width of the coexistence interval are in good agreement with the experimentally obtained data.

Zr-Cu alloys are similar to the Zr-Ni system with respect to their rather high glass forming ability and, correspondingly, with respect to the magnitude of $\Delta G_V(c)$ [10]. Thus, the experimental observation of similar values for d_{max} and d_{min} is in accordance with the implications of the model. In contrast, Zr-Al is not known to form glass easily. To the knowledge of the authors, specific numerical values for $\Delta G_V(c)$ are not available for this system. Yet, the low glass forming ability indicates, that the $\Delta G_V(c)$ values for this system are larger and the composition interval for amorphization smaller than for the Zr-Cu and Zr-Ni alloys. Therefore, and in qualitative accordance with Desré's model, even a grain size reduction to diameters of about 3 nm did not result in a solid state amorphization reaction.

On cold rolling a Zr-Cu-Ni-Al alloy a minimum grain size of about 50 nm was observed after 80 passes. Further rolling resulted in a complete amorphization of the sample; grains smaller than 50 nm have not been observed [12]. Reliable data for $\Delta G_V(c)$ is lacking for this quaternary system, yet, it has been shown that this alloy composition can be vitrified by melt quenching at a rate of only 10^2-10^3 K/s [13] or about 3-4 orders of magnitude lower than necessary to obtain amorphous binary Zr-Cu and Zr-Ni alloys by solidification processing techniques. This experimental result suggests a small ΔG_V, which, in turn leads to the prediction of a large critical nanograin size within the framework of the model.

CONCLUSIONS

Systematic investigations on the solid-state amorphization behavior of several binary and a quaternary Zr-rich alloys have indicated the existence of a system-specific critical grain size for amorphization. The existence of a size range where a nanocrystalline and an amorphous phase coexist, as well as experimental values for the critical grain diameter and for the lower limit of the nanocrystalline grain size, are in accordance with a recently proposed model. The comparison of the amorphization characteristics of the different alloys indicates that multi-component alloying can effectively enhance the glass forming ability because of a reduction of $\Delta G_V(c)$ which leads to amorphization at larger grain sizes. Within the context of this model it can also be understood that the composition range for glass formation by mechanical alloying is larger than for rapid solidification. Continued working can lead to grain sizes lower than the critical size and result in amorphization although the limit for supersaturation has not been exceeded. This processing route is not available for melt quenching techniques.

ACKNOWLEGMENTS

The support of the Alexander von Humboldt-Foundation via the Feodor-Lynen-Program (G.W., V-2.FLF-DEU/1052606) and ARO (DAAG 55-97-1-0261) is gratefully acknowledged.

REFERENCES

[1]   T.B. Massalski, Proc. 4th Int. Conf. on Rapidly Quenched Metals, (1982), 203.
[2]   B.C. Giessen and S. Wang, J. Phys. C, **41**, (1980) 8.
[3]   T Egami, Mat. Sci. Eng., **A 226-228**, (1997) 261.
[4]   P.J. Desré, Phil. Mag. A, **A74**, (1996) 103.
[5]   A. Sagel, H. Sieber and J.H. Perepezko, Phil. Mag. Lett., **77**, (1998) 109.
[6]   C. Ricollaeu, A. Loiseau and F. Ducastelle, Phase Transitions, **30**, (1989) 243.
[7]   C.C. Koch, Appl. Phys. Lett., **43**, (1983) 1017.
[8]   J. Eckert, J.C. Holzer, C.E. Krill and W.L. Johnson, J. Mat. Res., **8**, (1992) 1980.
[9]   G. Wilde, H. Sieber and J.H. Perepezko, J. Non-Cryst. Solids, in press.
[10]   N. Saunders, Calphad, **9**, (1985) 297.
[11]   J. Meudt, G. Wilde and J.H. Perepezko, in preparation.
[12]   H. Sieber, G. Wilde, A. Sagel and J.H. Perepezko, J. Non-Cryst. Solids, in press.
[13]   T. Zhang, A. Inoue and T. Masumoto, Mater. Trans. JIM, **32**, (1991) 1005.

THE SHEAR MODULUS OF GLASSY AND SUPERCOOLED LIQUID $Pd_{40}Ni_{40}P_{20}$

A. S. Bains, C. A. Gordon, A. V. Granato, University of Illinois, Department of Physics, Urbana, Illinois 61801, a-bains@uiuc.edu and R. B. Schwarz, Los Alamos National Labs, Center for Materials Science, Los Alamos, NM

ABSTRACT

We have measured the shear modulus and its temperature dependence of $Pd_{40}Ni_{40}P_{20}$ using an EMAT technique. The room temperature value of $3.92(10^{11})$ d/cm^2 is in fair agreement with that of $3.66(10^{11})$ d/cm^2 given earlier by He and Schwarz[1], using a resonant ultrasound spectroscopy technique. The relative change with temperature coefficient for T $<<$ T_g is $3.4(10^{-4})$ K^{-1}. For the heating rate of ~1 K/min used here, $T_g = 565$ K compared with a value of 575 K reported earlier[1]. The shear modulus is continuous at T_g, but its temperature coefficient is larger by a factor of 6.5 for T $>>$ T_g. During temperature cycling near but below T_g, irreversible aging effects are found showing that the amorphous state is not an equilibrium state. Near but above T_g, the cycling effects are reversible for time scales of the order of hours but not for time scales of the order of days, showing that metastable equilibrium states have not yet been fully attained. The results are in overall agreement with the predictions[2] of the Interstitialcy Theory.[3]

INTRODUCTION

The shear modulus G and its temperature dependence of $Pd_{40}Ni_{40}P_{20}$ has been measured using an EMAT technique from room temperature up to about 610 K, beyond the reported glass temperature of 575 K.[1] The results are in overall agreement with the predictions[2] of the Granato Interstitialcy Theory of Condensed Matter States[3].

EXPERIMENT AND THEORY

The shear modulus of a disk shaped sample of $Pd_{40}Ni_{40}P_{20}$ with a diameter of about 6 mm and a thickness of about 2 mm was measured using an EMAT technique based on a method by Lyall and Cochran.[4] Frequency modulation and phase sensitive detection by a lock in amplifier to automatically maintain the resonance was added to the circuitry, after Read and Holder.[5] Temperature measurements were made with a type N thermocouple in slight contact with the edge of the sample. A computer controlled power supply driving a chromel heating coil regulated the temperature of the sample. With this apparatus the resonant frequency f_R of the fundamental shear thickness mode of vibration of the sample can be continuously recorded as a function of temperature.

The resonant frequency f_R is related to G by

$$f_R^2 = G/\rho\lambda^2 \tag{1}$$

where λ is 2t, twice the thickness of the disk and ρ is the density of the sample. Small fractional changes in f_R are more simply related to fractional changes in G.

$$2\Delta f_R/f_R = \Delta G/G + \Delta V/3V \qquad (2)$$

Hence, neglecting the small $\Delta V/3V$ contribution, changes of G as a function of temperature can be as easily and accurately measured as changes of f_R.

According to the Interstitialcy Theory G depends on the concentration of interstitialcies, c_i.

$$G = G_x\exp(-\beta c_i) \qquad (3)$$

where G_x is the perfect crystal shear modulus, and $\beta \sim 20$.[3] Expanding c_i in a Taylor series around T_g and keeping only the first term[6],

$$G = G_g\exp \gamma(1 - T/T_g) \qquad (4)$$

where G_g is the shear modulus at the glass temperature, and

$$\gamma = \beta T_g(dc_i/dT)_g \qquad (5)$$

is the softening parameter. At low temperatures $T \ll T_g$ where interstitialcies are frozen in, G is expected to exhibit characteristic crystalline behavior, but at higher temperatures G is expected to decrease exponentially according to equation 4. For small temperature changes equation 4 predicts the relation

$$\Delta G/G = -\gamma(\Delta T/T_g) \qquad (6)$$

RESULTS

In Figure 1 the resonant frequency f_R of a disk shaped sample of $Pd_{40}Ni_{40}P_{20}$ vibrating in its fundamental shear thickness mode is plotted as a function of temperature. The curves beginning with the points 1, 2 and 3 are constant heating rate warm ups while the curves ending in points 3 and 4 are quick and slow cool downs respectively. From Figure 1 it is seen that G changes continuously with T up to and beyond T_g. At low temperatures from 300 K - 400 K, G displays a typically crystalline slope that is T - independent but beyond T_g the magnitude of the slope increases by a factor of about 6. In an intermediate temperature range from 400 K - 500 K, G increases above the value extrapolated from low temperatures in curve 1. A comparison of the realatively slower warm up curve beginning at point 3 with the faster warm up curve beginning at point 2 shows that this effect is more pronounced at slower heating rates. A quick cool down after the warm up to 612 K which began at point 2 reduced the room temperature f_R to point 3. The warm up beginning at point 3 was followed by a slower cooling and increased f_R at room temperature to point 4. Overall it can be seen that G increased by about 10% as f_R moved from point 1 to point 4.

The crystalline value of dG/dT at temperatures below 400 K regardless of the value of G is explained by the temperature independent diaelastic softening expected according to equation 3. At higher temperatures but below T_g the decreasing magnitude of dG/dT is caused by annealing of interstitialcies possible due to higher temperatures with a consequent increase of G. The rate dependence of this effect is explained since slower heating rates allow time for greater annealing and hence a greater recovery of G to take place. The marked increase in the magnitude of dG/dT

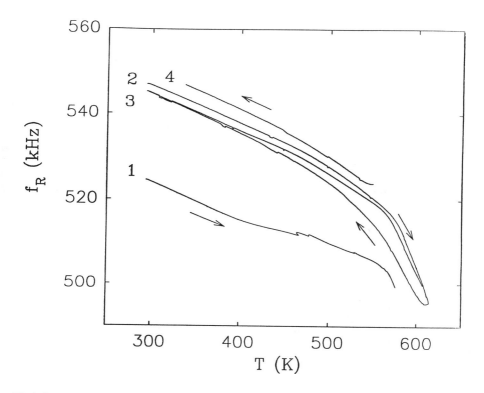

Fig 1. Resonant frequency f_R is plotted against temperature T. Curves beginning with points 1, 2 and 3 are warm ups at rates 1 K/min, 1.7 K/min, .4 K/min. Curves ending in points 3 and 4 are quick and slow cool downs respectively.

above T_g indicated a large shear softening due to an increase in c_i as required at high temperatures by the Interstitialcy Theory. A rapid cooling from these temperatures above T_g is expected to freeze in more interstitialcies as the sample cools too quickly to allow annealing and explains the drop in f_R from point 2 to point 3 after heating to 612 K. Conversely a slower cooling rate would allow more time for annealing and thus explains the increase in f_R from point 3 to point 4. The overall recovery of G by ~ 10% requires a decrease in c_i of only ~ 1/2% and is still less than the 20% - 30% reduction of G_x usually found for the amorphous state. A value for the softening parameter γ can be calculated from the slope for the second warm up curve near 610 K, and yields a value of $\gamma = 1.2$. This is less than the value of 2.7 found by Granato from data taken by Barlow[6].

In Figure 2, f_R and temperature T are plotted against time. As the temperature is cycled over 2 K near T_g, f_R cycles repeatably over 800 Hz on the time scale of the cycling. The magnitude of the ratio , $\Delta G/\Delta T$ is higher than at room temperature but is less than that taken from the constant heating rate curves at higher temperatures in Figure 1. This is evidence that the specimen is not yet in full equilibrium after 6000 s at this temperature.

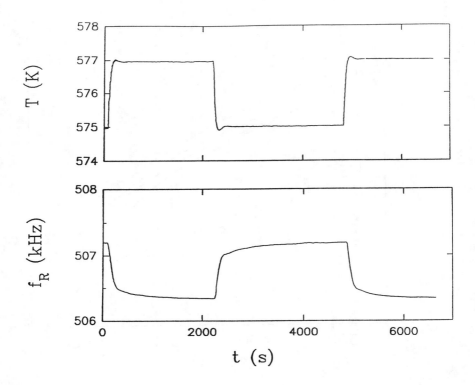

Fig 2. Resonant frequency f_R and temperature T are plotted against time t during 2 K temperature cycling.

CONCLUSION

This experimental technique is useful for accurate measurements of G near and above T_g permitting detailed quantitative comparisons with theory to be made. There is overall agreement with behavior expected from the Interstitialcy Theory. G is continuous at T_g and is comparable in magnitude to G_x. The interstitialcy concentration c_i decreases with aging, and increases with temperature increases as seen by the increasing magnitude of dG/dT above T_g, changes in G during temperature cycling and decreases in G at room temperature after annealing at 610 K and quickly cooling. Some aging ocurs as low as 400 K. From measurements of $\Delta G/\Delta T$ at the highest temperatures examined the softening parameter γ is found to have a value of 1.2 .

ACKNOWLEDGEMENTS

This work is supported by the National Science Foundation under grant DMR9705750.

REFERENCES

1. Y. He and R.B. Schwarz, Mat. Res. Soc. Symp. **455**, 495 (1997).

2. A.V. Granato, J. de Physique IV, **6**, C8-1 (1996)

3. A.V. Granato, Phys. Rev. Lett. **68**, p. 974-977 (1992).

4. K.R. Lyall and J.F. Cochran, Canadian Journal of Physics, **149**, p. 1075-1097 (1971).

5. D. Read and J. Holder, Rev. Sci. Instr. **43**, p. 932-935 (1972)

FRACTURE IN BULK AMORPHOUS ALLOYS

J. A. HORTON, J. L. WRIGHT and J. H. SCHNEIBEL
Metals and Ceramics Division, Oak Ridge National Laboratory
Oak Ridge, TN 37831-6115, hortonja@ornl.gov

ABSTRACT

The fracture behavior of a Zr-based bulk amorphous alloy, Zr-10 Al-5 Ti-17.9 Cu-14.6 Ni (at.%), was examined by transmission electron microscopy (TEM) and x-ray diffraction for any evidence of crystallization preceding crack propagation. No evidence for crystallization was found in shear bands in compression specimens or at the fracture surface in tensile specimens. In- situ TEM deformation experiments were performed to more closely examine actual crack tip regions. During the in-situ deformation experiment, controlled crack growth occurred to the point where the specimen was approximately 20 μm thick at which point uncontrolled crack growth occurred. No evidence of any crystallization was found at the crack tips or the crack flanks. Subsequent scanning microscope examination showed that the uncontrolled crack growth region exhibited ridges and veins that appeared to have resulted from melting. Performing the deformations, both bulk and in-situ TEM, at liquid nitrogen temperatures (LN_2) resulted in an increase in the amount of controlled crack growth. The surface roughness of the bulk regions fractured at LN_2 temperatures corresponded with the roughness of the crack propagation observed during the in-situ TEM experiment, suggesting that the smooth-appearing room temperature fracture surfaces may also be a result of localized melting.

INTRODUCTION

In 1989, amorphous alloy compositions based on La-Al-TM were found that could be produced in a bulk form by casting into water cooled copper molds.[1] These materials have some unusual mechanical properties that are attractive for a number of applications. One related alloy developed by Lin, Johnson and Rhim[2] with a composition of Zr-10 Al-5 Ti-17.9 Cu-14.6 Ni (at.%) exhibits a fracture stress[3] of 1680 MPa at an elastic elongation of ~2%. The Young's modulus is 89 GPa with fracture toughnesses that exceed 50 MPa√m. While previous reports have described the appearance of shear bands and the melted-appearing regions on fracture surfaces[3-6], no detailed analysis has appeared concerning the fracture process itself and what happens at the tips of propagating cracks. Transmission electron microscopy (TEM) was used here to examine shear bands in compression specimens, fracture surfaces in tensile specimens, and the actual crack tips in propagating cracks by performing in-situ deformation experiments.

EXPERIMENTAL PROCEDURE

Ingots of a bulk amorphous alloy with a composition of Zr-10 Al-5 Ti-17.9 Cu-14.6 Ni that was developed at Caltech[2] were cast at Oak Ridge National Laboratory. The alloys were prepared by arc melting in inert gas and drop casting into 7 mm diam by 72 mm long Cu molds. Further details regarding alloy preparation and bulk tensile and compressive tests are given in reference 3. Specimens for TEM examinations of bulk deformed tensile specimens were

Mat. Res. Soc. Symp. Proc. Vol. 554 © 1999 Materials Research Society

prepared by slicing perpendicular to the fracture surface, mechanical grinding and ion milling this fracture edge in a manner that would sharpen this edge while preserving the area of interest. Short cylinders were compressed up to 43% at a stress of 2.7 GPa and TEM specimens were prepared. TEM examinations and the in situ TEM deformation experiments at both room temperature and liquid nitrogen temperatures were performed at 300 kV using previously described deformation stages.[7] XRD was performed in a powder diffractometer with Cu K_α radiation. Bulk deformation experiments were performed on bars $5 \times 5 \times 40$ mm with a chevron notch[8] with the specimen immersed in liquid nitrogen.

RESULTS AND DISCUSSION

TEM specimens were prepared from bulk deformed material to examine the shear bands that form during bulk tensile or compressive deformation (see Fig. 10a in ref. 3). No evidence of any crystallization was found. In addition, specimens were prepared from the fracture surface and again no evidence of any crystallization was found. Because of the difficulty in preparing specimens that were certain to retain the areas of interest, TEM in situ deformation experiments were initiated to examine the region in front of a stressed propagating crack.

Previous successful in situ deformation experiments on brittle intermetallics such as the $L1_2$ trialuminide, Al-8 Cr-25 Ti, with toughnesses around 2 MPa√m, suggested that these experiments would be possible for amorphous alloys.[9] Several design features of the TEM stage were incorporated to make the stage as stiff as possible. The stiffness of the pull rod, relative to the cross sectional area of the tensile specimen where the crack initiates, maximizes the chance for controlled crack growth. The lack of triaxiality in the stress state could change some characteristics of the fracture process in thin sections. However, the TEM deformation experiments have followed bulk fracture behavior remarkably well for a wide range of material, from a low stacking fault energy material such as stainless steel, with a rather planar slip and narrow crack opening; to aluminum, with extensive dislocation cross slip and wide crack openings; to a bcc material such as molybdenum, with jerky dislocation motion ahead of the crack tips. In the case of the brittle trialuminide, dislocation generation was profuse with lots of cross slip and wide crack openings which agrees with the low indentation hardness. While the trialuminides are brittle with low toughness but are rather soft, the bulk amorphous alloys examined here are brittle in the sense of no measurable plastic deformation during a tensile test but with high toughness and high indentation hardness.

Figure 1 shows a sequence of TEM micrographs of an actual crack propagating during an in-situ deformation at LN_2 temperature. The arrow marks the same area in each micrograph. No evidence of crystallization was found at the crack tip or along the crack flank. The cracks started and propagated at approximately 90° to the tensile axis. However, on a microscopic scale the crack propagated in a zigzag manner leaving rough crack flanks. While the crack propagated in a controlled manner for quite some time, at some point the crack would propagate in an uncontrolled manner. The original electropolished hole provided the stress concentrator and the crack initiating site. The cracks start in a thin region and propagate into thicker material similar to the chevron notch toughness test.

A scanning electron microscope (SEM) examination of the fracture surface of a TEM specimen fractured at room temperature showed a distinct difference between the controlled crack growth region and the uncontrolled crack growth region. Figure 2 shows that the thicker area has a melted appearance on the ridges and veins on the surface as described by others.[3-6] The explanation for the prior observations concluded that the high level of stored energy in a

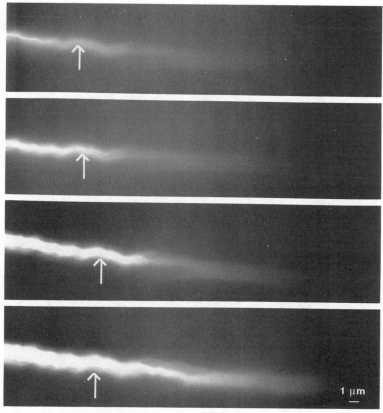

Fig. 1. TEM micrographs of a propagating crack during an in situ deformation at LN$_2$ temperature in a bulk amorphous alloy. The arrows point to the same spot in each image. No crystalline areas were found in front of or along the crack flanks. Tensile axis was also in line with the arrow. Crack propagation was slow and easily controlled.

material which elastically strains 2% is dumped into heat which results in partial melting of the surface. Bruck, Rosakis and Johnson[4] actually measured temperature rises during fracture of a similar bulk amorphous alloy and concluded that no heating was found before crack propagation and substantial heating occurred after failure. The crack growth was controlled until the specimen was approximately 20 μm thick. The serrations on the fracture surface correspond to the crack jumps during the in situ deformation. While the sequence shown in Fig. 1 was very smooth, as the material got thicker, the crack actually tended to propagate in small jumps and then self arrest.

If the released heat caused partial melting of the surface then does this acts as an aid to

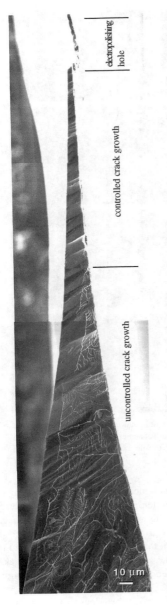

Fig. 2. Fracture surface of a specimen fracture inside the TEM at room temperature. The thinnest area underwent controlled crack growth and corresponds to the area here without the melted appearing ridges. The crack did often incrementally advance in the TEM and apparently this corresponds to the serations seen here.

allowing a crack to propagate along heated material? If true, then the toughness might be higher at lower temperatures. Both TEM in situ experiments and bulk chevron notch bend tests were conducted at liquid nitrogen temperatures. The smooth sequence shown in Fig. 1 was performed at LN$_2$ temperature. Some increase in the area of the fracture surface that did not have the melted ridges was observed in both TEM and bulk specimens but no quantifiable results came out of the chevron notch tests. Within the limits of the experimental setup for the bulk test, the crack growth was still largely uncontrolled.

However, differences were noted in the fracture surface appearance. Figure 3 shows two typical fracture surfaces from the in-situ TEM experiments. The cold fracture surface (Fig. 3b) is much rougher than that obtained at room temrperature (Fig. 3a). The roughness corresponds with the roughness suggested by the TEM in-situ deformation experiment. In Fig. 3b the particular specimen was fractured in the TEM and allowed to warm inside the ion-pumped-out, liquid-nitrogen-cold-trapped vacuum until the specimen reached room temperature. So no external surface water vapor reaction was expected. The heating phenomena that results in the ridges and veins of Fig. 3b may also "smooth" the surface between the ridges, and the rougher fracture surface in Fig. 3b is a more accurate representation of the actual crack propagation than the faster more uncontrolled fracture surfaces as seen in Fig. 3a.

A confirmation of this correlation between crack growth rate and appearance of the melted surfaces is shown in Fig. 4. In Fig. 4a, the crack starts at the apex of the triangle at c, is briefly controlled according to the load displacement curve, then in region d is uncontrolled. It finally becomes controlled again in region e as the relative compliance of the the speci-

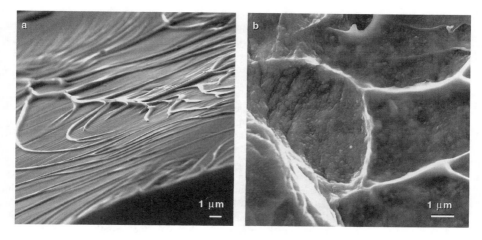

Fig. 3. Fracture surfaces of BAA-11, both from in-situ TEM experiments, (a) room temperature uncontrolled crack growth and (b) fracture at liquid nitrogen temperature in vacuum, also uncontrolled crack growth. The cold specimen was allowed to warm in the vacuum. While the supposed melted ridges looked similar, the room temperature specimen shows a smooth fracture surface otherwise while the LN_2 specimen shows a roughness on a similar scale to the TEM in situ observations.

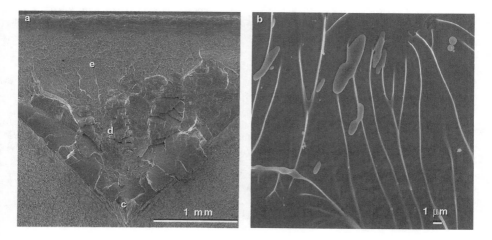

Fig. 4. (a) Fracture surface of a chevron notch bend test specimen. Region (c) and (e) showed controlled crack growth, while region (d) was uncontrolled crack growth. The crack starts at (c). Figure 4b shows an enlargement of region (d) showing how the uncontrolled crack region exhibits the melted appearance while the rougher controlled crack growth region (e) is similar to Fig. 3b.

men/load train increases relative to the specimen geometry. In the uncontrolled fast crack growth region, droplets with a melted appearance were seen on the surfaces (Fig. 4b).

X-ray diffraction was performed on these fracture surfaces with the melted ridges and veins and no evidence of crystallinity has been found, suggesting that the molten material resolidifies again as amorphous material.

CONCLUSIONS

In bulk amorphous alloys, post failure analysis of fracture surfaces have often found melted appearing features suggesting that the heat dump from the 2% stored elastic energy could be contributing to fast crack propagation by allowing the crack to propagate along a hotter or even molten path. No evidence was found for any crystallization in shear bands or at fracture surfaces in bulk deformed specimens. In situ TEM deformation tests also showed no evidence of any crystallization at the crack tip or along the crack flank. In specimen regions thinner than 20 μm, the cracks propagated in a controlled fashion and left fracture surfaces without any evidence of melting. In thicker areas the crack propagation was uncontrolled and the fracture surface contained the molten veins and ridges. Fracture at liquid nitrogen temperatures partially suppressed this phenomena and increased the amount of controlled crack growth. Furthermore, the smooth appearance to the fracture surface between the ridges is also apparently a result of localized melting since the fracture surface of the specimens fractured at liquid nitrogen temperatures were rough and corresponded to the rougher crack flanks produced during the in situ TEM deformation experiments.

ACKNOWLEDGMENTS

The authors wish to thank J. W. Jones for specimen preparation and C. G. McKamey and C.T. Liu for manuscript review. This research was sponsored by the Division of Materials Sciences, US Department of Energy, under contract DE-AC05-84OR21400 with Lockheed Martin Energy Research, Inc.

REFERENCES

1. A. Inoue, T. Zhang, and T. Masumoto, *Mater. Trans. JIM,* **30,** 177 (1989).
2. X.L. Lin, W. L. Johnson, and W. K. Rhim, *Mater. Trans. JIM,* **38,** 473 (1997).
3. C.T. Liu, L. Heatherly, D.S. Easton, C.A. Carmichael, J.H. Schneibel, C.H. Chen, J.L. Wright, M.H. Yoo, J.A. Horton, and A. Inoue, *Metall. and Mater. Trans. A.,* **29A,** 1811 (1998).
4. H.A. Bruck, A.J. Rosakis, and W.L. Johnson, *J. Mater. Res.* **11,** #2, 503 (1996).
5. A. Inoue and T. Zhang, *Mater. Trans. JIM,* **37,** #11, 1726 (1996).
6. C.J. Gilbert, J.M. Lippmann, and R.O. Ritchie, *Scripta Mater.,* **38,** #4, 537 (1998).
7. J.A. Horton, *Proc. of the 40th Ann. Mtg. of EMSA,* 748 (1982).
8. J.A. Horton and J.H. Schneibel, *Mat. Res. Soc. Symp. Proc.* **364,** 1107 (1995).
9. J.A. Horton, *J. Mater. Sci & Tech.,* **9** #1, 745 (1993).

MECHANISM FOR LIGHT EMISSION DURING FRACTURE OF A Zr-Ti-Cu-Ni-Be BULK METALLIC GLASS: TEMPERATURE MEASUREMENTS IN AIR AND NITROGEN

C. J. GILBERT,[1,2] J. W. AGER III,[1] V. SCHROEDER,[1,2] AND R. O. RITCHIE[1,2]
[1]Materials Sciences Division, Lawrence Berkeley National Laboratory
[2]Dept. of Materials Science & Mineral Eng., University of California, Berkeley, CA 94720-1760

ABSTRACT

Light emitted during rupture of $Zr_{41.2}Ti_{13.8}Cu_{12.5}Ni_{10}Be_{22.5}$ (at%) bulk metallic glass has been investigated. Charpy V-notch specimens fractured in a pendulum impact apparatus were used to excite light emission. Spectra acquired from rupture in air exhibited a single broad peak and were fit to a blackbody temperature of ~3175 K. In nitrogen, however, light emission was no longer visible to the eye. In this case, the captured light was at least four orders of magnitude less intense, and spectra were shifted to the red with an effective blackbody temperature of ~1400 K. Fracture surfaces generated in both air and nitrogen exhibited local melting, providing further evidence of intense heating during fracture. Based on these observations we argue that the intense light emission in air is associated with pyrolysis of fresh material exposed during rupture. Results were compared to preliminary observations of light emission from ribbons of a $Zr_{57}Nb_5Ni_{12.6}Cu_{15.4}Al_{10}$ (at%) glass.

INTRODUCTION

We investigate here intense light emission observed during rupture of the bulk metallic glass, $Zr_{41.2}Ti_{13.8}Cu_{12.5}Ni_{10}Be_{22.5}$ (at%), recently developed by Peker and Johnson [1]. Light emission of this intensity is unprecedented in conventional crystalline metals and remains poorly understood. Although recent work on the bulk amorphous alloys $Zr_{52.5}Al_{10}Ti_5Cu_{17.9}Ni_{14.6}$ and $Zr_{57}Cu_{20}Al_{10}Ni_8Ti_5$ report observations of light accompanying tensile rupture [2,3], the mechanistic basis of this behavior is still speculative. Furthermore, the spectral properties of the emissions have yet to be characterized.

The emission of photons, electrons, ions, and neutral particles following fracture or deformation of a range of materials has been studied for many years. The phenomenon is often referred to as fracto-emission [4]. Fracto-emission is associated with the release of mechanical energy applied to regions at or near a crack tip over very small time scales. Fracture surfaces undergo mechanical, chemical, and electronic relaxation from the non-equilibrium, high-energy states caused by rupture. These non-equilibrium states can induce the creation of defects at or near freshly created surfaces, the interaction of reactive species, charge separation between crack walls, and the production of heat and acoustic waves.

While there are numerous reports of fracto-emission in brittle insulating solids such as ceramics, oxide glasses, semiconductors, and polymers [4-7], fracto-emission in metals is less well studied and tends to be weaker (particularly emission of visible light). When fracto-emission is observed in metals, it is attributed to oxidation of freshly created surfaces [4], heating during deformation (as in polycrystalline Ti alloys) [5,6], rupture of surface oxides [6,7], or to recombination associated with dislocation or other defect formation [7].

To our knowledge, however, fracto-emission in amorphous metals has not been studied quantitatively. Consequently, the goals of the present work are to (*i*) measure the spectral properties of the emitted light, and (*ii*) identify the mechanisms responsible for this surprising behavior.

EXPERIMENTAL PROCEDURES

Fully amorphous $Zr_{41.2}Ti_{13.8}Cu_{12.5}Ni_{10.0}Be_{22.5}$ (at%) was machined from as-cast plates into standard Charpy V-notch impact specimens (50 × 10 × 8 mm) [8]. The processing methods are described elsewhere [1]. Samples were fractured at ambient temperature in both room air and nitrogen gas using a pendulum impact apparatus under dynamic loading conditions. Pendulum velocity was ~3.5 m/s at impact. Tests in nitrogen were performed in a plastic tent with a positive gas pressure, and gas flushed the chamber for at least 1 h prior to measurements. The optical path was oriented along a tangent of the pendulum swing, and the notch was imaged 1:1 with a 200 mm f/4 lens onto the entrance slit of a 0.25 m spectrometer. Beam alignment was performed using a HeNe laser directed at the notch. Spectra were collected over the visible range (350 to 925 nm) with a liquid nitrogen cooled CCD camera (Princeton Instruments), and in the near-infrared using a liquid nitrogen cooled HgCdTe detector (Rockwell, Inc.) equipped with a 1100 to 1300 nm filter. The spectral calibration was established using Hg and Ar atomic emission lines. Visible spectra collected with the CCD detector were corrected for variations of instrument throughput and detector quantum efficiency by normalizing the collected signal to that obtained from a NIST-traceable standard of spectral irradiance (quartz-halogen tungsten lamp operated at 39 W, Optronics Laboratories Model 245A).

Light emission was time-resolved by splitting off a portion of the light with a 50/50 beam splitter and focussing it onto a fast photodiode with a spectral range of 200 to 1100 nm. The transient signal was captured with a 100 kHz data collection card triggered immediately prior to impact. On one sample, the light was also captured with a digital camera using an exposure time of 1 s, again timed to coincide with the release of the pendulum. Fracture surfaces produced in both air and nitrogen were subsequently examined in the scanning electron microscope (SEM).

Results for the bulk $Zr_{41.2}Ti_{13.8}Cu_{12.5}Ni_{10.0}Be_{22.5}$ glass were also compared to limited experiments on a $Zr_{57}Nb_5Ni_{12.6}Cu_{15.4}Al_{10}$ glass in ribbon form (~0.6 mm thick and 5 mm wide).

Fig. 1: Light emission in the Zr-Ti-Cu-Ni-Be glass.

These samples were broken in bending (at non-dynamic loading rates), and the light was similarly captured using a digital camera. Resulting fracture surfaces were examined in the SEM.

RESULTS AND DISCUSSION

An optical photograph of light emission accompanying rupture in room air (a continuous exposure taken over the entire fracture event) revealed discrete streaks of visible light emanating from the fracture plane (Fig. 1). These flashes were easily visible to the eye, even in ambient light (these flashes were also visible when samples were cooled to liquid nitrogen temperature prior to fracture). The sample in Fig. 1 was loaded vertically into the pendulum apparatus, with the lower half rigidly clamped into position (the notch is oriented

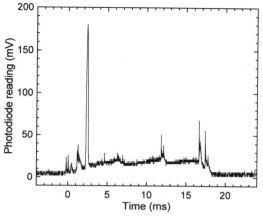

Fig. 2: Time-resolved light emission in the Zr-Ti-Cu-Ni-Be glass (from 200 to 1100 nm).

horizontally at the center of the photograph). The axis of the pendulum swing was perpendicular to the page, with the pendulum striking the sample ~10 mm below its top edge. The bright streaks suggest that light emission was associated with the ejection of hot particles from the fracture surface. In fact, closer inspection of the streaks indicates that several bounced off the sides of the testing apparatus, further supporting the notion that they are caused by heated particles flung from the newly-exposed surface.

Visible light was emitted over a time interval of ~18 ms (Fig. 2). Several discrete peaks of high intensity, each lasting ~1 ms, were observed within this time interval. Based on the streaks found in Fig. 1, we believe these are associated with individual heated particles crossing the photodiode's field of view.

In room air the emitted spectrum (Fig. 3) exhibited a broad peak with no sharp features. As such, spectral data were regression fit to Planck's blackbody relationship using weighted non-linear least squares fitting. This allowed for the determination of an effective blackbody temperature consistent with the measured spectral data. In this case, a fit was obtained to a blackbody temperature of 3175 K (measured using a total of three specimens, represented by different symbols in Fig. 3). In nitrogen, however, the signal intensity captured by the CCD detector was reduced by over four orders of magnitude, and a flash was no longer visible to the eye. Although a broad spectral peak was again observed within the range of our measurements (Fig. 3), it was red-shifted and a regression fit yielded a much lower blackbody temperature of 1400 K (observed on two specimens).

Along with a vein morphology typical of amorphous metals [9], evidence for local melting was observed on fracture surfaces developed in both air and nitrogen when examined in the SEM (Fig. 4). These melted features existed only near the root of the notch (within the first ~250 μm) where fracture initiated.

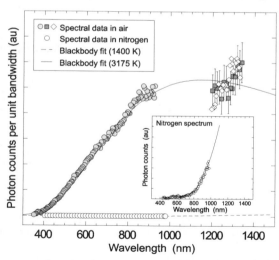

Fig. 3: Light emission spectra from fracture of the Zr-Ti-Cu-Ni-Be glass in air and nitrogen.

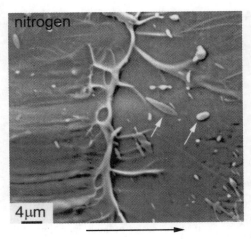

nitrogen

4μm

Fig. 4: SEM of the fracture surface of Zr-Ti-Cu-Ni-Be fractured in nitrogen. Similar features were found on surfaces ruptured in air.

Such apparently melted and re-solidified regions have been observed previously in this and many other amorphous alloys following rupture [9-13]. Given that the glass transition temperature is ~625 K (at a heating rate of 20 K/min) [1] and the liquidus temperature is ~923 K [1], the presence of such features places a lower bound on the temperatures achieved during fracture, consistent with our blackbody measurements. Furthermore, the "globular" morphologies of some features in Fig. 4 (marked by white arrows) suggest that they may be the source of what appear to be expelled particles in Fig. 1.

Based on the experimental observations discussed above, we conclude that in air the light emission and extreme blackbody temperature of 3175 K are associated with oxidation of freshly exposed material following fracture. Indeed, Zr and Ti both possess a strong affinity for oxygen, and their oxides have particularly large heats of formation (Table 1) [14]. The suppression of light emission in nitrogen, along with the large drop in temperature, was presumably associated with a decrease in oxygen partial pressure. Although others have speculated that oxidation may be the source of light emission [3], our observations provide the first direct evidence for the operation of such a mechanism. Moreover, we have successfully developed a technique to measure the associated temperatures.

Table 1: Heats of formation for selected oxides of Zr, Ti, Ni, Cu, and Be

Oxide	Heat of Formation (298 K) kJ/mole
TiO_2 (rutile/anatase)	-945/-939
Ti_3O_5	-2446
BeO (alpha/beta)	-608/-602
ZrO_2	-1097
CuO	-156
CuO_2	-171

Similar (qualitative) observations were also made in the $Zr_{57}Nb_5Ni_{12.6}Cu_{15.4}Al_{10}$ glass. In air, bright flashes of light were observed (Fig. 5), and fracture surfaces exhibited similar evidence for local melting (Fig. 6). Spectral measurements of the light, however, have not yet been made on this alloy (this is the subject of on-going experiments).

Of particular interest is the distinction between temperature rise due to oxidation and due to deformation. Deformation in these alloys is highly localized into narrow slip bands [2,3,10,13],

Fig. 5: Light emission during fracture of the Zr-Nb-Ni-Cu-Al glass in air.

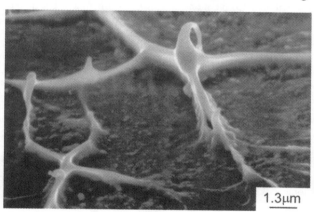

1.3μm

Fig. 6: SEM of the fracture surface of Zr-Nb-Ni-Cu-Al fractured in air (crack propagated top to bottom).

and such localization is known to generate intense heating associated with concentrated plastic work [5,13]. Strain and thermal localization are favored by many of the properties of this alloy, including a flow stress that drops rapidly with temperature near the glass transition, non-existent work hardening, and low thermal conductivity [10,13]. Indeed, estimates for uniaxial tensile tests based on the conversion of elastic strain energy into heat show that a temperature rise of ~900 K in a slip band is not unreasonable [3].

Tests in oxygen-free environments should allow for the direct measurement of deformation-induced temperatures (they will not be masked by the extreme temperatures accompanying pyrolysis). The temperature measured in nitrogen is presumably approaching this value. In fact, we speculate that the severe pyrolysis in air may be facilitated by initial deformation-induced heating which, upon rupture, exposes heated metal (as high as 1400 K according to our measurements) to an oxidizing environment. Improved measurements of this temperature (i.e., in cleaner inert atmospheres) are the subject of future experiments.

SUMMARY AND CONCLUSIONS

A novel light emission phenomenon observed during dynamic fracture of a bulk metallic glass, $Zr_{41.2}Ti_{13.8}Cu_{12.5}Ni_{10}Be_{22.5}$ (at%), has been investigated using standard Charpy V-notch specimens fractured in a pendulum impact apparatus. Spectra acquired from specimens ruptured in air exhibited a single broad peak, which could be fit to a blackbody temperature of ~3175 K. Emission from specimens fractured in nitrogen, however, was at least four orders of magnitude less intense. The spectrum was shifted to the red with an effective blackbody temperature of ~1400 K. Fracture surfaces of specimens ruptured in both air and nitrogen exhibited local melting, providing further evidence of intense heating during fracture. Based on these observations we argue that the intense light emission in air is associated with pyrolysis of fresh material exposed during rupture. Similar observations of light emission were also made on a $Zr_{57}Nb_5Ni_{12.6}Cu_{15.4}Al_{10}$ glass, although measurements on this alloy are still preliminary.

ACKNOWLEDGEMENTS

Work supported by the Director, U.S. Office of Energy Research, Office of Basic Energy Sciences, Materials Sciences Division of the U.S. Department of Energy, with additional support from Amorphous Technologies International, Corporation and Howmet Corporation. The authors also thank Mr. J. P. Lloyd and Prof. J. R. Graham of the Dept. of Astronomy, University of California, Berkeley for the use of their HgCdTe infrared detector.

REFERENCES

1. A. Peker, and W. L. Johnson, Appl. Phys. Lett. **63** (17), 2342-2344 (1993).
2. L. Q. Xing, C. Bertrand, J.-P. Dallas, and M. Cornet, Mater. Sci. Eng. **A241** (1-2), 216-225 (1998).
3. C. T. Liu, L. Heatherly, D. S. Easton, C. A. Carmichael, J. H. Schniebel, C. H. Chen, J. L. Wright, M. H. Yoo, J. A. Horton, and A. Inoue, Metall. Trans. **29A**, 1811-1820 (1998).
4. J. T. Dickinson, E. E. Donaldson, and M. K. Park, J. Mat. Sci. **16**, 2897-2908 (1981).
5. J. R. Rice, and N. Levy, in *The Physics of Strength and Plasticity*, ed. by A. S. Argon (MIT Press, Cambridge, MA, 1969), pp. 277-293.
6. K. B. Abramova, A. B. Pakhomov, B. P. Peregud, and I. P. Shcherbakov, Sov. Phys. Tech. Phys. **35** (6), 752-754 (1990).
7. J. T. Dickinson, P. F. Braunlich, L. A. Larson, and A. Marceau, Appl. Surf. Sci. **1**, 515-537 (1978).
8. ASTM E23, in *Annual Book of ASTM Standards* **3.01**, (American Society for Testing and Materials, West Conshohocken, PA, 1996), pp. 137-156.
9. C. A. Pampillo, and A. C. Reimschuessel, J. Mater. Sci. **9**, 718-724 (1974).
10. H. A. Bruck, T. Christman, A. J. Rosakis, and W. L. Johnson, Scripta Mat. **30**, 429-434 (1994).
11. C. J. Gilbert, R. O. Ritchie, and W. L. Johnson, Appl. Phys. Lett. **71** (4), 476-478 (1997).
12. P. Lowhaphandu, and J. J. Lewandowski, Scripta Mat. **38** (12), 1811-1817 (1998).
13. H. A. Bruck, A. J. Rosakis, and W. L. Johnson, J. Mater. Res. **11** (2), 503-511 (1996).
14. *JANAF Thermochemical Tables*, 3rd ed., American Chemical Society, 1986.

Part IV

Glass Forming Ability and Thermal Stability

CHARACTERIZATION OF GAS ATOMIZED Cu$_{48}$Ti$_{34}$Zr$_{10}$Ni$_8$ AMORPHOUS METAL POWDER

J. C. FOLEY, D. J. SORDELET, AND T. A. LOGRASSO
Ames Laboratory, Iowa State University, Ames, IA 50011

ABSTRACT

The advent of multi-component metallic alloys, which exhibit relatively good glass forming ability, has opened opportunities for processing metallic glasses into thick cross section components. The relatively good glass forming ability is important because conventional processing techniques (e.g., casting, extrusion and rolling) may be used to fabricate useful shapes while retaining the excellent engineering properties of an amorphous structure. In particular, the favorable processing characteristics of bulk amorphous alloys are the low cooling rates which can be exercised to yield an amorphous structure and the operating temperature range between the glass transition temperature (T_g) and the crystallization temperature (T_x). Current work is focused on developing a processing strategy that will allow us to fabricate even larger cross section amorphous alloys than are currently achievable by casting methods. The technique involves producing high pressure gas atomized (HPGA) Cu$_{48}$Ti$_{34}$Zr$_{10}$Ni$_8$ powders and consolidating them at temperatures above T_g, but below T_x. Thermal analysis of atomized powders by DSC provides details of the influence of powder particle size, which is related to cooling rate during atomization. The results of experiments characterizing the thermal and kinetic behavior of Cu$_{48}$Ti$_{34}$Zr$_{10}$Ni$_8$ powders indicate that short processing times are required to retain the amorphous structure during consolidation in the temperature regime between T_g and T_x.

INTRODUCTION

The first report of metallic glass formation during continuous cooling from the melt [1] sparked numerous experiments dealing with non-crystalline metallic solids. Many of these reports have involved systems that are capable of producing amorphous materials by various rapid solidification techniques that achieve very large cooling rates (10^5-10^8 K/s) such as splat-quenching or melt-spinning. It was not until 1974 that the first report of bulk metallic glasses, cylindrical rods 1 to 3 mm in diameter of Pd-Cu-Si, Pd-Ni-P and Pt-Ni-P, were reported by Chen [2]. More recently, Inoue and coworkers [3, 4] have reported the formation of Zr-based amorphous multicomponent alloys that do not contain relatively expensive Pd and Pt. In addition to greater glass formability, the Zr-based bulk amorphous alloys also exhibited desirable mechanical properties and a reported wide supercooled liquid range. The presence of a wide supercooled liquid range gave rise to the promise of novel near-net shape deformation processes to form relatively large structural components of metallic glasses [5]. Recently, a successful synthesis of glassy alloy compacts with full strength by warm extrusion of atomized glassy powder in the supercooled liquid state was reported [6].

In principle, powder metallurgy (P/M) processing of amorphous powder could be applied to any amorphous alloy that can be atomized and exhibits a wide supercooled region. The ability to fabricate rod and sheet stock of high strength, high fracture toughness amorphous alloys will open the possibility for many technical applications. However, before successful consolidation of amorphous metals becomes common place, a clear understanding of the formation and stability of powder amorphous alloys is needed. In general, crystallization of amorphous alloys occurs by nucleation and growth processes [7], but several decomposition pathways have been reported [7-10]. A determination of the decomposition pathways active in $Cu_{48}Ti_{34}Zr_{10}Ni_8$ is required to set consolidation processing parameters.

A bulk amorphous sample with the composition $Cu_{47}Ti_{34}Zr_{10}Ni_8$ is reported to exhibit a T_g at 398°C, T_x at 444°C and $\Delta T = 46$°C [11]. In addition, the critical cooling rate is reported to be less than 250°C/s. While the wide supercooled region and relatively low critical cooling rate should enable production of amorphous powder, it is unclear how the increased surface area of powder particle compared to bulk amorphous ingots and the change in cooling rate as a function of powder size will affect the decomposition pathways.

EXPERIMENT

An ingot of the composition $Cu_{48}Ti_{34}Zr_{10}Ni_8$ was made by induction melting 99.99% pure Ti, 99.99% pure Zr, 99.999% pure Cu, and 99.99% pure Ni in a graphite crucible. The ingot material was high pressure gas atomized (HPGA) at Ames Laboratory using a hard fired Al_2O_3 bottom pour crucible and 99.99% pure Ar gas. A portion of the powder obtained from HPGA was classified into the size categories of diameter < 5 μm, 5 μm-15 μm, 15 μm-25 μm, 25 μm-38 μm, 38 μm-45 μm, 45 μm-53 μm, 53 μm-62 μm, 62 μm-75 μm, 75 μm-90 μm, and 90 μm-106 μm. Select samples of the classified powder were examined with differential scanning calorimeter (DSC), x-ray diffraction (XRD) using a Cu K α source, and scanning electron microscopy (SEM). The starting ingot and HPGA powder in the range of diameter 75-106 μm and diameter < 75 μm was examined with inductively coupled plasma (ICP) and inert gas fusion (IGF) techniques. In addition, a small (diameter = 6.25 mm) arc-cast ingot was made for comparison to the HPGA powder.

RESULTS

Chemical analysis of the starting ingot indicated that the carbon content was well below 100 ppmw and that the alloy composition was similar to the desired nominal composition. Chemical analysis results of the powder in the range of diameter 75-106 μm and diameter < 75 μm are shown in Table 1. Although the chemical analysis results indicate that the copper content is slightly higher than the nominal composition and the titanium content is slightly lower than nominal, the compositions of the two powder samples are very similar.

Table 1
Chemical composition results of the HPGA powder

Sample	Cu (at%)	Ti (at%)	ZR (At%)	Ni (at%)	O (ppmw)	N (ppmw)
75-106μm	52.4	30	9.5	8.1	762	<1
< 75 μm	54.6	29.4	9.3	6.7	855	1

The X-ray diffraction trace of a slice of the arc-cast ingot is shown in Figure 1. The X-ray diffraction trace shows strong evidence of crystalline peaks with some evidence of amorphous scattering. In contrast, the X-ray diffraction traces of the HPGA powder shown in Figure 2 only exhibit amorphous scattering maximums. The location of the amorphous scattering maximums occurs at the same two-theta values. The presence of the amorphous scattering maximums in the same locations is an indication that the majority of the powder, regardless of the powder diameter, is amorphous.

The continuous DSC traces of select $Cu_{48}Ti_{34}Zr_{10}Ni_8$ HPGA powder sizes at 40°C/min are shown in Figure 3. The traces were normalized to the sample weights and overlaid for comparison. All of the continuous DSC traces exhibit a glass transition (T_g) signal around 440°C and crystallization temperature (T_x) around 480°C. In addition, all of the subsequent reactions are the same and occur at the same temperatures. The presence of the T_g signal around 440°C is strong evidence that all the powder sizes evaluated are completely amorphous. The isothermal DSC traces of the < 5 μm and 90-106 μm powder are shown in Figure 4. Again, the thermal signal from the very small powder is very similar to the signal of the very large powder. The differences in onset time, peak time and ΔH associated with the decomposition reaction are negligible.

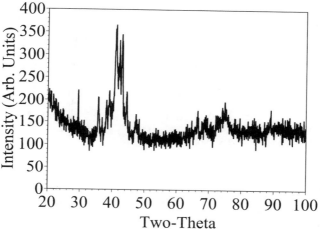

Figure 1. The X-ray diffraction trace of the arc-cast $Cu_{48}Ti_{34}Zr_{10}Ni_8$ alloy.

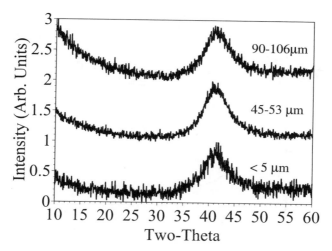

Figure 2. The X-ray diffraction traces of the $Cu_{48}Ti_{34}Zr_{10}Ni_8$ HPGA powder.

New samples of < 5 μm and 90-106 μm powder were annealed in an sealed quartz tube at 450°C for 3 minutes to partially crystallize the powder. Samples of the partially crystallized powder were examined with SEM and are shown in Figures 5A and 5B. The SEM micrographs show no evidence of a crystalline phase. The continuous DSC traces of the < 5 μm powder conducted at 5°C/min and 40°C/min are shown in Figure 6. From the traces shown in Figure 6, it is apparent that the thermal signal obtained at the different heating rates are quite different.

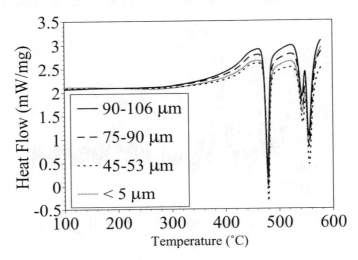

Figure 3. The continuous DSC traces of select $Cu_{48}Ti_{34}Zr_{10}Ni_8$ HPGA powder sizes at 40°C/min.

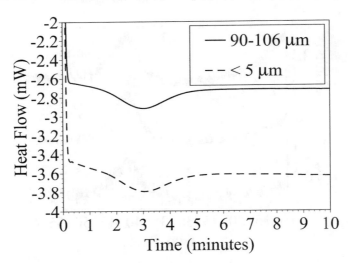

Figure 4. The isothermal DSC traces of the < 5 μm 90-106 μm $Cu_{48}Ti_{34}Zr_{10}Ni_8$ HPGA powder at 450°C.

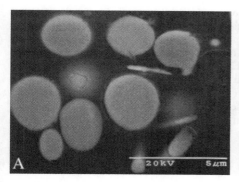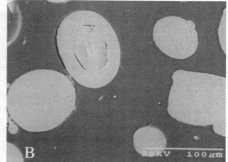

Figure 5. The SEM micrographs of the < 5 μm (A) and 90-106 μm (B) after annealing at 450°C for 3 minutes.

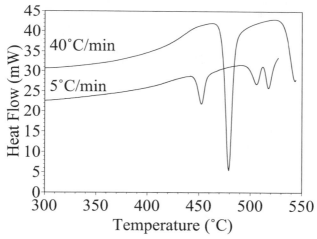

Figure 6. The continuous DSC traces of the < 5 μm $Cu_{48}Ti_{34}Zr_{10}Ni_8$ HPGA powder at 5°C/min and 40°C/min.

DISCUSSION

The $Cu_{48}Ti_{34}Zr_{10}Ni_8$ powders made by HPGA appear to be completely amorphous. The range of particle sizes examined exhibit the same T_g, T_x, and decomposition kinetics. It appears that the same nucleation site is active in both the large (90-106 μm) and small (< 5 μm) powder sizes. This is an indication that the powder surface of these powders is not a strong nucleation site for the crystallization of the amorphous powder. If the powder surface were a strong nucleation site, it would stand to reason that the smaller powder particles would have a much higher density of potent sites, which should cause a noticeable difference in the decomposition behavior. In contrast, if the powder contained a high density ($> 10^{20}$ m^{-3}) of internal sites, as other amorphous forming alloys seem to contain [12], the different powder sizes examined in this work would exhibit similar decomposition behavior.

While it is likely that internal heterogeneous nucleation is the rationale for the observed similar thermal decomposition signals, it is unclear what type of heterogeneous nucleant is active in the $Cu_{48}Ti_{34}Zr_{10}Ni_8$ HPGA powder. As mentioned previously, other amorphous forming alloys are reported to contain a high density of heterogeneous nucleants that can be bypassed with relatively high continuous cooling rates [12]. Another possible source of nucleants is a result of phase separation that has been reported to occur in some bulk amorphous alloys [8]. The current study can not resolve this issue, but if phase separation is occurring two T_g reactions should exist for each of the two amorphous phases [7]. Moreover, examination of the partially crystallized powders with transmission electron microscopy could elucidate the most likely decomposition route.

CONCLUSIONS

The formation of amorphous $Cu_{48}Ti_{34}Zr_{10}Ni_8$ powder was successfully made by HPGA. All of the powder diameters evaluated were completely amorphous and exhibited the same decomposition pathway and kinetics. It is likely that the same nucleation sites are active in both the < 5 μm and 90-106 μm powder. A wide supercooled region was only observed when the powder was heated at 40°C/min. Slower heating rates resulted in different decomposition reactions to occur. The relatively short crystallization delay time of the $Cu_{48}Ti_{34}Zr_{10}Ni_8$ powder at 450°C will make consolidation of the powder into a completely amorphous compact difficult. Further investigations of the decomposition pathway and kinetics are needed.

ACKNOWLEDGEMENTS

The authors would like to acknowledge the financial support of the Materials Science Division of DOE/BES under contract number W-7405-Eng-82 and a Special Research Initiation Grant from the Office of the Vice Provost, Iowa State University. Thanks also go to Amy Ross for her help in obtaining SEM micrographs, to John Wheelock for his help in the fabrication of arc-cast rods and to Chris Gross for the chemical analysis.

REFERENCES

1.   P. Duwez, R.H. Willens, and W. Klement, Appl. Phys. Lett., **31**, 1136-1150, (1960).
2.   H.S. Chen, Acta Met., **22**, 1505, (1974).
3.   A. Inoue, T. Zhang, and T. Masumoto, Mater. Trans. JIM, **31**, 177-183, (1990).
4.   T. Zhang, A. Inoue, and T. Masumoto, Mater. Trans. JIM, **32**, 1005-1010, (1991).
5.   T. Masumoto, Sci. Rep. RITU, **A39**, 98-101, (1994).
6.   Y. Kawamura, H. Kato, A. Inoue, and T. Masumoto, International Journal of Powder Metallurgy, **33**, 50-61, (1997).
7.   U. Köster and U. Herold, Topics in Applied Physics, **46**, 225-259, (1981).
8.   S. Schneider, P. Thiyagarajan, and W.L. Johnson, Appl. Phys. Lett., **68**, 493-495, (1996).
9.   J.C. Foley, D.R. Allen, and J.H. Perepezko, Scripta Materialia, **35**, 655-660, (1996).
10.   D.R. Allen, J.C. Foley, and J.H. Perepezko, Acta Mater., **46**, 431-440, (1998).
11.   X.H. Lin and W.L. Johnson, J. Appl. Phys., **78**, 6514-6519, (1995).
12.   J.C. Foley, H. Sieber, D.R. Allen, and J.H. Perepezko. "The Effect of Processing on the Microstructural Evolution During Solidification of Al-Y-Fe Glass Forming Alloys".*The 4th Decennial International Conference on Solidification Processing*1997. p. 602-605.

SIGNIFICANT UNDERCOOLED LIQUID REGION OF OVER 200K IN RARE EARTH BASED METALLIC GLASSES

Z.P. LU*, Y. LI*, S.C. NG**, Y.P. Feng**
*Department of Materials Science, National University of Singapore, Singapore 119260
**Department of Physics, National University of Singapore, Singapore 119260

ABSTRACT

Using a newly developed temperature modulated differential scanning calorimetry (TMDSC), glass transition has been detected in several RE-Fe-Al (RE = ND, Pr, Sm and Y) metallic glasses, which can not be observed by a conventional DSC. Our results show that the glass transition occurred together with a small fraction crystallization in the as-spun samples, the final main crystallization transition occurred at higher temperature (over 720K) in these rare-earth based alloys. Large undercooled liquid regions up to 200K at a underlying heating rate of 5K/min have been observed in these glasses.

INTRODUCTION

In DSC measurements, the sample temperature is either ramped linearly at a constant heating (or cooling) rate, or kept constant as in isothermal experiments. The measured heat flow can be described by the following equation:

$$\frac{dH}{dt} = -c_p \beta + f(T,t) \qquad (1)$$

where H is the amount of heat evolved, c_p the heat capacity, β the heating rate, T the absolute temperature and t the time. It states that the heat flow measured in a DSC cell, which is referred as total heat flow (HF), is made up of two terms. The first term is a function of the heat capacity of the sample and the heating rate, which is denoted as the thermodynamic or reversing heat flow (RHF). The second term is a function of absolute temperature and time, which governs the kinetics and is denoted as the non-reversing heat flow (NHF). The measured heat flow in a DSC is the total heat flow including both the reversing and the non-reversing heat flows. Recently, an enhanced version of the DSC method, temperature modulated differential scanning calorimetry (TMDSC) has been introduced [1,2]. One of advantages of the TMDSC is that it is able to separate the RHF and NHF from the total heat flow during a phase transition [1-6].

Glass transition and crystallization have been studied extensively by conventional differential scanning calorimetry (DSC). However for many metallic glasses, the glass transitions are closely followed by crystallization and often these glass transitions were not observed by conventional DSC. The glass transition upon heating is a reversible process in which heat is absorbed to accommodate the heat capacity increase during the transition, while crystallization is a non-reversible exothermic process which releases heats. Using TMDSC, glass transition in some metallic glasses has been separated from crystallization [6]. In this work, we present our latest results of TMDSC study on the glass transition for several rare-earth based $La_{55}Al_{25}Ni_{10}Cu_{10}$, $Nd_{60}Fe_{30}Al_{10}$, $Y_{60}Fe_{30}Al_{10}$, $Pr_{60}Fe_{30}Al_{10}$ and $Sm_{60}Fe_{30}Al_{10}$ metallic glasses.

EXPERIMENTAL

Alloy ingots of $La_{55}Al_{25}Ni_{20}$, $Nd_{60}Fe_{30}Al_{10}$ and $Zr_{60}Y_{30}Al_{15}Ni_{25}$ were first made by arc-melting 99.9% pure La, Y and Zr, 99.99% pure Ni and Fe, and 99.999% pure Al under a Ti-gettered argon atmosphere. The ingots were then melt-spun into ribbons using a single-roller melt-spinner under argon. The TMDSC experiments were carried out on a temperature-modulated DSC (TA Instruments Inc. USA) using a refrigerated cooling system or an argon-gas cooling system with a nitrogen-gas DSC cell purge. The modulated DSC regime was utilized to measure the modulated heat flow under continuous heating rates of 1 or 5 K/min up to 880 K. In order to obtain accurate and reproducible results three experimental parameters need to be controlled in TMDSC measurements, namely the underlying heating rate, the temperature oscillation amplitude and the oscillation period. The oscillation amplitudes and periods were determined so that at least four modulation runs could be performed during the transition. Oscillation amplitudes between 0.1 and 0.2 K and a modulation period of 60 second were used with sample weights of 8 to 15 mg. The experiments were repeated at least three times for each sample to eliminate other possible errors.

RESULTS AND DISCUSSION

Figure 1 shows typical TMDSC results for $La_{55}Al_{25}Ni_{10}Cu_{10}$ glass at an underlying heating rate of 1 K/min which illustrate total (HF), reversing (RHF) and non-reversing (NHF) heat flows curves. On the HF curve, the glass transition and crystallization are observed at temperature of 438 and 507K respectively, which are identical to the result obtained for a conventional DSC. On the RHF curve, there is a step change in the endothermic direction at an onset temperature of 451 K. The glass transition at best is a second-degree transition, only associated with a heat capacity change. Since the reversing heat flow in TMDSC is only related to the underlying heating rate and heat capacity, this step change clearly represents a glass transition in amorphous $La_{55}Al_{25}Ni_{10}Cu_{10}$. This glass transition temperature of 451 K on the reversing curve is about 13K higher than that obtained under a conventional DSC for the same alloy. Towards the end of crystallization shown on the HF curve, the reversing heat flow curve turns upward in a step change reaching the baseline, indicating the heat capacity changed back to that of the crystalline state.

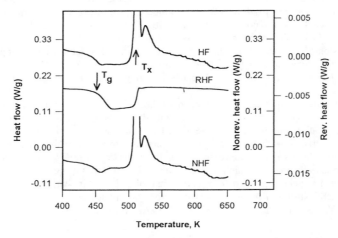

Figure 1 TMDSC results for $La_{55}Al_{25}Ni_{10}Cu_{10}$ amorphous phase.

Figure 2 shows the TMDSC results for $Nd_{60}Fe_{30}Al_{10}$ metallic glass obtained with an underlying heating rate of 5 K/min. There is small crystallization at the onset temperature of about 596 K and the main crystallization at the onset temperature of about 722 K on the NHF curve. No glass transition can be observed in the HF curve. However, there is a clear step change in the endothermic direction with an onset temperature of 591 K associated with a glass transition on the reversing heat flow curve. The glass transition temperature is about 5 K below the corresponding crystallization temperature of 596 K on the total heat flow curve. This weak glass transition could be related to the corresponding weak crystallization at 420 K. It is also noted that there is no further downward step change on the reversing heat flow curve near the temperature (722K) corresponding to the main crystallization reaction, indicating that there is no glass transition occurred here.

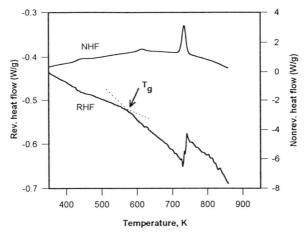

Figure 2 TMDSC results for $Nd_{60}Fe_{30}Al_{10}$ amorphous ribbon.

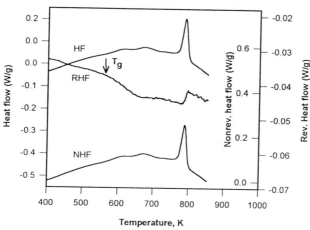

Figure 3. TMDSC results for $Y_{60}Fe_{30}Al_{10}$ amorphous ribbons

Similar results are obtained for $Y_{60}Fe_{30}Al_{10}$ (Figure 3), $Pr_{60}Fe_{30}Al_{10}$ and $Sm_{60}Fe_{30}Al_{10}$ glasses. As shown on the RHF curve in Figure 3 for $Y_{60}Fe_{30}Al_{10}$ glass, the glass transition occurred at temperature of 572 and there is no further glass transition corresponding to the main crystallization at around 773K showing a wide undercooled region. The detailed results on the temperatures of glass transition and crystallization for all the glasses are summarized in Table I. Figure 4 shows the conventional DSC results at heating rate of 40K/min for the $Y_{60}Fe_{30}Al_{10}$ ribbon as-spun and heat treated. The results show that the two small crystallization at 568 and 649 K were not present in the sample after heat treatment at 480°C for 10 mins, but the crystallization enthalpies measured for the main crystallization at temperature around of 790K are 38J/g and 37.7J/g for the as-spun and heat treated samples respectively. Figure 5 of TMDSC results shows that for the heat treated $Y_{60}Fe_{30}Al_{10}$ sample, the glass transition still can be observed at 590 K.

Table I. Characteristic temperatures (K) obtained from an TMDSC for $La_{55}Al_{25}Ni_{110}Cu_{10}$, $Nd_{60}Fe_{30}Al_{10}$, $Pr_{60}Fe_{30}Al_{10}$, $Sm_{60}Fe_{30}Al_{10}$, and $Y_{60}Fe_{30}Al_{10}$ glasses.($\Delta T = T_{x3} - T_g$).

Alloy	RHF	HF (K)			NFH (K)			$\Delta T(K)$
	T_g	T_{x1}	T_{x2}	T_{x3}	T_{x1}	T_{x2}	T_{x3}	
$La_{55}Al_{25}Ni_{110}Cu_{10}$	451	507			507			56
$Nd_{60}Fe_{30}Al_{10}$	591	402	596	722	402	595	722	131
$Y_{60}Fe_{30}Al_{10}$	572	568	649	773	571	649	772	201
$Pr_{60}Fe_{30}Al_{10}$	575	570	670	729	572	671	729	141
$Sm_{60}Fe_{30}Al_{10}$	593	536	627	734	535	626	734	151

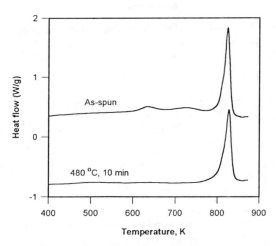

Figure 4 Normal DSC results at a heating rate of 40 K/min for $Y_{60}Fe_{30}Al_{10}$ as-spun and heat treated at 480°C for 10 mins.

The above results clearly exhibit that upon heating in a conventional DSC a small volume fraction of $Nd_{60}Fe_{30}Al_{10}$, and $Y_{60}Fe_{30}Al_{10}$ glasses underwent crystallization at temperatures below 700K and the majority of the glasses crystallized at a much higher temperature (over 720K). No glass transitions were observed in these as-spun samples by a conventional DSC. However, our present TMDSC results seem to point out that the glass transition occurred for this majority part of glasses at the temperature much lower than the temperature of the corresponding crystallization. It is pointed out that the signal from RHF for the glass transition was very weak when compared with that from NRF for the crystallization. The TMDSC results for the present as-spun samples show that despite a small fraction of sample crystallized, the remaining majority part of the sample was in the undercooled liquid region which is over 130K wide before they finally crystallized. The fact that the glass transition can still be observed at 590K by TMDSC for the $Y_{60}Fe_{30}Al_{10}$ ribbon heat treated at 480 °C for 10 mins shows that this partial crystallized sample has a glass transition temperature far below its crystallization temperature indicating the existence of a large supercooled liquid region.

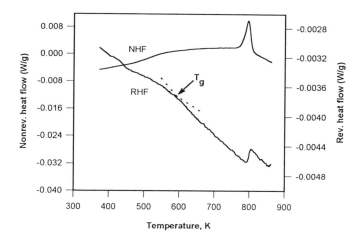

Figure 5 TMDSC results for $Y_{60}Fe_{30}Al_{10}$ ribbon heat treated at 480°C for 10 mins.

CONCLUSIONS

Glass transition in $La_{55}Al_{25}Ni_{10}Cu_{10}$, $Nd_{60}Fe_{30}Al_{10}$, $Y_{60}Fe_{30}Al_{10}$, $Pr_{60}Fe_{30}Al_{10}$ and $Sm_{60}Fe_{30}Al_{10}$ has been observed by TMDSC. Our results show that despite small fraction of crystallization in $Y_{60}Fe_{30}Al_{10}$,, $Pr_{60}Fe_{30}Al_{10}$ and $Sm_{60}Fe_{30}Al_{10}$ alloys, majority part of these rare earth based metallic glasses have large undercooled region up to 200K.

REFERENCES

1. M. Reading, Trends in Polymers Sci. **1,** p. 248 (1993).
2. M. Reading, D. Elliott, and V.L. Hill, J. Thermal. Anal. **40**, p. 949 (1993).
3. A. Boller, C. Schich, and B. Wunderlich, Thermochimica Acta **266**, p.97 (1995).
4. T. Wagner, and S.O. Kasap, Phil. Mag. B. **74**, p.667 (1997).
5. Y. Li, S.C. Ng, Z.P. Lu, Y.P. Feng, and K. Lu, Phil. Mag. Lett. **78**, p.37 (1998).
6. Y. Li, S.C. Ng, Z.P. Lu, Y.P. Feng, and K. Lu, Phil. Mag. Lett. **78**, p.213 (1998).
7. Y. Li, S.C. Ng, C.K. Ong, H.H. Hng and T.T. Goh, Scripta Mater. **36**, p.783 (1997).
8. A. Inoue, T. Zhang, A. Takeuchi and W. Zhang, Mater. Trans. JIM **37**, p. 636 (1997).

FORMATION OF GLASSY SPINODALS IN BULK METALLIC GLASSES

W.H. GUO, C.C. LEUNG, and H.W. KUI
Department of Physics, The Chinese University of Hong Kong, Shatin, N.T., Hong Kong, China

ABSTRACT

When a eutectic alloy melt is undercooled well below its liquidus T_l, liquid state phase separation occurs. The liquid morphologies can be frozen by subsequent crystallization. It was found that for $Pd_{82}Si_{18}$ when rapid quenching is also employed, one of the phase-separated liquids after solidification is amorphous. The undercooled specimen is in bulk form with a diameter of ~ 5 mm.

INTRODUCTION

In a eutectic alloy, the like species prefer to stay together in the solid state. If the interaction between the same species and the different species remains qualitatively the same in the liquid state, when the temperature is sufficiently low, liquid state phase separation by both nucleation and growth and spinodal decomposition can occur. Experience [1-5] indicates that the liquid phase separation occurs most often at temperatures below the liquidus T_l of alloys. To facilitate the discussions below, the extent below T_l is called undercooling ΔT defined as $\Delta T = T_l - T$, where T is the temperature of the undercooled melt.

When the discussion is extended to include composition, a metastable liquid miscibility gap is expected. Hong et al. [5] found that for the eutectic Pd-Si alloy, the tip of the metastable miscibility gap locates more or less directly below the eutectic point.

The morphology of an undercooled melt that has undergone liquid phase separation by the nucleation and growth process is characterized by island-like structures if grain coarsening has not been serious [6]. Also, when the composition of the original homogeneous melt is close to the eutectic composition, after liquid phase separation, the liquidus of the phase-separated liquids would be higher than that of the original melt. In other words, all the liquid-components are metastable/undercooled and are subject to crystallization. It turns out that the subsequent crystallization can freeze the island-like morphology.

The morphology of an undercooled melt that has undergone liquid phase separation by the spinodal decomposition process can be described as network-like [7]. For the same reasons mentioned in the last paragraph, the split liquid networks/spinodals are metastable and the subsequent crystallization can freeze the connected structure if the composition of the original homogeneous melt is close to the eutectic composition.

Schneider et al. [8] attributed the observation of nanocrystals in $Zr_{41.2}Ti_{13.8}Cu_{12.5}Ni_{10}Be_{2.5}$ to a chemical decomposition process in the liquid state prior to crystallization. Liu and Johnson [9] used the same reason to explain the appearance of nanocrystals in bulk Mg-Cu-Y amorphous alloys. Lee and Kui [4] were able to prepare undercooled $Pd_{80}Si_{20}$ specimens that crystallized at different (T. Microstructural analysis indicates that there are two morphological transitions. The first one dictates the transition from a eutectic structure to island-like morphology (characteristic of nucleation growth process) while the microstructure change observed in the second one can very well be described as a

Mat. Res. Soc. Symp. Proc. Vol. 554 © 1999 Materials Research Society

transition from liquid nucleation and growth to liquid spinodal decomposition. Similar microstructural evolution is also observed in undercooled $Pd_{40.5}Ni_{40.5}P_{19}$ specimens [1, 2].

All the as prepared specimens in Ref. [1, 4] are crystalline. This is not unexpected for they were prepared either under isothermal conditions or by slow quenching rates. In this article, we report microstructures of undercooled $Pd_{82}Si_{18}$ alloys that had been subject to fast quenching, mostly by water quenching, after the occurrence of liquid phase decomposition.

EXPERIMENTAL

$Pd_{82}Si_{18}$ ingots were prepared from elemental Pd (99.99% pure) and Si (99.999% pure) granules. After the right proportion of Pd and Si were weighed, they were put into a clean fused silica tube. Alloying was brought about by rf induction heating under Ar atmosphere. All the as prepared specimens had a diameter of ~ 5 mm.

It had been demonstrated repeatedly that by a fluxing technique, molten metals could be undercooled way below their T_l. For instance, a $Pd_{40} Ni_{40}P_{20}$ [10] melt can be undercooled to its glass state bypassing crystallization with a cooling rate of ~ 0.75 K s^{-1} [11]. Similarly, a Ge melt [12] can be undercooled by as much as 342 K below its T_l. Earlier, the fluxing method had been applied to $Pd_{80}Si_{20}$ with anhydrous B_2O_3 as the fluxing agent [4]. A maximum undercooling of ~300 K was recorded. In this work, the fluxing technique was again employed and anhydrous boron oxide B_2O_3 was chosen as the fluxing agent. The preparation of anhydrous B_2O_3 is described in Ref. [10].

Since the experiments were carried out at elevated temperatures, it was necessary to employ a furnace. A Transtemp furnace [11] was chosen for this purpose, which is mainly composed of a heating coil surrounded by a gold-plated fused silica tube. The Au coating is thin (serves to trap heat from radiating away from the chamber of the furnace) so that a molten specimen inside the furnace could be observed directly by us. The furnace is connected to a personal computer that can regulate the power into the furnace so that heating/cooling cycles (including isothermal annealing) can be prescribed in advance.

In the experiment, a $Pd_{82}Si_{18}$ ingot and anhydrous B_2O_3 flux were put into a clean fused silica tube of vacuum 10^{-3} Torr. The whole system was then heated up by a torch to a temperature of ~1350 K. Prolonged high temperature heat treatment was applied for 4 h to facilitate the removal of heterophase impurities from the molten specimen. After the heat treatment, the whole system was inserted into the Transtemp furnace (of initial temperature 1173 K) sitting on a thermocouple which served to read out the temperature of the molten specimen. We waited for 20 min. for the establishment of thermal equilibrium between the furnace and the system. Next the temperature of the furnace was lowered down at a rate of 10 K min-1 to a temperature T_i which is below the liquidus of $Pd_{82}Si_{18}$(= ~ 1133 K). As soon as T_i was reached , it was changed to an isothermal condition and the molten specimen was annealed at T_i for 10 min. As soon as the thermal annealing was over, whole system was removed from the furnace and quenched in water. as prepared for clarity. The experimental procedure is illustrated graphically in Fig. 1.

Microstructures of the undercooled $Pd_{82}Si_{18}$

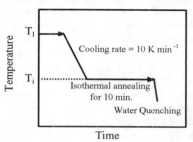

Fig.1 Experimental procedures taken in preparing phase-separated specimens.

specimens were mainly studied by scanning electron microscopy(SEM) and transmission electron microscopy (TEM) . Both SEM and TEM were equipped with EDX to carry out composition analysis. For SEM studies, the etching solution used was $HCl:HNO_3:H_2O = 5:1:3$.

RESULTS

The microstructures of undercooled $Pd_{100-x}Si_x$ specimens (x = 19, 20, 21, 22, 23) were studied earlier by Hong et al. [5]. It was found that for each value of x, there are three different reactions, namely, eutectic solidification, liquid phase nucleation and growth (followed by crystallization) and liquid phase spinodal decomposition (followed by crystallization), each prevailed over a different undercooling range.

In this work, the transition temperatures of undercooled $Pd_{82}Si_{18}$ alloy melt from eutectic crystallization to liquid phase nucleation and growth (denoted by T_{LNG}) and from liquid nucleation and growth to liquid phase spinodal decomposition (denoted T_{LSD}) are found to be $T_{LNG} \sim 883$ K and $T_{LSD} \sim 860$ K, respectively, in qualitative agreement with the work described in Ref. [5].

In the followings, the microstructures of three different undercooled specimens, one chosen from the regime of liquid nucleation and growth and the other two from the liquid spinodal decomposition regime are discussed in detail. It must be emphasized that each of these bulk specimens (diameter $\sim$ 5 mm) has its own characteristic microstructures. However, for the same undercooled specimen, its microstructure does not vary from one place to another.

Fig. 2a is a SEM micrograph displaying the microstructures of an undercooled specimen with $T_i = 873$ K. The microstructures are characterized by two kinds of island-like structures. The first type is heavily populated with dot-like particles (such an island is marked by H in the figure) while in the second kind, the density of particles is smaller (such an island is marked by L in the figure). To avoid confusion, lines are drawn to identify the boundaries that separate the two types of islands (Fig. 2b). In H, the compositions of the matrix and the dot-like particles are Pd_9Si_2 and Pd_3Si, respectively. In L, the compositions of the matrix and the particles are Pd_9Si_2 and Pd, respectively.

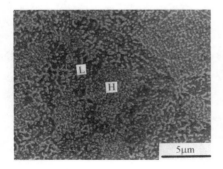 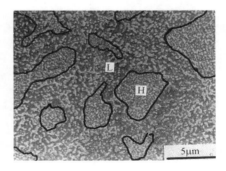

Fig. 2a A SEM micrograph showing the micro-structures of a specimen with T_i = 873 K. The undercooled specimen is occupied by two types of island- like structures.

Fig. 2b The boundaries between the two kinds of island-like structures shown in Fig. 2a are traced out for clarity.

The microstructures of an undercooled specimen with T_i = 853 K is shown in Fig. 3a. It depicts a network (made up of two subnetworks). Again, to avoid confusion, lines are drawn to identify the boundaries that separate the two subnetworks (Fig. 3b). Now it is obvious that one of the subnetworks denoted by H' has more dot-like particles embedded in it. The compositions of the matrix and the dot-like particles are Pd_9Si_2 and Pd_3Si, respectively. The compositions of the matrix and the dot-like particles in the other subnetworks denoted by L' are Pd_3Si and Pd,

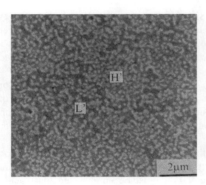

Fig. 3a The microstructures of a specimen with T_i = 853 K. It displays a connected structure made up of two subnetworks.

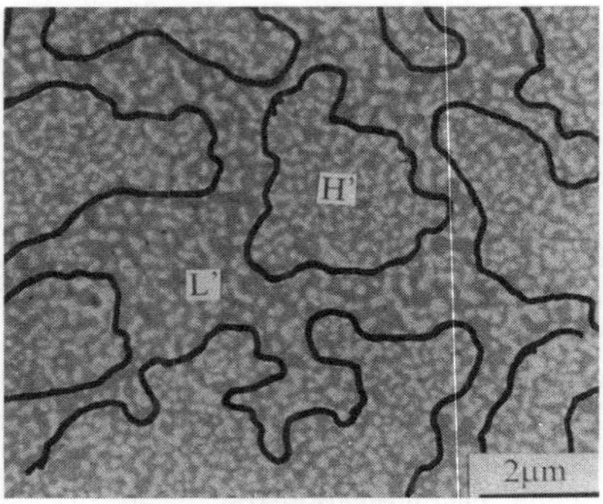

Fig. 3b The boundaries between the two sub-networks shown in Fig. 3a are traced out to be avoid confusion.

Fig. 4a A TEM micrograph showing the detailed microstructures of the region H' of the undercooled specimen shown in Fig. 3a.

Fig. 4b A TEM micrograph showing the detailed microstructures of the region L' of the undercooled specimen shown in Fig. 3a.

214

respectively. Therefore, the microstructures in H' are identical to that in H while the microstructures in L' are identical to that in L.

A TEM micrograph displaying the detailed microstructures of H' is shown in Fig. 4a. There are two phases. The areas covered with parallel straight lines have a composition of Pd_9Si_2. Those with smooth surfaces are Pd_3Si compounds. A TEM micrograph illustrating the detailed microstructures of L' is shown in Fig. 4b. Again the areas filled with parallel straight lines have a composition of Pd_9Si_2 while the parts with smooth surfaces are Pd crystals.

A TEM micrograph of an undercooled specimen with Ti = 813 K is shown in Fig. 5a. It depicts a two-phase connected morphology, similar to the one shown in Fig. 3b. The bright areas combine together to form a crystalline subnetwork The composition is close to Pd_4Si. The dark areas form the other subnetwork. An electron diffraction pattern taken from the dark area is shown in Fig. 5b. There are only halos indicating that this subnetwork is amorphous in nature. The average composition of the amorphous phase is $Pd_{84}Si_{16}$.

Fig. 5a   A TEM micrograph showing the microstructures of a specimen with T_1 = 813 K. A connected morphology is displayed.

Fig. 5b   An electron diffraction pattern taken from the area of the specimen shown in Fig. 5a.

DISCUSSIONS

According to Ref. [5], the solidification mechanism of the undercooled specimens shown in Fig. 2a and 2b can be postulated as follows: During the cooling cycle and the isothermal annealing, the original homogeneous liquid undergoes liquid phase separation by the nucleation and growth process. Since $Pd_{82}Si_{18}$ is a binary alloy, the phase-separated system consists of island-like structures of two different compositions,. As argued in the introduction section, both liquids would crystallize for they are metastable. Microstructural analysis indicates that eutectic crystallization prevails in both types of liquids. Moreover, the morphologies of the split liquids which also serves to preserve the morphologies of the split liquids. Similar arguments are applicable to the undercooled specimen shown in Fig. 3a that had undergone liquid phase spinodal decomposition.

Marcus and Turnbull [13] proposed that the glass forming ability (GFA) of a metallic alloy scales with its reduced glass transition temperature $T_{rg} = T_g/T_l$ where T_g is the glass transition temperature of the alloy. Indeed, experience indicates that those alloys with deep eutectics, i.e., smaller values of T_l, tend to form glass quite readily. In bulk metallic glass formation, it is always necessary to employ slow quenching rates. There is however one drawback for slow cooling. It allows time for liquid phase separation, esp. the liquid phase spinodal decomposition to take place. The transformation would reduce or even destroy the GFA of the system for it results in an increase of the reduced undercooling defined as $T_l = T/T_l$ of the liquid components as discussed in the introduction section. In the case of $Pd_{82}Si_{18}$, the bulk glass forming ability is indeed limited by the liquid phase separation for one of the split liquids crystallizes quite readily to a face-centered cubic Pd_4Si crystal.

Tanner and Ray [14] were able to prepare an alloy consisted of amorphous phase-separated components by rapid quenching technique. It is apparent that the quenching rate employed by them is just right for the occurrence of liquid phase separation and for the glass formation of the phase-separated liquid components. In this work, the experimental method developed enables us to have a better control of the microstructures.

CONCLUSIONS

An undercooled $Pd_{82}Si_{18}$ specimen consisted of two subnetworks, one crystalline and the other amorphous, is synthesized successfully in bulk form with a diameter of ~ 5 mm. The as observed morphology results from liquid phase spinodal decomposition, followed by crystallization. It is apparent that liquid phase separation reduces the glass forming ability of metallic alloys.

ACKNOWLEDGMENTS

We thank Hong Kong Research Grants Council for financial support.

REFERENCES

1. C.W. Yuen, K.L. Lee, and H.W. Kui, J. Mater. Res. 12, 314 (1997).
2. C.W. Yuen and H.W. Kui, J. Mater. Res., to appear in the Dec. 1998 issue.
3. C.W. Yuen and H.W. Kui, J. Mater. Res., to appear in the Dec. 1998 issue.
4. K.L. Lee and H.W. Kui, J. Mater. Res., accepted.
5. S.Y. Hong, W.H. Guo, and H.W. Kui, J. Mater. Res., accepted.
6. T.P. Seward, D.R. Uhlmann, and D. Turnbull, J. Am. Ceram. Soc. 51, 634 (1968).
7. J.W. Cahn, Trans. Met. Soc. AIME 242, 166 (1968).
8. S. Schneider, P. Thiyagarajan, and W.L. Johnson, Appl. Phys. Lett. 68, 493 (1996).
9. W. Liu and W.L. Johnson, J. Mater. Res. 11, 2388 (1996).
10. H.W. Kui, A.L. Greer, and D. Turnbull, Appl. Phys. Lett. 45, 615 (1984).
11. C.F. Lau and H.W. Kui, J. Appl. Phys. 73, 2599 (1993).
12. C.F. Lau and H.W. Kui, Acta Metall. et Mater. 39, 323 (1991).
13. M. Marcus and D. Turnbull, Mater. Sci. Eng. 23, 211 (1976).
14. L. Tanner and R. Ray, Scripta Metall. 14, 657 (1980).

EQUILIBRIUM THERMODYNAMICS NEAR THE GLASS TRANSITION - THE CONCEPTUAL APPLICATION OF THE LIMITING FICTIVE TEMPERATURE

G. WILDE, J.H. PEREPEZKO
University of Wisconsin - Madison, Department of Material Science and Engineering, Madison, 53706, WI, USA

ABSTRACT

The enthalpy, entropy, specific heat, specific volume and the equilibrium shear viscosity of the deeply undercooled melt of the bulk glass forming alloy $Pd_{40}Ni_{40}P_{20}$ have been determined as functions of temperature. The concept of limiting fictive temperature was applied to the entire set of measurements in order to allow for a valid comparison of the data based upon the respective equilibrium values. The comparison of the equilibrium properties shows that a proposed hierarchy of stability limits does not apply for this alloy. The results also indicate that the glass temperature as defined by the limiting fictive temperature does not depend on the property under observation.

INTRODUCTION

Thermodynamic quantities as well as properties related to the kinetics of a system like the shear viscosity, η become time dependent when the glass transition is approached. Nevertheless, metastable thermodynamic equilibrium can be approached in principle by relaxation at temperatures higher than the isentropic (Kauzmann-) temperature T_K [1]. Thus, the determination of equilibrium properties of the undercooled melt within the glass transition region requires a sample in a fully relaxed state during the entire measurement.

The concept of the limiting fictive temperature, introduced by Tool and Eichlin in 1931 [2], offers the possibility to determine reliable values for the equilibrium properties in the temperature range of the glass transition and allows for the extension of the accessible measurement interval towards temperatures close to T_K. The construction to determine the limiting fictive temperature of a sample can be expressed in terms of the enthalpy, H, as follows:

$$\Delta H^{a,b} = \int_{T_g^b}^{T_g^a} \left(C_p^l - C_p^{gl} \right) dT \tag{1}$$

with $\Delta H^{a,b}$ the enthalpy difference between two final glassy states obtained by different cooling rates or by different annealing treatments, respectively. The specific heats of the undercooled liquid, C_p^l, and the glass, C_p^{gl}, are extrapolated from the equilibrium liquid and the glassy state at lower temperatures which is on the experimental time scale stable with respect to relaxation. Even though this concept is well established, it has received only limited use for metallic glasses. Therefore, the aim of the present paper is to describe and compare the results of calorimetry, dilatometry and rheology applied to glassy and undercooled liquid bulk metallic samples, which have been obtained in compliance with this thermodynamic concept. For this study it is important to note, that the limiting fictive temperature is identical to the annealing temperature when the metastable equilibrium has been attained. Thus, C_p^l can be determined indirectly on equilibrated samples by comparing eq. (1) with this precondition [3].

217

EXPERIMENTAL DETAILS

$Pd_{40}Ni_{40}P_{20}$ samples were prepared by alloying premelted Ni_2P samples (99.5 % pure) with a Pd ingot (99.9 % pure) in a resistance furnace under argon flow. Glassy rods of 3.5 mm in diameter and up to 35 mm in length were produced by either flow casting or die casting in argon atmosphere. For density measurements the entire rod of about 3 g in weight was used. Small discs of about 100 mg were cut for the relaxation experiments and subsequent DSC measurements (Perkin Elmer DSC 2-C). The samples used for the creep measurements (Perkin Elmer DMA-7) were strip-shaped (0.5 mm x 0.5 mm) at a length of 10 mm. For these measurements, glassy samples of the dimensions 35 mm x 35 mm x 1.3 mm were produced by die casting. The momentary sample length, l, was measured for each specimen at a constant tensile stress level $\sigma_\perp = 2.16$ MPa as a function of time, t. The corresponding shear viscosity:

$$\eta = \frac{\sigma_\perp}{3 \cdot d\varepsilon/dt} \tag{2}$$

then is a function of the applied stress and the elongation rate $d\varepsilon/dt = l^{-1}dl/dt$, which can be taken directly from the isothermal creep data.

Measurements on the coefficient of thermal expansion (Perkin Elmer DMA-7) were performed on cylindrical samples of 3.5 mm in diameter and 10 mm in length. All samples were checked for the absence of crystallinity by X-ray diffraction prior to and after thermal aging. The DSC and DMA measurements and the applied calibration procedures are described in detail elsewhere [4,5].

Amorphous samples were relaxed in the temperature range between 537 K and 561 K using a high-precision isothermal vacuum furnace. Following each annealing segment, the apparent specific heat of the sample was measured. Equilibration at each temperature was defined to be complete if aging for twice the time changed the apparent specific heat by less than 1%. The annealing times sufficient to reach the metastable enthalpic equilibrium at the different temperatures are summarized in table I. It should be noted, that because of the method of doubling the annealing times until no further relaxation was observed, only sufficient times for equilibration could be determined. Thus, the actual average relaxation times will be shorter than the sufficient aging times.

Table I: Relaxation times to achieve metastable equilibrium based upon double annealing periods.

T [K]	537	540	545	553	558	560	565
t [10^3 s]	1900.8	1058.4	482.6	86.4	14.4	7.2	1.8

The density of structurally relaxed samples was determined at room temperature by the fluid displacement method in ethanol and in water using a microbalance (Sartorius 2504) with a resolution of ± 5 µg. During the weight measurement, the samples were kept in a Pt-basket which was constructed such that its weight amounted to only 1/30 of the weight of the sample. The measured absolute values were always checked by calibration measurements on a 2.5 g sapphire rod with a density of 3.9850 g/cm^3. The accuracy in the density measurements was determined to be about ± 0.0005 g/cm^3. From the experimental values of the room temperature density $\rho(T_R)$ and the coefficient of thermal expansion $\beta(T)$ which for isotropic materials like glasses equals three times the linear thermal expansion coefficient $\alpha(T)$, the specific volume $V(T)$ of the undercooled liquid at $T_A = T_g$ can be determined as:

$$V(T_g) = \frac{1}{\rho(T_R)} \cdot [1 + 3 \cdot \alpha(T) \cdot (T_g - T_R)] \tag{3}$$

By this method, specific volumes at different limiting fictive temperatures can be determined accurately and without any interference from macroscopic flow of the samples.

RESULTS

The results of the indirect C_p-determination using the concept of the limiting fictive temperature are shown in Fig. 1 together with the data obtained by dynamic calorimetry [6]. After a relaxation period of 22 days, the lowest value of T_g was reached at 537 K. In order to obtain a limiting fictive temperature of 537 K by isochronal cooling, a cooling rate of approximately 2×10^{-3} K/min would have to be applied. This rate is too low for calorimetric measurements.

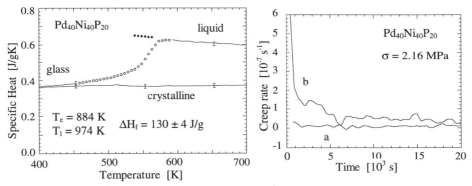

Fig. 1: The specific heat of $Pd_{40}Ni_{40}P_{20}$ obtained by direct calorimetry (o) and by enthalpy measurements on relaxed samples (♦).

Fig. 2: Creep rate at 540 K in dependence on time of samples, which have been equilibrated at 540 K (a) and 552 K (b).

The indirectly determined specific heat of the deeply undercooled liquid follows the course of a linear extrapolation of the data determined by direct calorimetry. Using the measured values of C_p^l, the specific heat of the crystalline state C_p^x and the melting enthalpy ΔH_f, [6] the enthalpy- and entropy difference between undercooled melt and crystal, consisting of the phases Ni-Pd, Ni_5P_2 and Pd_3P, were calculated.

Creep measurements have also been performed on equilibrated samples. For comparison, analogous experiments under equal experimental conditions were conducted on samples which had been equilibrated at higher and lower temperatures with respect to the temperature of the creep experiment. Figure 2 shows the time dependence of the elongation rate of two samples at T = 540 K. Sample (a) was equilibrated at 540 K, the temperature of the measurement. In accordance with the definition of the limiting fictive temperature, the elongation rate and thus the viscosity of sample (a) is constant. In contrast to this result, the sample (b), which has been equilibrated at 552 K, shows a pronounced time dependence of the elongation rate. Thus, the transient behavior of curve (b) reflects the *in situ* relaxation towards the metastable equilibrium state corresponding to a temperature of 540 K. Figure 2 shows that this relaxation is not completed after 20×10^3 s. Thus, equilibrium viscosity data according to eq. (2) were determined by isothermal creep measurements at the respective limiting fictive temperatures of the samples.

The linear thermal expansion coefficients of crystalline and of glassy alloy rods were measured in accordance with literature [7] as $\alpha_x = 13 \times 10^{-6}$ K^{-1} and $\alpha_g = 17 \times 10^{-6}$ K^{-1} in the glass transition region. Annealing of the glassy rods at different temperatures prior to the measurements did not result in any measurable changes of α_g. The density of large glassy samples of about 3 g in weight which had been annealed at different temperatures according to the enthalpy equilibration times, was measured at ambient temperature. Figure 3 shows the result on a Pd$_{40}$Ni$_{40}$P$_{20}$ sample. The numbers indicate the sequence of the annealing treatments. Equal densities e.g. at points (1) and (3) have been obtained after annealing samples of initially different thermal history. The specific volume of the alloy in the different states measured at ambient temperature can be determined as a function of the limiting fictive temperature by the use of eq. (3).

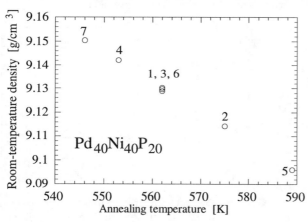

Fig. 3: Room-temperature density of samples in dependence on the equilibration temperature. The numbers indicate the sequence of annealing treatments. Three independent measurements have been performed at 562 K (1, 3, 6) yielding equal values for the density.

DISCUSSION

Equilibration times have been determined, which are sufficient to reach the enthalpic equilibrium for specific temperatures in the glass transition range of Pd$_{40}$Ni$_{40}$P$_{20}$. Creep measurements on samples, which have been annealed for equal time intervals at these temperatures revealed that the thermal treatment was also sufficient to attain the metastable equilibrium of the undercooled melt with respect to viscous flow. The measurements of the room temperature density of samples with different limiting fictive temperatures showed that reproducible values were obtained, although the initial state of the samples was altered by intermediate annealing treatments at different temperatures. The observed reproducibility of the density data indicates that metastable equilibrium has been attained during the initial annealing treatment i.e. within the same time interval as for the enthalpy and the shear viscosity. Otherwise, if the metastable equilibrium has not been achieved after step (1) in Fig. 3, annealing at T = 551 K (step(3)) after equilibration at 563 K (step (2)) would have resulted in an increase of the density towards equilibrium.

By definition, the limiting fictive temperature is independent of the thermal pathway by which the corresponding relaxational state is attained. Thus, the achievement of corresponding relaxational states with respect to different material properties within similar time intervals is equivalent to the achievement of equal limiting fictive temperatures on cooling with similar rates. Therefore, and in contrast to previous findings [9], the glass temperature defined as the limiting fictive temperature of the sample is independent of the property under study and depends only upon its thermal history. Obviously, this result only holds for the metastable equilibrium. Non-equilibrium effects like e.g. cross-over behavior [10] cannot be described by this method evolving from equilibrium thermodynamics.

The density measurements additionally show the importance of an accurate description of the thermal history of a glassy sample, since many physical properties are related to the specific volume [11]. Thus, giving numerical values for equilibrium properties obtained on "well relaxed" samples does not enhance the comparability of the data unless the annealing time and temperature and the initial state of the glassy sample are specified as well.

The normalized functions of ΔS, ΔH and ΔV as obtained by calorimetry, thermomechanical analysis and the density measurements are summarized in Fig. 4.

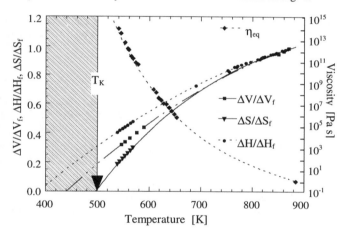

Fig. 4: Excess quantities of the enthalpy, entropy and the specific volume normalized to the values at the melting temperature. The diamonds indicate measured values for the equilibrium viscosity. The dashed curve represents the fit according to the free-volume model [8].

Following the arguments of Tallon [12], at least in the case of systems with a close-packed crystalline phase it is expected that the isochoric temperature $T(\Delta V = 0)$ should be higher than the isentropic temperature $T(\Delta S = 0)$ and the isenthalpic temperature $T(\Delta H = 0)$. Yet, this sequence of so called thermodynamic stability limits was derived under the assumption of a constant amount of free volume persistent in the liquid phase. It is uncertain if a model derived from such an assumption can be applied to glass forming systems. Obviously, this relation is not applicable in the case of the investigated bulk glass-forming alloy. From experimental data the sequence of stability limits for $Pd_{40}Ni_{40}P_{20}$ is: $T(\Delta H = 0) < T(\Delta V = 0) < T(\Delta S = 0)$.

Figure 4 additionally shows the equilibrium shear viscosity determined by isothermal creep measurements together with data obtained by parallel-plate rheometry [13] and by a capillary flow method [14]. The course of the viscosity follows a continuous increase with decreasing temperature. This result confirms a previous remark on the characterization of the glass transition [15] which states that the transition itself becomes imperceptible in measurements of equilibrium properties. The result also demonstrates that the arbitrary definition of the glass temperature as the temperature at which the viscosity amounts to $\eta_{eq} = 10^{12}$ Pa $\cdot$s is not applicable in the case of isothermal measurements performed on a time scale of the order of the sample relaxation time.

CONCLUSIONS

The rigorous application of the concept of the limiting fictive temperature for the characterization of the glass temperature has led to a reliable determination of the heat capacity, density and the viscosity. During these measurements effects due to the glass transition itself have been omitted by preequilibrating the samples at different temperatures. Thus, the temperature range accessible for determining equilibrium properties of the undercooled melt has been extended considerably. The equilibrium thermophysical property functions as well as their slopes are continuous and monotonous. Relaxation experiments on long time scales have shown, that samples which have been equilibrated with respect to the enthalpy equally correspond to the metastable equilibrium of the undercooled liquid state with respect to volume or shear viscosity. This result shows that the glass temperature is not dependent on the property under study given that the glass temperature is defined as the limiting fictive temperature of the sample.

The comparison of the excess thermodynamic quantities has shown that in the limit of slow cooling, the entropy difference between undercooled liquid and crystal would vanish first which indicates that the Kauzmann temperature sets a lower limit to the existence range of the liquid state of the alloy under study.

ACKNOWLEDGMENTS

The authors gratefully acknowledge the support by the Alexander von Humboldt-Foundation via the Feodor-Lynen-Program (G.W., V-2.FLF-DEU/1052606) and ARO (DAAG 55-97-1-0261).

REFERENCES

[1]   W. Kauzmann, Chem. Rev., **43** (1948) 219.
[2]   A.Q. Tool and C.G. Eichlin, J. Amer. Ceram. Soc., **14** (1931) 276.
[3]   C. Mitsch, G.P. Görler, G. Wilde and R. Willnecker, to be published
[4]   G. Wilde, G.P. Goerler, R. Willnecker, S. Klose and H.J. Fecht, Mater. Sci. Forum, **225-227** (1996) 101.
[5]   G. Wilde, S.G. Klose, W. Soellner, G.P. Görler, K. Jeropoulos, R. Willnecker and H.J. Fecht, Mat. Sci. Eng. A, **226-228** (1997) 434.
[6]   G. Wilde, G.P. Görler, R. Willnecker and G. Dietz, Appl. Phys. Lett., **65** (1994) 397.
[7]   H.S. Chen, J.T. Krause and E.A. Sigety, J. Non-Cryst. Solids, **13** (1973/74) 321.
[8]   M.H. Cohen and G.S. Grest, Phys. Rev. B, **20** (1979) 1077.
[9]   H.R. Sinning und F. Haessner, Mat. Sci. Eng., **97** (1988) 453.
[10]   C.A. Volkert und F. Spaepen, Acta Met., **37** (1989) 1355.
[11]   A. Bondi, *Physical Properties of Molecular Crystals, Liquids and Glasses*, Wiley, New York, 1968, pp. 370-404.
[12]   J.L. Tallon, Nature, **342** (1989) 658.
[13]   G. Wilde, G.P. Görler, K. Jeropoulos, R. Willnecker and H.J. Fecht, Mater. Sci. Forum, **269-272** (1998) 541.
[14]   K.H. Tsang, S.K. Lee and H.W. Kui, J. Appl. Phys., **70** (1991) 4837.
[15]   C.T. Moynihan, Rev. in Mineralogy, **32** (1995) 1.

VISCOSITY, RELAXATION AND CRYSTALLIZATION KINETICS IN ZR-TI-CU-NI-BE STRONG BULK METALLIC GLASS FORMING LIQUIDS

Ralf Busch*, Andreas Masuhr, Eric Bakke, T. Andy Waniuk, and William L. Johnson

Keck Laboratory of Engineering Materials, California Institute of Technology, Pasadena, CA 91125

ABSTRACT

The high thermal stability of bulk metallic glass (BMG) forming liquids in the undercooled state allows for measurements of thermophysical properties in a large time and temperature window. In this contribution, results on viscous flow, relaxation and crystallization of Zr-Ti-Cu-Ni-Be BMG forming alloys are presented. The data are compared with the kinetics of other metallic and non-metallic liquids. BMG formers are relatively strong liquids with melt viscosities that are about three orders of magnitude larger than in pure metals and other alloys. The strong liquid behavior of these alloys is also reflected by a small entropy of fusion and a weak temperature dependence of the thermodynamic functions upon undercooling. The high viscosity and small driving force for crystallization are major contributing factors to the high glass forming ability and low critical cooling rate. The upper portions of experimental time-temperature-transformation diagrams down to the crystallization nose can be described well using the kinetics deduced from the viscosity data. For lower temperature the viscosity can not describe the crystallization kinetics. The time scale for structural relaxation becomes larger than for diffusive hopping processes. Diffusion stays relatively fast, whereas viscosity and structural relaxation time upon undercooling follow a Vogel-Fulcher-Tammann relation.

INTRODUCTION

Viscosity is an important parameter to describe the kinetic slowdown when a melt is undercooled below its liquidus temperature. The increase of viscosity with undercooling reflects the increasingly longer time scale for structural rearrangements in the supercooled liquid state. If the viscosity reaches a value of approx. 10^{12} Pa·s upon undercooling, the intrinsic time scale for maintaining metastable equilibrium becomes comparable to the laboratory time scale, i.e. the cooling time. The liquid freezes to a glass, namely it vitrifies.

Viscosities of metallic supercooled liquids close to the glass transition have been previously measured for example by Chen and Turnbull [1], Tsao and Spaepen [2] and others [3,4]. The viscosities were determined in the glass transition region and crystallization did not allow measurements of the equilibrium viscosity below 10^9 Pa s or for long times to eliminate relaxation effects.

In recent years multicomponent alloy systems were found that are much more robust with respect to crystallization and form bulk metallic glass. Examples are the La-Al-Ni [5], the Zr-Al-Ni-Cu [6] and the Zr-Ti-Cu-Ni-Be [7] alloy systems of which the latter one is by far the best bulk metallic glass former with critical cooling rates as low as 1 K/s [8]. Amorphous ingots typically up to 50 mm in the smallest dimension can be produced.

---

* Present and permanent address: Department of Mechnical Engineering, Oregon State University, Rogers Hall 204, Corvallis, OR 97331

Mat. Res. Soc. Symp. Proc. Vol. 554 © 1999 Materials Research Society

The high resistance with respect to crystallization allows for measurements of the thermophysical properties of these undercooled metallic liquids in a broad time and temperature range. For example, results have been obtained for the specific heat capacity [9], diffusion coefficients [10] or emissivity [11].

In this contribution we present viscosity measurements performed on these metallic glasses using parallel plate rheometry and three-point beam bending and rotating cup viscosimetry. I addition the relaxation kinetics in the glass transition region and the influence of the kinetics on the glass forming ability and crystallization is discussed.

EXPERIMENTAL METHODS

Amorphous $Zr_{41}Ti_{14}Cu_{12.5}Ni_{10.0}Be_{22.5}$ (Vit1) and $Zr_{47}Ti_8Cu_{7.5}Ni_{10}Be_{27.5}$ (Vit4) alloys were prepared from the mixture of the elements of purity ranging from 99.5% to 99.9% by induction melting and subsequent water quenching in 6.35 and 10 mm inner diameter silica tubes. Cylindrical samples were cut from the 6.35 mm rods for use in parallel plate rheometry experiments. Beams of rectangular cross section were cut from the 10 mm rods for beam bending.

Parallel plate rheometry as described by Stephan [12] and Diennes and Klemm [13] was used to study the viscosity of Vit4 between 10^9 and 10^5 Pa s as a function of temperature. The experimental apparatus used for both experiments was a Perkin-Elmer Thermal Mechanical Analyser (TMA 7). A disc of the material flows in between two 3.7 mm diameter quartz penetration probes that were used as parallel plates. Measurements were performed with different heating rates and also isothermally using samples of different initial height. By measuring the height of the sample versus time, the viscosity at any temperature is given by the Stephan equation

$$\eta = -\frac{2Fh^3}{3\pi a^4 \dfrac{dh}{dt}} \tag{1},$$

where F is the applied load, a, is the radius of the plates, and, h, is the height of the sample. Since Eq. (1) was derived by neglecting the velocity normal to the plates, the measured value of the viscosity [i.e., the value calculated according to Eq. (1)] depends on the aspect ratio between the sample height and the radius of the sample. The equilibrium viscosity can be determined by measuring the viscosity for different aspect ratios x=h/a between height, h, and radius, a, of the plates and fitting the data with the parabolic function $\eta=b+cx^2$ (2). The value of this function that corresponds to an aspect ratio of zero is the true viscosity (see also [14]).

Three-point beam bending is another method that was applied to measure viscosity of Vit1 and Vit4. With this technique viscosities in the range from 10^7 to 10^{14} Pa s can be measured. A beam that is supported at the ends by sharp edges is deflected with a constant force applied to the center of the beam. From the deflection rate the viscosity is determined by the equation [16-19]

$$\eta = -\frac{gL^3}{2.4I_c v}\left[M+\frac{\rho AL}{1.6}\right] \tag{3},$$

where g is the gravitational constant (m/s²), I_c the cross-section moment of inertia (m⁴) [20], v the midpoint deflection rate (m/s), M the applied load (kg), ρ the density of the glass (kg/m³), A, the cross-sectional area (m²), and L the support span ($5.08 \cdot 10^{-3}$m for our apparatus). In the case of three point beam bending the measured apparent viscosity is independent of the geometry of the beam. Beams with square and rectangular cross sections were used and gave consistent results.

In addition, we have designed a Couette viscometer to measure the rheological properties of Zr-based glass forming alloys in the range from 10^{-1} Pa s to 10^3 Pa s. The viscometer's concentric cylinder shear cell is machined from graphite and designed to hold about 5 cm^3 of the liquid alloy. The cell is mounted vertically inside a vacuum induction furnace with the outer cylinder attached to a static torque sensor. Ingots of ca. 30 g of Vit1 are prepared by melting the elements in an arc-melting device under argon. The ingots are placed inside the shear cell and melted at 1200 K. Subsequently, the inner rotatable cylinder was inserted into the melt. We measure the temperature, T, with a type K thermocouple mounted inside the crucible wall of the outer cylinder. The torque and temperature signals are digitally processed and are linked to the temperature controller of the induction furnace. The liquid alloy wets the graphite and allows a straightforward calculation of the flow field between the cylinder [21]. A more detailed description of the experiment can be found elsewhere [22].

Isothermal measurements in the equilibrium liquid are performed by both, clockwise and counterclockwise rotation of the inner cylinder to improve the resolution of the apparatus. The static torque on the outer cylinder is proportional to the Newtonian viscosity, η, where the proportionality constant can be calculated from the angular frequency of the inner cylinder and the geometry of the shear cell [23]. We find the flow to be Newtonian for strain rates from 0.4 s^{-1} to 4 s^{-1} used in this study. Additional viscosity measurements of the supercooled liquid were performed with a cooling rate of 0.7 K s^{-1} at constant shear rate. The graphite containers do not alter the times to nucleation as compared to electrostatic levitation experiments [24].

Calorimetric measurements were performed in a Perkin Elmer DSC7 and a Seteram DSC 2000K using different heating rates and isothermal experiments. The temperature scale of the DSC's and the TMA were calibrated for each heating rate using the melting points of indium and zinc. In order to calculate the thermodynamic functions in the supercooled liquid, heats of fusion, heats of crystallization and specific heat capacities, c_p, were determined experimentally, as shown in ref. 9 for the $Zr_{41.2}Ti_{13.8}Cu_{12.5}Ni_{10.0}Be_{22.5}$ alloy. The enthalpy recovery after isothermal annealing below the calorimetrically observed glass transition was measured in experiments with constant heating rate in the DSC.

RESULTS

Viscosity measurements

The viscosity measurements were performed either with constant heating rate or in isothermal experiments. Figure 1 shows a viscosity measurement by parallel plate rheometry performed at a heating rate of 0.833 K/s and a force of 2.6 N. The viscosity decreases with increasing temperature. At 743 K and a viscosity of $6.0 \cdot 10^5$ Pa s the material starts to crystallize resulting in a rising apparent viscosity. Below about 680 K the sample is subject to relaxation processes due to the glass transition. Figure 2 summarizes the apparent viscosities measured by parallel plate rheometry at a temperature of 703 K. The flow between the plates was measured with different heating rates, isothermally and at different aspect ratios. The data were fitted with the parabolic eq. (2) yielding an equilibrium value of $4.1 \cdot 10^6$ Pa s. Equilibrium viscosity data were obtained by parallel plate rheometry in a viscosity range between 10^5 and 10^9 Pa s (see Fig.7). Viscosities in this range have not been measured in metallic supercooled liquids previously.

In order to measure the viscosity from 10^9 to 10^{14} Pa s, three-point beam bending was used. Typical three point beam bending data are shown in Fig.3. Both beams had a rectangular

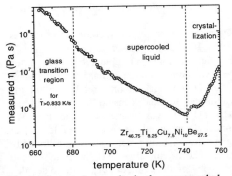

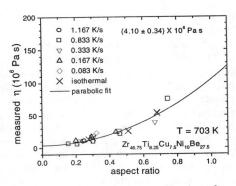

Fig.1: Measured viscosity in the supercooled liquid for a heating rate of 0.833 K/s.

Fig.2:Measured viscosity as a function of aspect ratio. Different heating rates and isothermal experiments are used.

cross section of about 0.7 by 0.35 mm. A force of 0.05 N and the respective heating rates of 0.033 K/s and 0.0833 K/s were used. Additionally shown is the equilibrium viscosity. The measured viscosities decrease with increasing temperature. The measured viscosity is found to be independent of force and cross section, provided the height of the beam is less than 1/5 of the span of the support knife edges.

Typical torque readings from a concentric cylinder shear cell containing liquid Vit1 are shown in Fig.4. The inner cylinder is alternately rotated in clockwise and counterclockwise direction for 60 s each. In this way, the determination of the torque baseline is straightforward and the resolution of the viscosity measurement is doubled compared to simple continuous motion. The torque meter signal (mV) is converted via a calibration curve into a torque reading (Nm). The analysis of the flow field between the two cylinders yields the viscosity, η, as a function of the temperature. The decrease in the torque signal with increasing temperature arises from the decrease in viscosity of the liquid alloy.

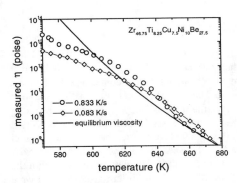

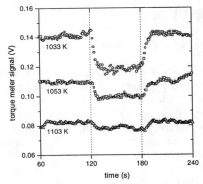

Fig.3: Non-equilibrium viscosity in the glass transition region measured by three-point beam bending for two different heating rates.

Fig.4:Isothermal torque reading at constant shear rate.Clockwise and counter clockwise rotation of the inner cylinder for 60s results in alternating torque signal.

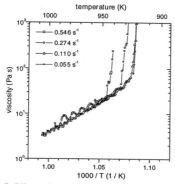

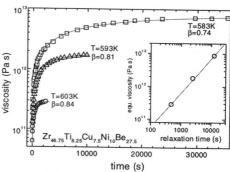

Fig.5: Viscosity measurement upon linear
cooling with a rate of 0.66 K/s. The
average shear rates are indicated.

Fig.6: Isothermal measurements of the
viscosity using three-point bending
and stretched exponential fits [Eq.(4)].

The viscosity of the supercooled liquid state of Vit1 was also measured upon cooling into the supercooled liquid state. The measured viscosity as a function of the temperature for various shear rates is plotted in Fig.5. No dependence of the viscosity on shear rate is observed indicating Newtonian flow behavior for the present temperature range. The sharp increase in the viscosity signal represents the onset of solidification.

Viscosity and enthalpy relaxation

In order to measure the equilibrium viscosity for low temperatures one has to use small heating rates or isothermal measurements. In the case of the beam bending experiments the samples were heated to a certain temperature at a rate of 0.833 K/s and the isothermal change of viscosity was monitored. In Fig. 6 three examples for the isothermal relaxation from the amorphous state into the supercooled liquid state are shown. The relaxation processes in the glass transition region are usually not found to follow an exponential law but obey a stretched exponential function [25]. The viscosity relaxation was fitted with a stretched exponential relaxation function

$$\eta(t) = \eta_a + \eta_{eq-a} \cdot (1 - e^{-(t/\tau_\eta)^\beta})$$ (4),

in which, τ_η, is an average internal relaxation time, β, a stretching exponent, t, the time and, η_a, the viscosity of the amorphous alloy before relaxation. η_{eq-a} is the total viscosity change during relaxation from the amorphous state into the equilibrium state. The fitted equilibrium viscosities reach values that are in good agreement with the extrapolation of the viscosities measured at higher temperatures by parallel plate rheometry as shown in Fig. 7.

In Fig.7 beam bending data measured with a constant heating rate of 0.833 K/s (◇) and viscosity values, after equilibrating the material in isothermal beam bending experiments (●), are depicted. Furthermore, the equilibrium viscosity data obtained by parallel plate rheometry (○) are included.

The viscosity is compared with the calorimetric glass transition that is studied in a DSC with the same respective heating rate as the viscosity measurement. A DSC scan is included in Fig.7. We can distinguish three different regions. For the heating rate of 0.833 K/min the material is amorphous below 623 K. Between 623 and 673 K it undergoes the calorimetric glass transition indicated by the gradual increase of the specific heat capacity (endothermic heat flow) in the DSC scan. Above 650K an overshooting is observed due to an enthalpy relaxation that is caused by the thermal history of the sample [26]. Beyond 673 K the material is a supercooled liquid.

The measured viscosity upon heating with constant rate undergoes the transition from the amorphous state to the supercooled liquid state in the same temperature range where the calorimetric glass transition is observed. Below 623 K the viscosity stays smaller than the equilibrium viscosity because of frozen-in free volume in the amorphous state. Above 673 K the measured viscosity corresponds to the equilibrium viscosity.

In general there is not one well-determined calorimetric glass transition temperature but a temperature interval in which the glass transition occurs. The location of this interval on the temperature axis depends on the heating or cooling rate. We measured the calorimetric glass transition in the DSC with different heating rates in the range between $8.33 \cdot 10^{-3}$ K/s and 5 K/s. The shift of the glass transition region with heating rate is shown in Fig.8. For each heating rate three temperatures are drawn, the onset (O), the point of the steepest ascent (△), and the temperature where the supercooled liquid is reached (□) (see also Fig.7). With increasing heating rate the glass transition is shifted to higher temperatures, because a shorter intrinsic relaxation time of the material is needed to reach the metastable equilibrium state of the supercooled liquid. In Fig.8 the temperatures are included where

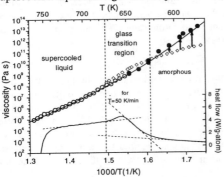
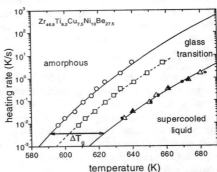

Fig.7: Viscosity in the glass transition region measured by beam bending with a constant rate (diamonds) and equilibrium viscosities (circles).

Fig.8: Temperature range in which the glass transition occurs as a function of heating rate measured in a DSC.

the measured viscosity reached the equilibrium value for each of the used heating rates (●). These temperatures are in good agreement with the values obtained from the DSC scans, where the specific heat capacity of the supercooled liquid is reached, indicating that the viscosity relaxation time equals the enthalpy relaxation time.

Enthalpy recovery

To investigate the relaxation of the enthalpy that accompanies the viscosity relaxation at low temperatures, alloys were heat treated isothermally at different temperatures below the calorimetrically observed glass transition. The annealing times were chosen to be ten times the relaxation time that was determined during viscosity relaxation (e.g. $1.2 \cdot 10^5$ s at 583 K). This guarantees that the samples reach states that are very close to metastable equilibrium. After the heat treatment the specimens were first cooled to room temperature with a rate of 3.33 K/s. Subsequently a DSC experiment with a constant heating rate of 0.083 K/s was performed.

Figure 9 shows the results of these enthalpy recovery experiments after annealing at four different temperatures as well as for an unrelaxed sample that was heated above the glass transition and cooled at the same rate (0.083 K/s) prior to the experiment. The annealed samples show a large endothermic heat recovery in the calorimetric glass transition region, whereas the unrelaxed reference sample does not exhibit this effect.

Van den Beukel and co-workers [27] observed enthalpy recovery in metallic glasses previously in the Ni-Pd-P system. They developed a model that describes the functional form of the DSC curves based on free volume theory. In the present study we use the enthalpy recovery to determine the change of the enthalpic state of the sample that occurred during the preceding relaxation. Enthalpy recovery increases with decreasing annealing temperature. The amount of enthalpy that was released during the isothermal heat treatment and that was recovered during reheating the sample, is the area between the curve of the relaxed sample and the unrelaxed material. In Fig.10 the measured recovered heats are plotted into an enthalpy diagram. The enthalpy, ΔH, of the supercooled liquid in reference to the crystalline state (solid line) was determined independently by DSC measurements [9].The enthalpy differences measured during enthalpy recovery are plotted, starting from the enthalpy of the supercooled liquid state in

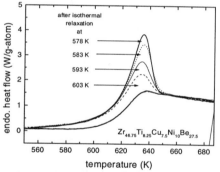
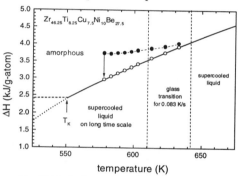

Fig.9: Enthalpy recovery experiments in the glass transition region after isothermal relaxation from the amorphous into the supercooled liquid state.

Fig.10:Enthalpy difference, ΔH, with respect to the crystalline mixture for the super-cooled liquid. The enthalpy changes from the frozen-in amorphous state into the supercooled liquid are added.

positive ΔH direction. The result is the enthalpy curve (solid circles) of the frozen-in amorphous alloy that was initially formed by cooling all samples with 0.083 K/s. We observe that the enthalpy of the amorphous alloy only increases slightly with temperature. This is due to the small specific heat capacity difference between the crystal and the amorphous alloy (see Fig.11). The measured enthalpy change with temperature can be attributed to a finite specific heat capacity difference between the amorphous alloy and the supercooled liquid. This specific heat capacity difference between supercooled liquid (sl) and amorphous (am) state at a temperature $T_1+(T_2-T_1)/2$ is calculated as

$$\Delta c_p^{sl-am} = (\Delta H^{sl-am}(T_1) - \Delta H^{sl-am}(T_2))/ (T_2-T_1) \quad (5),$$

where $\Delta H^{sl-am}(T_i)$ are the recovered enthalpies after annealing at the temperatures T_1 and T_2, respectively. The absolute specific heat capacity $c_p^{sl} = c_p^{am} + \Delta c_p^{sl-am}$ for the supercooled liquid on a long time scale (◇) is shown in Fig.11 in comparison with other c_p data of the crystal (□), the amorphous state (△), and the supercooled liquid at higher temperature (○). The data for the supercooled liquid on a long time scale are in good agreement with the extrapolation of the specific heat capacity data of the supercooled liquid obtained at higher temperatures.

It is important to note that the high c_p in the supercooled liquid at low temperatures can not be measured in the laboratory directly. A c_p measurement always involves a temperature change. The time to perform this temperature change (seconds) is always much shorter than the enthalpy relaxation time (days) in the studied temperature range. Therefore, the enthalpic state of the alloy will only change by a fraction of what is expected after full relaxation. Thus, the measured c_p only accounts for the amount of enthalpy that relaxed during this time period. In fact, when measured on laboratory time scale, the supercooled liquid exhibits an apparent c_p equal to that of the amorphous alloy [see (◆) Fig.11]. If c_p experiments could be performed on a much longer time scale (e.g. geological time scale) the c_p of the supercooled liquid could be measured at lower temperatures directly. In this sense, the observed location of the calorimetric glass transition on the temperature axis is merely a result of our very subjective feeling for time.

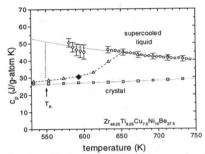

Fig.11: Specific heat capacity of the supercooled liquid (○), the crystal (□), and the glass(△). cp of th supercooled liquid on long time scale was obtained from the enthalpy recovery experiments (◇).

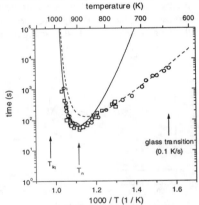

Fig.12: Time-temperature-transformation diagram for the primary crystallization of Vit 1 (○). Data from electrostatic levitation are included (□) [28].

Crystallization

We studied the temperature dependence of the onset of crystallization under isothermal conditions over the entire supercooled liquid range. Processing of ca. 0.05 cm³ of Vit1 in 4 mm diameter graphite crucibles in a computer controlled induction furnace allows cooling rates of up to 30 K/s from the equilibrium melt prior to the isothermal anneals. The logarithm of the onset times, t_x, of the first recalescence event as observed in the thermocouple signal are shown in Fig. 12. Results from electrostatic levitation experiments [28] are shown for comparison. The times to crystallization in Fig. 12 show a minimum at T_n=895 K, due to the competition between an increasing thermodynamic driving force for crystallization and decreasing atomic mobility upon supercooling. The shortest times to crystallization are as large as 60 s at this temperature, reflecting a crital cooling rate for glass formation of only 1 K/s. This rate is at least 6 orders of magnitude larger than values for simple metallic systems. At the temperature T_n, the viscosity of Vit1 is about 4 orders of magnitude larger than the viscosity of, e.g., Ni at a similar degree of supercooling [29]. As will be discussed below, the sluggish kinetics at elevated temperatures acccounts for a signifant part of Vit1's exceptional glass forming ability.

DISCUSSION

Strong liquid behavior

Figure 13 shows the viscosities for Vit 1 and Vit 4 in an Ahrrenius plot, obtained from the three described methods as well as one data point obtained by capillary flow [22]. The high viscosities are obtained after proper relaxation as described above. The data cover a range of 14 orders of magnitude and are fitted here by a Vogel-Fulcher-Tammann (VFT) equation

$$\eta = \eta_0 \cdot \exp[D \cdot T_0 / (T - T_0)] \qquad (6),$$

in a modification that was proposed by Angell [30] (dashed and dotted line). In this formulation D is the fragility parameter and T_0 is the VFT temperature. The best fits to the experimental data yield D=18.5 and T_0=412.5K for Vit 1 and D=22.7 and T_0=372 K for Vit 4. The value η_0 was set as $4 \cdot 10^{-5}$ Pa·s according to the relation $\eta_0 = N_A \cdot h/V$, with, N_A, Avogadro's constant, h, Planck's constant and, V, the molar volume [31]

Glass formation was observed and studied in a large variety of materials, mostly non-metallic systems. The temperature dependence of viscosity can differ substantially among different materials. The viscosity of SiO_2, for example, which is an open network glass, can be described well with an Ahrrenius law. Other substances such as materials with van der Waals bonds are best described by a VFT relation with a VFT temperature very close to the glass transition. A comprehensive concept to describe the sensitivity of the viscosity to temperature changes for different materials in the supercooled liquid state was developed by Angell (see, for example [30]. The viscosity is plotted normalized to the glass transition temperature, T_g.

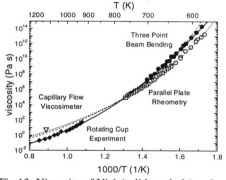
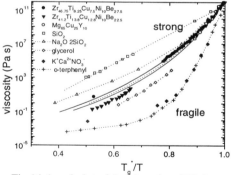

Fig.13: Viscosity of Vit1 (solid symbols) and Vit4 (open symbols). Included are the VFT fits for Vit1 (dashed) and Vit4 (dotted) and a fit according to Cohen and Grest [32]

Fig.14: Angel plot of the viscosity of Vit1, Vit4, Mg-Cu-Y and several non metallic "strong"and "fragile" glass forming liquids (see [30])

Figure 14 compares the data of the BMG forming Vit 1 and Vit 4 with a selection of some non-metallic liquids. Strong glasses, like SiO_2, are one extreme case. They exhibit very small VFT temperatures and very high melt viscosities. The other extreme are "fragile" glasses that show a VFT temperature close to the glass transition temperature, as well as low melt viscosities. The temperature dependence of the viscosity of the two BMG studied here is found to behave similar to that of the relatively strong sodium silicate glasses. We observe melt viscosities of the order of 2 Pa·s for both alloys. They are much more viscous (three orders of magnitude) than the melts of pure metals, where viscosities of typically $5 \cdot 10^{-3}$ Pa·s are found. The high viscosities of the BMG forming liquids is an important factor to understand the retarded nucleation and growth kinetics of crystals in these materials.

It is interesting to note that the data on Vit1 in Fig.13 can be better described by a Cohen and Grest fit [32] than by the VFT-fit. This is discussed elsewhere in detail [29,33].

Time scales for viscous flow and diffusion

The internal equilibration time, τ_η, of a liquid at a given temperature and pressure can be linked to the viscosity through $\tau_\eta = \eta/G$. The relaxation of the viscosity of Vit1 to its equilibrium values [34] below 650 K yields $G = 5.5 \times 10^8$ Pa. This value was determined by viscosity relaxation experiments in the same way as it is shown in Fig.6 for Vit4. We neglect the temperature dependence of G compared to η and τ_η in the following. Figure 15 shows the internal equalibration time as obtained from the equilibrium viscosity measurements (+) and the measured viscosity relaxation times(●). The total times for the calorimetric glass transition are directly linked to this internal relaxation time. Upon linear heating this time can be estimated from the calorimetric measurements as $\Delta T_g/R$, where R is the heating rate (see Fig.8). In Fig. 15 the resulting relaxation times from the calorimetric measurements (■) are added and show a good agreement with τ_η as measured by the viscosity. Note that τ_η is proportional but not equal to the shear stress relaxation time for small deformations as defined within Maxwell's model of viscoelasticity. The shear stress relaxation time is about two orders of magnitude shorter than the internal relaxation time.

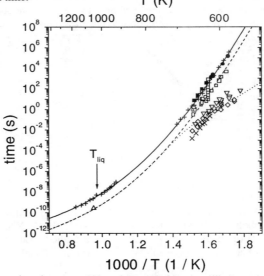

Fig.15: Structural relaxation times, τ_η, (●) and $\tau_\eta = \eta/G$ from equilibrium viscosities (+) of Vit1 and fit to Eq. (7) (-). Total times for calorimetric glass transition (■) from differential scanning calorimetry upon linear heating. Characteristic diffusion times, τ_D, for Be(x) [10], Ni (◇) [35], Co (▽), Al (□) [36], and Au (△) [37] in Vit1.

Recently, studies of Be interdiffusion and Ni, Co, and Al radiotracer diffusion in deeply supercooled Vit1 were performed [10,35,36]. Using the equation for random walk of the i-th tracer, $\tau_{D,i}=l^2/(6D_i)$, we consider times, $\tau_{D,i}$, for the successful displacement of an average atomic diameter, l. This definition of $\tau_{D,i}$ is not connected to a particular diffusion mechanism, e.g. it does not necessarily imply diffusion via thermally activated jumps. From the atomic volume, V_m, l is calculated to be 3.2×10^{-10}m and the corresponding $\tau_{D,i}$ are added in Fig. 15. Around 600 K the diffusivities of Al and Ni differ by 3 orders of magnitude, while they show a tendency to merge at higher temperatures. The temperature dependence of Al diffusion is similar to that of the viscosity: $\tau_{D,Al}\cong\tau_\eta/14$ (dashed curve). The proportionality factor implies that the mean displacement of an Al atom in the supercooled liquid during a typical relaxation time is roughly 4 interatomic diameters.

Ni and Co show significantly smaller absolute values of τ_D and activation energies of about 2.0 eV to 2.2 eV. Clearly, the time scales for viscous flow and diffusion of medium-sized atoms have different temperature dependencies in the deeply supercooled liquid state of Vit1. A collective hopping mechanism is likely to control the diffusion of Ni and Co in this temperature range as this behavior has been confirmed for Co in Vit4 by isotope effect measurements [38]. Similarily, we expect the migration of the smallest atom in the alloy, Be, to be also controlled by thermally activated jumps on a potential energy surface that fluctuates on a time scale given by τ_η.

For the equilibrium liquid, only preliminary results of Au interdiffusion with the equilibrium melt of Vit1 at 1050 K are available [37]. Quasielastic neutron scattering experiments on Vit4 revealed that above the liquidus temperature the differences in the diffusion coefficients of the various elements in the liquid alloy are less than one order of magnitude [39]. This finding is in accordance with the merging tracer diffusion coefficients in Fig. 15 and is well known for foreign diffusion in the equilibrium melts of Ag, Hg, Sn, Pb, and Bi [40].Guided by the differences in diffusion of the various elements at low temperatures and by the scaling between viscosity and diffusion in the equilibrium liquid, we consider a hybrid equation for the diffusion times, τ_D, of medium sized atoms in Vit1. Within this model, the time for a successful displacement, l, of Ni, for example, can either be limited by density fluctuations of the matrix or by a solid-like jump, where Eq. (7) takes into account the parallel development of both processes

$$\frac{1}{\tau_{D,Ni}} = \frac{g_{Ni}}{\tau_\eta} + \frac{1}{\tau_{Ni,0}}\exp\left(-\frac{Q_{Ni}}{kT}\right) \qquad (7)$$

At high temperatures $\tau_{D,Ni}$ is limited by the first term on the r.h.s in Eq. (7), i.e. by the time scale for changes in the surrounding matrix, τ_η. The value of $g_{Al}=14$ (see above) may serve as a lower limit and a first estimate for g_{Ni} as we do not expect Ni diffusion to be faster than Au near the liquidus (see Fig. 15). The corresponding times, $\tau_{D,Ni}$, are plotted in Fig. 15 with $Q_{Ni}\cong2.1$eV. Below about 800 K, thermally activated jumps of Ni with an average activation energy Q_{Ni} can occur as the potential energy barriers for Ni diffusion become fixed during the time between two successive jumps. Both processes in Eq. (7) act in parallel in a relatively wide intermediate temperature range between about 700 K to 800 K, which can be regarded as the glass transition regime for Ni diffusion in Vit 1. This temperature range is independent of the time scale of the experiment but may differ among the various components in the alloy. In fact, for the fast diffusing species (Ni, Cu, Be) in the alloy we expect a cascade of transitions. The jumps of a medium sized atom like Ni in the dense packed supercooled liquid will probably also affect the positions of the fastest diffusor, Be.

Viscosity, diffusion and crystallization kinetics

In this paragraph we want to discuss the crystallization of Vit 1 with a special emphasis on the changing kinetics as discribed by Eq. (7). For simplification we apply classical nucleation theory. One however must note that the real crystallization process (at least at temperatures below the crystallization nose) is likely to be very complex, involving phase separation and complicated diffusion fluxes that are affected by the diffusional asymetries between the different species.

Within classical nucleation theory the steady state nucleation rate, $I_s = A \cdot D_{eff} \cdot \exp(-\Delta G^*/kT)$ (8), is written as a product of an effective diffusivity, D_{eff}, and a Boltzmann factor of the Gibbs free energy, $\Delta G^* = 16\pi\sigma^3/3\Delta G^2$ (9), to form a critical nucleus. For the Gibbs free energy difference, ΔG, between the liquid and the crystalline phase we use results from differential scanning calorimetry [9] while the interfacial energy, σ, is treated as a fit parameter. The growth velocity, $u = f \cdot l^{-1} \cdot D_{eff}[1-\exp(-v_m \cdot \Delta G/kT)]$ (10), of the solid phase can similarly be expressed as a product of a kinetic and a thermodynamic factor. For simplicity, we use the same effective diffusivity, D_{eff}, as above and take the fraction of sites at the liquid-crystal interface where atoms are preferentially added or removed to be $f = 1$. With these assumptions, the time to crystallize a

$$t_x = \left(\frac{3x}{\pi I_s u^3} \right)^{\frac{1}{4}}$$

(11),

small volume fraction, x, is proportional to the inverse effective diffusivity: $t_x \sim D_{eff}^{-1}$ [40]. Except for D_{eff} one is left with A and σ as the only two fit parameters in Eq. 11. With the commonly used relation, $D_{eff} \sim \eta^{-1}$ the minimum in the nucleation time, t_x, at 895 K requires an interfacial energy of $\sigma = 0.040$ J/m^2. The corresponding temperature dependence of t_x is plotted in Fig. 12. While the low temperature part of the crystallization data cannot be described with the assumption $D_{eff} \sim \eta^{-1}$, for temperatures T>850 K satisfactory agreement between the experimental findings and classical nucleation and growth theory is found. In contrast, an Arrhenius-like effective diffusivity $D_{eff} \sim \exp(-Q_{eff}/kT)$ with $Q_{eff} = 1.2$ eV describes the crystallization times in the vicinity of the glass transition very well as shown in Fig. 12. We note that all viscosity and crystallization data were obtained on samples in their supercooled liquid state and that the failure of the viscosity to describe the onset times of crystallization is therefore not connected with the calorimetric glass transition.

The qualitative temperature dependence of D_{eff} is very similar to that of the tracer diffusion of medium-sized atoms discussed above. It is tempting to use a hybrid equation similar to Eq. 7 for D_{eff}. One cannot expect, however, that effective diffusivities and activation energies used to describe kinetics at liquid-solid interfaces during crystallization in a multicomponent system are directly comparable to tracer diffusion coefficients. Highly collective transport mechanisms proposed to control atomic transport in the bulk of the sample are likely to be altered in the immediate vicinity of a crystalline cluster.

CONCLUSIONS

The viscosity of Zr-Ti-Cu-Ni-Be bulk metallic glass forming liquids has been determined in the entire supercooled liquid region. This could be achieved due to their superior thermal stability with respect to crystallization compared to previous alloys. The novel alloys exhibit strong liquid behavior which means that they are much more viscous than previous metallic glass formers or pure metals. We have found that the viscosity of the particular Vit1 in the equilibirium melt is three orders of magnitude larger than for simple metallic systems. The sluggish kinetics reflected in the high viscosity in the temperature above T_n contribute significantly to the glass forming ability of the alloy. The time scales obtained from our viscosity measurements suggest that in the deeply supercooled a cascade of transitions from liquid-like to solid-like diffusion of small and medium sized atoms occurs. Similarly, the temperature dependence of the time to crystallization at low temperatures is best described with an Arrhenius-like effective diffusivity. For the present system we therefore conclude that the stability with respect to crystallization in the vicinity of the glass transition is not directly related to the sluggish kinetics at high temperatures that favor glass formation. For multicomponent metallic glass formers with large size differences among the constituents this is more likely to be the rule than the excemption.

ACKNOWLEDGEMENT

This work was supported by the U.S. Department of Energy (Grant No. DEFG-03-86ER45242)

REFERENCES:

1. H.S. Chen and D. Turnbull, J. Chem Phys. **48**, 2560 (1968).

2. S.S. Tsao and F. Spaepen, Acta Metall. **33**, 1355 (1985).

3. H.S. Chen, J. Non-Crystalline Solids **27**, 257 (1978).

4. C.A. Volkert and F. Spaepen, Acta Metall. **37**, 1355 (1989).

5. A. Inoue, T. Zhang and T. Masumoto, Mater. Trans. JIM 31, 425 (1991).

6. T. Zhang, A. Inoue and T. Masumoto, Mater. Trans. JIM 32, 1005 (1991).

7. A. Peker and W. L. Johnson, Appl. Phys. Lett. **63**, 2342 (1993).

8. Y.J. Kim, R. Busch, W.L. Johnson, A.J. Rulison, W. K. Rhim, Appl. Phys Lett. **65**, 2136 (1994).

9. R. Busch, Y.J. Kim, and W.L Johnson, J. Appl. Phys. **77**, 4039 (1995).

10. U. Geyer, S. Schneider, W.L. Johnson, Y. Qiu, T. A. Tombrello, and M. P. Macht, Phys. Rev. Lett. **75**, 2364 (1995).

11. R. Busch, Y.J. Kim, W.L. Johnson, A.J. Rulison, W. K. Rhim, and D. Isheim, Appl. Phys Lett. **66**, 3111 (1995).

12. M.J. Stephan, Akad. Wiss. Wien. Math.-Natur Klasse Abt. 2. **69**, 713 (1984).

13. G.J. Diennes and H.F. Klemm, J. Appl. Phys. **17**,458 (1946).

14. E. Bakke, R. Busch, and W.L. Johnson, Appl. Phys. Lett. **67**, 3260 (1995).

15. R. Busch, E. Bakke, and W.L. Johnson, Acta Mater. **46**, 4725 (1998).

16. E. Bakke, R. Busch, and W.L. Johnson, Mat. Sci. Forum **225-227**, 95 (1996).

17. H.E. Hagy, J. Amer. Ceram. Soc. **46**, 93 (1963).

18. F.T. Trouton, Proc. Roy. Soc (London) **77**, 426 (1906).

19. M. Reiner, Rheology, Vol. 1 (ed F. R. Eirich), Academic Press, New York, 9 (1956).

20. The cross section moment of inertia for a rectangular beam is $(a \cdot h^3)/3$, with, a, width and, h, height.

21. P.K. Kundu, Fluid Mechanics (Academic Press, San Diego, 1990).

22. A. Masuhr, PhD thesis, California Institute of Technology, 1998.

23. M.M. Couette, Ann. Chim. Phys. **21**, 433 (1880).

24. A. Masuhr, R. Busch, and W.L. Johnson, Mater. Sci. Forum **269-272**, 779 (1998).

25. J.C. Phillips, Rep. Prog. Phys. **59** (1996) 1133.

26. R. Busch, and W.L. Johnson Appl. Phys. Lett. **72**, 2695 (1998).

27. P. Tuinstra, R.A. Duine, J. Sietsma and A. van den Beukel, Acta metall. Mater. **43**, 2815 (1995).

28. Y.J. Kim, R. Busch, W.L. Johnson, A.J. Rulison, and W. K. Rhim, Appl. Phys Lett. **68**, 1057 (1996).

29. A. Masuhr, T.A. Waniuk, R. Busch and W.L. Johnson, Phys. Rev. Letters (submitted).

30. C.A. Angell, Science **267**, 1924 (1995).

31. S. Glasstone, K.J. Laidler and H. Eyring, The Theory of Rate Processes (McGraw-Hill, New York, 1941).

32. G.S. Grest and M.H. Cohen, Adv. Chem. Phys. **48**, 455 (1981).

33. A. Masuhr, R. Busch and W.L. Johnson, J. Non Cryst. Solids (in press).

34. T.A. Waniuk and R. Busch, unpublished.

35. F. Wenwer, K. Knorr, M.P. Macht and H. Mehrer, Defect and Diffusion Forum **143-147**, 831 (1997).

36. E. Budke, P. Fielitz, M.P. Macht, V. Naundorf and G. Frohberg, Defect and Diffusion Forum **143-147**, 825 (1997).

37. A. Masuhr and U. Geyer, unpublished.

38. H. Ehmler, A. Heesemann, K. Rätzke, F. Faupel and U. Geyer, Phys. Rev. Lett. **80**, 4919 (1998).

39. A. Meyer, J. Wuttke, W. Petry, O.G. Randl and H. Schober, Phys. Rev. Lett. **80**, 4454 (1998).

40. T.E. Faber, Introduction to the Theory of Liquid Metals (Cambridge University Press, 1972).

41. D.R. Uhlmann, J. Non-Cryst. Solids **7**, 337 (1972).

CRITICAL COOLING RATE VS. REDUCED GLASS TRANSITION: SCALING FACTORS AND MASTER CURVES

N. CLAVAGUERA*, M.T. CLAVAGUERA-MORA**
* Grup de Física de l'Estat Sòlid, Facultat de Física, Universitat de Barcelona, Diagonal 647, 08028-Barcelona, Spain.
** Grup de Física de Materials I, Facultat de Ciències, Universitat Autònoma Barcelona, 08193-Bellaterra, Spain.

ABSTRACT

The aim of the present paper is to analyse the glass formation and stability of bulk metallic glasses. Attention is focused to metallic alloys as systems which may develop a large glass-forming ability. Glass formation when quenching from the liquid state is discussed in terms of the thermodynamics and kinetics of the stable/metastable competing phases. Thermodynamics is required to relate glass transition temperature, T_g, to the energetics of the supercooled liquid. Kinetic destabilisation of equilibrium solidification and, consequently, glass forming ability are favoured by the high viscosity values achieved under continuous cooling. The relative thermal stability of the supercooled liquid depends on the thermodynamic driving force and interfacial energy between each competing nucleating phase and the molten alloy. It is shown that the quantities representative of the process, once scaled, have a temperature dependence that is mostly fixed by the reduced glass transition temperature, $T_{gr} = T_g/T_m$, T_m being the melting temperature. Based on the classical models of nucleation and crystal growth, the reduced critical cooling rate is shown to follow master curves when plotted against T_{gr}. Experimental trends for specific systems are compared to predicted values from these master curves.

INTRODUCTION

There are several empirical rules that have been successfully employed to predict the glass-forming ability. For systems which exhibit a glass transition at temperature T_g, one general accepted rule - usually called the "two thirds" rule - is that $T_{gr}=T_g/T_m \approx 2/3$. In particular, the achievement of bulk metallic glasses is viewed [1] as the result of application of the three following "golden" empirical rules. (1) Multicomponent systems consisting of more than three constituent elements; (2) significantly different atomic size ratios (above about 12%) among the main constituent elements; and, (3) negative heats of mixing among the constituent atoms. From the work of Uhlmann [2] the study of crystallisation has been enlarged to the theoretical evaluation of the transformation diagrams. With the goal of establishing criteria for bulk glass formation, the most relevant diagrams are the so called temperature *vs.* cooling rate transformation (T-CR-T) diagrams, in which are represented the loci of points where a certain crystalline fraction is achieved under continuous cooling conditions [3,4]. The value at T_g of the cooling rate for which a negligible fraction, x, of material crystallises (recognised typically as $x=10^{-6}$) is precisely the critical cooling rate, R_C, needed to avoid the transformation.

The aim of this paper is to analyse how specific material properties enhance glass-forming ability. The underlying statement all over the paper is the recognition that the R_C value of a supercooled liquid alloy is determined by an homogeneous nucleation frequency, I, and an interface controlled growth, u. This is irrespective of the crystallisation mode: polymorphous, primary or eutectic precipitation [5,6]. Thus, the procedure consists in the evaluation of the relationship between critical cooling rate and material properties.

Mat. Res. Soc. Symp. Proc. Vol. 554 © 1999 Materials Research Society

STABILITY OF THE SUPERCOOLED LIQUID VS. CRYSTALLISATION

Theoretically, T_g might be defined either as the temperature below which the relaxation time is too long for equilibrium to be reached in a finite time or as the temperature at which the probability of finding the critical local concentration of free volume for viscous flow vanishes. The viscosity of liquid metallic alloys increases rapidly with decreasing temperature to yield a glass, the viscosity of which is defined arbitrarily to be greater than 10^{13} poise (10^{14} Pa·s). Such behaviour is described by the phenomenological Vogel Fulcher equation:

$$\eta = \exp\{B/(T-T_o)\} \tag{1}$$

where B and T_o are empirical fitting parameters. T_o is found to lie near but somewhat below T_g.

Since the work of Chen and Turnbull [7], it has been accepted that the heat capacity, C_p, increases gradually as T decreases for a molten metallic alloy, as depicted schematically in Fig. 1a. The abrupt change of C_p, ΔC_p, to about half its value when the supercooled liquid becomes a glass occurs to avoid the entropy catastrophe [8]. The entropy change is shown in Fig. 1b.

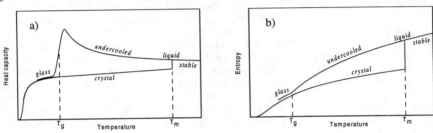

Fig. 1. Heat capacity and entropy for the stable and metastable phases *vs.* temperature.

SCALING FACTORS

Both the viscosity and the Gibbs energy of formation of a supercooled liquid, ΔG_f, are directly related to T_{gr}, when properly expressed in reduced units [4,9,10] as indicated in Fig. 2. The thermodynamic driving force depends also on the melting enthalpy, $\Delta H_m = T_m \cdot \Delta S_m$ (with ΔS_m the melting entropy). Also, the interfacial energy between the nucleating phase and the molten alloy, σ, plays an important role in the nucleation process. Consequently, apart from

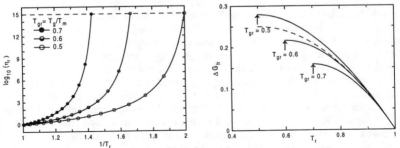

Fig. 2. Evolution of η_r *vs.* $1/T_r$ and ΔG_{fr} *vs.* T_r for different values of T_{gr}.

T_{gr}, a reduced nucleation frequency and a reduced crystal growth rate may be defined which do not dependent on specific material properties but do change with the Turnbull factors $\alpha=\sigma/\Delta H_m$ and $\beta=\Delta S_m/R$ [11]. The main reduced quantities already discussed are as follows.

Reduced viscosity:

$$\eta_r = \exp\left\{\frac{B_r \cdot \Delta T_r}{(T_r-T_{ro})(1-T_{ro})}\right\} \quad (2)$$

with $\eta_r=\eta(T)/\eta(T_m)$, $T_r = T/T_m$, $B_r = B/T_m$, $\Delta T_r = 1-T_r$ and $T_{ro} = T_o/T_m$.

Reduced ΔG_f:

$$\Delta G_{fr} = \Delta G_f/\Delta H_m$$
$$= \Delta T_r - \gamma \cdot \{\Delta T_r + T_r \cdot ln T_r\} \quad (3)$$

with $\gamma := \Delta C_p/\Delta S_m$.

The value of η_r in the supercooled liquid regime is only dependent on T_{gr} provided a decrease of 15 orders of magnitude is assumed from the glass transition to the melting point (See Fig. 2a). Equivalently, T_{gr} fixes the behaviour of ΔG_{fr}, since γ is determined imposing no residual entropy in the glassy state, as shown in Fig. 2b. Also, the reduced nucleation frequency, I_r, and crystal growth rate, u_r, are obtained through specific scaling factors. That is:

$$I_r = I \cdot \frac{V^2 \cdot \eta(T_m)}{N_A \cdot RT_m} \quad (4)$$

and

$$u_r = u \cdot \frac{V^{\frac{2}{3}} \cdot N_A^{\frac{1}{3}} \cdot \eta(T_m)}{RT_m} \quad (5)$$

resulting in

$$I_r = \frac{T_r}{\eta_r} \cdot \exp\left(-\frac{16\pi\alpha^3\beta}{3(\Delta G_{fr})^2 T_r}\right) \quad (6)$$

$$u_r = \frac{T_r}{\eta_r} \cdot [1 - \exp(-\beta \cdot \Delta G_{fr}/T_r)] \quad (7)$$

were V is the molar volume and N_A the Avogadro's number.

Since in metallic systems $\beta \cong 1$, only the combination $\alpha^3\beta$ will be specified in the following. The interplay in between the two main parameters, and, in two relevant reduced quantities, I_r and u_r, is presented in Fig. 3 and 4.

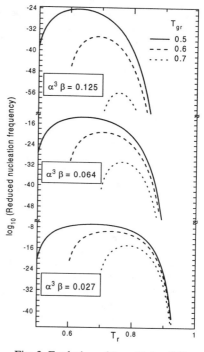

Fig. 3. Evolution of I_r vs. T_r for different values of both T_{gr} and $\alpha^3\beta$.

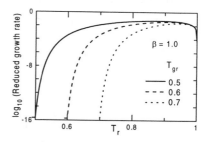

Fig. 4. Evolution of u_r vs. T_r for different values of T_{rg}.

The representation of these reduced quantities vs. the reduced glass transition temperature illustrates how the use of the scaling factors permits to omit specific properties apart from the parameters α and β. The importance of the interfacial energy in between the nuclei and the molten alloy, i.e. α, is clearly apparent in Fig. 3. The maximum value of the reduced nucleation frequency increases by about 8 orders of magnitude when $\alpha^3\beta$ passes from 0.125 to 0.064, or from 0.064 to 0.027. Another fact shown in Fig. 3 is the increasing temperature interval in which the nucleation rate is high for decreasing values of $\alpha^3\beta$, the effect being more pronounced when T_{gr}=0.5. The relative temperature interval in which the reduced growth rate stays close to its maximum value is also enlarged when T_{gr}=0.5, as shown in Fig. 4. Otherwise, all curves shown in this figure show a very sharp increase of u_r for low supercooling.

REDUCED T-CR-T DIAGRAMS

The solidification behaviour in terms of reduced quantities, by use of the Avrami formalism [12], is related to the fraction $x=x(T_r,R_r)$ crystallised at T_r after continuous cooling at the reduced rate R_r [4]:

$$x(T_r,R_r) = 1 - \exp\left\{-\frac{y(T_r)}{R_r^4}\right\} \tag{8}$$

with

$$y(T_r) = \frac{4\pi}{3}\int_{T_r}^1 I_r(x)dx\left\{\int_{T_r}^x u_r(x')dx'\right\}^3 \tag{9}$$

and

$$R_r = R\cdot\frac{V\cdot\eta(T_m)}{RT_m^2} \tag{10}$$

The reduced T-CR-T diagrams for a transformed fraction of $x=10^{-6}$ are presented in Fig. 5. At low reduced cooling rates the degree of supercooling needed to initiate the transformation increases with increasing values of both T_{gr} and $\alpha^3\beta$. However, the value at which the T-CR-T curves attain a R_r value independent of T_r (vertical slope) is much more sensitive to the reduced temperature at which I_r has its maximum. As a consequence, the reduced critical cooling rate, R_{Cr}, decreases successively also with increasing values of both T_{gr} and $\alpha^3\beta$.

CRITICAL COOLING RATE: MASTER CURVES VS. REAL VALUES

The main interest of the use of reduced quantities is to separate the representation of the

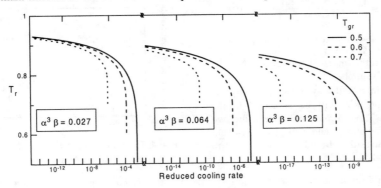

Fig. 5. Reduced T-CR-T curves for a fixed crystalline fraction $x=10^{-6}$.

glass-forming ability from specific material properties. According to the previous discussion, this is possible when considering the reduced critical cooling rate *vs.* both T_{gr} and $\alpha^3\beta$. A possible representation of the general trends in metallic systems is presented in Fig. 6. Such a figure shows that the value of R_{Cr} decreases by about 4 to 9 orders of magnitude when the reduced glass transition temperature evolves from 0.5 to 0.7, with a fixed $\alpha^3\beta$ value. The continuous reduction of R_{Cr} with decreasing $\alpha^3\beta$ may be interpreted as either a decreasing interfacial tension in the specific molten alloy or to the inclusion of preferred nucleation sites under heterogeneous nucleation.

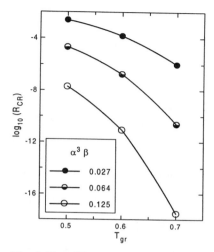

Fig. 6. Plot of R_{Cr} *vs.* T_{gr} for different values of $\alpha^3\beta$ in metallic systems.

For practical purposes one needs to pass from the reduced quantities to absolute values referred to the particular system under study. The scaling factor for transforming R_{Cr} into R_C is given by equation (10). Owing to the fact that the several quantities intervening in that equation are expected to vary in a limited range, the plot of the absolute value of R_C *vs.* T_{gr} is presented in Fig. 7 for three possible sets of specific material properties ranges. Superimposed to this figure are plotted the already published data [13,14] on the relationship between the critical cooling rate and T_{gr}. It may be recognised that the main kinetic factor which stabilises glass-formation comes from the viscosity, provided heterogeneous nucleation events are avoided, since for metallic systems no much spread occurs in the current values of both the melting entropy and the interfacial energy.

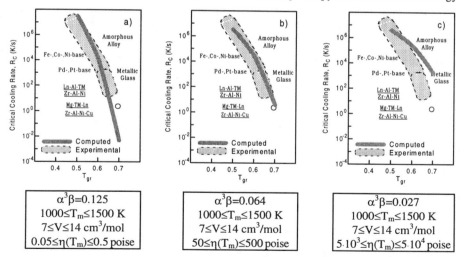

Fig. 7. Comparison between experimental and computed values of R_C.

CONCLUSIONS

The kinetics of crystallisation onset of a supercooled molten alloy under continuous cooling conditions has been analysed on the basis of homogeneous nucleation and three dimensional interface-controlled growth. Scaling factors used to define reduced quantities, cleared from some particular material characteristics, are proposed. The specific temperature dependence of these reduced quantities is analysed in terms of both the reduced glass transition temperature and the values of $\alpha^3\beta$.

The general trends of the reduced critical cooling rate are obtained in the form of masters curves. They can be converted to estimates of specific critical cooling rate values by use of quantities directly accessible from experiment, such as the viscosity of the stable molten alloy, its molar volume and its melting temperature.

The predictions resulting from the kinetic under continuous cooling, performed for generic metallic molten alloy systems, agree with the experimental observation that the ease of glass formation increases with the value of the reduced glass transition temperature.

ACKNOWLEDGEMENTS

Support form CICYT by Projects N° MAT96-0692 and MAT96-0769 is acknowledged.

REFERENCES

1. A. Inoue, Proc. Japan Acad. **73**, Ser. B, p. 19 (1997).

2. R. Uhlmannn, J. Non-Cryst. Solids **7**, p. 337 (1972).

3. N.Clavaguera, J. Non-Cryst. Solids **162**, p. 40 (1993).

4. N.Clavaguera, and M.T. Clavaguera-Mora,. Mater. Sci. & Eng. **A179/A180**, p. 288 (1994).

5. N.Clavaguera, and M.T. Clavaguera-Mora,.in: *Thermodynamics and Kinetics of Phase Transformations* edited by J.S. Im, B. Park, A.L. Greer, and G. B. Stephenson, (Mat. Res. Soc. Symp. Proc. **398**, Pittsburgh, PA 1996) p. 319-24.

6. M.T. Clavaguera-Mora and N.Clavaguera, J. Alloys & Comp. **247**, p. 93 (1997).

7. H.S. Chen, and D. Turnbull, J. Appl. Phys. **38**, p. 3,646 (1967).

8. W. Kauzmann, Chem. Rev. **43**, p. 305 (1948).

9. M.T.Clavaguera-Mora and N. Clavaguera, J. Mater. Res. **4**, p. 906 (1989).

10. M.T. Clavaguera-Mora, Ber. Bunsenges. Phys. Chem. **102**, p. 1,291 (1998).

11. D. Turnbull, Contemp. Phys. **10** (1969) p. 473.

12. J.W. Christian, *The theory of phase transformations in metals and alloys,* Pergamon Press, Oxford, 1965, pp. 525-48.

13. T. Masumoto, Mater. Sci. Eng. **A179/180**, p. 8 (1994)

14. A. Inoue, Mater. Sci. Eng. **A226-228**, p. 357 (1997)

Glass Formation and Phase Separation In The Vitreloy Type Bulk Metallic Glasses

Charles C. Hays, Paul Kim, and William L. Johnson
California Institute of Technology
Department of Materials Science & Engineering
W. M. Keck Laboratory for Engineering Materials 138-78
Pasadena, CA 91125 USA

Abstract

Results of calorimetric, differential thermal analysis, and structural measurements are presented for a series of bulk metallic glass forming compositions in the Zr-Ti-Cu-Ni-Be alloy system. The structural data identify the competing crystalline phases, formed on cooling from the liquid state, for various compositions in the Zr-Ti-X quasi-ternary phase diagram; with the Cu-Ni-Be ratio given by $X = Be_9Cu_5Ni_4$. For this region of the n-dimensional phase space (n=5), the bulk glass forming range is extensive and the calorimetric data exhibit thermal features associated with the occurrence of phase separation in the undercooled liquid state prior to primary crystallization. The topology of the composition manifold is complex; manifest by dramatic changes in crystallization behavior for small changes in the Zr-Ti-X ratio. Alloys with large supercooled liquid regions, ΔT, are observed; $\Delta T \approx 135$ K. Zr-Ti-Cu-Ni-Be alloy compositions with eutectic temperatures less than that of pure Al ($T_m = 933$ K) were also synthesized.

Introduction

The recently discovered bulk metallic glasses can be broadly described as pseudo ternary alloys. There are now a number of these multicomponent alloy systems; e. g., La-Ni-Al [1], Zr-Al-Cu-Ni [2], Zr-Ti-Cu-Ni-Be [3], and Zr-Ti-Cu-Ni [4]. The parent Be-containing bulk glass $Zr_{41.2}Ti_{13.8}Cu_{12.5}Ni_{10}Be_{22.5}$ (Alloy 1) developed at Caltech exhibits an exceptional glass forming ability (GFA) and has an excellent thermal stability with respect to crystallization. The crystallization behavior of $Zr_{41.2}Ti_{13.8}Cu_{12.5}Ni_{10}Be_{22.5}$ near the glass transition temperature T_g has been determined by electrostatic levitation and calorimetric methods [5, 6]. These results, when combined with the results of small angle neutron scattering (SANS) measurements, show that primary crystallization is preceded by phase separation in the supercooled liquid region (SLR) [7]. This phase separation process shows many of the features of a spinodal decomposition. The bulk glass forming compositions in the Zr-Ti-Cu-Ni-Be system are compactly written in terms of a Zr-Ti-X phase diagram, where X represents the Cu-Ni-Be ratio; $X = Be_9Cu_5Ni_4$, characteristic of the parent composition. In this paper preliminary results are presented for compositions that lie along various cuts in the Zr-Ti-X phase diagram. The bulk glass $Zr_{46.75}Ti_{8.25}Cu_{7.5}Ni_{10}Be_{27.5}$ (Alloy 4) also exhibits an excellent GFA. However, this alloy exhibits little phase separation prior to crystallization. In an effort to ellucidate this behavior, results of calorimetric measurements are presented for compositions that along the tie-line between the compositions of Alloy 1 and Alloy 4.

Experimental

The materials used in preparing the specimens were as follows: Cu (99.999%), Ni (99.995%), Ti (99.95%) from Cerac Inc., crystal bar Zr with < 300 ppm oxygen content from Teledyne Wah-Chang Inc., and Be (99.9%) from Electronic Space Products International. The master alloys were prepared in the form of 25g rods by arc melting in a Ti-gettered argon atmosphere on a water cooled copper crucible. The specimens were

Mat. Res. Soc. Symp. Proc. Vol. 554 © 1999 Materials Research Society

flipped and remelted several times to promote homogeneity. The alloy compositions along the tie-line between the compositions $Zr_{41.2}Ti_{13.8}Cu_{12.5}Ni_{10}Be_{22.5}$ (Alloy 1) and $Zr_{46.75}Ti_{8.25}Cu_{7.5}Ni_{10}Be_{27.5}$ (Alloy 4), were prepared in the form of thin strips, with dimensions of the order 1 x 5 x $\approx$ 25mm (thickness by width by length), by injection casting into a Cu metal mold. The specimens were cast in a bell-jar vacuum chamber evacuated to a pressure of P $\approx$ 10^{-5} mbar. The specimens were rapidly melted (τ_{heat} $\approx$ 1 sec) with a high frequency rf field in a fused quartz crucible and then injected into the mold with UHP argon gas (ΔP = 35 psi). The nominal compositions are used in this article.

The thermal properties of the alloys were measured in a Perkin-Elmer DSC-7 differential scanning calorimeter (DSC) under an ultra-high purity (99.9999%) argon gas blanket. The DSC scans were conducted in Al sample pans at a rate of 20 K/min between 40 to 580 °C followed by cooling to room temperature. A second scan was then immediately conducted under the same conditions to determine the thermal baseline. The melting points of the alloys were determined with a Perkin-Elmer DTA-7 differential thermal analyzer (DTA). The DTA scans were conducted under an ultra-high purity (99.9999%) argon gas blanket. in graphite crucibles at a rate of 10 K/min between 200 to 1100 °C followed by cooling to room temperature. The X-ray diffraction patterns were obtained with an INEL diffractometer using a CPS-120 position sensitive detector with a Co Kα radiation (λ = 0.179 nm) source. The detector scans over a 120° 2θ range (4096 channels). The diffraction patterns were calibrated against a silicon standard.

Results

The GFA of the Zr-Ti-Cu-Ni-Be quasi-ternary system was investigated using a rather simple method; the master alloys were prepared by arc melting on a watercooled Cu crucible. The $Zr_{41.2}Ti_{13.8}Cu_{12.5}Ni_{10}Be_{22.5}$ parent bulk glass has a GFA such that it can be prepared in the amorphous state by arc melting in the form of 25g rods. The alloy prepared in this manner is fully amorphous except for a thin crystalline skull ($\approx$ 1mm thick) present on the bottom of the as-cast rod. Upon removing the skull cap, the fully amorphous condition was confirmed by an x-ray diffraction pattern absent of any Bragg peaks.

The bulk glass forming range (GFR) of the Zr-Ti-Cu-Ni-Be phase diagram was examined by preparing a series of alloys along various cuts of the Zr-Ti-X phase diagram, where X represents the Cu-Ni-Be ratio; X = $Cu_5Ni_4Be_9$. The alloy series, and the respective composition limit of each bulk glass forming range, x(GFR), are presented in Table 1:

Series	Alloy Composition	x(GFR)
A)	$(Zr_{75}Ti_{25})_{100-x}((Ni_{45}Cu_{55})_{50}Be_{50})_x$	30
B)	$Zr_{41.2}Ti_x(Ni_{45}Cu_{55})_{50}Be_{50})_{100-41.2-x}$	20
C)	$Zr_x(Ti_{25}(Ni_{45}Cu_{55})_{50}Be_{50})_{100-x}$	50
D)	$(Zr_{100-x}Ti_x)_{55}((Ni_{45}Cu_{55})_{50}Be_{50})$	33

Table 1: Data for Zr-Ti-X Alloy Series

The calorimetric results show that the GFR in the Zr-Ti-Cu-Ni-Be system is quite extensive. All of the alloy compositions prepared within these ranges were fully amorphous by arc melting up to the x value specified. Upon exceeding these limits multi-phase, amorphous and crystalline, materials were formed on cooling. The multi-phase alloys are brittle due to the presence of the crystalline phases; which seriously degrade the ductility of the material. Examples of these crystalline phases are; the Ti_2Ni or Zr_2Cu ("E93" or "MoSi$_2$" structure types), Cu_2TiZr or Ti_2Cu ("MoSi$_2$ structure type"), NiTiZr (MgZn$_2$

{Laves phase} structure type). In the second series; $Zr_{41.2}Ti_x(Ni_{45}Cu_{55})_{50}Be_{50})_{100-41.2-x}$; for $30 < x < 35$, crystalline phases with icosahedral symmetry were found.

The DSC traces, conducted at 20 K/min. heating rate, for the above alloy series appear similar to that of Alloy 1 (see Fig. 1); namely they are characterized by a distinct endothermal heat effect due to the glass transition followed by three characteristic steps of heat release upon crystallization of the metastable supercooled liquid [6]. As discussed by Schneider *et al*, primary crystallization in $Zr_{41.2}Ti_{13.8}Cu_{12.5}Ni_{10}Be_{22.5}$ is not the result of homogeneous nucleation in the supercooled liquid state, but is preceded by a phase separation in the SLR that may occur via a spinodal decoposition [7, 8]. The decomposition process is followed by the nucleation of fcc-nanocrystals on a length scale correlated with the spinodal wavelength, $\lambda_s \approx 14$ nm. The first alloy series presented in Table 1, A = $(Zr_{75}Ti_{25})_{100-x}(Ni_{45}Cu_{55})_xBe_{22.5}$, fully amorphous to x = 30, exhibits DSC traces similar to the parent composition $Zr_{41.2}Ti_{13.8}Cu_{12.5}Ni_{10}Be_{22.5}$. However, the peak temperature at which the first exothermic heat release occurs; attributed to phase separation and primary crystallization, decreases rapidly with increasing x. The calorimetric results for the alloy series A to D, all demonstrate DSC traces that are indicative of phase separation prior to primary crystallization. Therefore, it might appear that phase separation prior to primary crystallization is a characteristic of the Zr-Ti-Cu-Ni-Be system; i. e., it may be global in nature, at least for the Cu-Ni-Be ratio, X = $Cu_5Ni_4Be_9$. The experimental results also indicate that the GFA in the Zr-Ti-Cu-Ni-Be system appears to be slowly varying in the bulk glass forming region.

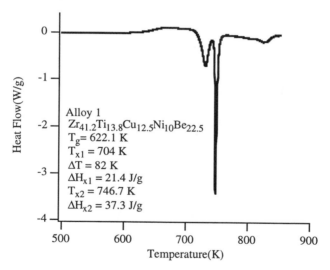

Fig. 1: DSC data for $(Zr_{75}Ti_{25})_{65}((Ni_{45}Cu_{55})_{50}Be_{50})_{35}$.

As stated before, there are Zr-Ti-Cu-Ni-Be alloys that exhibit little, if any, phase separation in the undercooled liquid region; e. g., $Zr_{46.75}Ti_{8.25}Cu_{7.5}Ni_{10}Be_{27.5}$. The GFA of this composition is less than that of the parent $Zr_{41.2}Ti_{13.8}Cu_{12.5}Ni_{10}Be_{22.5}$ composition. This manifest by the lack of an amorphous phase present in 25g specimens prepared by arc melting. However, the alloy composition is bulk glass forming, as fully amorphous rods may be prepared via water-quenching in evacuated quartz tubes, with diameters $\phi \approx 14$ mm. To examine the occurence of phase separation in the multi-component Zr-Ti-Cu-Ni-Be phase space, a series of five alloys were prepared (including the end members) that lie along the tie-line between the phase separating and non-phase separating compositions; labeled Alloys 1, 1a, 1b, 1c, and 4, and presented in Table 2. The specimens were first prepared in the form of 25g arc melted rods, and then as 1 mm thick strips by injection casting into a Cu metal mold. The results of x-ray diffraction measurements on the 25g arc melted rods show that the Alloys 1, 1a, 1b, and 1c were fully amorphous in this form.

Alloy Number	Alloy Composition
1	$(Zr_{75}Ti_{25})_{55}(Ni_{45}Cu_{55})_{22.5}Be_{22.5}$ $=Zr_{41.2}Ti_{13.8}Cu_{12.5}Ni_{10}Be_{22.5}$
1a	$(Zr_{77.5}Ti_{22.5})_{55}(Ni_{48}Cu_{52})_{21.25}Be_{23.75}$ $=Zr_{42.62}Ti_{12.38}Cu_{11.05}Ni_{10.2}Be_{23.75}$
1b	$(Zr_{80}Ti_{20})_{55}(Ni_{51}Cu_{49})_{20}Be_{25}$ $=Zr_{44}Ti_{11}Cu_{9.8}Ni_{10.2}Be_{25}$
1c	$(Zr_{82.5}Ti_{17.5})_{55}(Ni_{54}Cu_{46})_{18.75}Be_{26.25}$ $=Zr_{45.38}Ti_{9.62}Cu_{8.63}Ni_{10.12}Be_{26.25}$
4	$(Zr_{85}Ti_{15})_{55}(Ni_{57}Cu_{43})_{17.5}Be_{27.5}$ $=Zr_{46.75}Ti_{8.25}Cu_{7.5}Ni_{10}Be_{27.5}$

Table 2: Alloy 1 to Alloy 4 Tie-Line Compositions

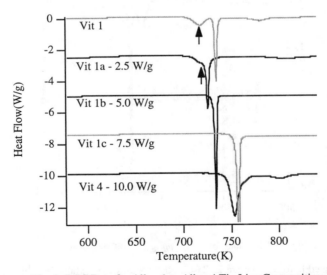

Fig. 2: DSC Data for Alloy 1 to Alloy 4 Tie-Line Compositions

The results of DSC measurements conducted on the 1mm injection cast strips of Alloys 1 to 4, at a heating rate of 20 K/min., are presented in Fig. 2. The data for each composition are shifted along the heat flow axis by a successive increment of -2.5 W/g.

The display of the DSC traces in this fashion shows the dramatic evolution of the thermal properties on going along the tie-line from Alloys 1 to 4. The results of analyses of the DSC and DTA data for these alloys are summarized in Fig. 3.

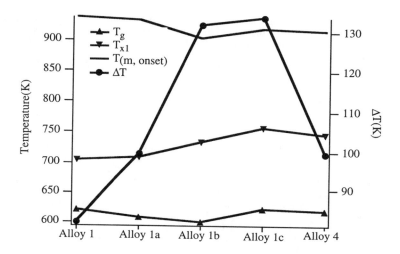

Fig. 3: DSC and DTA results for the Alloy 1 to Alloy 4 tie-line compositions

Disscussion and Conclusions

The calorimetric data for the alloy series A thru D (see Table 1), shown in Fig. 1, are consistent with the concept that phase separation in the SLR may be a global feature of the Zr-Ti-X compositions,with X = $Cu_5Ni_4Be_9$. The SANS measurements conducted by Schneider et al [7, 8] conclusively show that that primary crystallization is preceded by phase separation in the SLR for $Zr_{41.2}Ti_{13.8}Cu_{12.5}Ni_{10}Be_{22.5}$, and that the first exothermic heat release; obtained on heating via calorimetric measurements, may be associated with this decomposition process. The calorimetric results for the alloy series A thru D show for the first time that the occurrence of phase separation in the SLR is strongly influenced by controlled changes in alloy composition; the chemical driving force for this phase separation is stongly dependent on compositiion. This knowledge is vital for the design of improved BMG alloys.

The data for the Alloy 1 to Alloy 4 tie-line compositions (see Table 2), shown in Fig. 2, show a smooth transition from behavior characteristic of phase separation in the SLR; e. g., the first exothermic heat release shown in the DSC traces of compositions Alloy 1 and Alloy 1a (see arrows in Fig.2), to DSC traces that exhibit no features normally associated with phase separation prior to primary crystallization; Alloy 1b and Alloy 1c. The latter two alloys demonstrate a "eutectic-like" primary crystallization event. In an event of this type the crystalline phases are spontaneously formed upon primary crystallization. The width of the supercooled liquid region, for the Alloy 1c, $\Delta T = (T_x - T_g) \approx 135$ K, shows that this composition may be a highly processable bulk glass, with a low critical cooling rate. The magnitude of this ΔT value is the largest know reported for a bulk metallic glass. The onset of melting T_{onset} for these two alloys, 1b and 1c, are very low;

both demonstrate a near "eutectic-like" DTA trace, with a narrow melting interval. The values for the 1b and 1c compositions are $T_{onset}(1b)$ = 903 K and $T_{onset}(1c)$ = 918 K. Both are less than the melting point of pure alumimum (933 K). The reduced glass transition temperature, given by the ratio of the glass transition temperature to the melting temperature, for the Alloy 1c composition is T_{rg} = 0.679. The Alloy 4 composition, see Fig. 2, begins to display a high temperature shoulder beyond the primary crystallization event; this indicates that a small degree of phase separation may occur in this alloy. This is supported by unpublished three-point beam bending viscosity measurements conducted on Alloy 4 [9].

The thermal and structural properties of the alloys along the Alloy 1 to Alloy 4 tie-line will be further examined by calorimetric, and perhaps SANS measurements. Their excellent thermal stability will also enable the determination of their viscosity well into the supercooled liquid region via three-point beam bending measurements.

Summary

The calorimetric results for the alloys of Tables 1 and 2 show the dramatic influence of composition on the thermophysical properties. It has been proposed that the spinodal decomposition observed in the parent alloy $Zr_{41.2}Ti_{13.8}Cu_{12.5}Ni_{10}Be_{22.5}$ may result in the formation of coexisting strong and fragile liquids in the supercooled liquid region [10]. The SANS results of Schneider et al show that the wavelength of this decomposition is of the order ≈ 10 nm. This makes the determination of the actual compositions of these coexisting liquids experimentally intractable due to the small spatial scale. The dramatic thermal behviors observed in the DSC data for the alloys of Table 1; e. g., series A, give every indication that the spatial scale of the spinodal wavelength may be larger. We intend to measure the composition of these alloys via electron microprobe or atomprobe techniques together with SANS measurements, to verify the validity if this hypothesis. Overall, the data presented in this paper show that the topology of the glass forming ability, with regard to the composition manifold of the Zr-Ti-Cu-Ni-Be system is complex; manifest by dramatic changes in thermophysical behavior for small changes in the Zr-Ti, Cu-Ni, and Be ratios.

Acknowledgements- The authors would like to acknowledge the support of the U. S. Department of Energy (Grant No. DEFG-03-86ER45242). The authors would also like to thank Dr. John Haygarth of Teledyne Wah-Chang Inc. for providing the crystal bar Zr. used in the preparation of these alloys.

References

[1] A. Inoue, T. Zhang, and T. Masumoto, Mater. Trans., JIM, Vol. 31 (1990), pp. 425.
[2] A. Inoue, T. Zhang, N. Nishiyama, K. Ohba, and T. Masumoto, Mater. Trans. JIM Vol. 33 (1992), pp. 937.
[3]. A. Peker and W. L. Johnson, Appl. Phys. Lett. 63, 2342 (1993).
[4] X. H. Lin and W. L. Johnson, J. Appl. Phys. 78, 6514 (1995).
[5]. Y. J. Kim, R. Busch, W. L. Johnson, A. J. Rulison, W. K. Rhim, and D. Isheim, Appl. Phys. Lett. 65, 2136 (1994).
[6] R. Busch, Y. J. Kim, and W. L. Johnson, J. Appl. Phys. 77, 4039 (1995).
[7] S. Schneider, W. L. Johnson, and P. Thiyagarajan, Appl. Phys. Lett. 68, 493 (1996).
[8]. S. Schneider, U. Geyer, and P. Thiyagarajan, et al, Mat. Sci. Forum, in press (1997).
[9] Unpublished work of A. Wainuk, California Institute of Technology, 1998.
[10]. T. A. Wainuk, R. Busch, A. Masuhr, and W. L. Johnson, Acta Mater. 46, 5229 (1998).

Part V

Thermal Stability, Transport, and Magnetic Properties

SYNTHESIS AND PROPERTIES OF
FERROMAGNETIC BULK AMORPHOUS ALLOYS

A. INOUE *, T. ZHANG *, H. KOSHIBA ** AND T. ITOI ***
*Institute for Materials Research, Tohoku University, Sendai 980-8577, Japan
**INOUE SUPERLIQUID GLASS PROJECT, Exploratory Research for Advanced Technology,
    Japan Science and Technology Corporation, Sendai 982-0807, Japan
***Graduate School, Tohoku University

HISTORY OF Fe-BASED AMORPHOUS ALLOYS

Since an amorphous phase in Au-Si system was synthesized for the first time by rapid solidification in 1960[1], a large number of amorphous alloys have been prepared by various rapid solidification techniques. As the main amorphous alloy systems, one can list up the noble metal-, Fe-, Co-, Ni-, Ti-, Zr-, Nb-, Mo-, lanthanide(Ln)-, Al- and Mg-based alloys. Among these alloy systems, Fe-[2], Co-[2] and Al-[3]based amorphous alloys have been used in application fields of magnetic and high specific-strength materials. Thus, Fe- and Co-based amorphous alloys have gained the most important position as engineering amorphous alloys. When special attention is paid to Fe-based amorphous alloys, Fe-P-C alloys were synthesized in 1967[4] as the first Fe-based amorphous alloy. Subsequently, engineering important (Fe,Co)-Si-B amorphous alloys have been developed in 1974[5][6], followed by the formation of (Fe,Co,Ni)-(Cr,Mo,W)-C in 1978[7], (Fe,Co,Ni)-(Zr,Hf) in 1980[8] and then (Fe,Co,Ni)-(Zr,Hf,Nb)-B amorphous alloys in 1981[9]. The (Fe,Co)-Si-B amorphous alloys have been used in many application fields as soft magnetic materials[2]. However, after 1981, nobody have succeeded in finding a new amorphous alloy in Fe- and Co-based systems by rapid solidification from liquid phase. Besides, all these amorphous alloys have serious disadvantages that high cooling rates above 10^5 K/s are required for glass formation and the resulting sample thickness is limited to less than about 50 μm[10]. Great efforts have been devoted to find Fe- and Co-based amorphous alloys with a high thermal stability of supercooled liquid against crystallization and a high glass-forming ability (GFA). Very recently, we have succeeded in finding new ferromagnetic bulk amorphous alloys with critical sample thicknesses ranging from 1 to 15 mm in Fe-(Al,Ga)-(P,C,B,Si)[11]-[14], (Fe,Co,Ni)-(Zr,Hf,Nb)-B[15]-[17], (Fe,Co)-(Zr,Hf)-(Nb,Ta)-(Mo,W)-B[18], (Fe,Co)-Ln-B[19] (Ln=lanthanide metal) and (Nd,Pr)-Fe-Al[20]-[22] systems. In this review, we present the formation, thermal stability, mechanical strength and magnetic properties of these new ferromagnetic bulk amorphous alloys.

BASIC CONCEPT FOR THE ACHIEVEMENT OF HIGH GLASS-FORMING ABILITY FOR Fe-BASED AMORPHOUS ALLOYS

Figure 1 shows the relation among the critical cooling rate for glass formation (R_c), the maximum sample thickness (t_{max}) and the reduced glass transition temperature (T_g/T_m) for typical amorphous alloys reported up to date. Here, T_g and T_m represent the glass transition temperature and melting temperature, respectively. It is well known that Fe-, Co- and Ni-based amorphous alloys found before 1990 require high cooling rates above 10^5 K/s and the resulting t_{max} is limited to less than about 50 μm. As exceptional examples, Pd-Ni-P and Pt-Ni-P amorphous alloys have lower R_c of the other 10^3 K/s and the t_{max} is in the range of 1 to 3 mm in the non-fluxed liquid state[23]. More recently, we have succeeded in finding new multicomponent amorphous alloys with much lower R_c in Mg-[24], lanthanide(Ln)-[25], Zr-[26], Fe-[11], new Pd-[27] and Co-[28]based alloy systems. The lowest R_c is as low as 10^{-1} K/s[29] and the t_{max} reaches 80 mm[30]. Besides, there is a clear tendency for GFA to increase with increasing reduced glass transition temperature (T_g/T_m). Here, it is important to point out that the amorphous alloys with lower R_c and larger t_{max} have very high T_g/T_m values exceeding 0.7. Additionally, the GFA is strongly dependent on another factor of the temperature interval of a supercooled liquid region before crystallization which is defined by the difference between the crystallization temperature (T_x) and T_g, $\Delta T_x (=T_x-T_g)$. As summarized in Fig. 2, the R_c decreases and the t_{max} increases with increasing ΔT_x. It is to be noticed that the ΔT_x value exceeds 100 K for Zr-Al-Ni-Cu[31] and new Pd-Cu-Ni-P[32] amorphous alloys.

Table I summarizes typical bulk amorphous alloy systems reported hitherto and the calendar years when their amorphous alloys were reported. The bulk amorphous alloys are classified into two groups of nonferrous and ferrous alloy systems. As the nonferrous amorphous alloy systems, one can list up Mg-Ln-(Ni,Cu or Zn)[24][33], Ln-Al-TM[25][34] (Ln=lanthanide metal, TM=VI to VIII group transition metals), Zr-Al-TM[26], Zr-Al-TM-(Ti,Nd or Pd)[35], Zr-Ti-Ni-Cu-Be[36], Pd-Cu-Ni-P[27], Pd-Cu-B-Si[37] and Pd-Ni-Fe-P[38]. Ferrous alloy systems are composed of Fe-(Al,Ga)-metalloid[11]-[14], Fe-(Zr,Hf,Nb)-B[15]-[17] and Fe-(Zr,Hf,Nb)-(Mo,W)-B[18] alloys. Thus, the bulk amorphous alloys have been developed in the order of nonferrous and ferrous alloy systems. Besides, the ferrous bulk amorphous alloys were synthesized for the last two years. With the aim of developing the above-described Fe-based bulk amorphous alloys, we looked for a guide line to search for a new amorphous alloy with high GFA. Based on the nonferrous alloy compositions with high GFA found before 1993, we have noticed the existence of three empirical rules for achievement of high GFA for metallic alloys[39]-[43]. That is, (1) multicomponent alloy systems consisting of more than three elements, (2) significant difference in atomic size ratios above about 12 % among the main three constituent elements, and (3) negative heats of mixing among the main three constituent elements. The application of the three empirical rules to Fe-based amorphous alloys enabled the synthesis of Fe-based amorphous alloys with high GFA in several alloy systems shown in Table I.

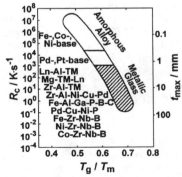

Fig. 1. Relation among the critical cooling rate for glass formation (R_c), the maximum sample thickness (t_{max}) and the reduced glass transition temperature (T_g/T_m) for typical amorphous alloys.

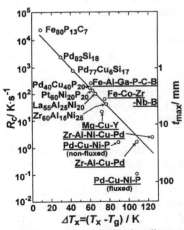

Fig. 2. Relation among the critical cooling rate for glass formation (R_c), the maximum sample thickness (t_{max}) and the temperature interval of the supercooled liquid region ($\Delta T_x=T_x-T_g$) for typical amorphous alloys.

Table I. Typical bulk amorphous alloy systems and calendar years when the alloys were discovered.

1. Nonferrous alloy systems	Years	2. Ferrous alloy systems	Years
Mg-Ln-M (Ln=lanthanide metal, M=Ni,Cu,Zn)	1988	Fe-(Al,Ga)-(P,C,B,Si,Ge)	1995
Ln-Al-TM (TM=VI~VIII group transition metal)	1989	Fe-(Nb,Mo)-(Al,Ga)-(P,B,Si)	1995
Ln-Ga-TM	1990	Co-(Al,Ga)-(P,B,Si)	1996
Zr-Al-TM	1990	Fe-(Zr,Hf,Nb)-B	1996
Ti-Zr-TM	1993	Co-(Zr,Hf,Nb)-B	1996
Zr-Ti-TM-Be	1993	Ni-(Zr,Hf,Nb)-B	1996
Zr-(Ti,Nb,Pd)-Al-TM	1995	Fe-Co-Ln-B	1998
Pd-Cu-Ni-P	1996		
Pd-Cu-B-Si	1996		
Pd-Ni-Fe-P	1997		
Ti-Zr-TM-Sn	1998		

FORMATION OF BULK AMORPHOUS ALLOYS

By choosing the above-described new multicomponent alloy systems, bulk amorphous alloys can be produced by two kinds of techniques of solidification and consolidation. As the solidification technique, one can list up water quenching[44], copper mold casting[45], high-pressure die casting[46], arc melting[47], unidirectional melting[48] and suction casting[49]. Besides, the bulk amorphous alloys are produced by hot pressing or warm extrusion of atomized amorphous powders in the supercooled liquid region before crystallization[50][51]. Table II summarizes the t_{max} by use of the solidification technique and the R_c of typical bulk amorphous alloys. The t_{max} is about 10 mm for the Ln-[52] and Mg-[53]based alloys, 30 mm[49] for the Zr-based alloys, 6 mm[18] for Fe-based alloys and 75 mm[30] for the new Pd-Cu-based alloys. As examples, Fig. 3 shows the formation of the bulk amorphous $Pd_{40}Cu_{30}Ni_{10}P_{20}$ and $Zr_{60}Al_{10}Ni_{10}Cu_{20}$ cylinders with diameters of 75 and 17 mm, respectively. The bulk amorphous alloys have smooth outer surface and good metallic luster. The R_c of the Pd-based amorphous alloy was determined to be 0.100 K/s in the B_2O_3 fluxed state and 1.58 K/s in a non-fluxed state from the continuous cooling transformation (C.C.T.) curve. The B_2O_3 flux treatment is effective for the further decrease in R_c. The effectiveness of the B_2O_3 flux treatment has been recognized for other Pd-based amorphous alloys in Pd-Ni-P[54]-[57] and Pd-Ni-Fe-P[38] systems and the bulk amorphous alloys with a maximum size of 10 mm and 30 mm, respectively, have been synthesized by the same water quenching technique. The origin of the effectiveness has been presumed[56]-[57] to be attributed to the suppression of heterogeneous nucleation by the increase in the degree of cleanness of the molten liquid.

Table II. Thermal stability and mechanical properties of typical nonferrous metal base bulk amorphous alloys.

System	t_{max}(mm)	R_c(K/s)
Ln-Al-(Cu,Ni)	$\cong 10$	$\cong 200$
Mg-Al-(Cu,Ni)	$\cong 10$	$\cong 200$
Zr-Al-(Cu,Ni)	$\cong 30$	$1 \sim 10$
Zr-Ti-Al-(Cu,Ni)	$\cong 30$	$1 \sim 5$
Zr-Ti-(Cu,Ni)-Be	$\cong 30$	$1 \sim 5$

Fig. 3. Outer surface morphology of bulk amorphous (a)$Pd_{40}Cu_{30}Ni_{10}P_{20}$ and (b)$Zr_{60}Al_{10}Ni_{10}Cu_{20}$ cylinders prepared by the water quenching technique.

FORMATION OF Fe-(Al,Ga)-(P,C,B,Si) BULK AMORPHOUS ALLOYS AND THEIR SOFT MAGNETIC PROPERTIES

As described above, the bulk amorphous alloys have useful properties of high GFA, high mechanical strength, high elastic energy up to yielding, good workability, good bondability etc. It has previously been reported that the Fe- and Co-based amorphous ribbons prepared by melt spinning exhibit good soft magnetic properties at room temperature. Based on the above-described three empirical rules for achievement of high GFA, we have searched for Fe- and Co-based amorphous alloys with a wide supercooled liquid region before crystallization because the high thermal stability of the supercooled liquid against crystallization enables the production of bulk amorphous alloys. The compositional dependence of the ΔT_x was examined for melt-spun $Fe_{80}(P,B,Si)_{20}$ amorphous alloys[58]. It have been reported that the amorphous alloys with glass transition are formed in the composition range of 9 to 13 at%P, 3 to 5 at%B and 3 to 7 at%Si and the largest ΔT_x value of 36 K is obtained for $Fe_{80}P_{12}B_4Si_4$. Subsequently, we examined the effect of additional Al and Ga elements on the increase of ΔT_x for the Fe-P-B-Si amorphous alloys because the addition of their elements causes the satisfaction of the three empirical rules. Figure 4 shows the compositional dependence of ΔT_x for the amorphous $Fe_{74}Al_4Ga_2(P,B,Si)_{20}$ alloys[58]. The simultaneous addition of Al and Ga elements extends the composition range of an amorphous phase with glass transition and increases the largest ΔT_x to 49 K. As a result, a truly internal equilibrium supercooled liquid state is obtained in the temperature range just below crystallization temperature. Besides, these amorphous alloys crystallize through a single exothermic reaction, accompanying the simultaneous precipitation of five crystalline phases of α-Fe, Fe_3P, Fe_3B, FeP and Fe_2B. This crystallization mode also indicates the necessity of long-range rearrangements of the constituent elements. However, the long-range rearrangements are not always easy in the Fe-based amorphous alloys with a higher degree of dense random packed structure, leading to a high thermal stability of supercooled liquid.

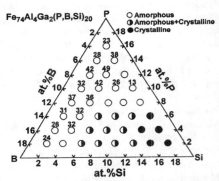

Fig. 4. Compositional dependence of ΔT_x for $Fe_{74}Al_4Ga_2(P,B,Si)_{20}$ alloys.

These Fe-based amorphous alloys exhibit good soft magnetic properties in an optimum annealed state at temperatures between Curie temperature (T_c) and T_g. Table III summarizes the

Table III. Soft magnetic properties of Fe-P-B-Si, Fe-Al-P-B-Si and Fe-Al-Ga-P-B-Si amorphous alloys ((a):As-quenched, (b)Annealed for 600 s at 723 K).

Composition		μ_e(1kHz)	H_c(A/m)	B_s(T)	B_r(T)	B_s/B_r	$\lambda_s(10^{-6})$	T_g(K)	T_x(K)	ΔT_x(T)
$Fe_{80}P_{12}B_4Si_4$	(a)	5800	1.3	1.10	0.32	0.29	31	753	789	36
	(b)	22000	1.1	1.34	0.46	0.34				
$Fe_{76}Al_4P_{12}B_4Si_4$	(a)	2600	12.7	0.96	0.30	0.31	30	738	780	46
	(b)	21000	2.6	1.24	0.43	0.35				
$Fe_{74}Al_4Ga_2P_{12}B_4Si_4$	(a)	1900	19.1	0.91	0.27	0.30	21	737	786	49
	(b)	19000	6.4	1.14	0.40	0.35				

saturated magnetization (I_s), residual magnetization (I_r), squareness ratio (I_r/I_s), coercivity (H_c) and effective permeability (μ_e) at 1 kHz for the Fe-P-B-Si, Fe-Al-P-B-Si and Fe-Al-Ga-P-B-Si amorphous alloys[58]. It is concluded that the Fe-based amorphous alloys have useful combined characteristics, i.e., good soft magnetic properties of high I_s of 1.1 to 1.3 T, low H_c of 1 to 5 A/m and high μ_e of 19000 to 22000 and high thermal stability of supercooled liquid, which have not been reported for any kinds of Fe-based amorphous alloys reported hitherto.

A wider supercooled liquid region before crystallization has also been recognized for another alloy system of Fe-Al-Ga-P-C-B[11] which satisfies the three empirical rules for high GFA. Figure 5 shows the DSC curves of the $Fe_{72}Al_5Ga_2P_{11}C_6B_4$[11] and $Fe_{72-x}Al_5Ga_2P_{11}C_6B_4M_x$ (M=Nb, Mo Cr or Co)[12] amorphous alloys. The wide supercooled liquid region exceeding 60 K is observed and the addition of a small amount of Nb, Mo or Cr is effective for further extension of the supercooled liquid region. The largest ΔT_x in the Fe-Al-Ga-P-C-B base system is 64 K for $Fe_{70}Al_5Ga_2P_{11}C_6B_4Nb_2$. It is also seen that these amorphous alloys also crystallize through a single exothermic reaction accompanying the simultaneous precipitation of five crystalline phases, in agreement with that for the Fe-Al-Ga-P-B-Si amorphous alloys. By using the Fe-Al-Ga-P-C-B alloys with ΔT_x above 55 K, a bulk amorphous alloy in a cylindrical form with a diameter of 1 mm is formed by the copper mold casting method[13], as shown in Fig. 6. The amorphous alloy cylinder has smooth outer surface and good metallic luster. The feature of the DSC curve of the bulk amorphous cylinder is just the same as that for the melt-spun amorphous alloy ribbon and no appreciable difference in T_g, ΔT_x, T_x and heat of crystallization is recognized between the cast bulk amorphous cylinder and the melt-spun amorphous ribbon. The cast amorphous cylinder also exhibits good soft magnetic properties. As an example, the hysteresis I-H loops of the $Fe_{72}Al_5Ga_2P_{11}C_6B_4$ amorphous cylinder in as cast and annealed (723 K, 600 s) states are shown in Fig. 7[13]. It is seen that the annealed bulk amorphous alloy exhibits high I_s of 1.1 T, low H_c of 3 A/m and rather low

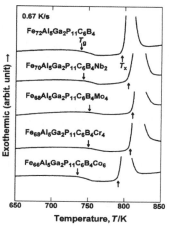

Fig. 5. DSC curves of $Fe_{72}Al_5Ga_2P_{11}C_6B_4$ and $Fe_{72-x}Al_5Ga_2P_{11}C_6B_4M_x$ (M=Nb, Mo, Cr or Co) amorphous alloys.

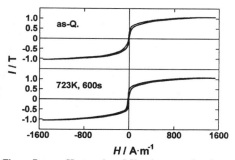

Fig. 6. Outer surface appearance of a cast amorphous $Fe_{72}Al_5Ga_2P_{11}C_6B_4$ cylinder with a diameter of 1 mm.

Fig. 7. Hysteresis I-H loops of the $Fe_{72}Al_5Ga_2P_{11}C_6B_4$ amorphous cylinder in as cast and annealed (723 K, 600 s) states.

saturated magnetostriction (λ_s) of 21×10^{-6}. Furthermore, the permeability at 1 kHz is measured to be about 7000. It has subsequently been found that the partial replacement of P by only 1 at%Si causes a further extension of the supercooled liquid region before crystallization to 67 K[59]. The use of the 1 %Si-containing alloy has enabled the production of bulk amorphous alloys in a cylindrical form with larger diameters of 2 to 3 mm and in a sheet form of 1 mm in thickness, 5 mm in width and 70 mm in length. No appreciable contrast revealing the precipitation of a crystalline phase is seen over the whole transverse cross section of the cast cylinder with a diameter of 2 mm even in an etched state. The $Fe_{72}Al_5Ga_2P_{10}C_6B_4Si_1$ amorphous cylinder with a diameter of 2 mm also exhibits better soft magnetic properties combined with a higher squareness ratio[59][60]. The bulk amorphous alloys in Fe-Al-Ga-P-C-B and Fe-Al-Ga-P-C-B-Si systems have been reported to have the T_c values of 580 to 610 K from the thermomagnetic data.

FORMATION OF $(Fe,Co,Ni)_{70}(Zr,Hf,Nb,Ta)_{10}B_{20}$ BULK AMORPHOUS ALLOYS AND THEIR SOFT MAGNETIC PROPERTIES

In addition to the Fe-(Al,Ga)-(P,C,B,Si) amorphous alloys, it was found[15] that (Fe,Co,Ni)-Zr-B amorphous alloys also exhibit a wide supercooled liquid region before crystallization and their ΔT_x values are larger than those for the previously reported Fe-based amorphous alloys. Figure 8 shows the compositional dependence of T_g for the $(Fe,Co,Ni)_{70}Zr_{10}B_{20}$ amorphous alloys. The glass transition phenomenon is observed over the whole composition range and the T_g increases monotonously from 800 to 850 K with increasing Fe and Co contents. The high T_g exceeding 850 K is believed to be the highest for all metallic amorphous alloys reported hitherto. Besides, large ΔT_x values above 65 K are obtained in the wide Fe-rich composition range of 3 to 20 at%Co and 3 to 28 at%Ni. In addition to the high stability of the supercooled liquid against crystallization, the Fe-rich amorphous alloys exhibit high Vickers hardness number (H_v) above 1000. The compositional dependence of H_v is analogous to that for T_g shown in Fig. 8.

The new Fe-based amorphous alloys were also found to exhibit good soft magnetic properties[15]. The ferromagnetic characteristics are obtained in the wide composition range except the Ni-rich corner for the $(Fe,Co,Ni)_{70}Zr_{10}B_{20}$ amorphous alloys. The H_c decreases gradually from 6 to 3 A/m with increasing Fe content and the Fe-rich amorphous alloys with ΔT_x above 65 K exhibit low H_c below 5 A/m. The compositional dependence of I_s, μ_e at 1 kHz and λ_s is also summarized in Fig. 9. The I_s increases from 0.3 to 0.9 T with increasing Fe content, while the λ_s shows zero in the Co-rich composition range and increases monotonously to 15×10^{-6} with increasing Fe content. The compositional dependence of μ_e is complicated and the maximum value of about 20000 is obtained in the Fe- and Co-rich composition ranges. From the compositional dependence of the above-described soft magnetic properties, it is clearly recognized that the new multicomponent amorphous alloys exhibit good soft magnetic properties of I_s above 0.9 T, H_c of 3 to 4 A/m, λ_s of 12-15×10^{-6} and μ_e of 20000 in the Fe-rich composition range and I_s of 0.5 T, H_c of 6 A/m, nearly zero λ_s and μ_e of 20000 in the Co-rich composition range.

It has subsequently been found that the replacement of Zr by 2 at%Nb or Ta for $Fe_{56}Co_7Ni_7Zr_{10-x}Nb_xB_{20}$ amorphous alloys causes a further increase in ΔT_x from 72 to 87 K[16][17]. Furthermore, the 2 %Nb-containing alloy exhibits an improved μ_e value of 25000 in the maintenance of nearly the same low H_c and λ_s values as those for the 10 %Zr alloy. Table IV summarizes the I_s, H_c, μ_e and T_c of the $Fe_{56}Co_7Ni_7Zr_8M_2B_{20}$ (M=Ti, Hf, V, Nb, Ta, Cr, Mo or W) amorphous alloys. It is seen that the best combination of soft magnetic properties is obtained for the Ti-, Nb-, Ta- and Cr-containing alloys and their I_s, H_c, μ_e and T_c are in the ranges of 0.75 to 0.82 T, 1.1 to 2.7 A/m, 10040 to 25000 and 503 to 531 K, respectively.

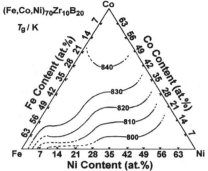

Fig. 8. Compositional dependence of T_g for $(Fe,Co,Ni)_{70}Zr_{10}B_{20}$ amorphous alloys.

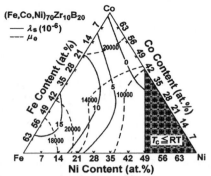

Fig. 9. Compositional dependence of I_s, μ_e at 1 kHz and λ_s for $(Fe,Co,Ni)_{70}Zr_{10}B_{20}$ amorphous alloys.

By choosing the Fe-Co-Zr-M-B (M=Nb or Ta) alloys, bulk amorphous alloys in a cylindrical form of 3 mm$^\phi$ x 110 mm and 5 mm$^\phi$ x 90 mm shown in Fig. 10 were prepared by the copper mold casting method[18]. These bulk amorphous alloys exhibit smooth outer surface and good metallic luster. No contrast revealing the precipitation of a crystalline phase is seen over the whole outer surface, indicating the formation of an amorphous phase. The t_{max}, T_g/T_m and ΔT_x values of the Fe-Co-Ni-Zr-M-B (M=Nb, Ta, Mo or W) amorphous alloys are plotted in the relation among R_c, t_{max}, T_g/T_m or ΔT_x in Fig. 11. The data of the new Fe-based alloys are located in the same relation as that for other bulk amorphous alloys, indicating the similarity of the mechanism for the achievement of high GFA between the Fe-based and other alloy base bulk amorphous alloys.

Table IV. Soft magnetic properties of $Fe_{56}Co_7Ni_7Zr_8M_2B_{20}$ (M=Ti, Hf, V, Nb, Ta, Cr, Mo or W) amorphous alloys.

M	I_s / T	H_c / Am^{-1}	μ_e(1kHz)	T_c / K
Ti	0.82	1.9	12470	528
Hf	0.80	2.2	9270	509
V	0.83	4.4	5210	530
Nb	0.75	1.1	25000	531
Ta	0.74	2.7	11970	503
Cr	0.75	1.9	10040	508
Mo	0.73	4.9	6490	490
Ta	0.70	5.7	8320	476

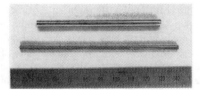

Fig. 10. Outer surface appearance of bulk amorphous $Fe_{61}Co_7Zr_{10}Mo_5W_2B_{15}$ cylinders with diameters of 3 and 5 mm.

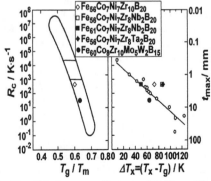

Fig. 11. Relation among R_c, t_{max} and T_g/T_m or ΔT_x for typical Fe-Co-Ni-Zr-M-B (M=Nb, Ta, Mo or W) amorphous alloys. The data of other typical bulk amorphous alloys are also shown for comparison.

FORMATION OF Co-BASED $(Co-Fe)_{70}(Zr,Nb,Ta,W)_{10}B_{20}$ AMORPHOUS ALLOYS WITH A WIDE SUPERCOOLED LIQUID REGION AND THEIR SOFT MAGNETIC PROPERTIES

In addition to the Fe-based bulk amorphous alloys, the combined properties of large ΔT_x and good soft magnetic properties have also been obtained for Co-based amorphous alloys in Co-Fe-Zr-B and Co-Fe-Zr-M-B (M=Nb, Ta or W) systems[61][62]. Although T_g of the $Co_{63}Fe_7Zr_{10-x}M_xB_{20}$ amorphous alloys decreases monotonously with increasing M content, T_x remains almost constant in the range up to 5 %M and then decreases gradually, resulting in the maximum ΔT_x value of 45 K at 4 %M. Besides, no glass transition is observed for the alloys containing less than 1 at%M. This indicates the effectiveness of the simultaneous addition of Zr and M elements for the increase in the stability of the supercooled liquid against crystallization. Figure 12 shows the changes with M content in I_s, H_c, λ_s and μ_e of the $Co_{63}Fe_7Zr_{10-x}M_xB_{20}$ (M=Nb, Ta or W) amorphous alloys annealed for 600 s at the temperatures between T_c and T_g. The Co-based amorphous alloys with the glass transition phenomenon exhibit good soft magnetic properties of low H_c of 3 A/m, low λ_s of 2×10^{-6} and high μ_e of 20000 at 1 kHz. Furthermore, the Co-based amorphous alloys

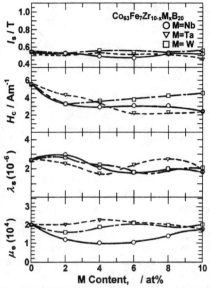

Fig. 12. Changes in I_s, H_c, λ_s and μ_e with M content for $Fe_{63}Co_7Zr_{10-x}M_xB_{20}$ (M=Nb, Ta or W) amorphous alloys annealed for 600 s at the temepratures between T_c and T_g.

with glass transition were found[62] to exhibit high-frequency μ_e, as is evidenced from the high μ_e value of 6500 at 1 MHz which is much superior to those for commercial Fe- and Co-based amorphous alloys, as shown in Fig. 13. The good high-frequency μ_e has also been recognized from the result that the imaginary part of μ_e shows a maximum near the high frequency of 0.8 MHz. The high-frequency μ_e is further improved by the simultaneous addition of Zr and Ta elements and the high μ_e value of 6700 is obtained at 1 MHz. These high-frequency μ_e characteristics have not been obtained for any kinds of soft magnetic materials reported hitherto. With the aim of investigating the reason for the achievement of the excellent high-frequency μ_e, the electrical resistivity (ρ) at room temperature was measured for the Co-based amorphous alloys. The ρ is in the range from 1.7 to 2.1 $\mu\Omega m$ which are much higher than those[2][63] for previously reported ferromagnetic amorphous alloys. It is therefore presumed that the high ρ causes a decrease of eddy current loss in the high frequency range, leading to the excellent high-frequency μ_e. However, it is rather difficult to consider that the appearance of the high frequency μ_e originates only from the decrease of the eddy current loss. The increase of homogeneity in the domain wall structure by the increase in the GFA as well as the low λ_s and H_c values seems to play an important role in the achievement of the excellent high-frequency μ_e.

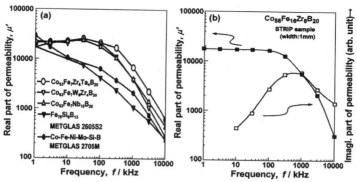

Fig. 13. High-frequency permeability of Co-Fe-Zr-B and Co-Fe-Zr-(Nb or Ta)-B amorphous alloys annealed for 600 s at the temperatures between T_c and T_g.

FORMATION OF $(Fe,Co,Ni)_{62}Nb_8B_{30}$ AMORPHOUS ALLOYS WITH A WIDE SUPERCOOLED LIQUID REGION AND THEIR SOFT MAGNETIC PROPERTIES

Figure 14 (a) and (b) show the compositional dependence of T_g and ΔT_x of the $(Fe,Co,Ni)_{62}Nb_8B_{30}$ amorphous alloys, respectively. Amorphous alloys were formed over the entire composition range including $Fe_{62}Nb_8B_{30}$, $Co_{62}Nb_8B_{30}$ and $Ni_{62}Nb_8B_{30}$. As seen in Fig. 2, the glass transition is observed in a wide composition range except the Ni-rich alloys containing more than about 50 at%Ni. The T_g shows a maximum in the Fe-rich composition range and decreases with increasing Co and Ni contents. The decrease in T_g is more significant for Ni. The resulting ΔT_x shows the maximum values above 80 K for the $Fe_{9-46}Co_{6-43}Ni_{0-10}Nb_8B_{30}$ alloys and decreases with increasing Co and Ni contents. However, it is noticed that the wide supercooled liquid region exceeding 60 K is obtained in the wide composition range of 8 to 62 %Fe, 0 to 53 at%Co and 0 to 28 at%Ni. Considering the previous data[62] that the ΔT_x value of a $Fe_{72}Nb_8B_{20}$ amorphous alloy is 30 K, the increase in B content from 20 to 30 at% is concluded to be very effective for the increase of the thermal stability of the supercooled liquid against crystallization.

Figure 15 shows the compositional dependence of I_s and H_c for the $(Fe,Co,Ni)_{62}Nb_8B_{30}$

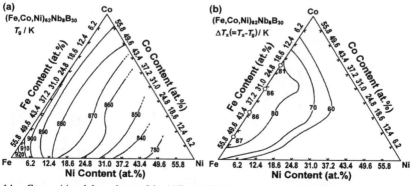

Fig. 14. Compositional dependence of the (a)T_g and (b)ΔT_x for the $(Fe,Co,Ni)_{62}Nb_8B_{30}$ amorphous alloys.

amorphous alloys. The I_s is 0.68 T for $Fe_{62}Nb_8B_{30}$ and decreases monotonously to 0.2 T with increasing Co and Ni contents and becomes zero because of the decrease of T_c down to room temperature. The H_c shows the lower values below 2.3 A/m for the Fe-Co-Nb-B alloys containing less than about 5 at%Ni and tends to increase slightly with increasing No content. Thus, the H_c values appear to be dependent on only Ni content. The low H_c values allow us to expect that the Fe-Co-Nb-B amorphous alloys exhibit high μ_e. Figure 16 shows the compositional dependence of μ_e at 1 kHz and λ_s for the $(Fe,Co,Ni)_{62}Nb_8B_{30}$ amorphous alloys. The highest μ_e exceeding 28000 is obtained for the Co-rich $Co_{40-50}Fe_{12-24}Ni_{0-3}Nb_8B_{30}$ alloys and the high μ_e above 20000 is also maintained over the whole composition range. Furthermore, the λ_s also shows very low values of $2.4-7.0 \times 10^{-6}$ in the whole composition range where the T_c is above room temperature. Table V summarizes the thermal stability, soft magnetic properties and ρ of the $Fe_{52}Co_{10}Nb_8B_{30}$ and $Co_{40}Fe_{22}Nb_8B_{30}$ amorphous alloys together with the data of the other amorphous alloys including commercial Fe-Si-B and Co-Fe-Ni-Mo-B-Si amorphous alloys. The features of the present Fe-Co-Nb-B amorphous alloys are summarized as follows; (1) high thermal stability which is evidenced from the higher T_x values and the wide supercooled liquid region above 80 K before crystallization, (2) high μ_e at 1 kHz comparable to the METGLASS 2705M[2] for the Co-rich Co-Fe-Nb-B alloy, (3) lower λ_s as compared with the METGLAS 2605S2 alloy[2], (4) much higher ρ of 232 to 237 $\mu\Omega$cm which is about 1.7 times higher than those for the commercial Fe- and Co-based amorphous alloys, and (5) much higher high-frequency μ_e exceeding 7000 at 1 MHz which is one order higher than those for the commercial Fe- and Co-based amorphous alloys. However, the λ_s values are considerably lower than that for the commercial Fe-based amorphous alloy, though they are nearly the same as that for the commercial Co-based amorphous alloy. It is thus concluded that the present B-containing alloys have higher thermal stability, higher high-frequency μ_e and higher ρ. The higher ρ is presumably because the present alloys contain a large amount (30 at%) of B. The high ρ is thought to cause the maintenance of extremely high high-frequency μ_e exceeding 7000 even at the high frequency of 1 MHz through the decrease in the eddy current loss. The structure analysis on the local atomic configurations in the high B-containing amorphous alloys is under investigation and the result will shed some insight on the clarification of the mechanisms for the high thermal stability of the supercooled liquid, high ρ and high μ_e.

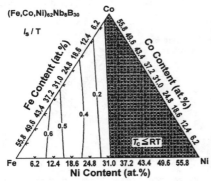

Fig. 15. Compositional dependence of I_s and H_c for the $(Fe,Co,Ni)_{62}Nb_8B_{30}$ amorphous alloys.

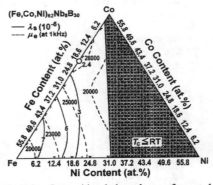

Fig. 16. Compositional dependence of μ_e at 1 kHz and λ_s for the $(Fe,Co,Ni)_{62}Nb_8B_{30}$ amorphous alloys.

Table V. Thermal stability and magnetic properties of the $Fe_{52}Co_{10}Nb_8B_{30}$ and $Co_{40}Fe_{22}Nb_8B_{30}$ amorphous alloys.

Alloy	T_g (K)	T_x (K)	ΔT_x (K)	I_s (T)	H_c (Am^{-1})	λ_s (10^{-6})	μ_e (at 1kHz)	μ_e (at 1MHz)	ρ_{RT} ($\mu\Omega$cm)
$Fe_{52}Co_{10}Nb_8B_{30}$	907	994	87	0.63	2.1	7.4	21000	4400	232
$Co_{40}Fe_{22}Nb_8B_{30}$	895	976	81	0.41	2.0	2.4	29300	7500	237

CONCLUSIONS

Tables IV and V summarize the magnetic characteristics of the Fe- and Co-based amorphous alloys with high stability of supercooled liquid against crystallization which enables the production of bulk amorphous alloys. As the features of magnetic properties for the present Fe- and Co-based amorphous alloys, one can list up the low H_c, low λ_s and high μ_e, though I_s is relatively lower because of the dissolution of a large amount of solute elements. In particular, the high frequency μ_e characteristics are excellent and cannot be obtained for any kinds of other soft magnetic materials. In addition to the good soft magnetic properties, the bulk amorphous alloys in the (Fe,Co)-TM-B (TM=Zr, Nb, Ta) systems also exhibit the following other features, i.e., (1) high dense random packed structure, (2) homogeneous mixing of the constituent elements, (3) high electrical resistivity, (4) high mechanical strength, (5) high corrosion resistance, (6) high T_g and T_x, (7) wide supercooled liquid region before crystallization, (8) large viscous flow, and (9) easy castability to near net shape. The unique and useful combination is expected to enable the future development of the present new Fe- and Co-based bulk amorphous alloys as a new type of soft magnetic material.

REFERENCES

1. W. Klement, R.H. Willens and P. Duwez, Nature, **187**, 869 (1960).
2. C.H. Smith in *Rapidly Solidified Alloys*, edited by H.H. Liebermann (Marcel Dekker, New York 1993), p. 617-663.
3. A. Inoue and L. Arnberg in *The Encyclopedia of Advanced Materials, Vol.1*, edited by R.W. Cahn (Pergamon 1994), p.217-227.
4. P. Duwez and S.C.H. Lin, J. Appl. Phys., **38**, 4096 (1967).
5. T. Masumoto, H.M. Kimura, A. Inoue and Y. Waseda, Mater. Sci. Eng., **23**, 141 (1976).
6. M. Kikuchi, H. Fujimori, T. Obi and T. Masumoto, Jpn. J. Appl. Phys., **14**, 1077 (1975).
7. A. Inoue, T. Masumoto, S. Arakawa and T. Iwadachi in *Rapidly Quecnhed Metals III*, edited by B. Cantor (The Metals Society, London 1978), Vo.I, p.265.
8. A. Inoue, K. Kobayashi, M. Nose and T. Masumoto, J. Phys. C-8, **41**, 831 (1980).
9. A. Inoue, K. Kobayashi, J. Kanehira and T. Masumoto, Sci. Rep. Res. Inst. Tohoku Univ., **29A**, 331 (1981).
10. H.A. Davies in *Amorphous Metallic Alloys*, edited by F.E. Luborsky (Butterwoths, London 1983), p.14.
11. A. Inoue and J.S. Gook, Mater. Trans., JIM, **36**, p.1282 (1995).
12. A. Inoue and J.S. Gook, Mater. Trans., JIM, **37**, p.32 (1996).
13. A. Inoue, Y. Shinohara and J.S. Gook, Mater. Trans., JIM, **36**, 1427 (1985).
14. A. Inoue, A. Murakami, T. Zhang and A. Takeuchi, Mater. Trans., JIM, **38**, 189(1997).
15. A. Inoue, T. Zhang and T. Itoi, Mater. Trans., JIM, **38**, 359 (1997).
16. A. Inoue, H. Koshiba, T. Zhang and A. Makino, Mater. Trans., JIM, **38**, 577 (1997).
17. A. Inoue, H. Koshiba, T. Zhang and A. Makino, J. Appl. Phys., **83** 1967 (1998).
18. A. Inoue, T. Zhang and A. Takeuchi, Appl. Phys. Lett., **71**, 464 (1997).

19. A. Inoue, W. Zhang and A. Takeuchi, Mater. Trans., JIM, to be submitted.
20. A. Inoue, T. Zhang, W. Zhang and A. Takeuchi, Mater. Trans., JIM, **37**, 99 (1996).
21. A. Inoue, T. Zhang and A. Takeuchi, Mater. Trans., JIM, **37**, 1731 (1996).
22. A. Inoue, T. Zhang and A. Takeuchi, IEEE Trans. Magn., **33**, 3814 (1997).
23. H.S. Chen, Rep. Prog. Phys., **43**, 353 (1980).
24. A. Inoue, K. Ohtera, K. Kita and T. Masumoto, Jpn. J. Appl. Phys., **27**, L2,248 (1988).
25. A. Inoue, T. Zhang and T. Masumoto, Mater. Trans., JIM, **30**, 965(1989).
26. A. Inoue, T. Zhang and T. Masumoto, Mater. Trans., JIM, **31**, 177 (1990).
27. A. Inoue, N. Nishiyama and T. Matsuda, Mater. Trans., JIM, **37**, 181 (1996).
28. A. Inoue and A. Katsuya, Mater. Trans., JIM, **37**, 1332 (1996).
29. A. Inoue and N. Nishiyama, Mater. Sci. Eng., **A226-228**, 401 (1997).
30. A. Inoue, N. Nishiyama and H.M. Kimura, Mater. Trans., JIM, **38**, 179 (1997).
31. T. Zhang, A. Inoue and T. Masumoto, Mater. Trans., JIM, **32**, 1005 (1991).
32. N. Nishiyama and A. Inoue, Mater. Trans., JIM, **37**, 1531(1996).
33. S.G. Kim, A. Inoue and T. Masumoto, Mater. Trans., JIM, **31**, 929 (1990).
34. A. Inoue, H. Yamaguchi, T. Zhang and T. Masumoto, Mater. Trans., JIM, **31**, 104 (1990).
35. A. Inoue, T. Shibata and T. Zhang, Mater. Trans., JIM, **36**, 1420 (1995).
36. A. Pecker and W.L. Johnson, Appl. Phys. Lett., **63**, 2342 (1993).
37. A. Inoue, T. Aoki and H.M. Kimura, Mater. Trans., JIM, **38**, 175 (1997).
38. R.W. Schwarz and Y. He, Mat. Sci. Forum, **235-238** 231 (1997).
39. A. Inoue, Mater. Trans., JIM, **36**, 866 (1995).
40. A. Inoue, Sci. Rep. Res. Inst. Tohoku Univ., **A42**, 1 (1996).
41. A. Inoue, Proc. Japan Acad., Ser.B,   **No.2**, 19 (1997).
42. A. Inoue, Mater. Sci. Eng., **A226-228**, 357(1997).
43. A. Inoue A. Takeuchi and T. Zhang, Metal. Mater. Trans., **29A** 1779 (1998).
44. A. Inoue, K. Kita, T. Zhang and T. Masumoto, Mater. Trans., JIM, **30**, 722 (1989).
45. A. Inoue, T. Zhang and T. Masumoto, Mater. Trans., JIM, **31**, 425 (1990).
46. A. Inoue, T. Nakamura, N. Nishiyama and T. Masumoto, Mater. Trans., JIM, **33**, 937 (1992).
47. Y. Yokoyama and A. Inoue, Mater. Trans., JIM, **36**, 1,398 (1995).
48. A. Inoue, Y. Yokoyama, Y. Shinohara and T. Masumoto, Mater. Trans., JIM, **35**, 923 (1994).
49. A. Inoue and T. Zhang, Mater. Trans., JIM, **36**, 1184 (1995).
50. H. Kato, Y. Kawamura and A. Inoue, Mater. Trans., JIM, **37**, 70 (1996).
51. Y. Kawamura, A. Kato, A. Inoue and T. Masumoto, Int. J. Powder Metall., **33**, 50 (1997).
52. A. Inoue, T. Nakamura, T. Sugita, T. Zhang and T. Masumoto, Mater. Trans., JIM, **34**, 351 (1993).
53. A. Inoue, A. Kato, T. Zhang, S.G. Kim and T. Masumoto, Mater. Trans., JIM, **32**, 609 (1991).
54. H.W. Kui, A.L. Greer and D. Turnbull, Appl. Phys. Lett., **45**, 615 (1984).
55. H.W. Kui and D. Turnbull, Appl. Phys. Lett., **47**, 796 (1985).
56. R. Willnecker, K. Wittmann and G.P. Gorler, J. Non-Cryst. Solids, **156-158**, 450 (1993).
57. R.W. Schwarz and Y. He, Mater. Sci. Forum, **235-238**, 231 (1997).
58. A. Inoue and R.E. Park, Mater. Trans., JIM, **37**, 1715 (1996).
59. T. Mizushima, A. Makino and A. Inoue, IEEE Trans. Magn., **33**, 3784 (1997).
60. A. Inoue and A. Makino, MMM, Grenoble, September (1997), in press.
61. A. Inoue, T. Itoi, H. Koshiba and A. Makino, Appl. Phys. Lett., **73** 744 (1998).
62. T. Itoi and A. Inoue, Mater. Trans., JIM, **39** 762 (1998).
63. *Materials Science of Amorphous Metals*, edited by T. Masumoto et al. (Ohmu, Tokyo 1982), p.97.

CRYSTALLIZATION OF SUPERCOOLED $Zr_{41}Ti_{14}Cu_{12}Ni_{10}B_{23}$ MELTS DURING CONTINUOUS HEATING AND COOLING

Jan Schroers, Andreas Masuhr, Ralf Busch and William L. Johnson

Keck Laboratory of Engineering Materials, California Institute of Technology, Pasadena, CA 91125

ABSTRACT

The crystallization behavior of the bulk glass forming $Zr_{41}Ti_{14}Cu_{12}Ni_{10}Be_{23}$ liquid was studied under different heating and cooling rates. Investigations were performed in high purity graphite crucibles since heterogeneous surface nucleation at the container walls does not effect the crystallization of the bulk sample. A rate of about 1 K/s is sufficient to circumvent crystallization of the melt while cooling from the equilibrium melt. In contrast, upon heating a rate of more than 150 K/s is necessary to avoid crystallization of $Zr_{41}Ti_{14}Cu_{12}Ni_{10}Be_{23}$ samples. The difference between the critical heating and cooling rate is discussed within classical nucleation theory and diffusion limited crystal growth. The calculated difference of the critical heating and cooling rate can be explained by the fact that nuclei formed during cooling and heating are expose to different growth rates.

INTRODUCTION

The ability to form a glass by cooling from the equilibrium liquid is equivalent to suppressing crystallization within the supercooled (undercooled) liquid. One of the central quantities in theoretical and experimental studies of glass formation is the critical cooling rate, R_c, to bypass crystallization upon cooling from the stable melt [1]. Besides its fundamental importance in quantifying the glass forming ability of a system, its practical importance stems from the fact that it is directly linked to the maximum thickness of a quenched amorphous ingot. Critical cooling rates for monoatomic metallic systems are typically on the order of 10^9 K/s. Recently discovered multicomponent alloys [2,3] exhibit excellent glass forming ability. The critical cooling rate for glass formation of the $Zr_{41}Ti_{14}Cu_{12}Ni_{10}Be_{23}$ (Vit 1) alloy, used in this work, is about 1 K/s [4].

In the past, much attention has been given to the onset of crystallization from amorphous Vit 1 [5,6,7,8]. The crystallization temperature appears to be strongly dependent on the heating rates [5]. Since these investigations were performed in a DSC the maximum heating rate was limited to about 5 K/s. In this paper, the onset of crystallization was investigated as a function of heating and cooling rate from the stable melt and amorphous Vit1, respectively. The experimental setup permits maximum heating rates of 350 K/s and maximum cooling rates of 40 K/s.

EXPERIMENTAL METHODS

Amorphous samples were prepared by arc-melting the constituents (purity ranges from 99.5 to 99.995 at.%) in a titanium gettered argon atmosphere. The graphite crucibles (machined from POCO Graphite, grade DFP-1, supplied by EDM Supplies Inc.) were heat treated at 1300 K for 30 min and 10^{-6} mbar prior to the experiment.

The samples were introduced into graphite crucibles and inductively heated in a titanium gettered argon atmosphere. The temperature was measured using a thermocouple (Type K) with an accuracy better than ±2 K. Maximum heating and cooling rates are as high as 350 K/s and 40 K/s upon heating and cooling, respectively. More details of the experimental setup can be found elsewhere [9].

The graphite crucible is designed to contain the Vit 1 sample as well as in a different slot a reference material to calibrate the temperature for the different heating rates. We choose tin with a melting temperature of $T_M^{Sn} = 505$ K for calibration. In conjunction with the eutectic temperature of Vit 1 of $T_{eut}^{Vit1} = 937$ K, taken from reference [5], we have two calibration temperatures in the temperature region of interest.

RESULTS

In Fig 1(a) various cooling rates as a function of temperature are shown for Vit 1 cooled from the equilibrium melt. The rates were calculated by differentiating the digitally recorded temperature-time profiles Prior to each cooling cycle the sample was overheated to 1175 K for 100 s. The cooling rates varies between 0.1 K/s and 2 K/s. The crystallization of the melt is detected by a temperature rise, so called recalescence, which has its origin in the release of the heat of fusion at the solid/liquid interface during crystallization. The recalescence leads to a decrease of the cooling rate. With increasing cooling rate the onset temperature of crystallization decreases. For rates larger than about 1 K/s no recalescence was detected. Therefore, cooling Vit 1 from the equilibrium melt faster than the critical cooling rate of about $R_c = 1$ K/s leads to the formation of a glass.

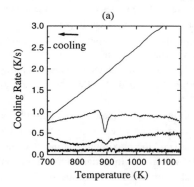 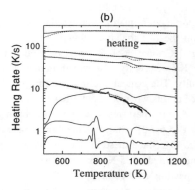

Fig 1:   Cooling (a) and heating (b) rates as a function of temperature for Vit 1 cooled from the equilibrium melt and heated from the amorphous state. In (b), solid curves (—) and dotted curves (- - -) correspond to heating curves from amorphous and crystalline Vit 1, respectively.

Fig 1(b) depicts the different heating curves for amorphous and crystalline Vit 1 samples heated with rates from 0.5 K/s up to 200 K/s. Prior to each heating procedure the sample was overheated to 1175 K and subsequently cooled to room temperature with a rate of 5 K/s which results in the formation of an amorphous sample. Crystalline samples were prepared by cooling from 1175 K with a rate of 0.2 K/s. A two point temperature calibration was performed with the melting peak of tin and the eutectic temperature of Vit 1. As an example, for an amorphous sample heated with 10 K/s the temperature was shifted by 20 K for T_M^{Sn} and by 28 K for T_{eut}^{Vit1}. After this temperature calibration the crystallization event of this particular sample is at $T_x = 816$ K. The heating of a crystalline Vit 1 sample with the same rate is denoted in Fig 1(b) by the dashed line. This branch only shows the melting of the crystalline Vit 1 sample. For low heating rates up to 7 K/s a single point calibration method was performed with the eutectic temperature of Vit 1. For heating rates above 60 K/s the melting of tin can not be determined from the heating curve since the large feature in the curve resulting from the setup itself superimpose it. Therefore, extrapolation of the temperature correction from the lower heating rates for T_M^{Sn} were performed.

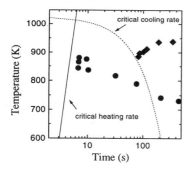

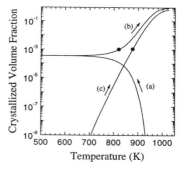

Fig 2: Continuos cooling and heating diagram of Vit 1. (♦) denotes the onset of crystallization for samples cooled from the equilibrium melt. The onset of crystallization for samples heated from the amorphous state is shown by (•) The critical cooling rate of about 1 K/s and the critical heating rate of approximately 200 K/s are denoted by (- - -) and (—), respectively.

Fig 3: Simulation for linear heating and cooling of Vit 1. Cooling from the equilibrium melt (a) with 1 K/s results in a volume fraction of crystallized material of $3.5 \cdot 10^{-4}$. Reheating this sample with 1 K/s (b) leads to the detection of the onset of the recalescence at 820 K. For heating a perfect amorphous sample with 1 K/s (c) the onset of the recalescence occurs at 880 K.

With increasing heating rate the crystallization temperatures shifted to higher temperatures. For the lowest heating rate of 0.5 K/s the onset of crystallization was detected at 730 K. A T_x of 880 K was observed for a heating rate of 170 K/s. Heating an amorphous Vit 1 sample with a rate of 200 K/s circumvents the crystallization of Vit 1 melts.

Fig 2 depicts the continuos heating and cooling diagram for Vit 1. In this diagram the time at which crystallization is detected for various linear cooling and heating rates is plotted versus the temperature. In addition, critical cooling (---) and heating rates (—) are shown.

DISCUSSION

According to our experiments the critical cooling rate for Vit 1 is about 1 K/s. The critical heating rate R_h of approximately 200 K/s is about two orders of magnitude larger. In the framework of steady state nucleation and crystalline growth, the onset of crystallization is somewhat arbitrarily defined as the point in time, where the crystalline volume fraction within the melt reaches some small but finite value. With the present setup a crystallized volume fraction of about 10^{-3} can be detected. The exact value of the detectable volume fraction has, however, marginal influence on the present discussion. In general, crystallization of a melt requires the formation of nuclei and subsequent growth of crystalline phase(s). Within classical nucleation theory, the steady state nucleation rate,

$$I_{ss} = A \cdot D \cdot \exp\left(-\frac{\Delta G^*}{kT}\right) \tag{1}$$

is written as the product of an effective diffusivity D times a constant A, and the thermodynamic Boltzmann factor of overcoming a nucleation barrier. T denotes the absolute temperature and k is Boltzmann's constant. Steady state nucleation can only serve as a first approximation of nucleation behavior of a complex system like Vit 1 [10] and other processes have to be included to describe the nucleation behavior [7,11]. The activation energy to form a critical nucleus is given by

$$\Delta G^* = \frac{16\pi \cdot \sigma^3}{3 \cdot \Delta G^2}. \tag{2}$$

Here σ denotes the energy of the interface between the melt and nucleus and ΔG the difference in Gibbs free energy of the solid and the liquid phase. The crystalline growth velocity can be described by

$$u = \frac{D}{a} \cdot \left[1 - \exp\left(-\frac{\Delta G}{kT}\right)\right] \tag{3}$$

with a as an interatomic spacing. Considering three dimensional growth and a steady state nucleation rate, the time-dependent volume fraction x, of crystallized material is obtained by integrating over all nucleation events:

$$x(t) = \frac{4\pi}{3} \cdot \int_0^t I(T,\tau)\left[\int_\tau^t u(T,t')\,dt'\right]^3 d\tau. \tag{4}$$

The double integral sums over all nucleation centers, appearing at time, τ, and their growth from time, τ, to time, t. The integral can be numerically solved for linear cooling with a rate R described by $T(t) = T_{liq} - R \cdot t$ (T_{liq} : liquidus temperature). Fig 3 depicts the crystallized volume as a function of temperature. Data for the effective diffusivity as well as for $\sigma = 0.04$ J/m^2 and $A = 10^{11.1}$ were taken from [12]. Differential scanning calorimetry results from [5] were taken as an estimate for ΔG. The crystallized volume fraction increases continuously upon cooling with 1 K/s from the liquidus temperature and becomes approximately constant at 600 K (curve (a) in Fig 3). The simulation ends at 450 K due to the freezing of the crystallization kinetics. The total crystallized volume fraction of $3.5 \cdot 10^{-4}$ would not be observed in the thermocouple signal of the experimental setup used. Continuing the simulation by heating the sample with the same rate of 1 K/s leads to the detection of the onset of crystallization ($x = 1 \cdot 10^{-3}$) at 820 K (curve (b) in Fig 3). For comparison, a simulation with a perfect amorphous sample heated with 1 K/s is also shown in Fig 3 by curve (c). Here, the sample 'crystallizes' at 880 K.

The number of nuclei formed during heating of a perfectly amorphous sample in this simulation is exactly the same as the number of nuclei formed during cooling from the liquidus temperature. However, the nuclei formed during cooling and heating are exposed to different growth rates. The growth rate calculated with the above mentioned parameters has a maximum at a relatively high temperatures of 985 K. Upon heating, nuclei formed at, e.g. the 'nose' temperature in the TTT-diagram of 880 K [13] will be exposed to these high growth rates before reaching the liquidus temperature. In contrast, nuclei formed upon cooling at the same temperature will experience lower growth rates. The result is that starting with a perfect amorphous sample the volume fraction crystallized during heating is larger than the volume fraction crystallized during cooling with the same rate from the liquidus temperature. If the effect of quenched-in nuclei is taken into account the difference is even larger. The critical cooling rate is therefore smaller than the critical heating rate to keep the crystallized volume fraction below a certain value. The calculations yields $R_c = 1$ K/s and $R_h = 9$ K/s. This suggest that this effect at least in part accounts for the experimental finding of a large asymmetry between R_c and R_h. Further contributing factors may be:

(1) Since the critical nucleation size decreases with undercooling, not only overcritical clusters (nuclei) are quenched-in upon cooling but also undercritical clusters. These undercritical clusters become overcritical at low temperatures. At reheating, their subsequent growth additionally contributes to the crystallization process, leading to an even stronger asymmetry between the critical cooling and the critical heating rate.

(2) There are indications, especially in the Vit 1 system, for a chemical decomposition process within the undercooled melt at low temperatures in the glass transition region [7]. These concentration modulations are likely to cause an increasing of the nucleation probability. The resulting high nucleation rate would also lead to a faster crystallization process upon reheating.

CONCLUSIONS

Linear heating and cooling experiments were performed on Vit 1 samples and critical heating and cooling rates were determined. Cooling Vit 1 from the equilibrium melt requires a rate of about 1 K/s to circumvent crystallization. By heating amorphous Vit 1 with a rate of approximately 200 K/s no crystallization event could be detected. The effect, that nuclei formed during cooling and heating are exposed to different growth rates at least in part accounts for the difference between critical cooling and heating rate. However, the observed microstructures [10] are much finer than would be expected after steady state nucleation. Therefore, additional contributions such as phase separation and quenched-in undercritical clusters have to be considered to explain the large asymmetry between the critical cooling and critical heating rate.

ACKNOWLEDGEMENT

This work was supported by the National Aeronautics and Space Administration (Grant No. NCC8-119), the Department of Energy (Grant No. DEFG-03086ER45242) and the ALCOA Technical Center.

REFERENCES

[1]   P.G. Debenedetti, Metastable Liquids, Princeton University Press, (1996).

[2]   A. Inoue, T. Zhang and T. Masumoto, Mater. Trans. JIM **31**, 177 (1990).

[3]   A. Peker and W.L. Johnson, Appl. Phys. Lett. **63**, 2342 (1993).

[4]   Y.J. Kim, R. Busch, W.L. Johnson, A.J. Rulison and W.K. Rhim, Appl. Phys. Lett. **65**, 2136 (1994).

[5]   R. Busch, Y.J. Kim, and W.L Johnson, J. Appl. Phys. **77**, 4039 (1995).

[6]   R. Busch, S. Schneider, A. Peker, and W.L. Johnson, Appl. Phys. Lett. **67**, 1544 (1995).

[7]   S. Schneider, P. Thiyagarajan, and W.L. Johnson, Appl. Phys. Lett. **68**,493 (1996).

[8]   A. Wiedemann, U. Keiderling, M. P. Macht, H. Wollenberger, Mat. Sci. Forum **225-227**, 71 (1996).

[9]   A. Masuhr, PhD Thesis, California Institute of Technology, (1998).

[10] J. Schroers, R. Busch, A. Masuhr and W.L. Johnson, J. Non-Cryst. Solids, in press.

[11] K.F. Kelton, Phil. Mag. Lett. **77**, 337 (1997).

[12] A. Masuhr, T.A. Waniuk, R. Busch and W.L. Johnson, to be published.

[13] A. Masuhr, R. Busch and W.L. Johnson, Mater. Sci. Forum **269-272**, 779 (1998).

SELF-DIFFUSION IN BULK METALLIC GLASSES

K. KNORR[1], M.-P. MACHT[2], H. MEHRER[1]
[1]Universität Münster, Institut für Metallforschung, D-48149 Münster, Germany,
knorrk@uni-muenster.de
[2]Hahn-Meitner-Institut Berlin GmbH, D-14109 Berlin, Germany

ABSTRACT

We have studied self-diffusion in the bulk metallic glasses $Zr_{46.75}Ti_{8.25}Cu_{7.5}Ni_{10}Be_{27.5}$ and $Zr_{65}Cu_{17.5}Ni_{10}Al_{7.5}$ by means of the radiotracer method. Diffusion of ^{63}Ni has been investigated as a function of temperature in both alloys and also as a function of hydrostatic pressure in $Zr_{46.75}Ti_{8.25}Cu_{7.5}Ni_{10}Be_{27.5}$. With the isotope ^{95}Zr diffusion studies of the major component Zr were performed in $Zr_{46.75}Ti_{8.25}Cu_{7.5}Ni_{10}Be_{27.5}$. The diffusivity of ^{95}Zr is much smaller than that of ^{63}Ni. The temperature dependence of ^{63}Ni self-diffusion into $Zr_{46.75}Ti_{8.25}Cu_{7.5}Ni_{10}Be_{27.5}$ and into $Zr_{65}Cu_{17.5}Ni_{10}Al_{7.5}$ cannot be described by a single set of Arrhenius parameters, breaks in the Arrhenius curves are observed. We attribute the non-linear Arrhenius behaviour to the transition from the glassy to the supercooled liquid state. For the first time activation volumes of diffusion in a supercooled melt have been determined. From the pressure dependence of ^{63}Ni diffusion in $Zr_{46.75}Ti_{8.25}Cu_{7.5}Ni_{10}Be_{27.5}$ we get activation volumes around one mean atomic volume favouring a diffusion mechanism via vacancy-like defects.

INTRODUCTION

In recent years, novel multicomponent Zr-based alloys like Zr-Ti-Cu-Ni-Be and Zr-Cu-Ni-Al were developed for which cooling rates of less than $100\,Ks^{-1}$ are sufficient for vitrification [1-5]. Hence the production of bulk metallic glasses has become possible. They show a fairly wide temperature range between the glass transition and the crystallization temperature and a remarkable resistance to crystallization. This enables studies of physical properties in the supercooled liquid region which is experimentally inaccessible for classical metallic glasses. Investigations concerning structural relaxation, phase separation, and crystallization were carried out [5-17]. Diffusion plays a major role in these processes. So far, mainly impurity diffusion of various diffusants was investigated [18-21]. Contrary to observations in many other less complex amorphous alloys a distinct non-linear Arrhenius behaviour of the diffusivities was observed in some cases [19,20]. Though several explanations of the break in the Arrhenius curve were proposed, this non-linearity is still a point of discussion.

We report a study of the temperature dependence of ^{63}Ni and ^{95}Zr self-diffusion in $Zr_{46.75}Ti_{8.25}Cu_{7.5}Ni_{10}Be_{27.5}$ and of ^{63}Ni self-diffusion in $Zr_{65}Cu_{17.5}Ni_{10}Al_{7.5}$. In $Zr_{46.75}Ti_{8.25}Cu_{7.5}Ni_{10}Be_{27.5}$ the non-linear Arrhenius behaviour of diffusion will be discussed including impurity diffusion data reported by other authors.

Activation volumes of diffusion in crystalline solids have contributed significantly to the understanding of diffusion mechanisms [22]. We report measurements of ^{63}Ni diffusion into $Zr_{46.75}Ti_{8.25}Cu_{7.5}Ni_{10}Be_{27.5}$ under hydrostatic pressure which provide activation volumes of diffusion. The results are discussed together with isotope effect data concerning the diffusion mechanism. A preliminary account of some results for $Zr_{46.75}Ti_{8.25}Cu_{7.5}Ni_{10}Be_{27.5}$ has recently been published [23].

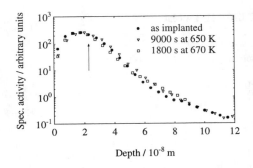

Figure 1: *Concentration-depth profiles in $Zr_{46.75}Ti_{8.25}Cu_{7.5}Ni_{10}Be_{27.5}$ of ^{95}Zr self-diffusion. The arrow marks the projected range for 60 keV ^{95}Zr tracer ions simulated with the program TRIM [27].*

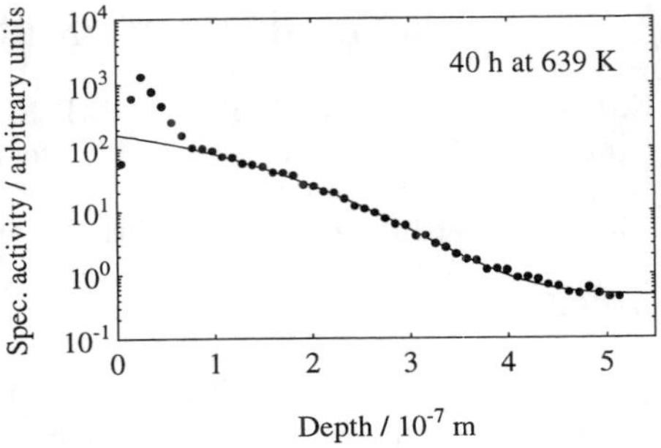

Figure 2: *Concentration-depth profile in $Zr_{46.75}Ti_{8.25}Cu_{7.5}Ni_{10}Be_{27.5}$ of ^{63}Ni self-diffusion.*

EXPERIMENT

Ingots with the nominal composition of $Zr_{46.75}Ti_{8.25}Cu_{7.5}Ni_{10}Be_{27.5}$ were produced by induction melting and quenching with a rate of about $10\,Ks^{-1}$. Samples of 1 to 2 mm height were cut and mechanically polished. From a $50\,\mu m$ thick melt-spun ribbon of $Zr_{65}Cu_{17.5}Ni_{10}Al_{7.5}$ samples of 9 mm in diameter were punched off. Some of these samples were mechanically polished while others were used without further polishing. The $Zr_{65}Cu_{17.5}Ni_{10}Al_{7.5}$ specimens were preannealed for 6 h at 478 K. X-ray diffraction patterns without any sharp Bragg peaks indicated the complete amorphous nature of both materials.

Self-diffusion coefficients were determined applying the radiotracer method: Special care was taken during tracer deposition to avoid diffusion barriers due to surface oxides. In case of Zr-diffusion samples the radiotracer ^{95}Zr was ion-implanted at the Bonn isotope separator. Ni diffusion specimens were prepared under ultra-high vacuum conditions by sputter depositing a ^{63}Ni layer of 1 to 2 nm thickness on a sputter-cleaned surface. The tracer layer was protected by a capping layer of the base material between 10 and 30 nm thick (see [24] for details). Diffusion annealings were carried out under vacuum in a temperature range from 555 to 685 K for times between 45 d and 10 min. Annealings longer than 90 min were performed in conventional resistance furnaces while for short-time annealing a radiation-heated furnace was used. High pressure annealing up to 0.7 GPa was carried out under Ar atmosphere in an inner-heated high pressure cell (see [25] for details). After the heat treatment each sample was checked for crystallization by X-ray diffraction. Concentration-depth profiles of the diffusing species were obtained by serial sectioning the samples applying ion-beam sputtering (see [26] for details). The eroded depth was determined from the total weight loss, the sputtered area and the density. The tracer concentration is proportional to the specific counting rate, which is the ratio of the counting rate and the section mass. The counting rate was detected in a Ge(Li) γ-detector for ^{95}Zr and in a liquid scintillation counter in case of the β-emitting radiotracer ^{63}Ni.

Temperature / K

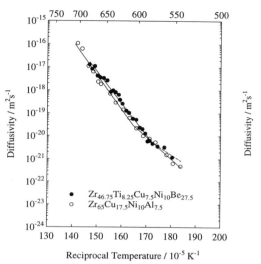

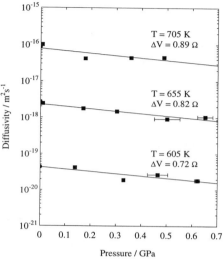

Figure 3: *Arrhenius diagram of* ^{63}Ni *self-diffusion. Error bars are within symbol size.*

Figure 4: *Pressure dependence of* ^{63}Ni *self-diffusion.*

RESULTS

Fig. 1 shows ^{95}Zr concentration depth profiles in $Zr_{46.75}Ti_{8.25}Cu_{7.5}Ni_{10}Be_{27.5}$ of an as implanted specimen and of two annealed samples. No significant changes in the profiles after diffusion annealing were observed. Unfortunately, the penetration depth cannot be increased by longer annealing times on account of the onset of crystallization. Since hold-up effects due to surface oxide layers can be excluded, we estimate that the diffusion coefficients of ^{95}Zr are smaller than $7 \cdot 10^{-21}\,m^2s^{-1}$ at 650 K and $3 \cdot 10^{-20}\,m^2s^{-1}$ at 670 K.

A typical penetration profile for ^{63}Ni diffusion into $Zr_{46.75}Ti_{8.25}Cu_{7.5}Ni_{10}Be_{27.5}$ is shown in Fig. 2. Profiles of $Zr_{65}Cu_{17.5}Ni_{10}Al_{7.5}$ samples look similar. A complementary error function was used to analyse the data, because the buried layer of radiotracer is not exhausted. The profiles were shifted by about 10 to 30 nm towards the surface on account of the capping layer. Furthermore, a constant background correction was applied. The resulting diffusion coefficients of ^{63}Ni self-diffusion in $Zr_{46.75}Ti_{8.25}Cu_{7.5}Ni_{10}Be_{27.5}$ respectively $Zr_{65}Cu_{17.5}Ni_{10}Al_{7.5}$ are plotted in an Arrhenius diagram presented in Fig. 3. A remarkable result is the non-linear Arrhenius behaviour of the ^{63}Ni diffusion. The temperature dependence of the diffusivity in $Zr_{46.75}Ti_{8.25}Cu_{7.5}Ni_{10}Be_{27.5}$ for temperatures higher than the kink temperature of 580 K can be described by

$$D = (4.3 \pm 0.2) \cdot 10^3 \exp\left(-\frac{(266 \pm 6)\,kJ\,mol^{-1}}{RT}\right)\,m^2s^{-1}. \qquad (1)$$

In case of ^{63}Ni diffusion into $Zr_{65}Cu_{17.5}Ni_{10}Al_{7.5}$ a break in the Arrhenius plot is observed at 608 K. The temperature dependence of the diffusivity above the kink temperature is given

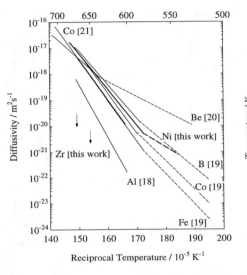

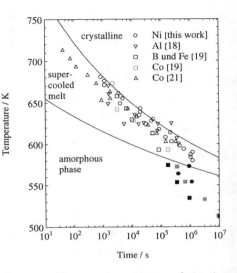

Figure 5: *Comparison of Al [18], B [19], Be [20], Fe [19], Co [19,21], and Ni [this work] diffusion in $Zr_{46.75}Ti_{8.25}Cu_{7.5}Ni_{10}Be_{27.5}$.*

Figure 6: *Time-temperature-transformation diagram of $Zr_{46.75}Ti_{8.25}Cu_{7.5}Ni_{10}Be_{27.5}$ with parameters of diffusion annealings of various diffusion studies.*

by

$$D = (1.2 \pm 0.1) \cdot 10^5 \exp\left(-\frac{(286 \pm 12)\,\text{kJ mol}^{-1}}{RT}\right)\,\text{m}^2\text{s}^{-1}. \tag{2}$$

Fig. 4 presents the diffusion coefficients of ^{63}Ni in $Zr_{46.75}Ti_{8.25}Cu_{7.5}Ni_{10}Be_{27.5}$ as a function of pressure for various temperatures. The activation volume can be obtained from the pressure dependence of the diffusivities according to

$$\Delta V \approx -kT\left(\frac{\partial \ln D}{\partial p}\right)_T. \tag{3}$$

For all investigated temperatures the resulting activation volumes given in Fig. 4 are around one mean atomic volume $\Omega = 1.63 \cdot 10^{-29}\,\text{m}^3$ of $Zr_{46.75}Ti_{8.25}Cu_{7.5}Ni_{10}Be_{27.5}$.

CONCLUSIONS

In Fig. 5 the ^{95}Zr and ^{63}Ni self-diffusion data in $Zr_{46.75}Ti_{8.25}Cu_{7.5}Ni_{10}Be_{27.5}$ are compared to Al, B, Be, Fe, and Co diffusion data reported by other authors recently [18-21]. Since the measurements of Be diffusion by Geyer et al. [20] represent interdiffusion and not Be tracer diffusion, they will not be discussed together with the other data on tracer diffusion in the following.

We note that Zr and Al are slow diffusors which can be attributed to their relatively large atomic radii. Diffuson of elements with smaller atomic radii like B and the late transition elements Fe, Co, and Ni is faster. A similar non-linearity in the Arrhenius behaviour like that found for ^{63}Ni self-diffusion in the present work was also observed for B, Fe, and Co diffusion by Fielitz et al. [19]. Linear Arrhenius plots are reported for Al diffusion by Budke et al. [18] and for Co diffusion by Ehmler et al. [21]. However, the data of Ehmler et al. [21] are limited to higher temperatures than those of Fielitz et al. [19]. To clarify the origin of the break in the Arrhenius function, we plotted in Fig. 6 the annealing times and temperatures applied in the above mentioned diffusion studies into the time-temperature-transformation (TTT) diagram of $Zr_{46.75}Ti_{8.25}Cu_{7.5}Ni_{10}Be_{27.5}$ published by Busch et al. [13]. Full symbols belong to time-temperature pairs yielding diffusion coefficients below the kink temperature and open symbols mark those pairs which lead to diffusion coefficients above this temperature. We note that the full symbols pertain to the amorphous phase whereas the open symbols pertain to the supercooled liquid state. According to these considerations the break in the Arrhenius plot of the diffusion coefficients reflects the transition from the glassy to the supercooled liquid state.

This explanation is supported by plotting the annealing times and temperatures applied in case of $Zr_{65}Cu_{17.5}Ni_{10}Al_{7.5}$ in a similar way into a pertaining TTT diagram [28]. Again, the non-linearity in the Arrhenius behaviour can be attributed to the transition from the glassy to the supercooled liquid state. Furthermore, ^{63}Ni self-diffusion is very similar to Co impurity diffusion. At 673 K Ehmler et al. [29] report $(6.8 \pm 1.7) \cdot 10^{-18} \, \mathrm{m^2 s^{-1}}$ for Co diffusion.

In crystals activation volumes of diffusion near one mean atomic volume indicate a diffusion mechanism via thermal vacancies [22]. To clarify the microscopic mechanism of long range atomic diffusion in the supercooled melt, we measured activation volumes of ^{63}Ni diffusion in $Zr_{46.75}Ti_{8.25}Cu_{7.5}Ni_{10}Be_{27.5}$ for temperatures above the kink temperature. According to our knowledge Fig. 4 represents the first measurements of activation volumes of diffusion in a supercooled melt. Sign and magnitude of the activation volumes around one mean atomic volume point to a vacancy-like diffusion mechanism. Additionally, the convergence of the diffusivities of all diffusants for higher temperatures and the weak mass dependence of Co diffusion found in isotope effect measurements by Ehmler et al. [21] indicate that diffusion occurs by collective processes. We propose a cooperative motion of atoms via vacancy-like defects in the supercooled liquid state of bulk metallic glasses.

ACKNOWLEDGMENTS

We thank Dr. F. Wenwer (now at Deutsche Bank AG, Eschborn, Germany) for assistance in parts of the experiments, Dr. K. Freitag (Universität Bonn, Germany) for implantation of ^{95}Zr produced by neutron activation of Zr in the Jülich reactor FRJ-2, and Dr. W. Ulfert (Max-Planck-Institut für Metallforschung, Stuttgart, Germany) for providing the $Zr_{65}Cu_{17.5}Ni_{10}Al_{7.5}$ ribbon. We acknowledge financial support by the Deutsche Forschungsgemeinschaft within the Schwerpunktprogramm "Unterkühlte Metallschmelzen: Phasenselektion und Glasbildung".

REFERENCES

1. A. Inoue, M. Kohinata, K. Ohtera, A.P. Tsai, and T. Masumoto, Mater. Trans. JIM

30, 378 (1989).

2. A. Inoue, T. Zhang, and T. Masumoto, Mater. Trans. JIM **31**, 177 (1990).

3. A. Peker and W.L. Johnson, Appl. Phys. Lett. **63**, 2342 (1993).

4. A. Inoue, Sci. Rep. RITU **A42**, 1 (1996).

5. W.L. Johnson, Mat. Sci. Forum **225-227**, 35 (1996).

6. R. Busch, Y.J. Kim, S. Schneider, and W.L. Johnson, Mat. Sci. Forum **225-227**, 77 (1996).

7. R. Busch, E. Bakke, and W.L. Johnson, Mat. Sci. Forum **235-238**, 327 (1997).

8. S. Schneider, U. Geyer, P. Thiyagarajan, and W.L. Johnson, Mat. Sci. Forum **235-238**, 337 (1997).

9. A. Wiedemann, U. Keiderling, M.-P. Macht, and H. Wollenberger, Mat. Sci. Forum **225-227**, 71 (1996).

10. N. Wanderka, Q. Wei, R. Dool, M. Jenkins, S. Friedrich, M.-P. Macht, and H. Wollenberger, Mat. Sci. Forum **269-272**, 773 (1998).

11. K. Ohsaka, S.K. Chung, W.K. Rhim, A. Peker, D. Scruggs, and W.L. Johnson, Appl. Phys. Lett. **70**, 726 (1997).

12. C.J. Gilbert, R.O. Ritchie, and W.L. Johnson, Appl. Phys. Lett. **71**, 476 (1997).

13. R. Busch and W.L. Johnson, Mat. Sci. Forum **269-272**, 577 (1998).

14. C. Nagel, K. Rätzke, E. Schmidtke, J. Wolff, U. Geyer, and F. Faupel, Phys. Rev. B **57**, 10224 (1998).

15. R. Busch, E. Bakke, and W.L. Johnson, Mat. Sci. Forum **225-227** 141 (1996).

16. J. Zappel and F. Sommer, J. Non-Cryst. Solids **205-207**, 494 (1996).

17. A. Meyer, H. Franz, B. Sepiol, J. Wuttke, and W. Petry, Europhys. Lett. **36**, 379 (1996).

18. E. Budke, P. Fielitz, M.-P. Macht, V. Naundorf, and G. Frohberg, Defect and Diffusion Forum **143-147**, 825 (1997).

19. P. Fielitz, M.-P. Macht, V. Naundorf, and G. Frohberg, J. Non-Cryst. Solids, in print.

20. U. Geyer, S. Schneider, and W.L. Johnson, Appl. Phys. Lett. **69**, 2492 (1996).

21. H. Ehmler, A. Heesemann, K. Rätzke, and F. Faupel, Phys. Rev. Lett. **80**, 4919 (1998).

22. H. Mehrer, Defect and Diffusion Forum **129-130**, 57 (1996).

23. K. Knorr, M.-P. Macht, K. Freitag, and H. Mehrer, J. Non-Cryst. Solids, in print.

24. F. Wenwer, K. Knorr, M.-P. Macht, and H. Mehrer, Defect and Diffusion Forum **143-147**, 831 (1997).

25. G. Rummel, Th. Zumkley, M. Eggersmann, K. Freitag, and H. Mehrer, Z. Metallkd. **85**, 131 (1995).

26. F. Wenwer, A. Gude, G. Rummel, M. Eggersmann, Th. Zumkley, N.A. Stolwijk, and H. Mehrer, Meas. Sci. Technol. **7**, 632 (1996).

27. J.P. Biersack and J.F. Ziegler, *Ion Implantation Techniques*, Series Electrophysics **10**, edited by H. Ryssel and H. Glaswischnig (Springer, Berlin, 1982).

28. R. Busch, private communication.

29. H. Ehmler, K. Rätzke, and F. Faupel, J. Non-Cryst. Solids, in print.

ATOMIC DIFFUSION IN BULK METALLIC GLASSES

ULRICH GEYER, ACHIM REHMET, SUSANNE SCHNEIDER
I. Physikalisches Institut der Universität Göttingen, Bunsenstr. 9, D-37073 Göttingen, Germany

ABSTRACT

Diffusion experiments in bulk metallic glasses are reviewed. Measurements carried out so far are focused on Zirconium based alloys and cover the amorphous and deeply undercooled liquid states as well as the equilibrium and slightly undercooled melts. Experimental findings are compared with theoretical predictions for atomic transport in glass forming systems.

INTRODUCTION

Diffusion experiments in amorphous metallic alloys have been carried out for several years. The nature of the diffusion mechanism, the role of thermal defects and the influence of structural relaxation, to name just a few, are some of the in part still controversely debated subjects of interest (for a recent review see e.g. [1]). The development of bulk metallic glass forming alloy systems, e.g. Pd-Ni-P [2, 3], Zr-Cu-Ni-Al [4], and Zr-Ti-Cu-Ni-Be [5], adds another exciting facet to this field of study: diffusion in the metastable equilibrium state of a metallic amorphous alloy, namely the undercooled liquid state.

It is well known that conventional metallic glasses transform into a "structrually relaxed state" upon annealing [6]. However, they never reach the undercooled liquid state during a typical diffusion experiment due to their limited thermal stability against crystallization. The discovery of bulk metallic glasses for the first time offers the possibility to study diffusion in the deeply undercooled liquid of a metallic system.

The dependences of diffusion in the undercooled liquid of bulk metallic glasses on temperature, mass, pressure and annealing time have already been studied. Methods applied include the standard radiotracer technique [7-9], secondary ion mass spectrometry [10], elastic backscattering (EBS) [11, 12], and nuclear magnetic resonance (NMR) [13]. Most of these experiments deal with self or impurity diffusion in the Zr-Ti-Cu-Ni-Be system; for this reason this article focuses mainly on the two most favored representatives of this alloy system, $Zr_{41.2}Ti_{13.8}Cu_{12.5}Ni_{10}Be_{22.5}$ (Vitreloy™ 1 or Vit 1) and $Zr_{46.7}Ti_{8.3}Cu_{7.5}Ni_{10}Be_{27.5}$ (Vit 4).

TEMPERATURE AND TIME DEPENDENCE OF DIFFUSION

In the first study of diffusion in the deeply undercooled liquid state of a bulk metallic glass interdiffusion experiments of Be and Vit 1 [11] were reported. In the case of Vit 1 only comparatively short diffusion anneals are feasible because this alloy has been found to undergo a decomposition which acts as a precursor to crystallization [14] and drastically influences the measured apparent diffusivities [15]. Therefore the choice of the diffusing species is limited to relatively fast diffusors. For these, measurable diffusion profiles can be produced by short annealing times, before effects of decomposition become important. Measurements of Be, B, and Fe in Vit 1 prooved to be reliable [10, 11].

Vit 4 is better suited for diffusion experiments since it does not show any sign of decomposition in the interesting temperature regime. The temperature dependence of diffusion in Vit 4 has

Mat. Res. Soc. Symp. Proc. Vol. 554 © 1999 Materials Research Society

been measured for a large variety of elements: Be [12], B, Fe, Al [10], Co [7, 10], Ni, and Zr [9] (Fig. 1).

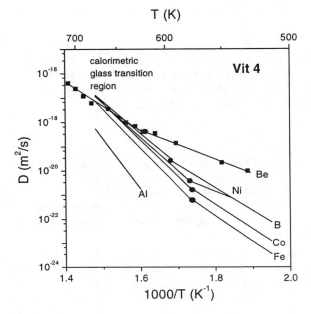

Figure 1:

Arrhenius plot for diffusivities in Vit 4. Bend temperatures are marked by dots.

The temperature dependences of all diffusion data in the deeply undercooled liquid can be described by Arrhenius fits. The activation enthalpy increases with increasing atomic size of the diffusing element from 1.9eV/atom (Be [12]) to 4.1eV/atom (Al, [10]). The diffusivities in the amorphous state are higher than expected from the extrapolation of the Arrhenius behavior in the undercooled liquid down to lower temperatures. They can be fitted with a second Arrhenius law giving a lower activation enthalpy and resulting in a bend in the Arrhenius plot at a temperature which is the higher the faster the species diffuses in the amorphous state (e.g. 625K in the case of Be). For the occurence of a bend in the Arrhenius plots of almost all diffusants studied basically two explanations have been suggested.

The first explanation attributes the higher diffusivities in the amorphous state to insufficient relaxation [9]. It assumes that the temperature dependence of the diffusivity in the undercooled liquid only extrapolates to lower temperatures if the duration of the diffusion anneals exceeds the average relaxation time of the alloy (see Fig. 2) at that temperature. Under this condition the matrix can relax isothermally into a state that corresponds structurally to an undercooled liquid at that temperature, even at temperatures below a typical calorimetric glass transition temperature. A resulting drop of the diffusivity should be observed when the sample "crosses" the isothermal glass transition line in Fig. 2 (time-temperature-transformation diagram taken from [16]). A collection of temperature/annealing time data for all diffusants except Be in Vit 4 can be found in [9]. These data seem to be consistent with the relaxation argument. Figure 2 contains the maximum annealing times for Be diffusion at several temperatures. According to the relaxation argument, the bend in the Be data is expected below 590K. Experimentally is has been found at 625K.

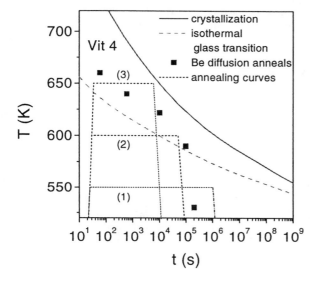

Figure 2:

Isothermal time-temperature-transformation diagram for Vit 4. Three diffusion anneals representing three different types of heat treatments are shown schematically. During anneal (1) the sample never reaches the undercooled liquid, during (2) it is mainly annealed in the undercooled liquid, and during (3) all the time. Squares indicate the maximum annealing times for Be.

The second explanation [10, 17, 18] claims a modification in the diffusion mechanism around the bend temperature to be responsible for the increased activation enthalpy in the undercooled liquid. The bend temperature is identified as the temperature at which the average attempt time of the diffusing atom,

$$\tau_D(T) = \frac{\lambda^2}{6\,D(T)} \tag{1}$$

(here, λ denotes the average hopping distance[1] and $D(T)$ the diffusion data in the amorphous state), approximately equals the matrix relaxation time $\tau_\eta(T)$ as given by a relation

$$\tau_\eta(T) = \frac{\eta(T)}{g \cdot G_\infty} \tag{2}$$

similar to the Maxwell-Relation, where η is the shear viscosity and $G_\infty = 33$GPa the high frequency shear modulus[2] of Vit 1 [19]. The hopping distance $\lambda=2.6$Å has been assumed to be a universal parameter for all diffusing species in Vit 4 and determined from the mean atomic volume of this alloy. This leaves the constant $g=30$ as the only free parameter. In Fig. 3 the competing times are compared.

---

[1] λ might have to be modified if the diffusion mechanism is not just single atom hopping but collective hopping of a group of atoms; however, the argument regarding two competing time scales remains the same.

[2] We expect no temperature dependence of the high frequency shear modulus in the temperature regime discussed here.

Below the crossover temperature, T_{cr}, the matrix is rigid on the time scale of the attempt time between two consecutive successful jumps, and thermally activated hopping is the only diffusion mechanism. Above the crossover temperature additional transport by shear flow starts and dominates at higher temperatures. Shear flow is a spontaneous process and therefore this contribution to atomic transport at higher temperatures is not thermally activated anymore.

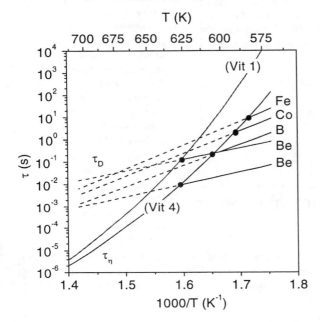

Figure 3:

Time scales for hopping and shear flow are compared for Vit 1 and Vit 4 as explained in the text. Crossover temperatures are marked by dots

Figure 3 gives a straightforward explanation why the observed bend temperatures decrease with increasing size of the diffusing species. With increasing size, the crossover temperature is simply shifted to lower values according to the individual Arrhenius parameters in the amorphous state. With the above mentioned values for λ, G_∞ and g the sequence of T_{cr} for Be, B, Co, and Fe can be reproduced. The argument also reproduces the observed identical T_{cr} for Be diffusion in Vit 1 and Vit 4: at low temperatures Be diffuses about one order of magnitude faster in Vit 4 than in Vit 1, but also the matrix relaxation time of Vit 4 is about one order of magnitude smaller than that of Vit 1. Both have been attributed to a higher amount of "free" volume in Vit 4 than in Vit 1 [12]. The relaxation argument would predict a bend lying at a lower temperature for Vit 4 than for Vit 1. The annealing times for both alloys were identical and the isothermal glass transition at the same temperature should occur earlier in Vit 4 than in Vit 1 due to the faster matrix relaxation in Vit 4.

In principal, it should be possible to discriminate between the two explanations given above by carefully measuring the time dependence in the vicinity of the isothermal glass transition using a sufficiently fast tracer.

In the case of Be as the diffusing species the proposed modification of the diffusion mechanism at a temperature that is characteristic for the diffusing species as presented above has been put into a more quantitative formulation. To account for the increased number of configurations that are sampled, a modified Arrhenius expression for $D(T)$ containing the configurational entropy of the undercooled liquid has been suggested [11]. In this context the high activation enthalpy of diffusion in the undercooled liquid state is only apparent since it is mimiced by the temperature dependent configuration entropy. A fit of this model to the EBS data describes Be diffusion in Vit 1 and Vit 4 well and gives reasonable values for the only fitting parameter, the number of neighboring atoms that take part in diffusion (N=13 for Vit 4 [12] and N=22 for Vit 1 [11]).

Recently performed measurements of the nuclear magnetic resonance of ^{9}Be in the amorphous and deeply undercooled liquid states of Vit 1 and Vit 4 corroborate this view [13]. No time dependence during isothermal annealing has been found which favors the second explanation. Furthermore, no bend in the Arrhenius plot has been found up to about 670K. Within the error of measurement the activation enthalpies determined by NMR and the low temperature values found by EBS are the same for both, Vit 1 and Vit 4 (around 1eV/atom). While NMR probes changes in the local environment of the Be atom, i.e. the short range atomic motion, EBS analyzes the long range effect of atomic transport. Most likely the flow events accelerating the long range diffusion above the crossover temperature do not significantly change the local environment of Be, such that NMR is not sensitive to flow in this temperature region. The NMR data also confirm the EBS results that Be diffusion in Vit 4 is one order of magnitude higher than in Vit 1 and that in both cases the migration entropy is significantly negative.

This significantly negative migration entropy suggests single atom hopping via a direct diffusion mechanism for Be in the amorphous states of Vit 1 and Vit 4. This is consistent with the common view that small atoms can perform single atom hopping in amorphous metals. In many cases of medium sized and larger atoms diffusion in amorphous metallic alloys has been attributed to a "collective" diffusion mechanism. Key experiments to elucidate the mechanism of diffusion in the undercooled liquid, in particular to probe the "collective" character of the mechanism, are measurements of the isotope effect.

MASS DEPENDENCE OF DIFFUSION

The isotope effect E is defined according to

$$E = \frac{\partial \ln D}{\partial \ln\left(1/\sqrt{m}\right)} \approx \left(\frac{D_\alpha}{D_\beta} - 1\right) \bigg/ \left(\sqrt{\frac{m_\beta}{m_\alpha}} - 1\right), \tag{3}$$

where $D_{\alpha,\beta}$ and $m_{\alpha,\beta}$ denote diffusivities and masses of tracer isotopes α and β. The isotope effect of cobalt (^{57}Co/^{60}Co) diffusion in Vit 4 and in $Zr_{65}Cu_{17.5}Ni_{10}Al_{7.5}$ has been measured [7, 8]. Its values in the deeply undercooled liquids are independent from temperature and almost vanish for both alloys. The average value is $E = 0.10 \pm 0.02$ in the case of Vit 4 and $E = 0.075 \pm 0.04$ in the case of $Zr_{65}Cu_{17.5}Ni_{10}Al_{7.5}$.

These values agree with those that have been obtained in relaxed conventional metallic glasses [20-22] and attributed to a highly cooperative hopping mechanism. This interpretation agrees with molecular dynamics simulations of supercooled liquids [23, 24] in which string- or chainlike displacements of groups of atoms are seen. In this picture the usual $m^{-1/2}$ mass dependence of the diffusivity that originates from the harmonic behavior of the attempt frequency is "diluted" by an effective mass M and has to be replaced by the relation

$$D \propto \frac{1}{\sqrt{m+M}} = \frac{1}{\sqrt{m+n\overline{m}}}. \tag{4}$$

The effective Mass $M = n\overline{m}$ accounts for all atoms that participate in the collective jump excluding the tracer atom; here $\overline{m}$ is the mean atomic mass of the alloy and n the number of participating neighbor atoms. From the measured isotope effects a value $n \approx 10$ can be estimated.

PRESSURE DEPENDENCE OF DIFFUSION

By measuring the pressure dependence of diffusion the role of thermal defects can be analyzed. If no pressure dependence can be observed (as e.g. for interstitial diffusion in crystals), a direct diffusion mechanism can be ruled out. The pressure dependence of ^{63}Ni diffusion in Vit 4 has been measured and reveals the activation volume,

$$\Delta V \approx -k_B T \frac{\partial \ln D}{\partial p}\bigg|_T, \tag{5}$$

for Ni diffusion in the deeply undercooled liquid [9]. ΔV is only slightly temperature dependent in the analyzed temperature interval, if at all, and averages to about 1.2 mean atomic volumes, 1.2Ω. An activation volume around Ω is usually attributed to thermally created vacancy-like defects or "quasi-vacancies". This term is motivated by the fact that activation volumes of about the same size are measured in crystalline densely packed metals and explained by the well established vacancy mechanism (see e.g. [25]). In case of Ni diffusion in Vit 4, the term "quasi-vacancy" is misleading; it suggests that a single defect of about the average size of the constituents of Vit 4 has to be created by fluctuations of the "free" volume in a location close to the tracer atom. A single jump of the tracer atom into the "quasi-vacancy" or the jump of an arbitrary atom into the "quasi-vacancy" as part of a chainlike motion of a group of atoms that contains the tracer atom would probably not be in agreement with the almost vanishing Co isotope effect. Since the diffusion mechanism most likely is the same for Co and Ni in Vit 4 (due to their similarity in size and chemistry), we suggest that the measured activation volume for ^{63}Ni is not the allocation of fluctuating "free" volume into a single located defect but is smeared out over all atoms that participate in the collective jump.

Results on the pressure dependence in conventional metallic glasses are non-uniform [26]. Even if one limits the comparison to Zirconium based glasses, the activation volume is concentration dependent and seems to reflect a diffusion mechanism that is sensitive to the Zirconium content. An experiment that is close to the one discussed here is a study on Co diffusion in relaxed $Zr_{58}Co_{42}$ [27] which revealed an activation volume of 0.66 mean atomic volumes and an activation enthalpy of 1.65eV/atom in combination with a vanishing isotope effect. These results are attributed to a collective diffusion mechanism that is mediated by delocalized thermal defects.

FREE VOLUME THEORY, SHEAR VISCOSITY AND DIFFUSION

The mass and pressure dependences for Co and Ni diffusion indicate the presence of delocalized thermal defects in Vit 4. Such defects are often described by the free volume theory

(FVT) which is restricted to the undercooled liquid where the system is in its metastable equilibrium. Any diffusivity is predicted to follow a temperature dependence like

$$D_{FVT}(T) = D_{0,i} \cdot exp\left(-\frac{b_i v_m}{v_f(T)}\right),$$
(6)

where $b_i v_m$ is the critical free volume for a diffusion event involving tracer i (in units of the mean atomic volume, v_m), and $v_f(T)$ denotes the temperature dependent average free volume per atom. The quantity $v_f(T)/v_m$ has recently been derived from shear viscosity measurements on Vit 1 over a range of fifteen orders of magnitude in shear viscosity [28], using the FVT equation

$$\eta(T) = \eta_0 \cdot exp\left(\frac{b_\eta v_m}{v_f(T)}\right)$$
(7)

and the expression for $v_f(T)$ given in [29]. The resulting value for the critical volume for flow is $b_\eta v_m = 0.105 \cdot v_m$. Equation (7) describes the experimentally found shear viscosity of Vit 1 better than the simpler Vogel-Fulcher-Tamann relation. With the $v_f(T)/v_m$ data from [28] we have fitted Eq. (6) to the Be diffusion data in the undercooled liquid of Vit 1, leaving $D_{0,i}$ and b_i as fit parameters (Fig. 4).

Figure 4:

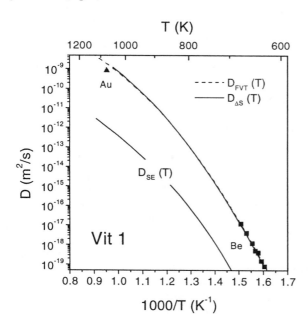

Be interdiffusion data in the deeply undercooled liquid of Vit 1 (squares, [12]) and Au interdiffusion with the equilibrium melt of Vit 1 (triangle, [30]). The fitting curve, $D_{FVT}(T)$, according to Eq. (6) describes the Be data very well and even extrapolates to the value for Au. It is not distiguishable from the fitting curve, $D_{\Delta S}(T)$, based on the modified Arrhenius expression discussed in [11]. Neither the low temperature nor the high temperature diffusion data follow the Stokes-Einstein relation, $D_{SE}(T)$.

Equation (6) describes the Be diffusion well with $D_{0,Be}=3.5 \cdot 10^{-5} m^2/s$ and $b_{Be}=0.097$. The resulting fitting curve, $D_{FVT}(T)$, even extrapolates well to an experimentally found value for interdiffusion of Au with the equilibrium melt of Vit 1 [30]. It is interesting to note that over the whole temperature regime shown and within the precision of the available data $D_{FVT}(T)$ cannot be distinguished from the fitting curve $D_{AS}(T)$ based on the modified Arrhenius expression mentioned above [11]. The critical volumes for Be diffusion, $b_{Be}v_m=0.097v_m$, as derived from our fit, and for flow, $b_\eta v_m=0.105v_m$ [18], are basically identical. This again supports our view that flow significantly contributes to long range diffusion in the undercooled liquid.

The Stokes-Einstein relation between shear viscosity and diffusivity,

$$D_{SE}(T) = \frac{k_B}{6\pi r} \cdot \frac{T}{\eta(T)} \tag{8}$$

(r denotes the atomic radius of the diffusing species), is neither fulfilled in the deeply undercooled liquid nor in the equilibrium melt (Fig. 4). This also holds in the deeply undercooled liquid [12] and equlibrium or slightly undercooled melts [31] of Vit 4 (Fig. 5). This is not surprising close to typical glass transition temperatures but has not been observed before in other glasses close to the liquidus temperature. The successful description of the shear viscosity and Be diffusion data in Vit 1 with Eqns. (6) and (7) rather indicates

$$D_{Be}(T) = \frac{D_{0,Be} \cdot \eta_0}{\eta(T)} = \frac{14 \cdot 10^{-10}\,Pa \cdot m^2}{\eta(T)}. \tag{9}$$

The time scale model presented above allows for the description of diffusion in the whole temperature range using a single expression. The interplay of shear flow and (single atom or collective) hopping of tracer i leads to an effective diffusivity D_{eff},

$$D_{eff}(T) = D_\eta(T) + D_i(T) = \frac{\ell^2 \cdot g \cdot G_\infty}{\eta(T)} + D_{0,i}^{SS} \cdot exp\left(-\frac{Q_i^{SS}}{k_B T}\right), \tag{10}$$

which interpolates between the low temperature Arrhenius behavior, characterized by $D_{0,i}^{SS}$ and Q_i^{SS}, and a high temperature contribution from flow. The length ℓ in Vit 1 can be extracted from Eq. (10), $\ell = 0.38\text{Å}$. This seems a very reasonable displacement in a flow process.

MODE COUPLING THEORY AND QUASIELASTIC NEUTRON SCATTERING MEASUREMENTS

Extrapolations of the data obtained in the undercooled liquid to higher temperatures suggest that all diffusivities merge to an universal $D(T)$ curve in the undercooled liquid (Fig. 5). Such behavior has been reported for molecular dynamics simulations of binary metallic glasses and discussed in terms of the mode coupling theory (MCT) [32].

For Vit 4 the atomic transport in the equilibrium melt above the liquidus temperature $T_{liq}=1050K$ and at moderate undercoolings has been studied by means of quasi-elastic neutron scattering [31]. The high frequency dynamics of this alloy has been found to be in accordance with the scenario predicted by MCT [33], a theory that has been successfully applied to interprete the dynamics of several non-metallic glass forming systems.

According to MCT a change in the diffusion mechanism from viscous flow to activated hopping is expected upon cooling below a critical temperature T_C which for Vit 4 has been derived as $T_C = 875 \pm 6K$ [31]. Unfortunately the temperature regime around T_C is not experimentally accessible. All available tracer diffusion data have been gained at temperatures far below this value.

From the broadening of the elastic neutron scattering signal average diffusivities of the constituents in the melt of this alloy have been determined (Fig. 5) [34]. Their temperature dependence is weak. A fit to the MCT prediction for $T > T_c$,

$$D(T) \propto (T - T_c)^\gamma \qquad (11)$$

with $\gamma = 1.71$ is added in Fig. 5. However, an independent analysis of the neutron scattering data yields $\gamma = 2.65$ [35].

Measurements of the isotope effect of diffusion, E, in the equlibrium and slightly undercooled melts would be very helpful for an independent check of MCT behavior in the undercooled liquid in these metallic systems. The isotope effect should approach unity at temperatures above T_c due to the expected increasing uncorrelated nature of binary atomic collisions in the liquid. However, the failure of the Stokes-Einstein relation as well as the unusual high viscosities around T_{liq} indicate a still highly cooperative atomic transport in this temperature regime. Therefore a relation $D(T) \propto T^n$ with n close to 2 which has been used to describe diffusivities in elemental melts [36] can only hold at temperatures far above T_c.

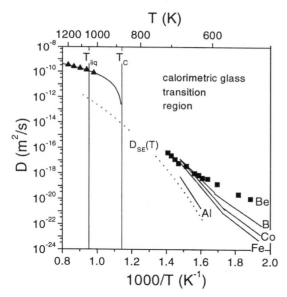

Figure 5:

A MCT fit to diffusion data derived from quasielastic neutron scattering (triangles) is shown in addition to the available low temperature diffusion data for Vit 4. The dotted line is the Stokes-Einstein prediction $D_{SE}(T)$ (Eq. 8) for Be diffusivity from shear viscosity data (r=1.1 Å).

SUMMARY

We have reviewed diffusion experiments in bulk metallic Zr-Ti-Cu-Ni-Be glasses which have been published during the past 4 years. These experiments cover the amorphous and deeply undercooled liquid states as well as the slightly undercooled and equlibrium melts. They focus on annealing time, temperature, mass and pressure dependence of diffusion with respect to the nature of the diffusion mechanisms in the various states. We have discussed the individual results and sketched some general ideas how they might fit together as well as into the framework of theories describing undercooled liquids.

More investigations of atomic diffusion and its mechanisms, particularly in the equilibrium and slightly undercooled melts are desirable and may be expected in the near future.

ACKNOWLEDGEMENTS

This work has been supported by Deutsche Forschungsgemeinschaft, Schwerpunktprogramm "Unterkühlte Metallschmelzen: Phasenselektion und Glasbildung". We gratefully acknowledge support by Professor W.L. Johnson (California Institute of Technology) through Grant No. DE-FG03-86ER-45242 (U.S. Department of Energy). We further thank the authors of [13, 18] for providing copies of their manuscripts prior to publication.

REFERENCES

[1]   H. Kronmüller, *Atomic transport and relaxation in rapidly solidified alloys*, in *Springer Series in Materials Science*, M.A. Otooni, Editor, 1998, Springer-Verlag: Berlin.

[2]   H.S. Chen, Act. Met. **22** (1974) 1505

[3]   K.W. Kui, A.L. Greer and D. Turnbull, Appl. Phys. Lett. **45** (1984) 615

[4]   T. Zhang, A. Inoue and T. Masumoto, Mater. Trans. JIM **32** (1991) 1005

[5]   A. Peker and W.L. Johnson, Appl. Phys. Lett. **63** (1993) 2342

[6]   W. Frank, J. Horváth and H. Kronmüller, Mat. Sci. Eng. **97** (1988) 415

[7]   H. Ehmler, A. Heesemann, K. Rätzke, F. Faupel and U. Geyer, Phys. Rev. Lett. **80** (1998) 4919

[8]   H. Ehmler, K. Rätzke and F. Faupel, Proceedings of the 10th International Conference on Liquid and Amorphous Metals, Dortmund 1998; to appear in J. Non-Cryst. Solids (1998)

[9]   K. Knorr, M.-P. Macht, K. Freitag and H. Mehrer, Proceedings of the 10th International Conference on Liquid and Amorphous Metals, Dortmund 1998; to appear in J. Non-Cryst. Solids (1998)

[10] P. Fielitz, M.-P. Macht, V. Naundorf and G. Frohberg, Proceedings of the 10th International Conference on Liquid and Amorphous Metals, Dortmund 1998; to appear in J. Non-Cryst. Solids (1998)

[11] U. Geyer, S. Schneider, W.L. Johnson, Y. Qiu, T.A. Tombrello and M.P. Macht, Phys. Rev. Lett. **75** (1995) 2364

[12] U. Geyer, W.L. Johnson, S. Schneider, Y. Qiu, T.A. Tombrello and M.P. Macht, Appl. Phys. Lett. **69** (1996) 2492

[13] X.-P. Tang, R. Busch, W.L. Johnson and Y. Wu, accepted for publication in Phys. Rev. Lett. (1998); Y.Wu, private communication

[14] S. Schneider, P. Thiyagarajan and W.L. Johnson, Appl. Phys. Lett. **68** (1996) 493

[15] E. Budke, P. Fielitz, M.P. Macht, V. Naundorf and G. Frohberg, Def. Diff. Forum **143** (1997) 825

[16] R. Busch and W.L. Johnson, Mater. Science Forum **269-272** (1998) 577

[17] U. Geyer, Habilitation thesis, Göttingen 1997

[18] A. Masuhr, T.A. Waniuk, R. Busch and W.L. Johnson, submitted to Phys. Rev. Lett. (1998)

[19] W.L. Johnson, Mater Science Forum **225-227** (1996) 35

[20] F. Faupel, P.W. Hüppe and K. Rätzke, Phys. Rev. Lett. **65** (1990) 1219

[21] K. Rätzke, A. Heesemann and F. Faupel, J. Non-Cryst. Solids **207** (1996) 607

[22] A. Heesemann, K. Rätzke, F. Faupel, J. Hoffmann and K. Heinemann, Europhys. Lett. **29** (1995) 221

[23] C. Donati, J.F. Douglas, W. Kob, S.J. Plimpton, P.H. Poole and S.C. Glotzer, Phys. Rev. Lett. **80** (1998) 2338

[24] H.R. Schober, C. Gaukel and C. Oligschleger, Progr. Theor. Phys. **126** (1997) 67

[25] H. Mehrer, Def. Diff. Forum **129-130** (1996) 57

[26] A. Grandjean, P. Blanchard and Y. Limoge, Phys. Rev. Lett. **78** (1997) 697

[27] P. Klugkist, *Aktivierungsvolumina der Diffusion in metallischen Gläsern*, Ph.D. thesis, Kiel 1998

[28] A. Masuhr, R. Busch and W.L. Johnson, Proceedings of the 10th International Conference on Liquid and Amorphous Metals, Dortmund 1998; to appear in J. Non-Cryst. Solids (1998)

[29] G.S. Grest and M.H. Cohen, Adv. Chem. Phys. **48** (1981) 455

[30] A. Masuhr and U. Geyer, unpublished

[31] A. Meyer, J. Wuttke, W. Petry, O.G. Randl and H. Schober, Phys. Rev. Lett. **80** (1998) 4454

[32] H. Teichler, Def. Diff. Forum **143** (1997) 717

[33] W. Götze and L. Sjögren, Transp. Theory Stat. Phys. **24** (1995) 801

[34] A. Meyer, J. Wuttke and W. Petry,Proceedings of the 10th International Conference on Liquid and Amorphous Metals, Dortmund 1998; to appear in J. Non-Cryst. Solids (1998)

[35] A. Meyer and A. Meyer, private communication (1998)

[36] G. Frohberg, K.H Kraatz, and H. Wever, Mater. Sci. Forum **15**, (1987) 529

HYDROGENATION AND CRYSTALLIZATION OF Zr-Cu-Ni-Al GLASSES

U. KÖSTER*, D. ZANDER*, H. LEPTIEN*, N. ELIAZ**, D. ELIEZER**
*Dept. Chem. Eng., University of Dortmund, D-44221 Dortmund, Germany
**Dept. Mat. Eng., Ben Gurion University of the Negev, Beer-Sheva 84105, Israel

ABSTRACT

Zr-Cu-Ni-Al belongs to the best glass forming systems known. Hydrogen charging was performed electrochemically in a 2:1 glycerin-phosphoric acid electrolyte. In comparison to binary Zr-Ni glasses the absorption kinetics in amorphous Zr-Cu-Ni-Al were found to be slower, but the storage capacity is similar. Desorption is hindered by the formation of thin ZrO_2 layers. Partial replacement of Ni by Pd in an amorphous $Zr_{68.5}Cu_{13}Ni_{11}Al_{7.5}$ alloy was found to change the absorption behavior to a faster kinetic and to improve the desorption.

The influence of hydrogen on the thermal stability was studied by DSC as well as by xray diffraction and TEM. Hydrogen was observed to play an important role. With increasing H/M ratio amorphous $Zr_{69.5}Cu_{12}Ni_{11}Al_{7.5}$ was found to transform during annealing above the glass transition temperature into a quasicrystalline structure with decreasing grain size until a nanocrystalline microstructure is achieved. Above a hydrogen content of H/M = 0.05 instead of quasicrystals a tetragonal phase with lattice parameters close to those of Zr_2Ni is formed. At very high hydrogen contents phase separation is assumed to take place followed by the formation of nanocrystalline ZrH_2 and other phases with reduced Zr content.

INTRODUCTION

Zr-based metallic glasses, in particular Zr-Ni glasses are known as hydrogen storage materials [1] and found some interest for application; alloying these glasses with B or Si was observed to deteriorate the hydrogen storage ability. Whether Al exhibits a similar effect is not known as yet. Whereas Zr-Ni has to be prepared by rapid quenching, Zr-Ni-Cu-Al alloys belong to the best glass forming systems known and can be cast at relatively low cooling rates in order to solidify as a glass [2]. Only recently it was shown that the glass transforms into a metastable icosahedral phase during annealing [3]. Since quasicrystals are often regarded as Hume-Rothery phases stabilized at a particular electron concentration, hydrogenation is expected to displace the electron concentration out of the stability range [4]. Therefore, hydrogenation of these bulk glasses is expected to hinder the formation of quasicrystals.

The aim of this paper is to study in detail the absorption kinetics and storage ability as well as the thermal stability of hydrogenated glassy $Zr_{69.5}Cu_{12}Ni_{11}Al_{7.5}$ alloys.

EXPERIMENTAL

Glassy $Zr_{69.5}Cu_{12}Ni_{11}Al_{7.5}$ ribbons were prepared by melt spinning as described elsewhere in detail [a]. Cathodically hydrogenation was carried out in a 2:1 glycerine-phosporic acid electrolyte at 25°C and a current density of i = 10 A/m². The hydrogen content was measured by a microbalance with an accuracy of ±1 μg.

The crystallization and the influence of hydrogen on the thermal stability were studied by means of differential scanning calorimetry (Perkin Elmer DSC7) as well as by microstructural investigations using x-ray diffraction (Cu-K$_\alpha$ radiation) and transmission electron microscopy (Philips CM200 operating at 200 kV).

RESULTS and DISCUSSION

Fig. 1 shows the hydrogen content versus charging time for an amorphous $Zr_{69.5}Cu_{12}Ni_{11}Al_{7.5}$ alloy in comparison with the binary $Zr_{60}Ni_{40}$ glass [1]. Whereas the kinetics is much slower for the quarternary glass, the maximum hydrogen content is probably similar. Hydrogen absorption as well as desorption of Zr-based alloys is known to be hindered even by a very thin layer of ZrO_2. Pd plating, often used to overcome this barrier, showed insufficient adhesion to the glassy ribbons. However, alloying with Pd (instead of the Ni) was found to improve the absorption kinetics significantly, probably due to Pd segregation at the surface of the ribbon. These Pd containing glasses, however, did not form quasicrystals upon annealing. With increasing hydrogen content the amorphous peak moves to smaller angles, thus reflecting the increase of the nearest neighbor distance from about 0.245 nm to 0.260 nm at a hydrogen content H/M = 1.0. The observed increase in length during hydrogen charging indicates a very surprising result: Instead of the continous increase of the specific volume per hydrogen atom known for example from Zr-Ni glasses [1], a decrease from a rather high value of about $4 \cdot 10^{-3}$ nm³/H-atom at very low hydrogen contents to about $2 \cdot 10^{-3}$ nm³/H-atom is observed; not understood at all as yet.

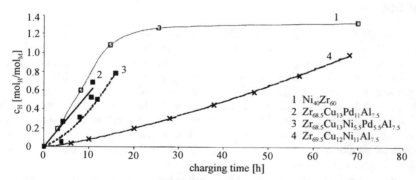

Fig. 1: Hydrogen charging of amorphous $Zr_{69.5}Cu_{12}Ni_{11}Al_{7.5}$ alloys (10 A/m², electrolyte: 2:1 glycerine-phosporic acid, 25°C).

The thermal stability was investigated by DSC of as-cast as well as hydrogenated Zr-Cu-Ni-Al metallic glasses. Desorption of hydrogen, as measured by TDA, was not observed to proceed at temperatures less than about 500°C. This allowed to investigate the influence of hydrogen on the crystallization of the glass. The DSC exhibits a two-step reaction in glassy $Zr_{69.5}Cu_{12}Ni_{11}Al_{7.5}$ (see Fig. 2): formation of icosahedral quasicrystals followed by decomposition of the quasi

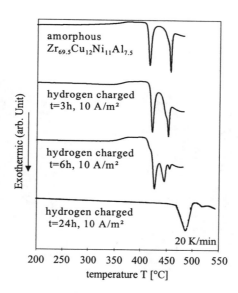

Fig. 2 DSC of uncharged and hydrogenated $Zr_{69.5}Cu_{12}Ni_{11}Al_{7.5}$ glasses.

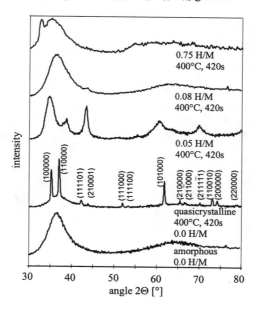

Fig. 3 X-ray diffraction ($\lambda = 0.154$ nm) of melt-spun as well as hydrogenated glassy $Zr_{69.5}Cu_{12}Ni_{11}Al_{7.5}$ ribbons after annealing at 400°C.

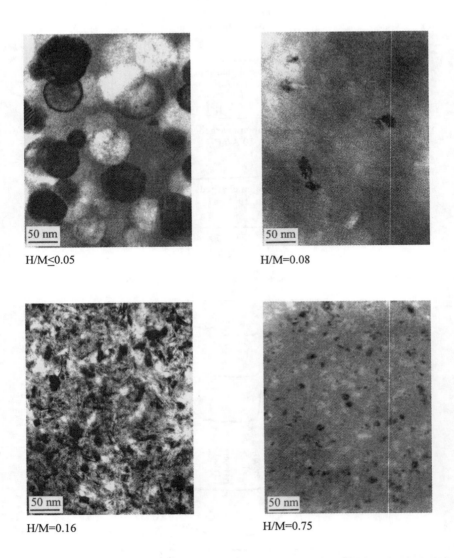

H/M≤0.05

H/M=0.08

H/M=0.16

H/M=0.75

Fig.4: Hydrogen charged amorphous $Zr_{69.5}Cu_{12}Ni_{11}Al_{7.5}$ after annealing for 7 min at 400°C

crystals into stable crystalline phases [5]. With increasing hydrogen content, the peak for the quasicrystal formation is shifted to higher temperatures. In addition we observe the crystallization of a small amount of the oxygen-stabilized fcc (big-cube) phase which itself transforms around 460°C (3rd peak) into the stable Zr_2Ni phase. The x-ray diffraction data show that these quasicrystals decompose around 450°C mainly into hexagonal Zr_6Al_2Ni, tetragonal Zr_2Ni and Zr_2Cu. Fig. 3 exhibits the x-ray diffraction pattern, fig. 4 the microstructure after annealing glassy ribbons with differing hydrogen contents for 420 s at 400°C. Hydrogen contents of $H/M < 0.05$ were found to reduce the number of quasicrystals formed during annealing. In the x-ray diffraction pattern all peaks of the icosahedral phase became weaker, thus indicating a smaller amount of quasicrystals. TEM investigations confirm that this change is due to a smaller number of quasicrystals; growth seems not to be affected. This might result from destabilization of the quasicrystalline structure by the hydrogen or more probably by the destruction of the quenched-in nucleation sites [6] for the quasicrystals.

At slightly larger hydrogen contents instead of the quasicrystalline phase a tetragonal one with lattice parameters close to Zr_2Ni is formed. At even larger hydrogen contents up to about $H/M = 0.6$, the amorphous state seems to be stabilized as also the number of these crystals decreased significantly. This might result from phase separation into two amorphous phases which leads first to elimination of quenched-in nucleation sites. The DSC (see fig. 2) exhibits also a quite different behavior as compared to less hydrogenated glasses.

Hydrogen contents of about $H/M = 1.0$, on the other hand, were observed to led to nanocrystallization at temperatures as low as 300°C, probably due to a progressive phase separation followed by easy nanocrystallization of ZrH_2 and phases with reduced Zr-content at relative low temperatures. A similar destabilizing phase separation was reported earlier for binary Ti-based metallic glasses [7].

CONCLUSIONS

With increasing hydrogen content the formation of icosahedral quasicrystals in amorphous $Zr_{69.5}Cu_{12}Ni_{11}Al_{7.5}$ is shifted to higher temperatures and then replaced by the formation of a tetragonal phase with lattice parameters close to Zr_2Ni. The quasicrystals decompose at higher temperature or longer annealing times mainly into the tetragonal Zr_2Cu phase, the tetragonal Zr_2Ni and the hexagonal Zr_6NiAl_2. At even higher hydrogen contents phase separation at much lower temperatures is assumed to be followed by the formation of nanocrystalline ZrH_2 and other phases with reduced Zr-content.

ACKNOWLEDGMENT

This work was supported by the Deutsche Forschungsgemeinschaft (DFG Ko 668/22-1). One of us (N. Eliaz) is grateful to the Israel Ministry of Science for the financial support through Eshkol Grant no. 2206-0587 for Scientific Infrastructures. The authors are indebted to cand.ing. H. Torwesten for his help in the hydrogenation experiments.

REFERENCES

1. H.-W. Schroeder, *Hydrogen in Zr-based Metallic Glasses*, in: Rapidly Quenched Metals, eds. S. Steeb, H. Warlimont, Elsevier Science Publ., Amsterdam 1985, p.1525;

2. A.Inoue, T.Nakamura, T.Sugita, T.Zhang, T.Masumoto, Mater.Sci.Eng.A (1994);

3. U. Köster, J. Meinhardt, S. Roos, H. Liebertz, Appl. Phys. Lett. **69,** 179 (1996);

4. P.A. Bancel, P.A. Heiney, Phys. Rev. **B33**, 7917 (1986);

5. U. Köster, J. Meinhardt, S. Roos, R. Busch, Mat.Sci.Eng. **A226-228** (1997), 995;

6. U. Köster, A. Rüdiger, J. Meinhardt, *Formation of Quasicrystals in Bulk Glass-Forming Zr-Cu-Ni-Al Alloys*, in: Proc. 6th International Conference on Quasicrystals, Tokyo1997, edited by S. Takeuchi, T. Fujiwara, (World Scientific, Singapore 1996), p.317;

7. D. Menzel, A. Niklas, U. Köster, Mat. Sci. Eng. **A133** (1991), 312.

ELASTIC AND DISSIPATIVE PROPERTIES OF AMORPHOUS AND NANOCRYSTALLINE Mg-Ni-Y ALLOY

Y.M.SOIFER, N.P.KOBELEV* , I.G.BRODOVA, A.N.MANUKHIN** , E.KORIN, L.SOIFER***

* Institute of Solid State Physics, RAC, Chernogolovka 142432, Moscow region, Russia
**Institute of Metal Physics, RAC, Kovalevskaya Str., Ekaterinburg 620219, Russia
***Department of Chemical Engineering, Ben-Gurion University of the Negev, P.O. Box 653 84105, Israel

ABSTRACT

The internal friction and Young's modulus of amorphous $Mg_{84}Ni_{12.5}Y_{3.5}$ alloy obtained by the melt spinning technique have been measured by a vibrating reed method at frequency of 250 Hz at heating and cooling runs in the temperature range from 300K to 625K.. The crystallization kinetics of the alloy was studied by the calorimetric methods (DSC and DTA). The structure of the samples was determined by the x-ray diffraction technique. The Young modulus measurements have revealed the irreversible multi-step changes (up to 50%) accompanied by the irreversible internal friction peaks. These changes were observed in the same temperature intervals where the anomalies of thermal properties were found out. The results obtained are explained by the structural rearrangement from amorphous to nanocrystalline state during the annealing.

INTRODUCTION

Interest in rapid solidification of magnesium alloys derives from the fact that poor strength, ductility, and corrosion resistance of conventional ingot metallurgy can be improved by microstructural refinement via rapid solidification processing. Recently, the development of new metallic materials with novel properties has been carried out through the formation of nonequilibrium phases such as amorphous, quasicrystalline and nanocrystalline phases[1,2,3]. Internal friction is known to be a structure-sensitive and has been used to investigate the structural relaxation, the glass transition and crystallization of amorphous alloys. In the present paper, we report the behavior of the internal friction and the Young's modulus of an amorphous Mg-Y-Ni alloy associated with the structural relaxation and crystallization of this alloy.

EXPERIMENTAL

$Mg_{84}Ni_{12.5}Y_{3.5}$ alloy has been prepared by melting the mixture of pure element in the alumina crucible under KCl-NaCl slag-cove at the temperature 1000 K. Rapidly quenched alloys have been produced by melt-spinning on the copper drum surface in vacuum. Cooling rates were 5×10^6 K/s from the temperature 880 K. The specimens for internal friction measurements were cut from amorphous ribbon and their typical dimensions were (10-15) x (1.5-3) x (0.025-0.030) mm. All the measurements of elastic and dissipative properties of the specimens were carried out by the vibrating reed method at the frequency of about 250 Hz in the temperature range from 300K to 650K. The heating rate was 2K/min. The measuring procedure is described in [4]. The internal friction Q^{-1} and resonant frequency f were determined. Young's modulus was determined from the expression $E = Kl^4 \rho f / h^2$, where K is a coefficient, l is the specimen length, h the thickness, ρ the density, f the resonant frequency.

The structures of the phases were determined by x-ray diffraction with a PW 1050/70 Philips X-ray Diffractometer with Cu-Kα radiation at a constant temperature of 298 K after annealing amorphous samples at the certain temperature during 5 min.

Differential thermal analysis in Ar atmosphere at the heating rate 4 K/min was used to study crystallization of $Mg_{84}Ni_{12.5}Y_{3.5}$ amorphous samples.

DSC analyses of 4-5mg samples were performed by Mettler Toledo DSC 820 System. The analyses were performed in an Al crucible under a nitrogen atmosphere (99% pure) at heating rates of 8, 30, and 120 K/min in a temperature range from 298 to 673 K.

RESULTS

Fig.1 shows the DSC curves of the as-quenched $Mg_{84}Ni_{12.5}Y_{3.5}$ alloy when heating at constant rates of 8, 30 and 120 K/min. The curves exhibit several large exotherms at the temperatures which are strongly dependent on a heating rate. At heating rates of 30 and 120 K/min one can also notice a weak endothermic peak at low temperatures. X-ray patterns of the as-quenched alloy after heating at heating rate of 30 K/min following the quenching to the oil-bath are given in Fig.2. The analysis of the data obtained shows: 1. Up to ~410K the alloy structure is remained amorphous; 2. An endothermic peak at ~410-430K is related to the glass transition; 3. A transformation (crystallization) from amorphous to crystalline state goes through several stages. Roughly the crystallization can be divided into two steps. The first one is origination and growth of metastable $Mg_xNi_yY_z$ nanocrystalline phase (the exact structural formula did not determined) and nanocrystalline Mg (solid solution of Y in Mg). The second one is a stabilization of the crystalline structure due to disintegration of metastable phase of $Mg_xNi_yY_z$ to more stable phases mainly Mg_2Ni and Mg.

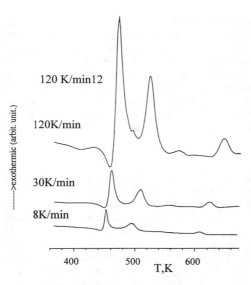

Fig.1. DSC curves of the amorphous ribbons $Mg_{84}Ni_{12.5}Y_{3.5}$ at heating rate of 8, 30 and 120 K/min.

DTA measurements confirm the conclusions on structure evolution during heating made on the base of analysis of the DSC curves. It is important to take attention to the fact that the DTA curves are obtained when heating at rate of 4K/min and the exotherms corresponding to the same crystallization stages are observed at the temperatures slightly lower than for DSC measurements.

Typical curves of the internal friction (IF) and relative Young's modulus change E/E_0 (E_0 is the Young's modulus of the as-quenched sample at 300K) measured during a heating-cooling run in the temperature range from 300K to 625K are given in the Fig.3. One can see two broad IF peaks at 425K and at 530K. In the cooling run the peaks disappear, i.e. they are irreversible ones, and the $Q^{-1}(T)$ curve becomes monotonous, the main decrease of the Q^{-1} value being occurred in the temperature range of 600-500K. (An irreversible change is a difference between the values of Q^{-1} and E/E_0 measured at the

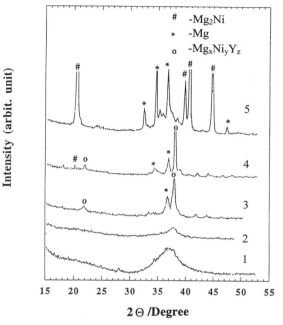

-Mg₂Ni

(legend in figure)

$\#$ -Mg$_2$Ni
$*$ -Mg
$\circ$ -Mg$_x$Ni$_y$Y$_z$

Fig.2. X-ray diffraction patterns of as-quenched Mg$_{84}$Ni$_{12.5}$Y$_{3.5}$ alloy before (1) and after heating at rate of 30 K/min up to 473, 533, 583 and 643K (curves 2 to 5, respectively).

same temperatures in the heating and cooling runs).

The Young's modulus dependence on the temperature has a very complicated behavior. In the heating run the monotonous decrease of the modulus takes place up to the temperatures of 350-360K, then a small peculiarity (a plateau on the E/E$_0$(T) curve) is observed at the temperatures from 360K to 390K, then again the Young's modulus decreases. However at the temperatures more than 415K and almost up to 490K the drastic Young's modulus rise occurs (the modulus increases by approximately 50%). After 490K the modulus begins to decrease again up to 615K. Cooling from this temperature to room temperature does not reveal any anomalies in the Young's modulus behavior, it increases monotonously, but its value at 300K is 20% higher than of the pristine one.

Compared together the IF data and the results of the study of structure evolution during the heating run, one can conclude that IF and the Young's modulus anomalies are observed approximately at the same temperatures which correspond to the exotherms in the DSC curves, that is at the temperatures corresponding to the different stages of Mg$_{84}$Ni$_{12.5}$Y$_{3.5}$ alloy crystallization from amorphous state. Thus, one can think that elastic and anelastic (Q^{-1}) phenomena observed in the temperature interval investigated are determined by the structural rearrangement occurred in the sample during the heating.

DISCUSSION

According to [5] in materials in which two phases connected by the phase transition can exit over a certain temperature range one should consider three main separate contributions to the total internal friction:

$$Q^{-1}_{tot} = Q^{-1}_{tr} + Q^{-1}_{pt} + Q^{-1}_{int} \qquad (1)$$

Q^{-1}_{tr} is the transient part of Q^{-1}_{tot} and it exits only during heating or cooling, i.e. (dT/dt)≠0. It depends on external parameters. Q^{-1}_{tr} depends on the transformation kinetics and is therefore proportional to the volume fraction α which is transformed per unit of time. Q^{-1}_{pt} is related to the mechanisms of the phase transition which are independent on the transformation rate, such as the movement of the phase interfaces. When the interface mobility is maximum Q^{-1}_{pt} exhibits a small peak. Q^{-1}_{int} is composed of the IF contribution of each phase and is strongly dependent on

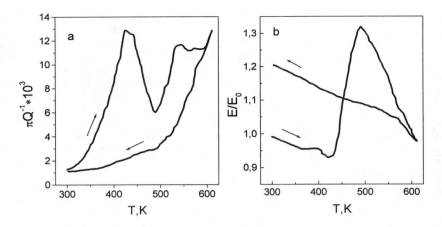

Fig.3. Internal friction (Q^{-1}) and Young's modulus (E/E_0) of $Mg_{84}Ni_{12.5}Y_{3.5}$ alloy measured during heating and cooling runs at rate of 2 K/min.

microstructural properties (interfaces , dislocations, point defects, grain and cluster boundaries, their density and mobility). Correspondingly, the Young's modulus can be given in the form:

$$E = E_{el} - \Delta E - \Delta E_{tr} \qquad (2)$$

$E_{el} = \alpha\, E_{el1} + (1-\alpha)\, E_{el2}$ is a pure elastic modulus, determined strictly by elastic stiffness matrixes of the phase1 (E_{el1}) and the phase2 (E_{el2}). α is a volume fraction of the phase1; ΔE is so called the modulus defect, determined by the anelastic behavior of solid (hysteretic, relaxation and resonance processes). $\Delta E = [\alpha \Delta E_1 + (1-\alpha)\Delta E_2]$;

Let us discuss the IF spectrum $Q^{-1}(T)$ and E/E_0 (T) changes obtained in the terms of definitions made above.

As mentioned above, according to the x-ray measurements no crystallization traces was observed up to 410K, i.e. in this temperature interval the specimen is in amorphous state ($\alpha=1$; $Q^{-1}_{tot} = Q^{-1}_{int}$; $E = E_{ela} - \Delta E_a$; here a denotes amorphous). Almost up to 350K the changes in Q^{-1} and E/E_0 are reversible. Decrease in E/E_0 is practically determined by the temperature dependence of E_{ela}. From 350K to 400K a slight increase in E/E_0 is associated with the topological ordering in amorphous matrix, i.e. a transition of amorphous system to a more stable amorphous state. Similar situation was discussed previously [6,7]. Such change in E/E_0 is an irreversible one and accompanied by the corresponding irreversible change in Q^{-1}.

Beginning from ~400K to ~500K the significant increase in E/E_0 is observed. This temperature range is characterized by the great changes in the sample structure. First of all at about 415K the glass transition in amorphous matrix occurs. According to [8] it leads to a reversible relaxation due to Q^{-1}_{int}, i.e. it should be manifested on the $Q^{-1}(T)$ curve as a relaxation maximum and on the E/E_0 (T) curve as a corresponding decrease of E/E_0 due to change in ΔE_a. However this decrease is overcovered by the sharp increase in E/E_0 caused by the first stage of $Mg_{84}Ni_{12.5}Y_{3.5}$ alloy crystallization from amorphous state (origination of $Mg_xNi_yY_z$ metastable crystalline phase and precipitation of nanocrystalline Mg). In accordance with (1) and (2) it means that α changes from 1 to 0 and the transition occurs from E_{el1} to E_{el2}. Q^{-1}_{tot} is determined mainly by the contribution of the Q^{-1}_{tr} component and therefore one should expect a peak behavior in $Q^{-1}(T)$ curve. Thus, the broad IF peak centered at 420-430K is supposed to be a

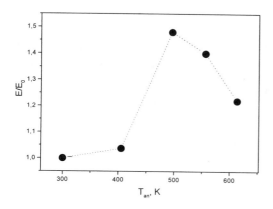

Fig.4. Young's modulus measured at 300K after successive heating-cooling runs versus the annealing temperature.

superposition, at least two internal friction peaks, one associated with glass transition and the other with the first stage of crystallization. Cooling from 500K shows that the changes in $Q^{-1}(T)$ and $E/E_0(T)$ after annealing at ~500K are irreversible, that confirms the assumption on transformation of amorphous state to crystalline one[10,11]. It is interesting to take notice that the maximal value of E/E_0 (490K) is attained at the lowest value of Q^{-1} on the high temperature side of the IF peak. One can assume that it means the end of the process of the high elastic phase formation. The further heating to temperatures above 500K leads to the essential change of the dE_{el}/dT derivative that points out to the onset (as one can conclude from the x-ray studies) of the disintegration of the $Mg_xNi_yY_z$ phase and an appearance a new phase with much lesser stiffness. The $Q^{-1}(T)$ measurements confirm the suggestion on the phase transition in the temperature range discussed. On $Q^{-1}(T)$ curve one can see IF peak centered at ~550K which disappears in the cooling run due to corresponding decrease of Q^{-1}_{tr} IF component. It should be pointed out in this connection that the Young's modulus proves to be a very sensitive characteristic of the phase composition of the sample. This statement can be illustrated by the change in the Young's modulus measured at 300K after successive annealings (Fig.4). Here the E magnitude is plotted against the annealing temperature which is defined as the maximum temperature attained in the heating cycle. It should be taken to notice to a very high magnitude of the Young's modulus of the structure obtained after annealing at 500K (nanocrystalline $Mg_xNi_yY_z$, nanocrystalline Mg and residual amorphous phase). Since after annealig at the temperatures above 550K $Mg_xNi_yY_z$ phase disappears one can conclude that peak value of the Young's modulus can be attributed chiefly to the $Mg_xNi_yY_z$ phase. It is clear that the determination of the exact chemical composition of the phase can be extremely important for the production of Mg alloys of high mechanical properties.

Thus, the results obtained demonstrate a strong influence of the structure of the amorphous and nanocrystalline materials on the formation their elastic and dissipative properties, that allow to use elastic characteristics and IF as a extremely informative instruments for the investigation of the kinetics of crystallization processes in amorphous alloys. The same results show the possibility of governing of Mg-Ni-Y alloy elastic and mechanical properties by the special heat treatment.

ACKNOWLEDGEMENTS

Ya.M.Soifer and N.P.Kobelev acknowledge Russian Foundation Foundation for Basic Research (Grant 98-02-16644 and Grant 96-15-96806) for a partial financial support.

REFERENCES

1. A.Inoe and T.Masumoto, Mater. Sci.Eng. **A173**, p.1-16 (1993)
2. A.Inoe, K.Ohetera, M.Kohinata et al., J.Non-Cryst. Solids **117/118**, p.712-719 (1990)
3. T.S.Srivatsan, Li Wei and C.F.Chang, J.Mater.Sci. **30**, pp.1832-1838 (1995)
4. N.P.Kobelev, Ya.M.Soifer, V.G.Shteinberg et al., Phys.Stat.Sol. (a),**102**, pp.773-780 (1987)
5. .J.E.Bidauh, R,Schaller and W.Benoit Acta Met. **37**, pp.803-811 (1989)
6. N.P.Kobelev and Ya.M.Soifer, Nanostr. Mater. **10**, pp.449-456 (1998)
7. A.van den Beukel, Acta Met.et Mater. **39**, pp.2709-2717 (1991)
8. He Yizhen and Li Xiao-Guang, Phys.Stat.Sol. (a) **99**, pp.115-120 (1987)
10. J.Perez, J.de Phys. **C10,** pp.427-430 (1985)
11 L.Kempen, U.Harms, H.Neuhauser et al., J.de Phys. IV **C8**, pp.643-646 (1996)

A RELATION BETWEEN FRAGILITY, SPECIFIC HEAT, AND SHEAR MODULUS TEMPERATURE DEPENDENCE FOR SIMPLE SUPERCOOLED LIQUIDS

A.V. Granato, University of Illinois, Dept. of Physics, Urbana, IL 61801, granato@uiuc.edu

ABSTRACT

It is often said that the fragility and specific heat discontinuity at the glass transition are related characteristic features of glasses. We give a relation between these quantities for simple liquids, and also find that the temperature dependence of the shear modulus should be closely related to these quantities. According to the Interstitialcy Theory, the liquid state shear modulus $G = G_o \exp(-\beta c) = G_o \exp\{-\gamma[(T/T_o)-1]\}$, where G_o is the shear modulus at a reference temperature T_o, which can be taken to be the glass temperature. In these relations, β is the diaelastic shear susceptibility, c is the interstitialcy concentration, T is the temperature, and $dc/dT = \gamma/\beta T_o$. It has been proposed by Dyre, Olsen and Christensen that U in the viscosity $\eta = \eta_o \exp(U/kT)$ be given as the work done in shoving aside particles during a diffusion step, and therefore be proportional to the (liquid state) shear modulus. If so, then combining the above relations, the fragility $F = [d \log \eta/d (T_g/T]_{T=T_g}$ becomes $F=(1+\gamma) \log[\eta (T_g)/\eta_o]$, where $\gamma = -(T_g/G)(dG/dT)$. Then, since $\delta C_v = U_F(dc/dT)$, where U_F is the interstitialcy formation energy, one obtains further with reasonable simplifying approximations $\delta C_v/C_v \sim \gamma G(T)/G_{oo}$, where G_{oo} is the zero temperature crystalline shear modulus. For a typical fragile glass $\gamma \sim 2$ (F $\sim$ 50) with $G(T_g)/G_{oo} \sim 1/2$, then $\delta C_v/C_v \sim 1$, but close to zero for strong glasses, in fair agreement with available experimental results. The perspective given by the Interstititialcy Theory is then that the fragility is given phenomenologically by the temperature dependence of the shear modulus, or microscopically by the rate of increase of the equilibrium interstitialcy concentration with temperature.

INTRODUCTION

The magnitude and temperature dependence of the specific heat and the non-Arrhenius behavior of the viscosity are usually regarded as the two most characteristic features of supercooled liquids and the glass transition. These effects are related with large discontinuities in the specific heat at the glass transition temperature T_g correlated with large deviations from Arrhenius behavior.[1]

We give a quantitative relation between these effects as well as with the shear modulus, based on the Interstitialcy Theory of Simple Condensed Matter.[2] We focus on the specific heat because a comprehensive theory of condensed matter, including liquids and glasses, should lead to a calculation of the specific heat with a minimum of assumptions, thereby providing a direct and immediate test of any theoretical model. We are accustomed to having a simple, quantitative, universal relation for the specific heat of gaseous and crystalline states, but the Interstitialcy Theory is the only theory that predicts the magnitude and temperature dependence of the specific heat of liquids.

PREDICTIONS OF THE INTERSTITIALCY THEORY

According to the Interstitialcy Theory, liquids are crystals containing a few percent of interstitials in thermal equilibrium and glasses are frozen liquids.

The shear modulus G depends on the interstitialcy concentration c as

Mat. Res. Soc. Symp. Proc. Vol. 554 © 1999 Materials Research Society

$$G = G_o \exp(-\beta c) \tag{1}$$

where $\beta = -d \ln G/dc$ is the shear susceptibility. With $\beta \sim 20$ to 30, a concentration of a few percent of interstitials is expected to greatly reduce G. The volume and bulk modulus are much less affected.

If c is expanded in a Taylor series in the liquid state, the first term is a good representation for $\Delta c = (dc/dT)(T - T_o)$, where T_o is any reference temperature. A convenient choice for T_o is the glass temperature so that the temperature dependence of G (in the liquid state) is given by

$$G^1 = G_g \exp \gamma (1-T/T_g) \tag{2}$$

where

$$\gamma = \beta T_g (dc/dT)_g \ . \tag{3}$$

The specific heat increase over the crystalline value is given by[2]

$$\delta C_v = E^F(T)dc/dT$$

$$= (\alpha G(T)\Omega)(\gamma/\beta T_g) + \delta C_{v2}, \ T > T_g \tag{4}$$

$$= \delta C_{vg} \exp \gamma(1-T/T_g) + \delta C_{v2}(\text{const.})$$

where E^F is the formation energy, Ω is the atomic volume, δC_{v2} is a small constant which can be absorbed into the definition of δC_v,[3] and

$$\delta C_{vg} = E_g^F(\gamma/\beta T_g) = (\alpha G_g \Omega)(\gamma/\beta T_g) \tag{5}$$

where E_g^F is the defect formation energy at $T = T_g$. In Eq. (4), the first factor is the formation energy of the interstitial, while the second factor is the rate of increase of the interstitial concentration with temperature. The value of C_v is thus given by the work necessary to create the increasing equilibrium concentration with increasing temperature. For $T < T_g$, $dc/dT = 0$ and $\delta C_v = 0$.

The magnitude and temperature dependence of the liquid state shear modulus G, specific heat C_p, diffusivity D, and viscosity η should all be closely related, according to the interstitialcy model, if a recent proposal by Dyre et al.[4] is generally true. They suppose that the viscosity is given by $\eta = \eta_o \exp(W/kT)$, where η_o is a reference viscosity and W is given by the work required to shove aside neighboring particle in a diffusion process, where $W = GV_c$ and V_c is a characteristic volume.

The fragility F may be defined[5] by

$$F = \frac{d \log \eta}{d(T_g/T)} \bigg|_{T=T_s}, \tag{6}$$

where $\log \eta$ falls from a value of 13 at T_g to -4 at $T = \infty$. F is sometimes called the steepness parameter m.[6,7] A normalized fragility may be defined as

$$f = F/\log(\eta_g/\eta_o) = F/17 \qquad (7)$$

Then, using Eqs. 2 and 6, one finds that the softness parameter γ is related to the normalized fragility parameter f by

$$f = \gamma + 1 \qquad (8)$$

Most supercooled liquids have fragilities which lie in the range from 17 to 68 called strong to fragile in Angell's[1] classification. A few may have larger values. The corresponding typical range of the differing fragility parameters is then

$$(strong) \ 17 < F < 68 \ (fragile)$$
$$1 < f < 4$$
$$0 < \gamma < 3$$

EXPERIMENTS

The heat capacities for some extreme cases where crystal, glass, and liquid state Cp's are compared are shown in Fig. 1, collected by Angell[1], and described in the Figure caption.

In Fig. 2 the log of the shear modulus, normalized to the value at the glass temperature, is plotted as a function of temperature for three frequencies from data given by Barlow, et al.[8] for sec-butyl benzene. At the lowest temperatures, the curves agree for all frequencies and give G(T). The fact that the data are the same for higher frequencies shows that this is the equilibrium curve, and the fit shows that the temperature dependence of G(T) is exponential, as predicted by Eq. 2, with $\gamma = 2.7$.

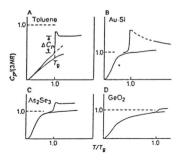

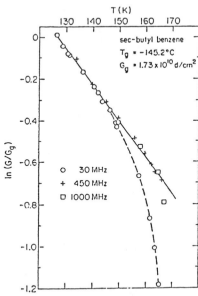

Fig.1. Heat capacity forms for liquid and crystal phases of different substance. (A) Molecular systems like toluene where the glass transition occurs over a range where the crystal heat capacity is not classical; (B) metallic systems like Au-Si where crystal and glass reach the classical regime before the glass transition occurs; (C) covalent systems like As_2Se_3 where the jump in liquid heat capacity occurs on a classical background and ΔC_p remains large above T_g; (D) open network systems like GeO_2 where ΔC_p is small and occurs on a classical background. The T_g is usually defined by the construction made at 10 K/min shown in (A) because this fixes T_g as the temperature where the average structural relaxation time is 100 s for scanning at 10 K/s; N is the number of atoms per formula unit and R is the gas constant. (After Angell[1])

Fig. 2. The shear modulus of sec-butyl benzene as a function of temperature for three frequencies, taken from data by Barlow, et al.[8] O, 30MhZ; + 450 MhZ; and ☐, 1000 MhZ.

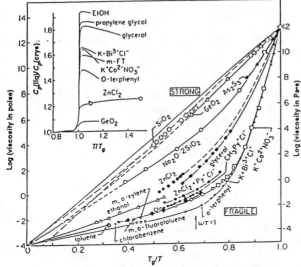

Fig. 3. Arrhenius plots of the viscosity data scaled by values of T_g showing the "strong-fragile" pattern of liquid behavior on which the liquid's classification of the same name is based. As shown in the insert, the jump in C_p at T_g is generally large for the fragile liquids and small for strong liquids, although there are a number of exceptions, particularly when hydrogen bonding is present. (After Angell[1])

Fig.3 is a standard Angell plot for the temperature dependence of the viscosity for a number of supercooled liquids. The inset gives the temperature dependence of the jump in specific heat. An Arrhenius behavior would be a straight line on a log η as T_g/T, and is called strong. This shows the correlation between C_p and η.

DISCUSSION

The change in specific heat according to Eqs. 4 and 5 is given by

$$\delta C_v = (\alpha G_g \Omega)(\gamma/\beta T_g) \exp[-\gamma(T-T_g)/T_g], \quad T>T_g \qquad (9)$$
$$= 0 \qquad T<T_g$$

For a simpler rough estimate for a typical glass, we may take

$$\alpha G_{oo}\Omega \sim 37kT_m^2, \ \beta \sim 20, \ G_g/G_{oo} \sim 1/2, \ T_g/T_m \sim 2/3$$

to obtain

$$\frac{\delta C_v}{3R} \approx \frac{\gamma}{2} \exp[-\gamma(T-T_g)/T_g],$$

$$(10)$$

where 3R is the DuLong Petite crystalline value. We thus expect a discontinuity in $C_v/3R$ at $T_g \sim \gamma/2 = (f-1)/2$.

For the case of strong GeO_2, we obtain from Fig. 3, $f \sim 1.14$, $\gamma \sim 0.14$ and $\delta C_v/3R \sim .07$, in reasonable agreement with Fig. 1. For a more typical value of $\gamma \sim 2$, we expect $\delta C/3R \sim 1$. We notice also, by comparing the cases of AuSi and As_2Se_3 in Fig. 1, that when γ is small, the temperature dependence of δC_p is weak. These estimates are of the right order of magnitude although there seem to be some exceptional cases. A more quantitative survey, comparing values of γ obtained from C_p, G and η for different materials, will be given elsewhere.

It is somewhat surprising to find that the theory for simple materials seems to apply as well for complex materials, such as covalent GeO_2. The basic phenomenological requirements of the theory are that there exist thermally accessible defects sensitive to shear distortion with low frequency resonance modes and a degeneracy of energy with orientation. These properties come close to defining interstitials, whose properties are well known for simple metals. There is an urgent need for investigations of the basic properties of interstitials in both metallic compounds and covalent materials, where very little is known about interstitials.

CONCLUSIONS

Eq. 9 giving $\delta C_v(T)$ has some notable features:
1. It is the first quantitative theory of the specific heat of liquids predicting its magnitude and temperature dependence.
2. The specific heat provides an immediate and direct test of any theory of condensed matter, but as yet, no other theory predicts $\delta C_v(T)$.
3. Eq. 9 (and 10) is simple, with a character mainly determined by γ. But it is unexpected, and different from all proposed phenomenological relations.
4. It has a universal character. The height is proportional to γ, but the width is inversely proportional to γ, so that all data could in principle be plotted on a single curve.
5. Since γ is the key parameter for each, the specific heat is simply related to the shear modulus and the viscosity, or fragility.

The perspective given by the Interstitialcy Theory is that the fragility is given phenomenologically by the temperature dependence of the shear modulus, or microscopically by the rate of increase of the equilibrium interstitialcy concentration with temperature.

ACKNOWLEDGMENTS

This work is supported by the National Science Foundation under grant DMR9705750.

REFERENCES

1. C.A. Angell, in *Relaxation in Complex Systems*, K. Ngai and G.B. Wight, Eds. (National Technical Information Service, U.S. Department of Commerce, Springfield, VA, 1985) p.1-11.

2. A.V. Granato, Phys. Rev. Lett. **68**, p. 974-977, (1992).

3. A.V. Granato, Met. and Mat. Trans. A, **29A**, p. 1837-1843 (1998).

4. J.E. Dyre, N.B. Olsen and T. Christensen, Phys. Rev. **B53**, p. 2171 (1996).

5. D.M. Zhu and H. Chen, J. Non-Cryst. Sol., **224**, p. 97-101 (1998).

6. D.J. Plazek and K.L. Ngai, Macromolecular **24**, p. 1222 (1991).

7. R. Bohmer, K.L. Ngai, C.A. Angell and D.J. Plazek, J. Chem. Phys. **99**, p. 4201-4209 (1993).

8. A.J. Barlow, J. Lamb, A.J. Matheson, P.R.K.L. Padmini, and J. Richter, Proc. Roy. Soc. London Ser. **1298**, pp. 467-480 (1967).

Effect of small amounts of B and C additions on glass formation and mechanical properties of a Zr-base alloy

C. T. Liu, L. M. Pike, and N. G. Chen
Metals and Ceramics Division
Oak Ridge National Laboratory
Oak Ridge, TN37831-6115

Abstract

The effect of B and C additions up to 0.4 at. % on glass formation and mechanical properties of a Zr-base alloy Vitreloy 105 was studied using various techniques. All alloys were prepared by arc melting and drop casting. Boron additions increase the glass forming ability by lowering T_m and increasing T_g. Carbon additions only lower T_m but not affect T_g. B and C additions occupy free space and do not harden the glass phase.

Introduction

Metallic glasses in amorphous states possess many attractive properties for structural and functional use [1]. Recent studies show that it is possible to produce the glass state in bulk metallic glass (BMG) alloys containing multiple components at critical cooling rates of 1 to 10^2 /s [2-6]. In comparison, high critical cooling rates of 10^5 to 10^6 K/s are generally required for formation of the glass state in metallic elements or simple binary alloys. These observations demonstrate that BMG alloys have a much superior glass forming ability. In the recent review paper, Inoue [5] pointed out that the high glass forming ability of BMG alloys can be simply rationalized by three empirical rules: (1) BMGs contain multiple elements, ≥3, (2) constituent elements have a large difference in atomic size, usually ≥12 %, and (3) these elements possess negative heats of mixing. It should be pointed out that no single intermetallic phase is stable with the multi-component composition. In view of these rules, it may be possible to further stabilize BMG alloys by adding small amounts (within solubility limits) of elements with small atomic sizes, such as boron and carbon. Boron and carbon could easily occupy open space in the amorphous states, resulting in enhancement of random packing density and stablization of the supercooled liquid state. Also, the formation of strong Zr-B or Zr-C atom pairs in Zr-base BMG alloys is expected to further lower the heat of mixing and increase the glass forming ability [6,7]. On the other hand, excessive amounts of B and C additions may cause the formation of Zr boride and carbide embryos, which may serve as nuclei for crystallization.

In this study, a Zr-base alloy, Zr-10Al-5Ti-17.9Cu-14.6Ni alloy (Vitreloy 105) was selected as the base composition [8,9], where additions of B and C up to 0.4 at. % were added to the base alloy. The B- and C-doped alloys were prepared by arc melting and drop casting, and their metallurgical and mechanical properties were characterized using various techniques. The addition of 0.3 % B or C lowers the melting by ~6°K and causes no significant hardening of the base alloy.

Experimental procedures

Three series of BAM alloys were prepared based on Vitreloy 105, with their compositions listed in Table 1. The first and second series were doped with up to 0. 4 at. % B and C, respectively, and the third was doped with equal amounts of B and C up to a total of 0.3 %. All the alloys were synthesized by arc melting in an inert gas, followed by drop casting into 6.2 mm diam. Cu molds chilled with water. Zone-refined Zr bars (containing 12 appm O and 10 appm Hf) together with pure metal elements, electron-grade boron, and pyrolytic graphite were used as charge

305

materials. Cast alloy ingots were then sectioned for microstructural analyses and property evaluation. Metallographic specimens were polished on a syntron machine and etched in a solution of 40-ml HNO_3 plus 10 drops of HF. The microstructure and phase compositions were analyzed by wavelength dispersive spectroscopy and energy dispersive spectroscopy using an electron microprobe. The phase transformation behavior was studied at a heating rate of 0.67 K/s by differential scanning calorimetry (DSC).

Table 1. Nominal Compositions

Composition (at. %)	0	0.05	0.1	0.2	0.3	0.4
Alloy number			B-Doped			
	BAM-11	BAM-31	BAM-23	BAM-24	BAM-25	BAM-26
			C-Doped			
	BAM-11	BAM-32	BAM-27	BAM-28	BAM-29	BAM-30
			(B+C)-Doped*			
	BAM-11		BAM-33	BAM-34	BAM-35	

*Equal amounts of B and C.

Tensile specimens with a 3.19-mm gage diameter were fabricated by centerless grinding. Tensile tests were all conducted on an Instron testing machine at room temperature at a strain rate of 3.3 x 10^{-3} s^{-1}. Microhardness was measured on metallographic samples using 25 to 200 g, dependent on the size of alloy phases. Sectioned alloy ingots were crushed in a hydraulic press to produce coarse powder for bulk density measurements. The density was measured using a helium pycnometer with an accuracy of approximately 0.01 vol. %. The use of powdered samples is intended to eliminate microporosities in alloy ingots, resulting in more accurate density measurements.

Results

The microstructure of BAM alloys was examined by both metallography and electron microprobe analysis. Figure 1 shows a back-scattered electron image (BSEC) of BAM-35 containing 0.15 B and 0.15 C. No microstructural features were detected in the matrix, indicating the formation of the glass phase in the alloy [9]. In contrast, massive crystalline phases are observed in BAM- 30 doped with 0.4 C (Fig. 2), and their amount increases steadily with the distance away from the ingot surface. This increase is due to a continuous decrease in cooling rate from ingot surface to center. Unmelted B (dark contrast) was observed in BAM-26 containing 0.4 B, which was surrounded by crystalline phases (light gray) and B-rich phase particles (dark gray) in Fig.3a. In comparison, unmelted C (dark contrast) was first detected in 0.3 C alloy (BAM-29), with no observation of carbon-containing phases around it. A particle of unmelted carbon and its surrounding structure in BAM-30 containing 0.4 C is shown in Fig. 3b.

Since the BMG alloys were produced by drop casting that induced turbulent flows, oxide particles (dark contrast) and isolated crystalline phase particles (gray contrast) were occasionally observed in turbulent-flow regions in cast ingots, as shown in Fig. 4 at a high magnification. The dendritic growth of large oxide particles is also visible in Fig. 4. Table 2 summarizes the composition of phases in BAM-35 by electron microprobe analyses. The glass matrix has its composition similar to the nominal alloy composition. Crystalline particles are enriched slightly with Ni and depleted with Al. On the other hand, oxide particles are enriched with Al and oxygen, indicating the formation of aluminum oxide particles. Since the volume (~2 µm) excited by the electron beam is larger than the size of oxide particles in dendritic shapes, the exact oxide composition can not be determined here.

Fig. 1 BSEI of BAM-35

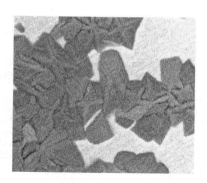

Fig.2 BSEI of BAM-30

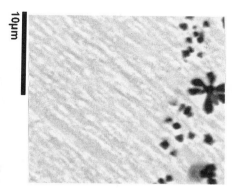

(a)

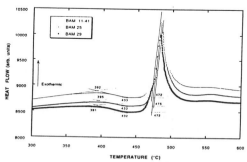

(b)

Fig. 3 BSEI of (a) BAM-26 showing unmelted B, and (b) BAM-30 showing unmelted C

Fig. 4 BSEI of BAM-35

Fig. 5 DSC curves of BAM-11, -25, and 29

The Glass transition (T_g) and crystallization temperatures (T_x) of BAM-25 containing 0.3 B and BAM-29 with 0.3 C together with the base alloy BAM-11-41 were determined by DSC, and the Table 2. Phase Compositions in BAM-25 (Containing 0.3 at. % B) Determined by Electron Microprobe Analyses

Table 2. Phase Compositions in BAM-25 (Containing 0.3 at. % B) Determined by Electron Microprobe Analyses

Phase	Compositions (at. %)				
	Cu	Ni	Al	Ti	O
Alloy nominal	17.9	14.7	10.0	5.0	
Glass matrix	18.0	14.7	10.0	5.4	
Crystalline phase	18.2	15.2	9.6	5.2	
Oxide particle	16.6	9.5	17.0	4.0	7.2

results are shown in Fig. 5. Noticeably, the B addition increases T_g and T_x by 3°K, whereas these temperatures are not affected by the C addition. Figure 6 shows the effect of B and C additions on the melting point of the base alloy. As indicated in the figure, both B and C lower the melting point, T_m, of the base by 5-6°K.

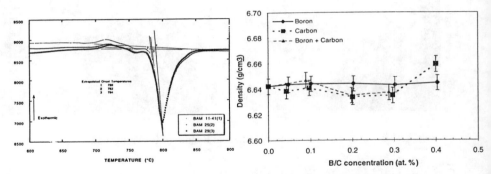

Fig. 6 DSC curves for BAM-11, -25, -29   Fig. 7 Plot of density vs B/C concentration

The measured density of BAM alloys is plotted as functions of B and C concentrations in Fig. 7. The density of the alloy in the glass state is essentially not affected by B and C additions at levels to 0.4 %. For the C and (B+C) doped alloys, their density exhibits a small decrease at 0.2-0.3 %. A relatively sharp increase in density is observed for 0.4 C alloy, mainly due to the massive formation of crystalline phases (see Fig. 2).

The mechanical properties of BAM alloys were determined by hardness and tensile testing at room temperature. Figure 8 shows that the hardness is essentially not sensitive to B and C additions. Note that the hardness measurements reported here were all made on the glass matrix. The fracture strength determined by tensile testing at room temperature is shown in Fig. 9. The strength of the base alloy is as high as 1650 MPa. Adding with a combination of B and C up to 0.3 % does not affect the tensile fracture strength. The strength is not sensitive to B or C at levels up to 0.2 %, and above that level it decreases with increasing B or C addition. A sharp decrease in strength at the C level above 0.3 % is due to a massive formation of crystalline phases, as observed in BAM-30 doped with 0.4 C (Fig. 2).

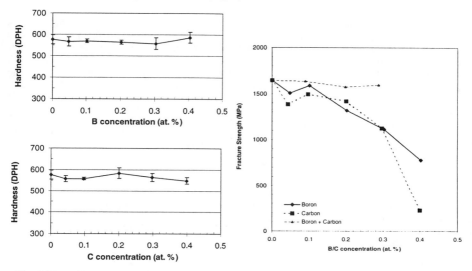

Fig. 8 Plot of hardness as functions of B & C Concentrations

Fig. 9 Plot of tensile fracture strength vs B/C concentration

Discussion

Recently, Inoue et al. [6,7] found that additions of B significantly increase the thermal stability of the supercooled liquid region in Zr- and Pd-base alloys prepared by melt spinning. The addition of 3 at. % B extends the supercooled liquid region, T_{xg} ($=T_x-T_g$), of Pd-6Cu-18Si alloy from 47 to 70 K [6]. In this case, the increase in T_{xg} is due to a combination of a small increase in T_x and a large decrease in T_g. For the Zr-base BMG alloy, Zr-27.5Cu-7.5 Al, the addition of 4 % B extends T_{xg} from 72 to 100 K, and this increase is mainly due to a significant increase in T_x [7]. In the present study, all BAM alloys were prepared by arc melting and drop casting, and their molten states could not be superheated and soaked extensively above T_m. The finding of unmelted elements suggests that the solubility of B and C appears to be ~0.4 % for B and ~0.3 for C in drop-cast BAM ingots. These small amounts apparently do not extend the supercooled liquid region (=80 K) of Vitreloy 105. Boron at 0.3 % increases both T_x and T_g by 3 K, resulting in no change of T_{xg}. Carbon at 0.3 % virtually does not alter T_x and T_g. However, in terms of the reduced glass transition temperature (T_g/T_x), the 0.3 % C and B additions increase the value from 0.626 to 0.633 for B and 0.626 to 0.628 for C.

Small atoms such as boron enhance the stability of glass forming ability; however, oxygen, on the other hand, lowers the glass forming ability and dramatically increases the critical cooling rate required for glass forming in Vitreloy 105. Lin et al. reported that the increase in the oxygen level from 0.025 to 0.525 at. % increases the critical cooling rate by several orders of magnitude [8]. Their study suggests that oxygen-induced precipitates are catalytic sites for heterogeneous nucleation for intermetallic phases and control the crystallization behavior of the glass forming alloy. It has been reported that small silicon additions between 0.5 to 1 % enhance the glass forming ability of Cu- and Ni-base alloys [10,11]. Choi-Yim et al. [10] suggest that the beneficial effect of silicon comes mainly from passivating oxygen impurities that promote heterogeneous nucleation in the molten state.

It is interesting to examine the hardening behavior of the glass state by B and C additions. As shown in Fig. 8, the addition of up to 0.4 % B and C causes no significant hardening in the glass matrix. This is very different form interstitial hardening in crystalline phases [12,13]. For instance, doping with 0.2 % of B results in a twofold increase in the yield strength of Ni3Al [12]. This suggests that B and C with small atomic sizes mainly occupy free space, resulting in no significant increase in lattice strain and hardness in the glass state. Adding 0.4 % C produces massive crystalline phases in BAM-30, whose hardness is 695 DPH as compared with 545 DPH for the glass phase. The tensile fracture strength, similar to the hardness, showed no increase with B and C additions. The significant decrease in fracture strength of BAM alloys with ≥0.3 % B or C is due to the existence of defects, such as unmelted B/C elements and crystalline phases. In this Zr-base alloy, the formation of crystalline phases causes severe embrittlement [14].

Acknowledgment

The authors would like to thank W. D. Porter, J. L. Wright, C. A. Carmichael, and L. R. Walker for their technical assistance and Connie Dowker for preparation of final manuscript. This research was sponsored by the Division of Materials Sciences, US Department of Energy under contract number DE-AC05-96OR22464 with Lockheed Martin Energy Research Corporation.

References

1. F. E. Luborsky, ed., "Amorphous Metallic Alloys," Butterworths, Boston (1983).
2. A. Inoue, T. Zhang, and T. Masumoto: *Mater. Trans., JIM* 30, 965-72 (1989).
3. A. Inoue, T. Zhang, N. Nishiyama, K. Ohba, and T. Masumoto: *Mater. Trans., JIM* 34, 1234-37 (1993).
4. A. Peker and W. L. Johnson, *J. Appl. Phys. Lett.,* 63, 2342-44 (1993).
5. A. Inoue, A. Takeuchi, and T. Zhang, *Metall. Trans. A* 29A, 1779-93 (1998).
6. A. Inoue, T. Aoki, and H. Kimura, Mat. Trans., JIM 38, 175-78 (1997).
7. A. Inoue, T. Negishi, H. Kimura and T. Aoki, Mat. Trans. JIM 38, 185-88 (1997).
8. X. H. Lin, W. L. Johnson, and W. K. Rhim, Mat. Trans. JIM 38, 473-77 (1997).
9. C. T. Liu, et al., Metall. Trans. A 29A, 1811-20 (1998).
10. H. Choi-Yim, R. Busch, and W. L. Johnson, J. Appl. Phys. 83, 7993-97 (1998).
11. A. Volkert, Ph.D. Thesis, Harvard University, Boston, MA, 1998.
12. C. T. Liu, C. L. White, and J. A. Horton, *Acta Metall.* 33, 213-29 (1985).
13. A. I. Taub, S. C. Huang, and K. M. Chang, *Metall. Trans. A* 15A, 399-402 (1984).
14. C. T. Liu, unpublished results, Oak Ridge National Laboratory, Oak Ridge, TN, 1998.

BULK GLASS-FORMING METALLIC ALLOYS: SCIENCE AND TECHNOLOGY
[1998 MRS MEDAL AWARD LECTURE, PRESENTED AT SYMPOSIUM MM]

William L. Johnson, Keck Laboratory of Engineering, 138-78, California Institute of Technology, Pasadena, CA   91125

ABSTRACT

The paper begins with some brief remarks about the history and background of the field of bulk glass-forming metallic alloys. This is followed by a discussion of multicomponent glass-forming alloys and deep eutectics, the chemical constitution of these new alloys, and how they differ from metallic glasses of a decade ago or earlier. The development of bulk glass forming alloys has led to interesting studies of the deeply undercooled liquid alloys, which are made possible by the exceptional stability with respect to crystallization. Recent advances made in this area will be illustrated by several examples. The paper continues with a discussion of some of the physical properties of bulk metallic glasses. Mechanical properties are specifically discussed. Some interesting potential applications of bulk metallic glasses will be mentioned.

HISTORICAL BACKGROUND AND DEVELOPMENT OF BULK METALLIC GLASSES

The first liquid metal alloy vitrified by cooling from the molten state to the glass transition was Au-Si as reported by Pol Duwez at Caltech in 1960 [1]. Duwez made this discovery as a result of developing rapid quenching techniques for chilling metallic liquids at very high rates of 10^5 –10^6 K/s. The work of David Turnbull and his group in the early 1960s was another critical contribution to the subject. This work illustrated the similarities between metallic glasses, ceramic glasses, silicates, etc. Specifically, Turnbull, Chen, and other later collaborators [2,3] clearly demonstrated the existence of a glass transition in rapidly quenched Au-Si glasses as well as other Pd-Si and Pd-Cu-Si glass forming alloys synthesized initially by the Duwez group. Earlier, Turnbull had predicted that as the ratio of the glass transition temperature to the melting point or liquidus temperature of an alloy (referred to as the reduced glass transition temperature, t_{rg}) increased from values near 1/2 to values near 2/3, the homogeneous nucleation of crystals in the undercooled melt should become very sluggish on laboratory time scales [4]. Work on the Au-Si, Pd-Si, and Pd-Cu-Si confirmed this prediction. This "Turnbull" criterion for the suppression of crystallization in undercooled melts remains today one of the best "rules of thumb" for predicting the glass forming ability of any liquid.

The field of metallic glasses gained momentum in the early 1970's when researchers at Allied Chemical Corporation developed continuous casting processes for commercial manufacture of metallic glass ribbons and sheets [5]. During the same period, Chen (then at Bell Laboratories) and collaborators used simple suction casting methods to form millimeter diameter rods of ternary Pd-Cu-Si alloys at significantly lower cooling rates in the range of 10^3 K/s [6]. If one arbitrarily defines the millimeter scale as "bulk", then these ternary glasses were perhaps the first examples of "bulk" metallic glasses. Beginning in 1982, Turnbull, Drehman, Kui, Greer, and other collaborators [7, 8], carried out experiments on Pd-Ni-P alloy melts using a boron oxide fluxing to dissolve heterogeneous nucleants into a glassy surface coating. The fluxing experiments showed that when heterogeneous nucleation was suppressed, this ternary alloy, with a reduced glass transition temperature of ~2/3, would form bulk glass ingots of centimeter size at cooling rates in the 10 K/s range. At the time, this work was perceived by many to be a laboratory curiosity. During the late 1980s, Akihisa Inoue and his coworkers in Sendai, Japan investigated the fabrication of amorphous aluminum alloys. In the course of this work, Inoue's research team considered ternary alloys of rare earth materials with Al and ferrous metals. While studying rapid solidification in these pseudo-ternary alloy systems, they found exceptional glass forming ability in the rare-earth rich alloys [9], e.g. La-Al-Ni and La-Al-Cu. By casting the alloys into copper molds, the researchers fabricated fully glassy rods and bars with cast thicknesses of several mm's. From there, the researchers studied similar quaternary and quinary materials (e.g. La-Al-Cu-Ni) and developed alloys that formed glass at cooling rates of under 100 K/s with critical casting thicknesses ranging upward toward 1 cm [10]. A similar family of alloys with the rare-earth metal (Yittrium here is considered to be a rare earth or Lanthanide alloying element) partially replaced by the alkali earth metal Mg were also developed (Mg-Y-Cu, Mg-Y-Ni, etc.) [11, 12] along with a parallel family of multicomponent Zr-base alloys (e.g. Zr-Cu-Ni-Al) [13, 14]. These multicomponent glass-forming alloys illustrated that bulk glass formation was far more ubiquitous than previously thought, and not confined to exotic Pd-base alloys (e.g. the Pd-Ni-P family). The work opened the door to the development of other broad classes of bulk metallic glasses. Building on the Inoue work, Peker and the author [15-17] developed a family of ternary, quinary, and higher order glass formers based on higher order alloys of Zr, Ti, Cu, Ni, Be (also combined with other transition metals). One extensively studied example, referred to as Vitreloy 1, has a composition which can be usefully represented in the form $(Zr_3Ti)_{0.55}(Cu_5Ni_4)_{0.225}$ $Be_{0.225}$. Direct measurements of the TTT-diagram of this alloy [18] placed the "nose" of the nucleation curve for bulk samples at time scales of the order of 100 seconds with critical cooling rates for glass formation of 1 K/s. The alloys were be cast in the form of fully glassy rods with diameters ranging up to 5-10 cm. The alloys require no fluxing or special processing treatments and form bulk glass by conventional metallurgical casting methods. The glass forming ability and processability is comparable to that of many silicate glasses. Metallic glasses could now be processed by common methods available in a foundry.

To put the development of metallic glass forming alloys in perspective, let us refer to Figure [1]. The figure shows a schematic time-temperature-transformation diagram for crystallization of the undercooled liquids. The Duwez glasses had nucleation kinetics in the undercooled region (between the melting point and glass transition temperature T_g) such that the time scale for crystallization was in the hundreds of microseconds or millisecond range at the "nose" of the nucleation curve. Chen's Pd-Cu-Si glass, for example, would have fallen in the range where crystallization was occurring in tens of milliseconds. The Pd-Ni-P glass studied by

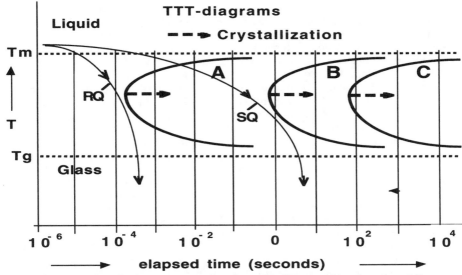

Fig. 1 Schematic TTT-diagrams comparing the crystallization kinetics of rapidly quenched (A), slower quenched (B), and bulk glass forming alloys (C). The two curves with arrows show the cooling history during rapid quenching at ~10^6 K/s (RQ) and relatively slow quench at ~10^2 K/s (SQ).

the Turnbull group exhibited a TTT-diagram with a "nose" in the range of 1-10 seconds. The new "Inoue" glasses also exhibited a crystallization "nose" in the range of 1-10 seconds. Finally, the alloys of the Vitreloy family exhibit a nose in the TTT-diagram at time scales on the order of 100 seconds. Very recent extensions of the earlier work on Pd-base glasses have also shown a TTT-nose at time scales of 100 seconds or more [19,20] for quaternary alloys of the form Pd-Cu-Ni-P etc. The development of such multicomponent alloys with exceptional glass forming ability will likely continue.

GLASS FORMATION IN MULTICOMPONENT ALLOYS – THE ROLE OF COMPLEXITY AND FRUSTRATION

To make bulk glasses, one must frustrate the process of crystallization. Simple fcc, hcp, and bcc crystals with one (or two in the case of hcp) atom per unit cell are formed typically by the pure elements and simple solid solutions of elements with extended solubility. If we put two species of atoms on a crystal lattice with one atom per unit cell, we "frustrate" the crystal by chemical disorder. This chemical disorder is associated with local atomic level strains arising from atomic size differences as well effects arising from differences in the valence electron configuration of the two species. Takeshi Egami [21,22] and his collaborators have studied the effects of random atomic level strains on the stability of solid solutions. They have demonstrated that for two atoms of differing size, there exists a critical maximum solubility in both of the terminal solid solutions. Outside these critical concentrations, the solid solutions become topologically unstable with respect to transformation to the glassy phase. Li et. al. [23] used molecular dynamics studies to show that such saturated solutions develop shear instability. This destabilization is reflected in the phase diagrams by a falling liquidus curve of the terminal solid solutions. The liquid phase, due to its configurational entropy, is better suited to accommodate atomic level strains arising from the size difference of two atomic species. With alloying, the liquid enthalpy of mixing is lowered vis a vis that of the crystalline solid solution, resulting in a lower melting point.

Consider a more complex crystal structure with two non-equivalent atomic positions per unit cell, and a preferred stoichiometry for the two atomic species. When we introduce a third atomic species, we can again frustrate the crystal. If the third species is chemically and topologically (e.g. atomic size or valence) different than the first two, we will once again create frustration in the crystal. The third species must substitute for one of the first two species on the lattice. The crystal will, once again, be subject to chemical disorder and accompanying local atomic strains etc. Thus, a third species can destabilize a "binary" ordered crystalline phase. Looking through the list of known crystal structures, one observes that as the order of a structure goes from elemental, to binary, ternary, quaternary, etc., the number of new structure types tends to diminish. Going from one component elemental structure (of which there are a relatively small handful – e.g. fcc, hcp, and bcc in metallic systems) to binary alloys, we find a relatively large number of new phases having two or more non-equivalent crystallographic sites. We identify still more new crystal structures in true ternary alloy phases (containing three or more non-equivalent crystallographic sites occupied by three or more atomic species) although the number of new ternary crystals is of the same order as the number of binary crystals. Extension to quaternary crystals introduces relatively fewer new crystal structure types. Further, the typical complexity and size of the unit cell in these higher order crystal structures implies that the energetic advantages of forming a periodically ordered structure becomes progressively more marginal. For example, in a quinary crystal where we have five nonequivalent atoms in the five nonequivalent positions in a unit cell, and 200 atoms in the unit cell, the long-range periodicity of the unit cell is over such a large scale compared to the range of the atomic interactions that the marginal benefit of crystallizing (from the melt) is significantly diminished. Such crystals tend to have marginally lower energy than the corresponding liquid, and thus relatively lower melting points. In an alloy, if we introduce multiple atoms that are chemically different (in size, valence, etc.), which do not "fit" into a finite number of nonequivalent sites of relatively simple crystals, then we will "frustrate" the tendency of the alloy to crystallize at all. Thermodynamically, the liquid, being better able to accommodate the chemical disorder, gains a negative enthalpy and entropy of mixing and a free energy advantage vis a vis the competing crystalline phases. The liquid phase is stabilized relative to the crystalline phase and the melting point is substantially depressed.

Turnbull's work on nucleation of crystals from the melt teaches us that the reduced glass transition temperature, $t_{rg} = (T_g/T_l)$, is the key parameter to consider in glass formation. The homogeneous crystal nucleation rate in the undercooled melt is a strong function of this dimensionless parameter. Bulk glass formation is predicted if this parameter reaches a value of about 2/3 or larger. In the real world, it is found that certain complex molecular liquids (e.g. polymer blends) have t_{rg} in excess of this 2/3 value. These liquids do not crystallize at all on laboratory time scales and always form glass. Other examples of exceedingly good glass formers are soda-silicate (Na_2O-SiO_2) and lime-soda-silicate glasses (CaO-Na_2O-SiO_2), which have t_{rg} of 2/3 or slightly larger. Incidentally, if $t_{rg} = 1$, the glass would be in the equilibrium state (at T_g) and would never crystallize on any time scale.

How do we combine the above discussion of "frustration" in crystalline alloy physics with the Turnbull criterion to develop bulk glass forming alloys? To illustrate the principles, consider the series of progressively higher order alloys which led to the development of Zr-Ti-Ni-Cu-Be bulk metallic glasses. On going from a one component metal to a two component alloy, one must first consider the destabilization of the parent elemental phases. The solubility of two atoms in a simple elemental crystal structure depends significantly on their relative atomic "size" as shown by Hume-Rothery [25] who suggested that mutual solubility of elements is very restricted when the atomic radii differ by more than 15%. This is reflected in the binary phase diagrams. Using the regular solution model for both the crystalline solid solutions and the liquid phase, one can compare the variation in the calculated phase diagrams as the heat of mixing parameter Ω_{mix} is varied for the liquid and crystalline solution phases. In this model, the heat of mixing is modeled by

$$\Delta H^{L,x}_{mix} = \Omega^{L,x} c (1-c)$$

where L and x refers to liquid and crystalline phases and c is the alloy composition. One obtains a "deep eutectic" phase diagram, and consequently a high reduced glass transition temperature, t_{rg}, for $\Omega^L \ll \Omega^x$. This condition is met, for example, in an alloy with atomic size differences lying outside the 15% Hume-Rothery rule. We want the liquid heat of mixing to be significantly negative compared to that of the crystal because the difference of the two controls the relative stability. Having a large negative heat of mixing itself does not necessarily provide a good glass former. Searching through binary phase diagrams, one finds the best options are binary alloys such as Pd-Si, Pd-P, and Pd-P. These exhibit very "deep binary eutectics". Binary metal alloys such as Zr-Ni and Nb-Ni also show such deep eutectic compositions.

Two interesting phase diagrams that we noticed in 1992 are Ti-Be and Zr-Be. The Zr-Be diagram is shown in Fig.2. This is a phase diagram of the very "small" atom Be, a simple metal, with metallic radius (0.112 nm). This compares with nickel (0.124 nm), copper (0.128), and the larger radii of Zr (0.160 nm) and Ti (0.147 nm). Both Ti-Be and Zr-Be show deep eutectics on the transition metal rich side of the diagram and only one new crystalline intermetallic and the 1:2 stoichiometry. $Zr-Be_2$ has the AlB_2 –type structure, while $TiBe_2$ is a laves phase of the $Mg-Zn_2$ type. There are no titanium-zirconium-rich compounds in either diagram, or in the ternary Zr-Ti-Be diagram. The ternary diagram shows an even deeper eutectic structure near the ZrTiBe equiatomic composition. Lee Tanner [26] of Allied Signal rapidly quenched Zr-Be, Ti-Be, and Zr-Ti-Be alloys in the late 1970s and demonstrated excellent glass forming ability, particularly in the ternary system. Early work also showed that binary alloys of Zr and Ti with ferrous metals (Fe, Co, Ni, and Cu) were good glass formers. Figs. 3 and 4 show examples, the binary Ti-Cu and ternary Zr-Ti-Cu phase diagrams. In Fig.3, the T_o-curves (polymorphic melting curves) for the competing fcc and bcc solid solution phases are also shown for comparison. Also shown is the composition range where amorphous (and fcc solid solution) phases are formed by rapid solidification. In both the Cu-Ti and Cu-Zr systems, one can form glasses by rapid quenching in the central portions of the diagrams. The reduced glass transition temperatures of the best binary glass formers are about ~0.56-0.58. The competing crystalline phases [e.g. fcc solid solutions, Zr_2Ni, Ti_2Cu, etc.] limit the glass forming ability. Ternary alloys of Zr-Ti-Cu and Zr-Ti-Ni are even better glass formers than the binaries. This is essentially because the competing binary crystals (e.g. $Zr_2Cu/ZrCu$ and $Ti_2Cu/TiCu$) have limited mutual solubility yielding a more frustrated ternary system. In Zr-Ti-Cu, there is only one ternary phase (ZrTiCu) which forms in the central region of the ternary system as seen in Fig.4. Around this ternary phase, we find a "circle of low melting alloys" with three true ternary eutectics (see E1, E2, and E3 in the diagram). The situation is further improved by taking quaternary alloys of Cu-Ni-Ti-Zr [see. ref. 17]. Near the quaternary eutectic compositions, one forms even better glasses having reduced glass transition temperatures. By example, one quaternary $Ti_{34}Zr_{11}Ni_8Cu_{48}$ alloy with $t_{rg} = 0.6$, forms glasses at cooling rates of about 50-100 K/s [17,27] and is based on the ternary eutectic E3 of Fig.4. The optimum compositions of several quaternary glass formers are illustrated in Fig. 5 and are found to lie in two distinct regions related to ternary eutectic points E2 and E3 of the liquidus projection shown in Fig.4.

Now recall that for the case of Zr-Be and Ti-Be, transition metal-rich compounds do not exist. This implies, for example, that Be will have limited solubility in the intermetallic phases of the Ti_2Cu, Zr_2Cu, Ti_2Ni-type etc. A Zr-Ti-rich higher order alloy of Zr-Ti-Ni-Cu-Be can be made for which there exists no single crystalline phase, which can form near the overall average alloy composition. In the Zr-Ti-Ni-Cu-Be system, five elements essentially act independently, each species preferring different non-equivalent crystallographic sites in the competing crystalline phases. The alloy is exceedingly "frustrated"! The competing crystals have 1–3 nonequivalent crystallographic positions which must accommodate 5 differing species. The crystalline phase will be forced to accommodate substantial atomic strains, etc. The thermodynamic competitiveness of the crystalline phases is diminished relative to the more accommodating liquid phase, resulting in a very deep higher order "eutectic" structure. Combining Cu-Ni-Ti-Zr with beryllium into a quinary system, we obtain still deeper eutectic features in virtually the entire central portion of the quinary phase diagram. The region of lowest lying liquidus features is illustrated in one type of ternary projection of the system in Fig.6. Alloys in this central shaded region of Fig.6 exhibit melting transitions (in a DTA) with a solidus as low as 630°C and liquidus temperature as low as 700C. Compared to the glass transition temperature (typically in the range of 360-400 C, this gives t_{rg} values

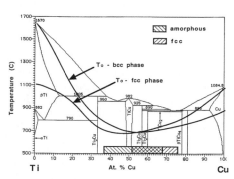

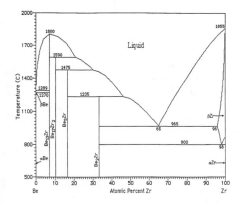

Fig. 2 The binary phase diagram of the Be-Zr system. Taken from ASM Binary Alloy Phase Diagrams

Fig.3 The binary Cu-Ti phase diagram showing T_o-lines of the competing bcc and fcc simple solid solution phases. Also shown are regions of the diagram where amorphous and fcc phase form by rapid quenching methods

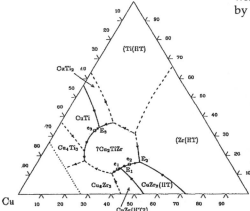

Fig.4 Liquidus projection of the Cu-Ti-Zr system. From ASM Compendium of ternary phase diagrams

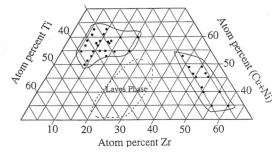

Fig.5 Pseudo-ternary diagram of the Cu-Ni-Ti-Zr system showing regions of "bulk" (>1mm thick strips) glass formation (dots) lying in two areas related to E2 and E3 of Fig.4. Laves phase region enclosed by dashed line. See refs. 17 & 27 for details.

ranging from 0.65- 0.7. According to Turnbull, we predict that homogeneous nucleation will be very difficult in these highly frustrated multicomponent alloys!

As discussed later, such a frustrated alloy will explore other ways to escape its chemically frustrated composition. The liquid can, for example, phase separate into two liquids of differing composition, in order to escape the "frustrated" composition range. As will be seen, the Zr-Ti-Ni-Cu-Be alloys ultimately crystallize by first undergoing phase separation to two liquids. At the new compositions, the liquidus curves of the competing crystals are higher, the reduced glass transition temperatures lower, and homogeneous nucleation becomes possible once again.

In 1992, Peker and the author developed and experimented with the Zr-Ti-Ni-Cu-Be alloys described above. It was ultimately shown that some of the quinary alloys could be cast into bulk glassy rods with thicknesses ranging between 5 and 10 cm! Glass formation is so easy that large glassy ingots (~100g) are generally formed when the original metals are alloyed in a plasma or induction melting system. Fig. 7 shows a commercial plate of Vitreloy 1 cast at Howmet Metal Mold (Howmet Corp., Whitehall, MI) in a facility used to cast bulk metallic glasses for commercial applications.

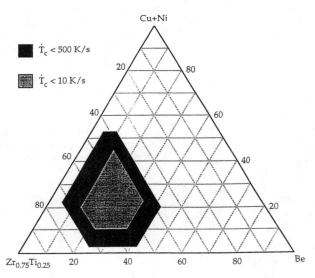

Fig. 6 Pseudo-ternary projection of the quinary Be-Cu-Ni-Ti-Zr   system where the ratio Zr:Ti is held at 3:1 while the overall Ni content varies from 5-15 at. %. Within the shaded regions, t_{rg} is approximately 2/3. Critical cooling rate for glass formation is less than 500 K/s (dark area)   and less than 10 K/s (light area). Taken from ref. 16

Fig.7 Commercially cast plate of Vitreloy 1. Cast by Howmet Corp. , Whitehall, MI., using commercial Howmet process. Plate approx. 4 mm thickness.

STUDIES OF THE DEEPLY UNDERCOOLED METALLIC MELTS, CRYSTALLIZATION, AND PHASE SEPARATION

To better characterize the bulk glass-forming ability of metallic alloys, one needs to study crystallization kinetics in detail. An important step was to develop a controlled method to examine nucleation and growth of crystals in the undercooled liquid. Containerless processing under high vacuum conditions provides a means of diminishing the influence of heterogeneous nucleation and permits processing a chemically reactive high temperature liquid under environmentally controlled conditions. In particular, the High Vacuum Electrostatic Levitation (HVESL) method developed by Rhim at Jet Propulsion Laboratory in the early 1990's provides an excellent platform for such studies [28]. The Rhim system employs active feedback control for positioning an electrostatically levitated liquid drop in a high vacuum chamber. The levitated drop is heated with a focussed quartz lamp or a YAG laser. Non-contact temperature measurements are made with a pyrometer.

Using the HVESL, a drop of roughly 3mm diameter can be heated, melted, then undercooled. Cooling is by free radiative cooling. By control of the heating input power, one can produce well-defined cooling histories. For example, the sample can be melted, then "free cooled" (Stephan-Boltzmann radiative decay) to a predetermined temperature (at a cooling rate of typically 5-10 C/s), then held isothermally by toggling the heating power back up to match the radiative loss at the temperature of interest. Crystallization is observed by following the recalescence behavior of the sample. In this manner, one can directly measure a TTT-diagram (time-temperature-transformation) for crystallization. It should be first noted that samples allowed to free cool from the equilibrium melt to below T_g form metallic glass! Radiative cooling rates are sufficient to produce glass!

The results of several "isothermal" runs are shown in Fig. 8 for Vitreloy 1 ($Zr_{41.2}Ti_{13.8}Ni_{10}Cu_{12.5}Be_{22.5}$). The alloy is preheated to temperatures above the liquidus ($T_l = 720$ C), free cooled to a preset temperature and held there isothermally until recalescence is observed. The time to recalescence is measured for each degree of undercooling. The radiative cooling rate is about 7°C/s for a 3-mm drop so that a transient cooling period precedes the isothermal plateau (the cooling is not instantaneous). This transient period (generally relatively short) is neglected in determining the time to crystallization at the "isothermal" plateau.

Figure 9 shows the TTT-diagram of Vitreloy 1 obtained from results of many runs carried out by Y.J. Kim et. al. [18,29]. The "nose" of the TTT-diagram, defined as the shortest time to crystallization at any undercooling, occurs at relatively shallow undercoolings at a time of about 70 s. From the position of the nose, we can estimate the critical cooling rate of the alloy to be of order 1-2 K/s. In fact, continuous cooling curves can be done on the ESL and place the critical cooling rate between 1 and 2 K/s. Simple theories of "polymorphic crystallization" using a Vogel-Fulcher viscosity law, normally used to describe crystal nucleation in one-component systems, predict the "nose" should occur at deeper undercoolings. In addition, the form of the TTT-curve is broader than expected form simple theories [30,31] and exhibits extensive asymmetry in the upper and lower branches of the curve. It is noteworthy that later experiments [32] showed that the curve is also independent of the size of the samples used. In other words, the time to the "nose" is at ~70 s (and the TTT-curve as a whole) are the same irrespective of the size of the sample studies. This suggests that simple homogeneous nucleation cannot be the rate limiting factor in crystallization. These features together provide limits for experimental models. Kim et. al., and others [18,33] tried to fit the data with simple steady state nucleation and got rather poor fits. The nucleation and growth of crystals in this system is apparently more complex than the simple one-component models would predict.

Kim et. al. noticed [18] that samples free cooled below about 800 K always exhibit a small anomaly in the cooling curve. This can be seen in the free cooling part of several curves in Fig. 8. X-ray diffraction and TEM analysis revealed that free cooled samples which exhibit this anomaly show no evidence of crystallization. Later work by Busch et. al. [34], and Miller et. al. [35] using Field Ion Microscopy and Atom Probe techniques revealed that free cooled samples showed liquid-liquid phase separation into a Beryllium rich and Be poor liquids. The image of the FIM sample tip shows regions of clear constrast. Atom probe FIM can be used to locally sample the compositions of the decomposed regions, Busch et. al. found that the dark/light regions correspond to areas of low/high beryllium concentration.

In a five-component system, there are four compositional degrees of freedom; a 4x4 matrix G [i, j] describes the stability of the liquid and is expressed in terms of the second partial derivatives of the Gibb's free energy with respect to the four degrees of freedom, i.e.

$$G''[i, j] = \partial^2 G/\partial x_i \partial x_j$$

where the x_i's are the compositional degrees of freedom. Stability is determined by the 4 eigenvalues of this matrix λ_i ($i=1...4$), which must be positive for stability. When an eigenvalue passes through zero on lowering of the temperature T, we have a critical point and spinodal instability results along one principal axis (compositional eigenvector) of the stability matrix. On cooling, 1 or more eigenvalues may change sign resulting in a "cascade of critical points". The change of sign of each eigenvalue opens a miscibility gap along a different direction in composition space. The resulting phase separation can be complex since the kinetic response to each unstable decomposition mode depends on the chemical interdiffusion constants along the corresponding principal axis in composition space. In general, the decomposition may evolve along a different direction than the driving force (the principal axis of G'' associated with a negative eigenvalue λ_1). The decomposition actually observed at high temperatures (the small anomaly, I, in Fig. 8) corresponds to a principal axis in composition space which lies primarily along the Be composition axis [34]

Susanne Schneider et. al. [36,37] investigated the phase separation in Vitreloy 1 using small-angle neutron scattering (SANS) and obtained a classic spinodal decomposition behavior in samples annealed at low temperature.

The decomposition that Schneider saw by aging the sample near the glass transition differs from that seen by Busch (decomposition primarily along the Be-axis of the phase diagram). The SANS is particularly sensitive to variations in the Ti content of the alloy (due to the negative scattering length of the Ti nucleus) The SANS spectra show a low angle interference peak which increases in intensity with annealing and exhibits a shift in the scattering vector of the peak (Q_{max}) with the annealing temperature. Cahn's theory predicts that the square of the dominant spinodal wavelength $\lambda^2 = [2\pi Q_{max}]^{-s}$ should be a linear function of T with an intercept at the spinodal temperature. Fig 10. shows such a plot from the SANS data [36,37] and suggests a spinodal temperature of about 670°C, about 30-40 °C above the T_g of the glassy alloy. Schneider determined that this low temperature decomposition involves primarily the Ti and Cu content of the alloy. Further, she determined that decomposition triggers the nucleation of fcc nanocrystals which are enriched in Ti and Cu within the decomposed zones. The growth of the fcc nanocrystals is then limited to the size of the domain in which they nucleated. Since the spinodal decomposition is a spatially correlated phenomenon, if we nucleate the fcc crystals below the spinodal, we obtain spatially correlated crystals. The same fcc crystal can be nucleated in V1 by annealing above the spinodal. In this case, we obtain random crystals that are spatially uncorrelated as shown in Schneiders' experiments.

By choosing a new alloy having the same composition as the Ti- and Cu-poor decomposed regions, one can prepare a bulk glass which is stable with respect to this type of spinodal decomposition near Tg. Such an alloy is stable with respect to nucleation of the FCC nanocrystals. Vitreloy 4, having a composition $Zr_{46.2}Ti_{8.8}Cu_{7.5}Ni_{10}Be_{27.5}$ (5 at. % reduction in the Ti- and Cu-content compared with V1), in fact has a spinodal temperature well below T_g. Above T_g, V4 is outside the spinodal gap and shows no evidence of phase separation in SANS [38]. V4 does not exhibit nucleation of the fcc phase at all! The V4 alloy is ideally suited for studies of

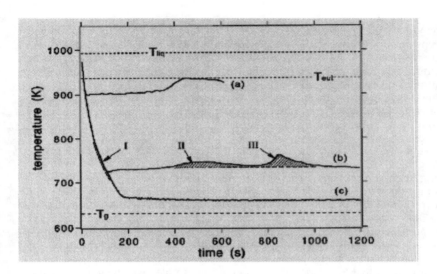

Fig.8 A series of "isothermal" undercooling curves obtained on a 3mm diameter sphere of Vitreloy 1 by radiatively cooling from the equilibrium melt then holding at an isothermal plateau. (a) shallow undercooling, (b) deep undercooling, and (c) deep undercooling to near Tg. The temperature rise observed following an elapsed isothermal hold time corresponds to recalescence and crystallization (see ref. 18). I, II, and III corresponds to phase separation, nucleation and growth of fcc nanocrytalline phase, and nucleation of Zr_2Cu-type phase respectively.

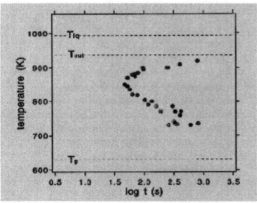

Fig.9 Time-temperature-transformation diagram showing the time to the onset of crystallization as a function of isothermal undercooling temperature. Solid symbols represent the first crystallization event observed. Below 800K, two crystallization events are observed. The first is denoted by solid symbols, the second is denoted by the open symbol with an x.

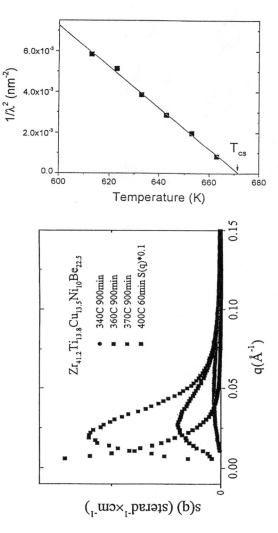

Fig.10 Left: A series of small angle neutron scattering curves following annealing of the glass at the indicated temperatures and times, Right: A plot of $1/\lambda^2$ vs. annealing temperature for annealed samples of Vitreloy 1. Here λ is the dominant spinodal wavelength which grows at the given annealing temperature (as determined by the maxima in the SANS curves). See refs. 36 and 37 for details.

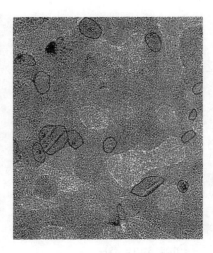

Fig.11 A high resolution TEM image of a glassy Mg-Y-Li-Cu sample exhibiting phase separation. See refs. 40 and 41 for details.

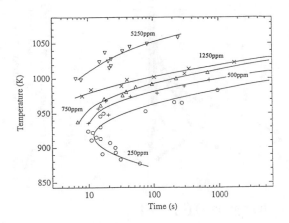

Fig.12 TTT-diagrams of Vitreloy 105 containing varying amounts of oxygen impurity. Data from ref. 44.

viscosity and atomic diffusion above T_g since the absence of phase separation and nanocrystallization simplifies the interpretation of these measurements.

Figure 11 shows a high resolution TEM image of a Mg-Y-Cu-Li alloy of the Inoue type [11,12] obtained by Liu et. al. [39]. The image shows evidence of phase separated domains. SAXS (small angle x-ray scattering) and ASAXS (anomalous small angle x-ray scattering) studies of this alloy also give evidence for phase separation [40] into Cu-rich and Cu-poor regions. Again, these phase-separated domains appear to be preferred sites for nucleation of bcc (Mg_6Li) nanocrystals [40,41]. The local composition shifts during decomposition towards compositions having a rising liquidus curve for the bcc phase in much the same manner as observed for the fcc-phase in V1. Clearly the precipitation of primary nanocrystalline phases in both systems is governed by the initial chemical decomposition of the undercooled liquids. Chemical decomposition of the undercooled melt controls crystallization in these systems.

In further assessing the role of phase separation and composition fluctuations in nucleation of crystals, the work of P. Desre has been particularly interesting [42], Desre investigated the role of composition fluctuations in determining the nucleation rate for a system which is "stable" with respect to such fluctuations (above the spinodal temperature). Even when all the eigenvalues of the G'' matrix are positive, Desre argues that the evolution concentration fluctuations in a higher order multicomponent system can become the rate limiting step to form a critical nucleus. Using the Landau theory for thermodynamic fluctuations, Desre derives an expression for the probability of a correlated fluctuation in m compositional variables and shows that this probability decreases by roughly an order of magnitude for each additional component. Desre's conclusion, when the time scale nucleation is dominated by the time scale for a correlated concentration fluctuation to develop, then addition of each independent component to the alloy increases this time scale by roughly one order of magnitude. Here, the term "independent" refers to the fact that the competing crystalline phase is chemically ordered with respect to each independent component. Glass formation should thus be favored in higher order multicomponent systems. This observation seems to agree with the experimental trend of bulk glass formation in progressively more complex alloys. The Desre work shows that this effect is operant even when phase separation does not occur! In a separate paper [43] Desre argues that this effect is particularly effective in suppressing heterogeneous nucleation. This may explain the absence of heterogeneous nucleation in bulk glass forming alloys in contact with crystalline solids (see discussion of BMG composites in the last section of the paper).

The HVESL method can be used to study other aspects of crystal nucleation and growth. Fig. 12 shows TTT-curves of another bulk glass forming alloy, V105 with the composition $Zr_{52.5}Ti_5Cu_{17.9}Ni_{14.6}Al_{10}$ studied by Lin and the author [44]. A series of TTT-curves corresponding to the addition of oxygen impurity to the parent alloy were measured. The oxygen content ranges from about 300 to 5000 atomic ppm. Over the range of oxygen contents studied, the TTT-curves vary by roughly two orders of magnitude along the time axis. In other words, oxygen contamination ranging up to 0.5 atom % can alter the critical cooling rate for glass formation by two

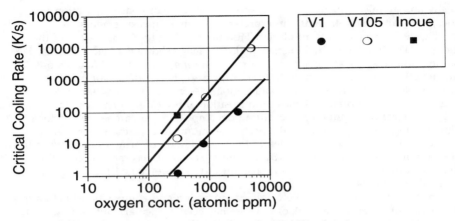

Fig. 13 The critical cooling rate for glass formation of Vitreloy 1 and Vitreloy 105 alloys as a function of the oxygen content of the alloys. Data on V1 and V105 are taken from HVESL work and other measurements of critical casting thickness. The alloy V105 has the composition $Zr_{52.5}Ti_5Cu_{17.9}Ni_{14.6}Al_{10}$ while the Inoue alloy is $Zr_{65}Cu_{15}Ni_{10}Al_{10}$.

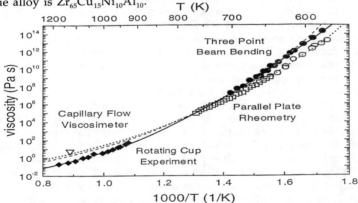

Fig14 Summary of Viscosity data on V1 and V4 taken using a variety of experimental methods as described in the text. Filled symbols are data for Vitreloy 1 ($Zr_{41.2}Ti_{13.8}Cu_{12.5}Ni_{10}Be_{22.5}$) while the open symbols are date for Vitreloy 4 ($Zr_{46.75}Ti_{8.8}Cu_{7.5}Ni_{10}Be_{27.5}$). The various methods used to measure the viscosity are listed on the figure. The fits to the data (solid and dashed curves) were obtained using the Vogel-Fulcher-Tamann relation $\_\_\_{\circ}exp[DT_{o}/(T-T_{o})]$ where the parameter D is the so-called "fragility" parameter. The parameters are $T_{o} = 412K$ (372K) and D = 18.5 (22.7) for Vitreloy 1 (Vitreloy 4). See references 46,49 and 52 for details.

orders of magnitude! This dramatic "impurity" effect quantitatively showed, for the first time, the extreme importance of alloy purity on crystallization behavior. Similar behavior has been found for the Vitreloy alloy series and the Inoue Zr-Ni-Cu-Al alloys. The effects of oxygen impurity levels on critical cooling rates of the V1 and V105 glass formers are summarized in Fig. 13. Note that the logarithm of estimated critical cooling rate varies roughly linearly with log (x_{oxygen}). The slope is about 2 indicating that $dT/dt]crit \sim x_{oxygen}^2$. For Vitreloy 1, a clean sample (300 ppm oxygen) has a critical cooling rate of about 1 K/s. With 1500 or 2000 ppm oxygen the critical cooling rate is increased to hundreds of K/s. With 1 atom % O, glass formation actually requires rapid solidification. Based on the above data and the observation of an "overheating effect", Lin et. al. [44,45] argued that the addition of oxygen induces nucleation of an oxide phase with low oxygen content. Oxygen impurities raise the liquidus temperature of this oxide phase and thus induce nucleation at progressively higher temperatures. This oxide nucleation event then heterogeneously catalyzes the nucleation of the other crystalline phases resulting in complete crystallization of the liquid.

DIFFUSION, VISCOSITY, AND TRANSPORT PROPERTIES

The stability of bulk metallic glass forming liquids in the undercooled region has permitted far more extensive investigation of the temperature dependence of the Newtonian viscosity for a deeply undercooled metallic system than possible previously. A variety of techniques can be applied to measure viscosity from the equilibrium melt down to the deeply undercooled range near T_g. Parallel plate rheometry, beam bending, and capillary flow methods have all been applied to the study of rheological properties [46-50]. For example, Masuhr et al. [48], have developed a rotating cup viscometer to measure liquid viscosity at high temperature in the equilibrium and somewhat undercooled melt. Masuhr also used capillary flow measurements in this range [50]. Busch, Bakke, et. al. used parallel plate rheometry and beam bend measurements for high viscosities found near and above T_g (deep undercoolings) [46,47]. Since the undercooled liquid alloys are relatively stable with respect to crystallization on laboratory time scales, one can do measurements throughout most of the undercooled liquid region, with the exception of a small gap covering a few orders of magnitude. This is seen in Fig. 14, which summarizes viscosity measurements of V1 and V4 ($Zr_{46.2}Ti_{8.8}Ni_{10}Cu_{7.5}Be_{27.5}$). The latter alloy exhibits better stability against crystallization in the low temperature region near T_g (as discussed above) and was thus chosen for rheological studies in this range. These data form what is probably the most complete set of viscosity data available for any liquid metal system in the undercooled region. The viscosity data extend from the equilibrium melt over much of the undercooled range. Such data were inaccessible in earlier metallic glass forming alloys because nucleation kinetics prevented experimental access to most of the undercooled region. For example, H.S. Chen [51], who pioneered rheological studies of glass forming metals at Bell Labs in the 1970s, was limited in measuring viscosity to temperatures of only 20-30 K above T_g (by the lack of stability against crystallization of the Pd-Cu-Si glasses which he studied). Essentially, earlier alloys crystallized when viscosity values fell below about 10^{10} Pas-s. In other early work, Spaepen and Taub [52], and Spaepen and Volkert [53] studied flow and relaxation of Pd-Ni-P glasses near and slightly above T_g. They found that small additions of Si to Pd-Ni-P alloys improved stability above T_g permitting extension of their viscosity data to slightly higher temperatures. These studies were nevertheless limited to minimum viscosities in the 10^{10} or 10^9 Pas-s range.

With the new bulk glasses, measurements have extended the data to much lower viscosities. As seen in Fig.14, the data obtained by several techniques over about 15 orders of magnitude in viscosity are well fit by the Vogel-Fulcher expression

$$\eta = \eta_o \, \exp[DT_o/(T-T_o)]$$

where T_o is called the Vogel-Fulcher temperature and D is often referred to as the "fragility" parameter. The D parameter will be discussed further below.

The surprise is that the Vogel-Fulcher temperature, contrary to what researchers had expected from earlier work, was not near the experimentally observed T_g, (observed for example in calorimetry experiments), but far below it. The apparent singularity in the viscosity occurs at a temperature that is roughly 60% of the temperature (in K) where we measure the laboratory glass transition by calorimetry, for example. Following Angell, the fragility can be assessed by plotting viscosity data as a function of Tg/T (where Tg is the calorimetric glass transition or the temperature where the viscosity is 10^{12} Pas-s). A comparison of the Vitreloy alloys with several other types of glasses such a silica, silicates, and organic glasses (O-terphenyl) is shown in Fig. 15. The bulk metallic glass formers are relatively strong glasses (much like soda-silicates). The results are surprising in view of earlier work on amorphous metals where researchers modeled crystal nucleation kinetics by assuming that T_0 and T_g were very close [56,57]. In fact, it has been common to replace T_g by T_0 in modeling kinetics in amorphous metals. Bulk glass formers are, in fact, very different from elemental metals and simple binary alloys. As seen in Fig. 15, the rheological behavior is similar to that of many silicate glasses. Silica itself is the "canonical example" of a strong glass where T_o approaches 0K and the viscosity has an Arrehnius form. In Fig.16, we compare the fragility of a series of glass forming alloys in which glass forming ability increases with increasing number of components. Here the fragility has been determined by using the heating rate dependence of T_g itself. The inverse of the heating rate is proportional to the relaxation time of the glass which can then be plotted vs. Tg/T as described in refs. 49,50. On going from rapidly quenched binary glasses (e.g. Zr-Ni), to ternary (Zr-Al-Ni), etc. to the best bulk glass forming alloys, one sees a steady increase in the fragility parameter D. The figure suggests that the ease of bulk glass formation in metals is correlated with rheological "strong glass behavior". This correlation should not be surprising. Strong glass behavior leads to higher values of viscosity at high temperatures where crystallization rates are greatest (near the nose in the simpler binary alloys. Note that the viscosity of V1 is of the order of 10 Pas-s at the equilibrium liquidus temperature of the alloy (720 C)! This value is 3-4 orders of

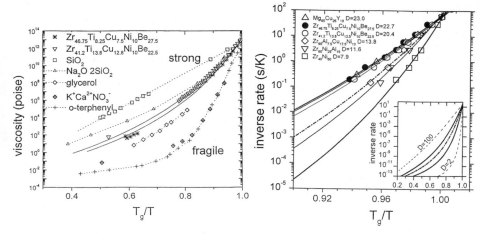

Fig.15 Angell plot comparing the rheological behavior of several liquids. The plot shows that the bulk metallic glass forming alloys V1 and V4 exhibit relatively "strong" liquid behavior. See text for details.

Fig.16 Angell plot comparing a series of progressively more complex alloys with increasing glass forming ability. Here, the dependence of Tg itself on inverse heating rate was used to characterize the relaxation time dependence on temperature of the glass [see ref.49]. The plot suggests that glass forming ability is correlated with increasing fragility index of the glass, i.e. "stronger" alloy liquids are better glass formers.

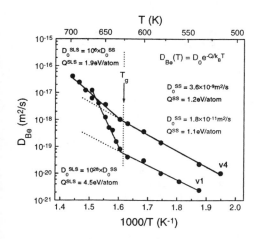

Fig. 17 Chemical diffusion of Be in Vitreloy 1 as measured by Geyer et. al. [58,59]. Note the break in the temperature dependence in the vicinity of the glass transition. The reader is referred to ref. 58 for details of the diffusion experiments.

329

magnitude greater than the viscosity of elemental metals or simpler binary alloys at their respective melting points! Indeed, this will result in a relative "slow down" of crystallization. This effect is likely an important contributing factor to the glass forming ability.

In addition to fluidity, it is of clear interest to study atomic diffusion in these systems. The first studies of atomic diffusion were carried out by Geyer et. al. [58]. Fig 17 shows results of measurements of Beryllium diffusion in the glassy and undercooled liquid states of glassy V1. One sees a clear break in the diffusion curve, which has been explained by the authors in terms of the supercooled liquid/glass transition. Here, the T_g is defined with respect to the time scale of the diffusion experiments. The apparent activation energy for diffusion in the glass (low T) is much smaller than that of the undercooled liquid. Geyer et. al. [58,59] discussed this in terms of a crossover in behavior from Be hopping in an essentially solid environment to Be transport by cooperative shearing in the liquid state (as reflected in viscosity measurements). Mehrer's group [60.61] have reported tracer diffusion studies of various elements in V1 and V4 at temperatures above and below T_g. They see similar behavior to Be where the break in the diffusion curve depends on the diffusing species. While "small" atoms such as Ni, and Co, show a pronounced break (though less pronounced than the still "smaller" atom Be), large atoms such as Al (or Zr), show almost no break at all. In the later case, the atomic diffusion constants follow a temperature dependence much like that of the viscosity. Masuhr et. al. [62] have used the relation $\tau_A = l^2/6D$ where l is a diffusive jump distance (of the order of the interatomic spacing) to define a characteristic "jump time" for atomic diffusion. From viscosity data, they define a Maxwell stress relaxation time $\tau_\eta = \eta/G_\infty$. These relaxation times can be plotted and compared on a plot vs. T_g/T as shown in Fig. 18. The relaxation times tend to follow the same temperature dependence at high temperatures. On cooling to towards T_g, one observes that the atomic diffusion curves "decouple" from the viscosity curve. The decoupling occurs at higher temperatures and in a more dramatic fashion (greater change in apparent activation energy) for the "smallest" atom (eg. Be). At low temperatures, the small atoms diffuse predominantly by an individual atom-hopping mechanism within a solid glassy matrix. The apparent activation energy in this range is much lower than expected based on the slope of the Vogel-Fulcher viscosity curve. The crossover is dramatic. The decoupling is barely seen in the diffusion of the large atoms (e.g. Al and Zr). The various elements undergo an apparent glass transition at different temperatures. If one assumes that chemical diffusion at high T is governed by the Vogel-Fulcher law, and at low temperature by the "decoupled" chemical diffusion of the smaller atoms, one can obtain a hybrid model which nicely describes the effect of atomic transport on the form of the TTT-diagram for nucleation of crystals [62]. Such a model provides an excellent description of the form of the TTT-diagram for V1 as measured in the HVESL (see previous discussion).

Decoupling behavior has previously been observed in ionic glasses and discussed by Angell and coworkers [63-64]. It implies that relaxation in the undercooled liquid state occurs by different processes depending on which atomic species is involved or whether one observes the collective shear behavior of the liquid as a whole. This means, for example, that atomic transport, stress relaxation,

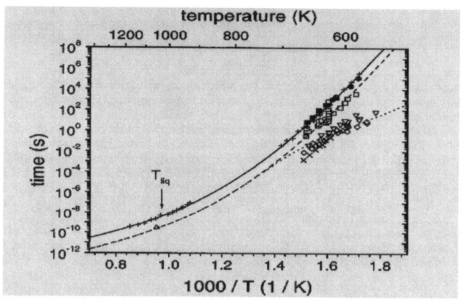

Fig. 18 Plot of the temperature dependence of both the Maxwell stress relaxation time τ_η (●) and $\tau_\eta = \eta/G_\eta$ (+) from equilbrium viscosity measurements of V1. Calorimetric relaxation times, (■) from DSC heating experiments. Diffusion relaxation times, τ_D, from intrinsic atomic diffusion measurements on Be (✖), Ni(◊), Co(∇), Au(Δ), and Al( ). Dashed line is a model from ref. 62.

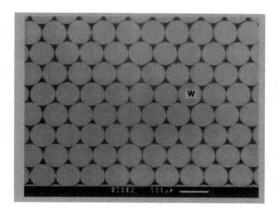

Fig. 19 SEM micrograph of metallic glass matrix composite containing 80 volume % W wire in a matrix of Viteloy 1. Light areas are the wires and dark areas the metallic glass matrix. Scale bar is 100 μm. Taken from ref. 77.

and other dynamic processes are each characterized by a unique relaxation time behavior. Obviously, the Stokes-Einstein relation will not hold in this case. Geyer, for instance, has suggested that Be diffusion can be related to viscosity by a more general relation proposed by van den Beukel [65].

Summarizing, the experimental elucidation of the details of transport and relaxation throughout the undercooled melt is of fundamental interest. The observed behavior demonstrates a new paradigm for understanding shear response and atomic diffusion in higher order liquid metal alloys. Much work remains to be done in this area.

The ability to experiment on deeply undercooled metallic liquids has opened other opportunities for testing theories of relaxation in liquids. For example, the Meyer group [66] has been able to study quasi-elastic and inelastic neutron scattering in the undercooled liquid state of the V1 and V4 alloys. This has permitted a test of Mode Coupling Theory. They find that the dynamic relaxation behavior at high temperatures as determined by quasi-elastic scattering is consistent with the predictions of the mode coupling theory. They were able to estimate the critical temperature for mode coupling from the experimental data providing the first experimental test of mode coupling theory on metallic liquids.

SOME PROPERTIES AND APPLICATIONS

What are metallic glasses good for? They have very high yield strength and high elastic limit, roughly a 2% elastic strain limit in tension or compression. The yield strength in tension is thus typically $\sigma_y = 0.02\ Y$, where Y is Young's modulus. For Vitreloy 1, with $Y = 95$ GPa and $\sigma_y = 1.9$ GPa [67]. The Hookian elastic strain that a metallic glass can support in tension or bending is at least double those of ordinary commercially useful crystalline materials. Metallic glass is the premier "spring " material. Upon yielding, metallic glasses tend to form localized shear bands. The localization of shear is associated with the absence of strain hardening (work hardening) mechanisms, possible strain softening mechanisms, and thermal softening during adiabatic heating of the material. Within the bands, one observes large local plastic strains. Unfortunately, without geometrical confinement, failure often occurs along a single band, which traverses the sample. Despite the tendency to localized deformation, metallic glasses are "tough" by many measures. For example, Vitreloy 1 has a plane strain fracture toughness of about $K_{1c} \sim 20$- 55 MN-$m^{1/2}$ [68,69,70]. This value suggests a significant "plastic zone" which screens a crack tip. On the other hand, under tensile load, there is little global plasticity of the sample as a whole.

By contrast to metallic glasses, useful crystalline metals exhibit substantial plastic strains following yielding under tension. This results in high fracture toughness, impact resistance, etc. For example, commercial steels and Ti-alloys have plane strain fracture toughness values of $K_{1c} \sim 50$-100 MN-$m^{1/2}$. This behavior is associated with dislocation activity, work hardening behavior, and a large plastic zone around crack tips. For reference, it should be mentioned that, "brittle" crystalline intermetallic compounds and ceramics, like metallic glasses, also frequently exhibit a high elastic limit (compared with ductile metals), but exhibit essentially no plasticity. For these materials, the Peierle's barrier is so high that dislocations are immobile. "Brittle" intermetallics and ceramics tend to fail by mode I (less frequently mode II) catastrophic crack propagation with little plastic deformation. They exhibit values of $K_{1c} \sim 1$ MN-m$^{1/2}$ or less.

The role of geometry in deformation of metallic glasses is of particular interest. Geometrical confinement of shear bands can dramatically enhance overall plasticity. For example, axial compression of a mm dimension cylindrical sample with an aspect ratio h/d (height/diameter) less than 1, results in deformation to 10-μm-thick foil - plastic strains of 1000- 10,000%. In this geometry, a single shear band cannot result in failure. For a cylinder with h/d = 2-3, failure occurs when a one or two planar shear bands cut across the sample at an angle of 45° (direction of maximum resolved shear stress) with respect to the compressive axis. Plastic strain to failure is less than 1%.

In a uniaxial tensile test, shear bands are always unconfined. Failure is invariably along a single or small number of bands and is catastrophic. The possibility of enforcing shear band confinement under tension, was nicely demonstrated by Courtney [71], who fabricated laminated composite specimens consisting of a layer of metallic glass bonded between two ductile metal layers. He deformed the laminated composite under tensile loading (tensile axis parallel to the layers). Shear bands formed in the glassy layer are "blunted" by the ductile layers. Stress is redistributed at the blunting site and multiple shear bands are formed. A single shear band no longer leads to failure. Courtney observed a high density of multiple shear bands "trapped" between the ductile metal layers. Overall plastic tensile strain of order 10% could be achieved.

In an effort to overcome the problem of limited plasticity and develop engineering applications, recent efforts have focussed on fabrication of metallic glass composites. A variety of composite materials have been formed by the direct introduction of a crystalline solid phase as reinforcement into the glass forming melt [72-78]. The melt is then cooled in an effort to produce a metallic glass matrix composite structure. Perhaps, the most noteworthy aspect of this work is the ease with which such composites can be fabricated. The introduction of a second crystalline phase into the melt would be expected to induce heterogeneous nucleation and crystallization of the glass forming liquid on undercooling. In many cases, this does not occur. The glass forming ability of the remaining liquid is not compromised by contact with the crystalline solid! It appears, for many cases studied, that heterogeneous nucleation does not occur, even when substantial chemical reaction occurs at the liquid/reinforcement interface. For example, Vitreloy 1 has been used as a matrix for particulate diamonds, particulate and filamentary SiC, TiC, tungsten metal particles, tungsten wires, etc. [73-75]. The reinforcement is directly introduced in the molten glass former at high temperatures (typically 800-1000 C) and the mixture is cooled at rates from 1-100 K/s. The result is a "clean" two phase metallic glass matrix composite with no evidence for nucleation and growth of third phases.

Figure 19 shows a composite of tungsten wires that was infiltrated with molten V1 at about 900 C. The tungsten wires are bundled and then infiltrated by the capillary flow of the melt under gravity alone. Similar results have been obtained for SiC fiber bundles, steel wires, and molybdenum wires. In these composites, we see basically no difference between the glass-forming ability of this melt when it is in contact with tungsten or silicon carbide than in the case of making a monolithic glass. In other words, the presence of a second crystalline phase does not affect the processibility. This is true even when significant chemical reaction occurs at the interface during processing. For instance, when SiC is used as reinforcement, a reduction reaction occurs in which SiC is reduced to ZrC along the fiber interface. A reduced layer of several microns thickness is formed between the remaining SiC and the glassy matrix. Elemental Si, a reaction product is dissolved in the glass forming matrix. Choi-Yim et. al. have found that the additional Si acquired by the matrix actually enhances the glass forming ability of some bulk glass forming alloys [75,76]

In W composites, tungsten nanocrystals were found to precipitate near the glass/W interface following processing. Tungsten dissolved in the glass forming melt at the infiltration/processing temperature. It diffuses out from the tungsten wires over a relatively narrow zone owing to an apparently very low diffusion constant. In a several micron thick layer near the glass /W-wire interface, the glass matrix is found to contain about 12 at. % W following processing. This dissolved W precipitates as nearly pure bcc W on cooling forming a dispersion of W nanocrystals in the glass matrix. The glass forming ability of the matrix is otherwise unaffected! The big surprise is that no other crystalline phases form. The metastable phase diagram of Fig. 20 illustrates what has happened. The bcc tungsten phase has a very high liquidus (near the W-end of the diagram) which falls to about 1000C (our processing temperature) with a tungsten concentration of about 12 at. %. If one were to process the composite at higher temperature, one would expect get more tungsten dissolution in the liquid. On cooling the nearly pure W precipitates out of the undercooled melt (undercooled with respect to the W-liquidus curve) and the remaining liquid/glass matrix is left with little or no W! Its glass forming ability is uncompromised. Apparently, the W-glass interface does not act as a heterogeneous nucleation site for any third phase! The glass forming system remains very "robust".

We now turn briefly to a discussion of some selected applications of bulk metallic glasses and BMG–composites. The mechanical properties and deformation behavior of the above-mentioned W/V1 matrix composite has been studied. When the composite is subjected to uniaxial compression and tension, one finds the overall composite exhibits enhanced plastic stain to failure (compared with monolithic glass) [74,78]. Deformation occurs, as in the glass, by the propagation of localized shear bands (at 45° to the compressive axis). But the interaction of shear bands with the W-wire results in a high density of multiple shear bands, which are confined by their interaction with the wires in much the same manner seen by Courtney on laminate [71].

The failure of the W containing composites in compression by the formation of localized shear bands is of practical interest. This mode of failure is greatly desired in the design of W-base kinetic energy penetrators. Ballistic testing of the W/BMG (bulk metallic glass) composites has indeed shown that the performance is greatly enhanced by comparison with ordinary crystalline W-alloys. This can be directly attributed to influence of the metallic glass matrix in inducing localized shear

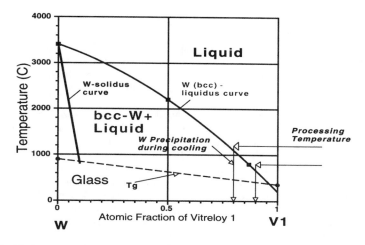

Fig.20 A metastable phase diagram showing the metastable equilibrium of the Vitreloy 1 glass forming liquid with the bcc tungsten-base solution. The figure shows the metastable solidus and liquidus curves of the bcc-W phase. Note that the solubility of W in liquid vitreloy at composite processing temperatures of ~1000C is ~ 12-14 at. %.

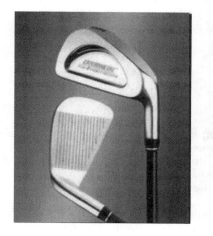

Fig.21 Golf clubs with heads ("irons" –left, "drivers"-right) fabricated from monolithic Vitreloy 1. The golf club heads are cast to net shape by Howmet Corp. of Whitehall, MI and sold commercially by Liquid Metal Golf (LMG) Inc. of Laguna Niguel, CA. Photo compliments of LMG.

deformation. This type of deformation, in turn, leads to a "self sharpening" behavior of the penetrator.

Another recent and interesting application of BMG exploits the very high elastic strain limit of metallic glasses. As mentioned above, BMG is essentially the "premier" material for the design of springs which can store high densities of elastic energy. The basic idea is that the maximum stored elastic energy is given by

$$E_{el} = 1/2\, M\, \varepsilon^2$$

where M is a suitable modulus (Youngs Modulus in tension, a bending modulus in bending, etc.) and ε the elastic strain limit. Since the elastic strain limit of metallic glasses often exceeds 2% compared with useful crystalline metals where the maximum strain is invariably less than 1%, one expects the maximum stored elastic energy density to be roughly 4x that of crystalline materials. This is indeed found to be the case. This property of metallic glasses finds utility in the design of certain types of sporting equipment (e.g. a baseball bat, bicycle spoke, or golf club to name a few). Vitreloy 1 has been used in the design of golf club heads. The BMG golf club exploits the high strength, favorable density, and perfectly elastic behavior of the metallic glass to very high strains to improve the performance of the golf club. Vitreloy 1 golf clubs are currently being manufactured at Howmet Corporation, Whitehall, MI and sold by Liquid Metal Golf Inc., Laguna Niguel, CA. Fig. 21 shows a photo of an fully glass head of "iron" –type club (here, "iron" is a type of golf club) [79]. The "iron" head is a roughly 1-lb. net shape metallic glass casting. The golf clubs represent the first commercial components fabricated from bulk metallic glass.

Given the unique and unusual properties of BMG, one can expect that this new engineering material will make it's way into a variety of other specific commercial products as the science and technology of this new field undergo further development.

ACKNOWLEDGMENTS

I'd like to acknowledge all the members of my research group at Caltech who have contributed to the research associated with this MRS medal award. In particular, I acknowledge Atakan Peker, who pioneered the development of the Vitreloy alloys as part of his thesis research. Ralf Busch, Uli Geyer, Susanne Schneider, Xian-Hong Lin, Eric Bakke, Konrad Samwer, and others, some of whom are in the audience, have made substantial contributions. Dr. Y.J. Kim, whose premature death in 1998 saddened all of us at Keck Lab, pioneered the use of the HVESL to study glass forming liquids.

I wish to acknowledge the continuous financial support of the U.S. DOE, Office of Basic Energy Science, Division of Metallurgy and Ceramics – particularly Drs. John Mundy, Joe Darby, and Robert Gottshall. Without their support, this work would not have progressed to its present state. Support for the ground-based and microgravity studies of containerlessly processed liquids was provided by NASA. Special thanks to Mike Wargo at NASA Headquarters, to Jan Rogers and the supporting staff at Marshall Space Flight Center. The Army Research Office and the Air Force Office of Scientific Research have supported the recent work on composite materials. Finally, I would like to express my extreme gratitude to the staff and management of Amorphous Technologies International and Liquid Metal Golf, especially James Kang and Mike Tenhover. Together with the outstanding technical team at Howmet Corporation (particularly N. Paton and D. Larson), they have pioneered the commercial development of bulk metallic glasses as a new engineering material.

REFERENCES

1. W. Clement, R.H. Willens, and P. Duwez, Nature, **187,** 869 (1960)
2. H.S. Chen and D. Turnbull, J. Chem. Phys., **48,** 2560 (1968)
3. H.S. Chen and D. Turnbull, Acta Metall., **17,** 1021 (1969)
4. D. Turnbull and J.C. Fisher, J. Chem. Phys., **17,** 71 (1949); also D. Turnbull, J. Chem. Phys., **18,** 198 (1950)
5. for a review of melt spinning and planar flow casting technology, see S. Kavsesh, in "Metallic Glasses", (ASM International, Metals Park, Ohio, 1978), Chapter 2
6. H.S. Chen, Acta Metall., **22,** 1505 (1974)
7. A.L. Drehman, A.L. Greer, and D. Turnbull, Appl. Phys. Lett., **41**716 (1982)
8. H.W. Kui, A.L. Greer, and D. Turnbull, Appl. Phys. Lett., **45,** 615 (1984)
9. A. Inoue, H. Yamaguchi, T. Zhang, et. al., Mater Trans., JIM, **31,** 104 (1990); also A. Inoue, T. Zhang, and T. Masumoto, Mater. Trans., JIM, **30,** 965 (1990)
10. A. Inoue, T. Nakamura, N. Nishiyama, and T. Masumoto, Mater. Trans., JIM, **33,** 937 (1992); also A. Inoue, T. Nakamura, T. Sugita, T. Zhang, and T. Masumoto, Mater. Trans., JIM, **34,** 351 (1993)
11. A. Inoue, A. Kato, T. Zhang, et. al., Mater. Trans., JIM, **32,** 609 (1991)
12. A. Inoue, Mater. Trans., JIM, **36,** 866 (1995)
13. A. Inoue, T. Zhang, and T. Masumoto, Mater. Trans., JIM, **31,** 177 (1990)
14. A. Inoue, T. Zhang, N. Nishiyama, et. al., Mater. Trans., JIM, **34,** 1234 (1993)
15. A. Peker and W.L. Johnson, Appl. Phys. Lett., **63,** 2342 (1993)
16. A. Peker and W.L. Johnson, U.S. Patent No. 5,288,344, issued Feb. 1994; also see A. Peker, "Formation and Characterization of Bulk Metallic Glasses", Ph.D. Thesis, Calif. Inst. of Tech. (1994)
17. X.H. Lin and W.L. Johnson, J. Appl. Phys., **78,** 6514 (1995)
18. Y.J. Kim, R. Busch, W.L. Johnson, et. al., Appl. Phys. Lett., **65,** 2136 (1994) ; also Y.J. Kim, R. Busch, W.L. Johnson, et. al., Appl. Phys. Lett., **68,** 1057 (1996)
19. R.B. Schwarz and Y. He, Mater. Sci. Forum, **235,** 231 (1997)
20. N. Nishiyama and A. Inoue, Mater. Trans., JIM, **37,** 1531 (1996)
21. T. Egami and Y. Waseda, J. Non Cryst. Sol., **64,** 113 (
22. T. Egami, Mat. Sci. and Eng. A, **226,** 261 (1997)
23. M. Li and W.L Johnson, Phys. Rev. Lett., **70,** 1120 (1993)
24. H.J. Fecht, P. Desre, and W.L. Johnson, Philos. Mag.,**59,** 577 (1989
25. W. Hume-Rothery, "The Structure of Metals and Alloys", (The Inst. of Metals, London, 1956) p.101
26. L.E. Tanner, Acta Metall., **27,** 1727 (1979)

27. X.H. Lin, "Bulk Metallic Glass Formation and Crystallization of Zr-Ti-based Alloys", Ph.D. Thesis, Calif. Inst. of Tech., (1997))

28. W.K. Rhim, Rev. Sci. Instr., **64**, 2961 (1993)

29. R. Busch, Y.J. Kim, S. Schneider, and W.L. Johnson, Mater. Sci. Forum, **225**, 77 (1996)

30. H.A. Davies, 3rd Int. Conf. On Rapid Quenching, ed. B. Cantor (The Metals Society, London, 1978) p.1

31. H.A. Davies, Phys. Chem. Glasses, **17**, 159 (1976)

32. A. Masuhr, "Viscous Flow and Crystallization of Bulk Metallic Glass Forming Liquids, Ph.D. Thesis, Calif. Inst. of Tech., (1999)

33. A. Masuhr, T.A. Waniuk, R. Busch, and W.L. Johnson, Phys. Rev. Lett., **82**, 2290 (1999)

34. R. Busch, S. Schneider, A. Peker, and W.L. Johnson, Appl. Phys. Lett., **67**, 1544 (1995)

35. M.K. Miller, K.F. Russel, P.M. Martin, R. Busch, and W.L. Johnson, J Phys. IV, **6**, 217 (1996)

36. S. Schneider, U. Geyer, P. Thiyagarajan, and W.L. Johnson, Mater. Sci. Forum, **225**, 59 (1996)

37. S. Schneider, U. Geyer, P. Thiyagarajan, and W.L. Johnson, Mater. Sci. Forum, **235**, 337 (1997)

38. R. Busch and W.L. Johnson, Appl. Phys. Lett., **72**, 2695 (1998)

39. R. Busch, W. Liu, and W.L Johnson, J. Appl. Phys., **83**, 4134 (1998)

40. W. Liu, W.L. Johnson, S. Schneider, U. Geyer, and P. Thiyagarajan, Phys. Rev. B, **59**, 11755 (1999)

41. Wenshan Liu, "Formation and Characterization of Mg-based Bulk Metallic Glasses and Nanocrystalline Materials", Ph.D. Thesis, Calif. Inst. of Tech., (1998)

42. P.J. Desre, Mater. Trans. JIM, **38**, 583 (1997)

43. P.J. Desre, MRS Symp. Proceeding, Symposium MM "Bulk Metallic Glasses", in press (1999)

44. X.H. Lin, W.L. Johnson, and W.K. Rhim, Mater. Trans. JIM, **38**, 473 (1997)

45. X. Lin, "Bulk Glass Formation and Crystallization of Zr-Ti-based Alloys, Ph.D. thesis, Calif. Inst. of Tech., (1997)

46. E. Bakke, R. Busch, and W.L. Johnson, Appl. Phys. Lett., **67**, 3260 (1995)

47. E. Bakke, R. Busch, and W.L. Johnson, Mater. Sci. Forum, **225**, 95 (1996)

48. R. Busch and W.L. Johnson, Mater. Sci. Forum, **269**, 577 (1998); also R. Busch, A. Masuhr, E. Bakke, and W.L. Johnson, Mater. Sci. Forum, **269**, 547 (1998)

49. R. Busch, E. Bakke, and W.L. Johnson, Acta Mater., **46**, 4725 (1998)

50. T.A. Waniuk, R. Busch, A. Masuhr, and W.L. Johnson, Acta Mater., **46**, 5229 (1998)

51. H.S. Chen and M. Goldstein, J. Appl. Phys., **43**, 1642 (1971)

52. A.I. Taub and F. Spaepen, Acta Metall. Mater., **28**, 633 (53. C.A. Volkert and F. Spaepen, Acta Metall. Mater., **37**, 1355 (1989)

54. C.A. Angell, Science, **267**, 1924 (1995)

55. C.A. Angell, B.E. Richards, and V. Velikov, Phys. Cond. Mater., **11**, A75-A94, (1999)

56. F. Spaepen and D. Turnbull, in Rapidly Quenched Metals II, ed. by N.J. Grant and B.C. Giessen, (MIT Press, Boston, 1976), p. 20557. H.A. Davies, in Rapidly Quenched Metals III, ed. by B. Cantor, (Chameleon Press, London, 1978) p.

58. U. Geyer, S. Schneider, W.L. Johnson, et. al. , Phys. Rev. Lett., 75, 2364 (1995)

59. U. Geyer, W.L. Johnson, S. Schneider, et. al., Appl. Phys. Lett., 69, 2492 (1996)

60. F. Wenwer, K. Knorr, M.P. Macht, et. al., Defect Diffusion Forum, 143, 831 (1997).

61. W. Dorner and H. Mehrer, Phys. Rev. B, 44, 101 (1991).

62. A. Masuhr, T.A. Waniuk, R. Busch, and W.L. Johnson, Phys. Rev. Lett., 82, 2290 (1999): also A. Masuhr, R. Busch, and W.L Johnson, MRS Symp. Proc., Symp. MM, "Bulk Metallic Glasses" (MRS, Boston, 1999)

63. C.A. Angell, Chem. Rev., 90, 523 (1990)

64. M.E. Ediger, C.A. Angell, and S.R. Nagel, J. Phys. Chem., 100, 13200 (1996)

65. U. Geyer, W.L. Johnson, S. Schneider, Y. Qiu, T.A. Tombrello, and M.P. Macht, Appl. Phys. Lett., 69, 2492 (1996

66. A. Meyer, J. Wuttke, W. Petry, et. al., Phys. Rev. Lett., 80, 4454 (1998)

67. H.A. Bruck, T. Christman, A.J. Rosakis, and W.L. Johnson, Scripta Metall., 30, 429 (1994).

68. R.D. Conner, A.J. Rosakis, W.L. Johnson, et. al., Scripta Mater., 37, 1373 (1997)

69. C.J. Gilbert, R.O. Ritchie, and W.L Johnson, Appl. Phys. Lett., 71, 476 (1997

70. P. Lowhaphandu, and J.J. Lewandowski, Scripta Meter., 38, 1811 (1998)

71. Y. Leng and T.H. Courtney, J. Mater. Sci., 26, 588 (1991)

72. R.B. Dandliker, R.D. Conner, and W.L. Johnson, J. Mater. Res., 13, 2896 (1998)

73. H. Choi-Yim and W.L. Johnson, Appl. Phys. Lett., 71, 3808 (1997) 74. R.D. Conner, R.B. Dandliker, and W.L. Johnson, Acta Meter., 46, 6089 (1998)

75. H. Choi-Yim, "Synthesis and Characterization of Bulk Metallic Glass Matrix Composites", Ph.D. thesis, Calif. Inst. of Tech., (1999) 76. H. Choi-Yim, R. Busch, and W.L. Johnson, J. Appl. Phys., 83, 7993 (1998)

77. R.B. Dandliker, "Bulk Metallic Glass Matrix Composites: Processing, Microstructure, and Applications as a Kinetic Energy Penetrator", Ph.D. Thesis, Calif. Inst. of Tech., (1998)

78. R.D. Conner, "Characterization and Mechanical Behavior of Bulk Metallic Glass Matrix Composites, Ph.D. Thesis, Calif. Inst. of Tech., (1998) Conner thesis

79. see article by S. Ashley, "Metallic Glasses Bulk Up", in Mechanical Engineering, June 1998, p. 73

Part VI

Mechanical and Other Properties I

FRACTURE AND FATIGUE IN A Zr-BASED BULK METALLIC GLASS

C. J. GILBERT, V. SCHROEDER, and R. O. RITCHIE
Materials Sciences Division, Lawrence Berkeley National Laboratory, and Department of Materials Science and Mineral Engineering, University of California, Berkeley, CA 94720-1760

ABSTRACT

The fracture and fatigue properties of the $Zr_{41.2}Ti_{13.8}Cu_{12.5}Ni_{10}Be_{22.5}$ (at.%) bulk metallic glass alloy have been examined. The plane-strain fracture toughness of the fully amorphous alloy was found to exceed 50 MPa√m, although results were sensitive to strain rate, showed significant variability and were influenced by the presence of residual stresses following processing. Fracture surfaces exhibited a characteristic vein morphology, consistent with micromechanical models for meniscus instabilities. Local melting was evident, consistent with the emission of light during rupture and very high local temperatures (>1000 K) measured during fracture. Upon partial or complete crystallization, the alloy was severely embrittled, with toughnesses dropping to ~1 MPa√m and the hardness increasing by ~10%. Under cyclic loading, crack-propagation behavior in the amorphous structure was similar to that observed in polycrystalline metals; the crack-advance mechanism was associated with alternating crack-tip blunting and resharpening, as evidenced by presence of fatigue striations. Conversely, the (unnotched) stress-life (S/N) properties were markedly different. Crack initiation and subsequent growth occurred quite readily due to the lack of microstructural barriers that would normally provide local crack-arrest points. This resulted in a very low fatigue limit of ~4% the ultimate tensile strength.

INTRODUCTION

Amorphous metallic alloys represent an interesting class of potential structural materials. Their properties include near theoretical strength, large elastic deflections, high hardness, excellent wear properties, and good potential for forming and shaping. Indeed, unlike ceramic glasses, metallic glasses can be quite ductile [1-4], with inhomogeneous deformation highly localized into slip bands. Although the precise flow mechanisms are unclear, bubble raft and computational studies suggest that they are associated with localized atomic shear-rearrangements correlated to regions of either excess free volume [3,5,6], or extreme shear-stress concentration [7].

Early versions of these materials required very high cooling rates (>10^5 K/s) to prevent crystallization; corresponding shapes were thus confined to very thin ribbons or wires (~10 to 100 μm), making the evaluation of many mechanical properties difficult. However, in recent years, several new multicomponent alloys have been developed with very high resistance to crystallization in the undercooled liquid state; correspondingly, as relatively low cooling rates (typically <10 K/s) are required to form a fully amorphous structure, these metallic glass alloys can be processed in bulk form. These include Mg-based alloys like Mg-Cu-Y [8], some recently discovered Fe-based alloys [9], and the Zr-Ti-Ni-Cu, Zr-Ti-Ni-Cu-Be, and Zr-Ti-Ni-Cu-Al alloys [10,11].

The first commercial alloy, $Zr_{41.2}Ti_{13.8}Cu_{12.5}Ni_{10}Be_{22.5}$ (at.%) (Vitreloy™), requires cooling rates of only ~1 K/s, such that fully amorphous rods several centimeters in diameter have been produced [10]; indeed, it is now used to fabricate golf club heads. Although this alloy appears to display ideal properties for such applications, i.e., high strength-to-stiffness ratio and low damping characteristics, studies have indicated that whereas it exhibits reasonable fracture toughness in the amorphous state, severe embrittlement can occur on partial or full crystallization [12-16]. Moreover, its comparative fatigue properties appear to be quite different depending upon whether they are assessed using crack propagation or stress/life (S/N) approaches. Accordingly, the objective of this study is to characterize fracture and fatigue in this bulk metallic glass alloy in order to develop a mechanistic understanding of crack initiation and growth in this class of materials.

EXPERIMENTAL PROCEDURES

Experiments were performed on as-cast plates (7 mm thick, 40 × 40 mm) of fully amorphous $Zr_{41.2}Ti_{13.8}Cu_{12.5}Ni_{10}Be_{22.5}$ (at.%), produced by Hitchener Manufacturing Co. (Milford, NH) and supplied by Amorphous Technologies International (Laguna Niguel, CA). This alloy was originally developed by Peker and Johnson [10]. The chemistry, measured using mass spectroscopy, was (wt.%) 62.5% Zr, 10.9%Ti, 12.9% Cu, 10.1% Ni, 3.8% Be, 0.10% O, and 0.005% N. Typical mechanical and thermal properties are listed in Table 1 [10,12,17,18].

Table 1: Selected Properties of $Zr_{41.2}Ti_{13.8}Cu_{12.5}Ni_{10}Be_{22.5}$ Bulk Metallic Glass

Density (g/cm^3)	Young's Modulus (GPa)	Shear Modulus (GPa)	Poisson's Ratio	Yield Strength (GPa)	Vickers Hardness (GPa)	Glass Transition[#] (K)	Toughness[*] (MPa√m)
5.9	95	35	0.35	1.9	5.4	~625	~55

[*]Measured at a $\dot{K}$ of 0.3 MPa√m s^{-1} on a 7 mm-thick, fatigue-precracked compact-tension sample [12].
[#]Measured at a scan rate of 0.33 K/s [10].

Using transmission electron microscopy (TEM), as-received plates were found to be fully amorphous (Fig. 1a) [19]. Partial crystallization was achieved by heat treating *in vacuo* (~10^{-6} Torr) at 633 K for 0.1 to 24 hr; the structure after 12 hr at 633 K consisted of <5 vol.% of ~3 to 5 nm crystallites of a Cu-rich, Ti-rich (*fcc*) phase (mean spacing ~20 nm) in an amorphous matrix (Fig. 1b) [19-22]. The fully crystallized multiphase microstructure, obtained by heat treating *in vacuo* for 24 hr at 723 K, consisted of a Laves phase with a "MgZn$_2$-type" (*hcp*) structure [21], a phase with the "Al$_2$Cu-type" structure, and at least one additional unidentified phase [20,23,24].

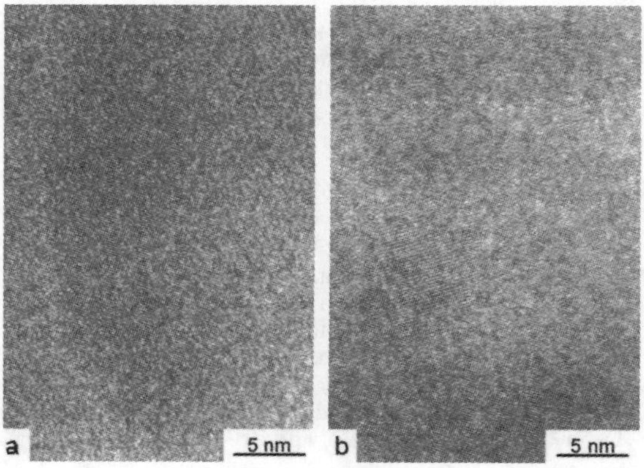

Fracture toughness and fatigue-crack growth rates were determined in room air (22°C, ~45% relative humidity) on 7 mm thick, 38 mm wide compact-tension C(T) specimens. Samples were cut from the as-received plates using electrode discharge machining; some samples were cut with the thickness essentially unchanged from the original casting, others were thinned (~1.5 mm removed from *each* surface to leave a thickness of 4 mm) due to concerns about surface residual stresses. Specimens were cycled on computer-controlled, servo-hydraulic mechanical testing

Fig. 1: Bright-field TEM of (a) as-received, amorphous Zr-Ti-Cu-Ni-Be, and (b) a partially crystallized microstructure (annealed 12 hr at 633 K). Courtesy W. L. Johnson [19].

machines under stress intensity, K, control, with a frequency, ν, of 25 Hz (sine wave), under constant load ratio (ratio of minimum to maximum load, i.e., $R = K_{min}/K_{max}$) conditions. R was varied from 0.1 to 0.5. Samples were first cycled with a decreasing stress-intensity range (at normalized K-gradients of 0.08 and 0.2 mm^{-1}) until measured growth rates were less than 10^{-10} m/cycle. The value of ΔK at this point was used to operationally define the fatigue threshold stress intensity (ΔK_{TH}), below which microstructurally large cracks are essentially dormant.

Crack lengths were continuously monitored using unloading elastic-compliance measurements with a back-face strain gauge; readings were checked periodically using a traveling microscope. Optical and compliance measurements were found to be within 2%. Data are presented in terms of the growth rate per loading cycle, da/dN, as a function of the alternating stress intensity range, $\Delta K (= K_{max} - K_{min})$, the latter being computed using standard linear-elastic handbook solutions.

Fatigue-crack closure, defined as premature contact of mating crack faces on unloading, was also monitored using the back-face strain gauge. The closure stress intensity, K_{cl}, was approximately defined at the point where the elastic unloading line initially deviated from linearity. Under plane-strain conditions, the prime origin of such closure is asperity contact behind the crack tip; this reduces the local stress-intensity range actually experienced at the crack tip to an effective level, $\Delta K_{eff} = K_{max} - K_{cl}$, provided $K_{cl} > K_{min}$.

During and after fracture and fatigue experiments, the crack profiles and fracture surfaces of selected specimens were analyzed both optically and in the scanning electron microscope (SEM).

Fracture toughness values in the fully amorphous structure were determined by monotonically loading the fatigue precracked specimens to failure at specified loading rates, $\dot{K}$. Due to their extreme brittleness, corresponding toughnesses in the partially and fully crystalline structures were obtained using indentation methods. In the latter case, measurements were averaged from at least five Vickers hardness indents under an indentation load, P_{ind}, of 49 N, with K_{Ic} values calculated from $K_{Ic} = \chi(P_{ind}/c^{3/2})$, where the $2c$ is the total surface crack length and χ is a constant dependent upon indent geometry and material properties. For brittle solids, χ can be estimated as $0.016(E/H)^{1/2}$, where E is the elastic modulus, and H is the hardness [25].

Fatigue lifetimes, N_f, were measured over a range of cyclic stresses by cycling $3 \times 3 \times 50$ mm rectangular beams in four-point bending, with an inner span, S_1, and outer span, S_2, of 10.2 mm and 20.3 mm, respectively. Polished specimens (to ~1 μm surface finish) were cycled under load control at $\nu = 25$ Hz (sine wave) with $R = 0.1$ in room air on a servo-hydraulic mechanical test frame. Stresses were calculated at the tensile surface within the inner span from $\sigma = [3P(S_2-S_1)]/2bh^2$, where P is the applied load, b is the specimen thickness, and h is the specimen height. A total of 21 beams were tested at maximum stresses ranging from 100 MPa to 1800 MPa (just below the tensile failure stress), with multiple measurements made at each stress when possible. Fracture surfaces of selected beams were examined after failure via both optical and scanning electron microscopy in order to discern the origin and mechanisms of failure. Stress/life data are presented in terms of the stress amplitude, $\sigma_a = \frac{1}{2}(\sigma_{max} - \sigma_{min})$, normalized by the uniaxial tensile strength, $\sigma_u = 1.9$ GPa, plotted as a function of the number of cycles to failure, N_f.

RESULTS

Fracture Toughness Behavior

Significant variability was observed in the fracture toughness data. The highest measured K_{Ic} value was ~68 MPa√m (measured on a 7 mm thick sample at a $\dot{K}$ 0.012 MPa√m s^{-1}); the lowest was ~30 MPa√m (on a 4 mm thick sample at the same $\dot{K}$). Sources of this variability may be associated with residual stresses at the surface of castings, compositional variation (particularly oxygen), crack branching and ligament bridging, and sensitivity to loading rate (these issue are discussed more fully below). For example, although both the 7 and 4 mm thick C(T) samples exceeded the ASTM E399 plane-strain thickness requirement ($B > 2.5(K_{Ic}/\sigma_Y)^2 \sim 2$ mm), significant crack branching was observed in the thicker specimens where the surface residual stresses had not been removed.

Thermal exposure, resulting in partial or full crystallization, led to a dramatic reduction in fracture toughness to ~1 MPa√m, with a corresponding small increase (~10%) in hardness (Fig. 2). Similarly, severe embrittlement was observed in a melt-spun ribbon (57 μm thick) of amorphous $Ni_{78}Si_{10}B_{12}$, where fracture toughness dropped from 67 MPa√m to 14 MPa√m after annealing at 713 K for 7 min [26].

Fracture surfaces in the fully amorphous alloy exhibited a vein-like morphology (Fig. 3a-c), typical of many metallic glasses [2,4], with additional evidence of local melting during fracture (Fig. 4). At low magnifications, ridges could be seen on the surfaces that were quite large and ran nominally parallel to the direction of crack propagation (Fig. 3a-b). Stereo-photogrammetric investigations of matching fracture surfaces reveal that the tips of the ridges match across the fracture plane [27] and show evidence of substantial local necking and plastic deformation. The vein-like features have been likened to that found after the separation of grease or adhesive films [1,2,28]. In marked contrast, fracture surfaces in the partially and fully crystallized structures were relatively featureless at comparable magnifications (Fig. 3d).

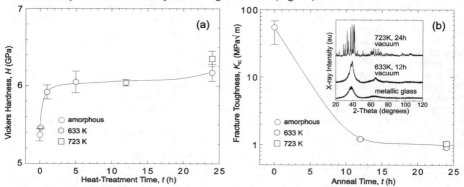

Fig. 2: (a) Vickers hardness and (b) fracture toughness plotted as a function of annealing time. Toughness tests were conducted at loading rates of 0.01-0.30 MPa√m s^{-1}. Insert shows X-ray diffraction patterns corresponding to anneals at 723 K and 633 K.

Fatigue Behavior

Fatigue-crack growth behavior: In the fully amorphous metal, stable fatigue-crack growth was readily characterized under cyclic loading. Growth rates, da/dN, are plotted against ΔK (Fig. 5a) for a specimen tested in plane strain at $R = 0.1$ and $\nu = 25$ Hz, and are compared to those for an ultrahigh strength steel (300-M, quenched and 200°C-tempered) [29] and an age-hardened aluminum alloy (2090-T81) [30] (Fig. 5b). It is apparent that the cyclic crack-growth rates in the amorphous metal are comparable to those observed in traditional polycrystalline metallic alloys. Indeed, when these data are regression fit to a simple Paris power-law equation [31]:

$$\frac{da}{dN} = C'\Delta K^{m},$$ [1]

(where m is the crack growth exponent, and C' a scaling constant), the exponent m in the mid-range of growth rates ($\sim 10^{-10}$ to 10^{-7} m/cycle) lies in the range of $m \sim 2$-5, typical of ductile crystalline metals in this regime. These values of m are in marked contrast to the fatigue properties of ceramics [32] and oxide glass [33], where m typically lies in the range of $m \sim 15$-50. Values of ΔK_{TH} in the metallic glass ranged from ~ 1 to 3 MPa√m, again comparable to many high strength steels and aluminum alloys.

It should be noted that, whereas fatigue-crack growth rates (at 25 Hz and $R = 0.1$) approach catastrophic failure at $\Delta K \sim 12$ MPa√m ($\dot{K} \sim 600$ MPa√m s^{-1} at instability), under monotonic loading fracture toughnesses as high as ~ 68 MPa√m ($\dot{K} \sim 0.012$ MPa√m s^{-1}) are measured. Preliminary studies suggest that fracture toughness may depend on loading rate, $\dot{K}$. In Fig. 6, K_{Ic} values are plotted in terms of $\dot{K}$ for all samples tested, together with the points of criticality from fatigue tests at 25 Hz. While there may be a trend toward lower toughness with higher loading rate, scatter in the data make any conclusions difficult. This scatter was far worse in the 7 mm thick samples, and was reasoned to be associated with the presence of compressive residual stresses in the surface layers.

As the load ratio was increased to $R = 0.5$, cyclic crack growth rates were accelerated and fatigue thresholds, ΔK_{TH}, reduced, similar to trends observed in traditional metals [29,30]. These data can be expressed in terms of a modified Paris power-law relationship that includes the effect of both ΔK and K_{max} [34]:

$$\frac{da}{dN} = C(K_{max})^n (\Delta K)^p, \qquad [2]$$

where n and p are experimentally determined crack-growth exponents and C is a scaling constant independent of K_{max}, ΔK, and R. In this form, $m = n + p$ and $C = C'(1 - R)^n$. A regression fit to Eq. 2 yields values of $C = 4.1 \times 10^{-12}$, $n = 2.3$ and $p = 1.2$ (units: m/cycle, MPa√m).

The effect of load ratio changes on fatigue-crack propagation in crystalline metals below

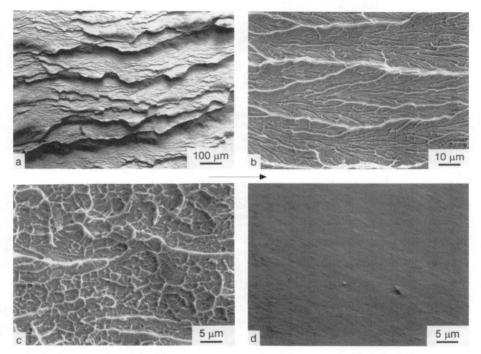

Fig. 3: SEM of overload fracture surfaces in (a-c) the fully amorphous alloy, and (d) a partially crystallized specimen annealed at 633 K for 12 hr. Arrow indicates direction of crack propagation.

~10^{-6} m/cycle is generally associated with crack closure [35-37]. In the metallic glass, significant levels of closure were also observed at $R = 0.1$, with no closure detected at $R = 0.5$. Using a normalization common to polycrystalline metals [35], crack-growth data were replotted in terms of $\Delta K_{eff} = K_{max} - K_{cl}$ (for $K_{cl} > K_{min}$). Such an approach is justified since in the presence of closure, ΔK_{eff} is a more appropriate measure of the *near-tip* stress-intensity range. However, this normalization was only partially successful when applied to the amorphous alloy.

Due to embrittlement upon crystallization, fatigue cracking was not seen in the partially or fully crystallized samples; attempts to grow stable cracks always led to catastrophic failure.

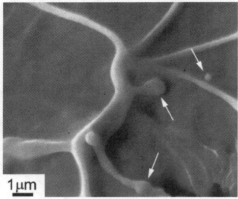

1 μm

Fig. 4: SEM of an overload fracture surface in a fully amorphous specimen. Features marked by arrows are suggestive of local melting. The crack propagated left to right.

Fractography: SEM examination of fatigue surfaces in the amorphous metal revealed that the surface roughness scaled directly with local crack-growth rates. Near-threshold regions exhibited a unique mirror-like surface finish. In Fig. 7a, the boundary between such a region and overload fracture is shown. The surface is relatively featureless, even at higher magnification (Fig. 7b). However, at higher ΔK levels, fatigue striations could be seen (Fig. 8). The striation spacings were found to scale directly with growth rates (Fig. 5b), consistent with previous studies on rapidly quenched Ni-Fe [38] and Ni-Si-B [39] metallic glasses.

Crack profiles revealed that fatigue cracks did not generally propagate continuously in the 7 mm thick specimens. Substantial crack deflection and branching were observed. Flaws frequently nucleated ahead of the main crack, creating regions of material that spanned the crack faces, resulting in crack bridging and in some cases multiple crack branches. The large degree of slip-band formation observed ahead of fatigue cracks in thin sheets of rapidly quenched amorphous metals [39,40] was not seen in the present alloy, presumably because deformation was constrained under plane-strain conditions. The source of the crack closure appeared to be the tortuous crack paths and rough fracture surfaces. Such branching was minimized in the thinner samples, where the surface layers (and hence residual stresses) had been machined off.

Stress-life behavior: Normalized stress amplitude, σ_a/σ_u, is plotted against cycles to failure, N_f, for the metallic glass in Fig. 9. Results are again compared with those for 300-M ultrahigh strength steel ($\sigma_u = 2.3$ GPa) [41] and 2090-T81 aluminum-lithium alloy ($\sigma_u = 0.56$ GPa) [30]. The S/N properties of amorphous and polycrystalline metals are very different despite similarities in crack-growth properties. At a given value of σ_a/σ_u fatigue lifetimes were significantly shorter in the metallic glass, and exhibited a markedly lower dependence on the stress amplitude. By fitting the S/N data to a Basquin equation $(\sigma_a)^k N_f = $ constant, fatigue lives in the metallic glass are proportional to $(\sigma_a)^{-3.4}$, compared to $(\sigma_a)^{-10}$ in steel and aluminum. Moreover, whereas crystalline alloys generally display a fatigue limit or 10^7-cycle endurance strength at values of σ_a/σ_u between 0.3 and 0.5, no fatigue limit was detected in the metallic glass until σ_a/σ_u dropped below ~0.04. Examination of fatigue fracture surfaces indicated that cracking originated from a corner on the tensile face of the beam, the extent of stable fatigue-crack growth increasing with life. Beams subjected to high bending stresses exhibited extensive slip-band formation at the specimen edges, suggesting that slip bands were the precursor of cracking during the S/N experiments.

DISCUSSION

Mechanisms Controlling Fracture in Bulk Metallic Glass

In brittle solids, fracture or cleavage is initiated when the cohesive strength of the solid is reached at the tip of a pre-existing sharp crack, prior to the onset of extensive inelasticity. Such behavior is generally observed in ceramics and oxide glasses at low homologous temperatures. In crystalline ductile solids, fracture events are dominated by crack-tip plasticity. Progressive separation proceeds by the linkage of voids initiated via internal cleavage within brittle particles or by interfacial decohesion at microstructural inhomogeneities.

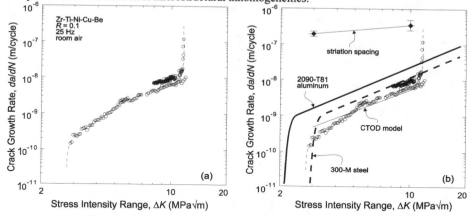

Fig. 5: Growth rates, da/dN, plotted as a function of ΔK are shown in (a) for the Zr-Ti-Cu-Ni-Be glass in air ($R = 0.1$ and $\nu = 25$ Hz), and are compared in (b) with behavior in a high strength steel (300-M) and an age-hardened aluminum alloy (2090-T81). Also shown in (b) are the striation spacing measurements and prediction from the CTOD model (Eq. 4).

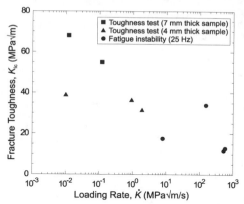

Fig. 6: Variation in fracture toughness of the fully amorphous alloy from toughness tests and the instability in fatigue-crack propagation tests, as a function of the loading rate.

In contrast to these other classes of materials, the micromechanisms that control tensile fracture and toughness properties of amorphous metals are poorly understood. The fundamental differences in both atomic structure and observed deformation behavior (e.g., extreme slip instability in tension, near-theoretical strength, distinctive overload fracture surface morphologies) make it clear that such mechanisms are quite distinct from tensile fracture in crystalline metals, ceramics, or oxide glasses. Indeed, the vein morphology commonly observed on fracture surfaces in metallic glasses [2,4] (Fig. 3c) has been suggested to be a variant of the Taylor instability [2,28]. This instability is associated with the tendency of a fluid meniscus (under a positive pressure gradient) propagating in the direction of its convex curvature to break-up into a series of fingers which penetrate into the fluid meniscus [42]. The process is considered to generate the ubiquitous vein morphology as the material between the fingers necks down to failure. The vein markings are comparable to those associated with intergranular fracture by diffusional flow at high homologous temperatures in some metals [43] and ceramics [44]. When this process dominates, the critical fracture event is associated with the onset of this instability, governed by the surface tension of the fluid and the

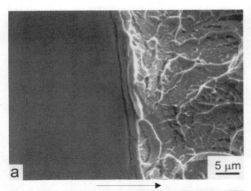

a ⟶ 5 μm

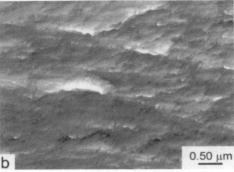

b 0.50 μm

Fig. 7: SEM in (a) of the boundary between near-threshold fatigue (left) and final fracture (right). The near-threshold fatigue region ($\Delta K < 3$ MPa√m) has a mirror-like appearance, and is shown at higher magnification in (b). Arrow indicates direction of crack propagation.

applied pressure gradient. The notion that material local to the crack tip is softened, possibly by adiabatic heating [4,45] or a strain-softening phenomenon [5,28], is supported by the fracture surface appearance (Fig. 4). In addition, recent measurements have detected significant local temperature rises (>1000 K) during fracture of this material [45,46].

A model [28] for the fracture toughness of metallic glasses, based on the resistance of a blunt crack to this instability, gives an expression for K_{Ic} in terms of the surface tension, Γ:

$$K_{Ic} = 24\pi^3 \sqrt{3}\,\frac{\beta \Gamma E}{\alpha}, \qquad [3]$$

where β is a scaling constant dependent on the work hardening behavior (for a solid with near ideal plastic behavior, $\beta \sim 4$) and $\alpha \sim 2.7$ [28]. Using values for the modulus, $E = 95$ GPa, and surface tension, $\Gamma = 1$ J/m^2 (the surface tensions of the component metals ranges from ~1 to 2 J/m^2) [47], Eq. 3 predicts a K_{Ic} value of merely ~13 MPa√m, far lower than measured values. It is speculated that this discrepancy may be associated with a strain rate effect, residual stresses and/or with extensive crack branching and ligament bridging observed along the fatigue crack prior to the measurement of toughness. Such mechanisms are potent in promoting toughness [48], and are obviously not accounted for in Eq. 3. In fact, Lowhaphandu, et al. [16] observed no crack branching in an alloy of nearly identical composition (containing much less oxygen) and measured a fracture toughness of ~18 MPa√m.

The severe embrittlement observed upon full crystallization is presumed to be associated with the complex intermetallic crystal structure that forms (five separate solid solutions in equilibrium in the present alloy) [20-22,24]. However, it is surprising that the toughness plummets with the precipitation of as little as 5 vol.% nanocrystals in an otherwise amorphous matrix (after annealing at 633 K for 12 h). At 5 vol.%, such nanocrystals cannot yet form a continuous network. The severe embrittlement must therefore also involve changes in the remaining amorphous material (such as local concentration of certain elements).

Mechanisms for Fatigue Degradation in Bulk Metallic Glass

Fatigue-crack growth behavior: The mechanism of striation formation in crystalline alloys is associated with irreversible crack-tip shear that alternately blunts and resharpens the crack during the fatigue cycle. Models for striation formation [e.g. 49] indicate that growth rates should scale with the range of crack-tip opening displacement, $\Delta\delta$. Using simple continuum mechanics [49]:

$$\frac{da}{dN} \propto \Delta\delta \sim \beta' \frac{\Delta K^2}{\sigma_Y E'} ,$$ [4]

where σ_Y is the flow stress, $E' = E$ (Young's modulus) in plane stress and $E/(1-v^2)$ in plane strain, and β' is a scaling constant (~0.01 to 0.1 for Mode I crack growth) which is a function of the degree of slip reversibility and elastic-plastic properties of the material. In amorphous metals, tensile experiments indicate that slip bands also form readily [1,17]. Moreover, Eq. 4 provides a reasonable description of the experimentally measured growth rates with $\beta' \sim 0.01$ (Fig. 5b). As striations are seen on the fracture surfaces (Fig. 8), the mechanism for cyclic crack advance in metallic glasses also appears to involve repetitive blunting and resharpening of the crack tip.

Fig. 8: SEM of fatigue fracture surfaces in the fully amorphous alloy at $R = 0.1$, $v = 25$ Hz, showing fatigue striations at (a) $\Delta K \sim 3$ MPa√m with da/dN $\sim 10^{-10}$ m/cycle, and (b) $\Delta K \sim 10$ MPa√m with *da/dN* $\sim 10^{-8}$ m/cycle. Arrow indicates direction of crack propagation.

Stress/life and crack initiation behavior: The results of this work present an interesting contrast between crystalline and amorphous metals. Whereas the crack-propagation properties (Fig. 5) are similar in the two classes of materials (i.e., the similar presence of fatigue striations and dependence of growth rates on ΔK), the S/N behavior (Fig. 9) is markedly different. In the S/N measurements, total life in the amorphous structure is far less dependent upon the applied stress amplitude and the fatigue limit is smaller by an order of magnitude. This implies that mechanistically the fatigue properties of crystalline and amorphous metals differ significantly with respect to crack *initiation*, or more precisely in the nucleation of crack growth. Such a difference may be associated with the easier natural initiation of a fatigue crack, e.g., via slip-band formation [17] although very little is known about the initiation mechanisms at this time. However, since the fatigue limit can be equated with the critical stress for crack initiation or more generally for an initiated (small) crack to overcome some microstructural barrier (e.g., a grain boundary) [50], we may presume that the markedly lower fatigue limits in the amorphous alloys are associated with a lack of microstructure which would normally provide local arrest points to newly initiated or pre-existing cracks. Indeed, low endurance limits appear to be typical of metallic glass [39,51,52].

SUMMARY AND CONCLUSIONS

1. Significant variability was seen in measured fracture toughnesses for the fully amorphous $Zr_{41.2}Ti_{13.8}Cu_{12.5}Ni_{10}Be_{22.5}$ alloy; measured K_{Ic} values ranged from 30 to 68 MPa√m. Sources of this variability are likely associated with surface residual stresses, compositional variation (particularly oxygen), crack branching and ligament bridging, and a sensitivity to loading rate.
2. In the partially or fully crystallized condition, the alloy displays a marginal (~10%) increase in hardness and a severe (30-50 fold) drop in fracture toughness.
3. Preliminary attempts to rationalize the fracture toughness of the amorphous structure indicate that a model based on the Taylor instability underpredicts the measured K_{Ic} values. The higher measured toughness is considered to result from strain rate effects and/or extensive branching, deflection and bridging of the crack associated with surface residual stress.

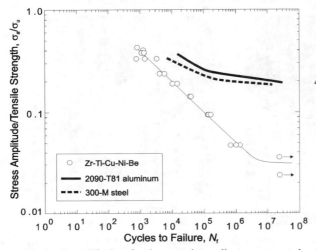

Fig. 9: Stress/life data for the amorphous alloy are presented as the stress amplitude, $\sigma_a = \frac{1}{2}(\sigma_{max} - \sigma_{min})$, normalized by the tensile strength, σ_u, plotted as a function of the number of cycles to failure, N_f. Data are compared to a high-strength steel (300-M) and an aluminum-lithium (2090-T81) alloy.

4. The amorphous Zr-Ti-Cu-Ni-Be alloy is susceptible to fatigue degradation. Crack-growth behavior is similar to that observed in traditional ductile crystalline alloys in terms of (*i*) the dependence of growth rates on the applied stress-intensity range, (*ii*) the role of load ratio and crack closure, and (*iii*) the presence of ductile striations on fatigue fracture surfaces.
5. Conversely, the stress/life behavior of the metallic glass is very different from that observed in crystalline alloys. Fatigue lifetimes were shorter in the amorphous alloy and exhibited a much lower dependence on the applied stress range. A fatigue limit could not be detected until the stress amplitude dropped to ~4% of the tensile strength, presumably because no micro-structural features exist to act as "barriers" to the growth of pre-existing or initiated cracks.

ACKNOWLEDGMENTS

This work was partially supported by the U.S. Air Force Office of Scientific Research under Grant No. F49620-98-1-0260. Thanks are also due to Drs. A. Peker and M. Tenhover (Amorphous Technologies International, Corp.) for their support and for supplying the material, to Howmet Research Corp. for additional financial support, to Prof. W. L. Johnson (Cal Tech.) for microscopy and numerous helpful discussions, and to Prof. R. Pippan and A. Tatschl (Erich-Schmid-Institut, Leoben) for stereo-photogrammetric measurements of fracture surfaces.

REFERENCES

1. J. J. Gilman, J. Appl. Phys. **46**, 1625-33 (1975).
2. C. A. Pampillo and A. C. Reimschuessel, J. Mater. Sci. **9**, 718-24 (1974).
3. A. S. Argon, Acta Metall. **27**, 47-58 (1979).
4. H. J. Leamy, H.S. Chen, and T. T. Wang, Metall. Trans. **3**, 699-708 (1972).
5. F. Spaepen, Acta Metall. **25**, 407-15 (1977).
6. H. Kimura and T. Masumoto, Scripta Metall. **9**, 211-22 (1975).
7. D. Srolovitz, V. Vitek, and T. Egami, Acta Metall. **31**, 335-52 (1983).
8. A. Inoue, T. Nakamura, N. Nishiyama, T. Masumoto, Mater. Trans. JIM **33**, 937-45 (1992).
9. A. Inoue, T. Zhang, and A. Takeuchi, Appl. Phys. Lett. **71**, 464-66 (1997).
10. A. Peker and W. L. Johnson, Appl. Phys. Lett. **63**, 2342-44 (1993).
11. X. H. Lin and W. L. Johnson, J. Appl. Phys. **78**, 6514-19 (1995).
12. C. J. Gilbert, R. O. Ritchie, and W. L. Johnson, Appl. Phys. Lett. **71**, 476-78 (1997).
13. C. J. Gilbert, J. M. Lippmann, and R. O. Ritchie, Scripta Mater. **38**, 537-42 (1998).
14. R. D. Conner, A. J. Rosakis, W.L. Johnson, D.M. Owen, Scripta Mater. **37**, 1373-78 (1997).
15. C. T. Liu, L. Heatherly, D. S. Easton, C. A. Carmichael, J. H. Schniebel, C. H. Chen, J. L. Wright, M. H. Yoo, J. A. Horton, and A. Inoue, Metall. Trans. A **29**, 1811-20 (1998).
16. P. Lowhaphandu and J. J. Lewandowski, Scripta Mater. **38**, 1811-17 (1998).
17. H. A. Bruck, T. Christman, A. J. Rosakis, and W. L. Johnson, Scripta Metall. Mater. **30**, 429-34 (1994).
18. W. L. Johnson and A. Peker, in *Science and Technology of Rapid Solidification and Technology*, edited by M. A. Otooni (Netherlands: Kluwer Acad. Publ.), pp. 25-41 (1995).
19. W. L. Johnson, California Institute of Technology, unpublished research (1997).
20. R. Busch, S. Schneider, A. Peker, and W. L. Johnson, Appl. Phys. Lett. **67**, 1544-46 (1995).
21. S. P. Schneider, P. Thiyagarajan, and W. L. Johnson, Appl. Phys. Lett. **68**, 493-95 (1996).
22. A. Peker and W. L. Johnson, Mater. Sci. Eng. A **179/180**, 173-75 (1994).
23. H. J. Fecht, Philos. Mag. B **76**, 495-503 (1997).
24. S. Spriano, C. Antonione, R. Doglione, L. Battezzati, S. Cardoso, J. C. Soares, and M. F. da Silva, Philos. Mag. B **76**, 529-40 (1997).
25. B. R. Lawn, *Fracture of Brittle Solids*, 2nd ed, Cambridge University Press (1993).
26. A. T. Alpas, L. Edwards, and C. N. Reid, Mater. Sci. Eng. **98**, 501-04 (1988).
27. A. Tatschl and R. Pippan, Schmid Institute, Leoben, private communication (1997).
28. A. S. Argon and M. Salama, Mater. Sci. Eng. **23**, 219-30 (1976).
29. R. O. Ritchie, J. Eng. Mater. Tech., Trans. ASME Series H **99**, 195-204 (1977).
30. K. T. Venkateswara Rao and R. O. Ritchie, Int. Mater. Rev. **37**, 153-85 (1992).
31. P. C. Paris and F. Erdogan, J. Basic Eng. **85**, 528-34 (1963).
32. C. J. Gilbert, R. H. Dauskardt, R. W. Steinbrech, R. N. Petrany, and R. O. Ritchie, J. Mater. Sci. **30**, 643-54 (1995).
33. S. J. Dill, S. J. Bennison, and R. H. Dauskardt, J. Am. Ceram. Soc. **80**, 773-76 (1997).
34. R.H.Dauskardt, M.R. James, J.R.Porter, R.O.Ritchie, J. Am. Ceram. Soc. **75**, 759-71 (1992).
35. W. Elber, Eng. Fract. Mech. **2**, 37-45 (1970).
36. R. A. Schmidt and P.C. Paris, in *Progress in Flaw Growth and Fracture Toughness Testing* (Philadelphia, PA: American Society for Testing and Materials), 79-94 (1973).
37. S. Suresh and R.O. Ritchie, in *Fatigue Crack Growth Threshold Concepts*, edited by D. L. Davidson and S. Suresh (Warrendale, PA: TMS-AIME), 227-61 (1984).
38. A. T. Alpas, L. Edwards, and C. N. Reid, Eng. Fract. Mech. **36**, 77-92 (1990).
39. A. T. Alpas, L. Edwards, and C. N. Reid, Metall. Trans. A **20**, 1395-409 (1989).
40. T. Ogura, K. Fukushima, and T. Masumoto, Mater. Sci. Eng. **23**, 231-35 (1976).
41. W. F. Brown, Jr., *Aerospace Structural Metals Handbook*, vol. code 1224, Metals and Ceramics Information Center, 1-30 (1989).
42. G. I. Taylor, Proc. R. Soc. London **A201**, 192-96 (1950).
43. R. J. Fields and M. F. Ashby, Philos. Mag. **33**, 33-48 (1976).

44. R. L. Tsai and R. Raj, Acta Metall. **30**, 1043-58 (1982).
45. H. A. Bruck, A. J. Rosakis, and W. L. Johnson, J. Mater. Res. **11**, 503-511 (1996).
46. C. J. Gilbert, J. W. Ager, III, V. Schroeder, and R. O. Ritchie, this volume.
47. M. A. LaMadrid, S. D. O'Connor, A. Peker, W. L. Johnson, and J. D. Baldeschwieler, J. Mater. Res. **11**, 1494-99 (1996).
48. R. O. Ritchie, Mater. Sci. Eng. A **103**, 15-28 (1988).
49. F. A. McClintock and R. M. N. Pelloux, *Boeing Scientific Research Laboratories Document D1-82-0708* (1968), cited in R. M. N. Pelloux, Trans. ASM **62**, 281-85 (1969).
50. K. J. Miller, Fatigue Fract. Engng. Mater. Struct. **10**, 93-113 (1987).
51. L. A. Davis, J. Mater. Sci. **11**, 711-17 (1976).
52. T. Ogura, T. Masumoto, and K. Fukushima, Scripta Metall. **9**, 109-14 (1975).

Environmental and Stress State Effects on Fracture and Fatigue Crack-Growth in Zr-Ti-Ni-Cu-Be Bulk Amorphous Metals

Katharine M. Flores, Daewoong Suh, and Reinhold H. Dauskardt
Department of Materials Science and Engineering, Stanford University, Stanford, CA 94305

ABSTRACT

Our work has focused on the fracture and fatigue crack-growth mechanisms of a $Zr_{41.25}Ti_{13.75}Ni_{10}Cu_{12.5}Be_{22.5}$ bulk metallic glass. These alloys exhibit failure strengths on the order of 2 GPa and toughnesses of 10-30 MPa√m with remarkably little plastic deformation. The effect of stress state on strength properties was studied in an effort to model fracture mechanisms. While fracture surfaces suggest significant plastic flow, failure strains are quite small and appear to be independent of stress state. Other methods of energy dissipation, including adiabatic heating, are discussed. Temperature increases of more than 20°C have been noted on the side face of the fracture sample at the crack tip during fracture. The micromechanics of fatigue crack growth are also considered and related to the batch chemistry, environment, and microstructure.

INTRODUCTION

The recent development of *bulk* metallic glasses offers the potential for metallic material systems with dramatically improved mechanical properties. Tensile strengths of 2 GPa [1] and toughnesses of up to 55 MPa√m [2] have been reported. These high strength and toughness values are accompanied by remarkably little plastic deformation. Until recently, little work has been undertaken to elucidate the fracture mechanisms of these alloys. In this work, we initially examine the effect of stress state on the tensile strength of a $Zr_{41.25}Ti_{13.75}Ni_{10}Cu_{12.5}Be_{22.5}$ bulk metallic glass. Early work has noted the presence of voids and vein patterns on the fracture surfaces, indicating the activation of flow processes and suggesting that failure occurs via microvoid coalescence. To model such behavior in other high strength metals, McClintock [3] and Mackenzie [4] used a stress-modified critical strain criterion, in which failure occurs when the local equivalent plastic strain exceeds a critical fracture strain over a characteristic distance ahead of the crack tip. The critical fracture strain is a function of stress state. However, it will be shown that for the metallic glass, the failure strain remains constant. Heating during plastic flow in metallic glass is another important variable that may affect the extent of plasticity. Local adiabatic heating is thought to significantly reduce the glass viscosity to produce local softening [5-10]. We report on detailed measurements of the surface temperature increase around the crack tip in a bulk metallic glass during fracture [11].

Finally, environment plays a central role in the ability of a material to perform in structural applications. Previous evidence from metallic glasses in ribbon form suggest that it is susceptible to the same deleterious effects of hydrogen embrittlement as crystalline metals [12]. Again, little is understood about the diffusion behavior of various environmental species in the metallic glass and their effect on micromechanisms of failure. Accordingly, the fatigue crack-growth behavior of the bulk metallic glass was examined in air and dry nitrogen, as well as after ex situ hydrogen charging.

THEORETICAL BACKGROUND

Notched Tensile Bars

The stress state in a tensile bar may be altered by introducing a notch [4, 13]. Varying the radius of the notch relative to the diameter of the bar allows a range of stress states to be examined. An estimate of the stress state parameter, the ratio of the mean stress to the effective stress, at the center of the bar is given by the Bridgman analysis as [14]:

$$\frac{\sigma_m}{\overline{\sigma}} = \frac{1}{3} + \ln\left(\frac{a}{2R} + 1\right), \tag{1}$$

where a is the minimum bar radius and R is the radius of the notch. The effective plastic strain is given by:

$$\overline{\varepsilon}_p = 2\ln\frac{a_0}{a}, \tag{2}$$

where a_0 is the initial minimum bar radius. Note that by varying the notch radius or the bar radius, the stress state may be varied from uniaxial ($a/2R \rightarrow 0, \sigma_m/\overline{\sigma} \rightarrow 1/3$) to nearly triaxial ($a/2R \rightarrow \infty, \sigma_m/\overline{\sigma} \rightarrow \infty$). The rate of void growth is strongly dependent on the stress state. A material which fails by void coalescence may therefore show a similar strong dependence of failure strain on stress state. Most metals which fail by void coalescence show a dramatic increase in the effective plastic strain as the loading is changed from triaxial to uniaxial due to the loss of constraint.

Heating at a Propagating Crack Tip

Rice and Levy's [15] model for local heating at a propagating crack tip assumes that all plastic work is dissipated as heat and employs a non-hardening plasticity model. This is a reasonable assumption for the present metallic glass alloy, since it exhibits virtually no work hardening [1]. For large propagation velocities, the maximum temperature rise at the crack tip approaches:

$$\Delta T = \sqrt{2}\frac{(1-v^2)K\sigma_0\sqrt{v}}{E\sqrt{\rho ck}} \tag{3}$$

where v is Poisson's ratio, K the applied stress intensity factor, σ_0 the yield stress, v the crack velocity, and E is Young's Modulus. This represents an upper bound for the temperature rise at the crack tip; lower propagation velocities and locations more remote from the crack tip experience only a fraction of this temperature rise. Post fracture, heat dissipation can also be predicted using a heat conduction model [11].

EXPERIMENTAL WORK

Notched bars with a 44 mm long, 10 mm diameter gauge section and 2.50 mm notch depth were machined from $Zr_{41.25}Ti_{13.75}Ni_{10}Cu_{12.5}Be_{22.5}$ bulk metallic glass rods. The notch radius was varied between 0.25 and 2.50 mm, with all other dimensions held constant. Testing was performed on an electro-servo-hydraulic system with a crosshead displacement rate of 5 μm/s. This resulted in strain rates near the notch of $\sim 8 \times 10^{-5}$ s^{-1}. Elongation across the notch was monitored with a clip gage. The initial and final bar diameters were carefully measured at the notch root for the effective plastic strain calculation.

To examine the fracture and fatigue crack-growth behavior of the bulk metallic glass, compact tension samples were machined from plates and were fatigue precracked. The samples had dimensions scaled in general accordance with the ASTM-E399 Standard for plane strain fracture toughness testing. During fatigue testing, crack lengths were continuously monitored using compliance techniques, and were confirmed optically. Specimens were loaded at 25 Hz with a load ratio of 0.1 under automated load-shedding schemes. Hydrogenation of some samples was performed by ex situ cathodic charging for selected times.

Fracture testing was performed under displacement control with a displacement rate of 1 μm/s. Load-displacement data were recorded and crack lengths were monitored optically. Thermal imaging was performed with an Amber Galileo IR camera with a 256 x 256 array and a 27 mm pixel size. Images of the sample surface were captured at 50 frames/sec. The sample was lightly coated with colloidal graphite to lower its reflectivity. Since the onset of crack growth was difficult to accurately predict, the frame capture could not be synchronized with incipient crack extension. The camera was therefore set to begin capturing images slightly before the fracture event.

RESULTS AND DISCUSSION

All tensile bars failed at the notch with a cup and cone type morphology similar to that found in ductile metals. Micrographs of the failure surface from the 0.25 mm notch radius bar are shown in Figure 1. These micrographs clearly illustrate the different morphologies of the shear lip and the center region. Voids are visible in both areas, but they have different characteristics. Most notably, the shear lip region appears to exhibit localized melting. The voids are smeared, and there are several droplets on the surface. The central region may also have melted slightly, although it does not appear to have softened to the same extent.

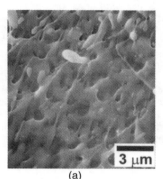 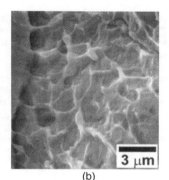

(a)   (b)

Figure 1. Micrographs of the failure surface of 0.25 mm notch radius bar with a cup and cone morphology. The void morphologies for the (a) shear and (b) central regions are shown. Note the molten droplets clearly visible in (a).

The variation in effective plastic strain with stress state parameter is shown in Figure 2a. A high strength steel is also shown for comparison. Although the fractography shows evidence of plastic flow processes, the metallic glass shows no significant change in effective plastic strain with stress state. In fact, there is almost no measurable plastic strain in any of the bars ($\bar{\varepsilon}_p \leq 0.01$). However, there is a variation in strength and the size of the shear lip with stress state parameter, as shown in Figure 2b. The shear lip ratio is defined as:

$$\text{shear lip ratio} = \frac{\text{shear lip width}}{\text{radius of failure cross section}}. \qquad (4)$$

Both the failure stress and the size of the shear lip decrease with increasing triaxiality.

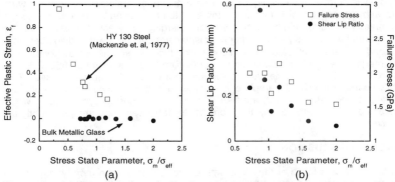

Figure 2. (a) The variation of effective plastic strain with stress state. The results for a pressure vessel steel are shown for comparison. (b) The variation of failure stress and shear lip ratio with stress state.

During the fracture toughness experiment, the bulk metallic glass exhibited stable crack growth over ~14 mm of crack extension during repeated fracture experiments. The resulting crack growth resistance data are shown in Figure 3. Several crack growth sequences were captured using the thermal imaging system as the crack propagated and arrested. A representative image is presented in Figure 4. A maximum temperature of 22.5 K above ambient was measured near the initial crack tip. The crack extended ~0.9 mm by unstable fracture until arrest in a time of less than 20 ms before the image shown. Using a heat dissipation model [11], we estimate that the fracture occurred 5 ms before the image shown, and the lower bound crack velocity was 175 mm/s. Additionally, the dissipation model indicates that the crack tip temperature change at initiation was ~54.2 K. This is in excellent agreement with the Rice and Levy model (Equation 3) which indicates that the temperature rise at a crack propagating at 175 mm/s is 56.5 K for an applied stress intensity of 21 MPa√m. While these temperature rises are insufficient to cause localized melting, it is evident from the fractography that softening did occur, causing the familiar vein patterns.

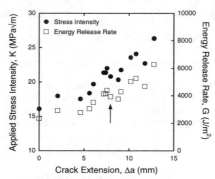

Figure 3. Crack growth resistance data obtained during repeated fracture events during the thermal imaging experiment. The point corresponding to the thermal image in Figure 4 is indicated with an arrow.

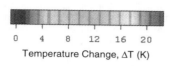

Temperature Change, ΔT (K)

Figure 4. Thermal image after a fracture event in which the crack extended by 0.875 mm. The image is 7 mm x 7 mm, and the stress intensity at initiation was 20.8 MPa√m. Temperature increases behind the crack tip are probably associated with the fracture of bridging ligaments.

Fatigue crack-growth data are presented in Figure 5. The behavior in ambient air and dry nitrogen is compared for two sample batches in Figure 5a. It is apparent from these results that the environmental sensitivity is dependant on the batch chemistry, since the environment had a marked effect on one batch and very little effect on the other. The batch chemistries are given in Table 1. These subtle variations in chemistry have a significant effect on the interaction of the metallic glass with its environment. The effect is a subject for further study. Hydrogen charging experiments were performed only on batch two, as shown in Figure 5b. Increased charging time resulted in increased retardation of crack growth. The crack path for the charged samples was quite tortuous, suggesting that the slower crack growth may be due to crack deflection and closure. These crack shielding effects may mask any embrittlement of the glass due to hydrogen. Work involving in situ hydrogen charging, as well as to determine the actual hydrogen concentration due to charging, is in progress.

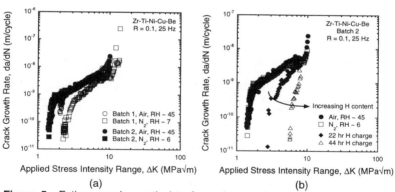

(a)   (b)

Figure 5. Fatigue crack-growth data for various environments showing (a) the batch to batch variation in behavior and (b) the effect of increasing hydrogen content.

Table 1. Batch Chemistries

Batch #	Zr	Ti	Ni	Cu	Be	H	O	N	C
			(wt. %)				(ppm)		
1	62.67	11.04	9.80	12.41	4.07	17	730	20	53
2	63.86	10.73	9.49	11.91	4.00	33	660	20	62

CONCLUSIONS

The fracture morphology of the metallic glass strongly suggests the activation of flow processes. Tensile bars exhibit cup and cone type fracture, and veins and voids are evident on the fracture surface. However, varying the stress state had no effect on the failure strain, indicating that the typical ductile failure criterion is inapplicable. Further, there appears to be no plastic strain at failure for any stress state. Examination of the shear lip surface strongly suggests that local melting occurred, indicating that plastic work was dissipated as heat.

Thermal images of a propagating crack tip provide direct, spatially resolved evidence of localized adiabatic heating thought to occur during inhomogeneous flow in amorphous metals. The temperature increase of at least 22.5 K is consistent with models for the heat generated by plastic work ahead of a moving crack tip. While this temperature rise is insufficient to cause melting, it does suggest that significant reductions of the glass viscosity may occur in localized shear bands, giving rise to the characteristic veined fracture surface. Further, higher strain rates, such as those associated with the tensile test, should result in larger temperature increases.

Environmental effects on the fatigue crack-growth behavior of the metallic glass appear to be batch dependant. Subtle variations in batch chemistry may cause significant changes in the interaction with environmental species. However, it is clear that increasing hydrogen charging time causes a decrease in crack-growth rates. We associate this with increasing crack deflection and closure effects masking the expected hydrogen embrittlement.

ACKNOWLEDGMENTS

The work was supported by the Air Force Office of Scientific Research under AFOSR Grant No. F49620-98-1-0260. Material for this work was provided by Amorphous Technologies International, Laguna Niguel, California, and by Howmet Research Corporation, Whitehall, Michigan. Thermal imaging equipment was provided and operated by Chris Johnston, Sierra Olympic Technologies, Inc., Issaquah Washington.

REFERENCES

1. H. A. Bruck, T. Christman, A. J. Rosakis, and W. L. Johnson, Scripta Metall. **30**, 429 (1994).
2. R. D. Conner, A. J. Rosakis, W. L. Johnson, and D. M. Owen, Scripta Metall. **37**, 1373 (1997).
3. McClintock in *Fundamental Aspects of Structural Alloy Design*, edited by R. I. Jaffee and B. A. Wilcox (Penum, New York, NY, 1977), p 147.
4. A. C. Mackenzie, J. W. Hancock, and D. K. Brown, Eng. Frac. Mech. **9**, 167 (1977).
5. F. Spaepen, Acta Metall. **25**, 407 (1911).
6. H. A. Bruck, A. J. Rosakis, and W. L. Johnson, J. Mater. Res. **11**, 503 (1996).
7. A. Leonhard, L. Q. Xing, M. Heilmaier, A. Gerbert, J. Eckert, and L. Shultz, Nanostruct. Mater. **10** (1998), in press.
8. H. Kato and A. Inoue, Mater. Trans. JIM **38**, 793 (1997).
9. R. Doblione, S. Spriano, and L. Battezzati, Nanostruct. Mater. **8**, 447 (1997).
10. V. Z. Bengus, E. D. Tabachnikova, S. E. Shumilin, Y. I. Golovin, M. V. Makarov, A. A. Shibokov, J. Miskuf, K. Csach, and V. Ocelik, Int, J. Rapid Solid. **8**, 21 (1993).
11. K. M. Flores and R. H. Dauskardt, J. Mater. Res. **14** (1999), in press.
12. R. K. Viswanadham , J. A. S. Green, and W. G. Montague, Scripta Metall. **10**, 229 (1976).
13. J. W. Hancock and A. C. Mackenzie, J. Mech. Phys. Solids **24**, 147 (1976).
14. P.W. Bridgman, *Studies in Large Flow and Fracture*, McGraw-Hill, New York (1952).
15. J. R. Rice and N. Levy, in *Physics of Strength and Plasticity*, edited by A. S. Argon (MIT Press , Cambridge, MA, 1969), p. 277.

HIGH DYNAMIC MECHANICAL STRENGTH
OF ZIRCONIUM-BASED BULK AMORPHOUS ALLOYS

TAO ZHANG AND AKIHISA INOUE
Institute for Materials Research, Tohoku University, Sendai 980-8577, Japan

ABSTRACT

A bulk amorphous $Zr_{55}Al_{10}Ni_5Cu_{30}$ alloy prepared by squeeze casting was found to exhibit high mechanical strength values, i.e., uniaxial tensile fracture strength (σ_t) of 1850 MPa, three-point bending strength (σ_b) of 3200 MPa, bending fatigue strength (σ_f) of 1100 MPa, Charpy impact fracture energy (E_f) of 135 kJ/m^2 and fracture toughness of 68 MPa$\sqrt{m}$. The σ_b, σ_f and E_f are about two times higher than those for the corresponding bulk amorphous alloys prepared by unidirectional solidification and powder consolidation techniques, though the σ_t of the squeeze cast sample is higher by about 15 % than those for the other samples. The remarkable increases in the σ_b, σ_f and E_f are presumably due to the introduction of high compressive residual stress of about 1240 MPa in the outer surface region only for the squeeze cast sample. The finding of the effectiveness of the compressive residual stress on the increase in the mechanical strength under the bending stress mode is important and expected to be widely used as a new strengthening mechanism for bulk amorphous alloys, as is the case for reinforced oxide glasses subjected to strengthening treatment.

INTRODUCTION

Recently, bulk amorphous alloys have attracted rapidly increasing attention because of the importance of basic science and engineering application[1][2]. Bulk amorphous alloys with diameters above several milimeters had been prepared in Pd-Ni-P system by repeated melting of their molten alloys fluxed with B_2O_3[3]-[5]. However, the bulk amorphous alloys had been limited to the noble metal base system for a long period of almost three decades after the finding of Pd-Ni-P glassy alloy in 1969[6]. The remarkable attention to bulk amorphous alloys seems to result from the findings of a series of bulk amorphous alloys in Mg-Ln-TM[7], Ln-Al-TM[8] and Zr-Al-TM[9] (Ln=lanthanide metal, TM=transition metal) systems for several years between 1988 and 1990. Subsequently, the alloy systems in which bulk amorphous alloys are formed by copper mold casting have been extended to Zr-Ti-Ni-Cu-Be[10], Zr-(Ti,Nb,Pd)-Al-Ni-Cu[11], Fe-(Al,Ga)-(P,C,B,Si)[12], Pd-Cu-Ni-P[13], Pd-Ni-Fe-P[14], (Fe,Co)-(Zr,Hf,Nb,Ta)-B[15] and Ti-Ni-Cu-Sn[16] systems. The bulk amorphous alloys in Zr-Al-Ni-Cu, Zr-Ti-Al-Ni-Cu and Zr-Ti-Ni-Cu-Be systems have already been used as sporting goods materials[17]. Furthermore, the Pd-Cu-Ni-P bulk amorphous alloys have also gained an application as an electrode material for high efficient generation of chloride gas[18]. The use of the Zr-based bulk amorphous alloys as sporting goods has been attributed to the high tensile fracture strength (σ_t), high Charpy impact fracture enenrgy (E_f), high bending flexural strength (σ_b) and high bending fatigue strength (σ_f). There has been no distinct difference in their mechanical properties among the above-described three types of Zr-based alloys prepared by unidirectional solidification[19] and powder consolidation[20] techniques. The values of σ_t, E_f, σ_b and σ_f have been reported in the range of 1500 to 1800 MPa, 65 to 70 kJ/m^2, 1700 to 1900 MPa and 500 to 600 MPa, respectively. Besides, these bulk amorphous alloys have another feature of much lower

Mat. Res. Soc. Symp. Proc. Vol. 554 © 1999 Materials Research Society

Young's module as compared with the crystalline alloys with the same σ_t(21).

It is well known that oxide glasses can be reinforced by the introduction of compressive residual stress in the region near outer surface. The compressive residual stress can be introduced by cooling the surface region at a higher cooling rate as compared with the cooling rate in the inner region(22). It is difficult to introduce the compressive residual stress in the outer surface region of bulk amorphous alloys in the use of conventional copper mold casting, unidirectional solidification and powder consolidation methods. If a compressive residual stress is introduced in the outer surface region of a bulk amorphous alloy, the resulting bulk amorphous alloy is expected to exhibit significant increases in σ_b, E_f and σ_f. With the aim of introducing the compressive residual stress only in the outer surface region, we have constructed a new liquid-forming equipment named as squeeze casting for the production of a bulk amorphous alloy. In this equipment, a molten metal is squeezed with water-cooled copper plates and then cooled to room temperature between the copper plates under an applied load of 60 MPa. The liquid-forming technique leads to the production of a bulk amorphous alloy sheet in which higher cooling rates are achieved only in the outer surface region.

EXPERIMENTAL PROCEDURE

An amorphous $Zr_{60}Al_{10}Ni_{10}Cu_{20}$ sheet with a thickness of about 3 mm was produced by the squeeze casting. From the alloy sheet, mechanical testing specimens were made by mechanical cutting in the condition where the surface region in the cast sheet is not eliminated. The dimension of the gauge part in the specimen and its total length are 20 x 3 x 2 mm and 50 mm, respectively, for tensile test and 40 x 5 x 2 mm and 80 mm, respectively, for three-point bending test, 30 x 10 x 2.5 mm and 80 mm, respectively, for bending fatigue test, 30 x 5 x 2.5 mm and 55mm, respectively, having a U-shape notch corresponding to JIS-No.3 (2.5 mm in thickness) type for Charpy impact fracture test and 20 x 30 x 2 mm and 3 mm, respectively, having a V-shape precrack corresponding to ASTM E-399 for fracture toughness. The tensile strength, bending flexural and fracture toughness tests were made with an Instron-type testing

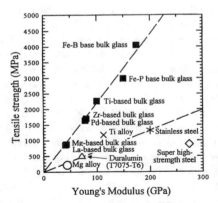

Young's Modulus (GPa)

Fig. 1 Relation between tensile fracture strength (σ_t) and Young's modulus (E) for bulk amorphous alloys. The data of conventional crystalline alloys are also shown for comparison.

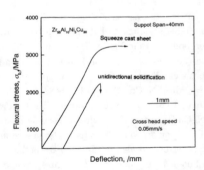

Fig. 2 Bending flexural strength-deflection curve of the bulk amorphous $Zr_{55}Al_{10}Ni_5Cu_{30}$ sheet prepared by squeeze casting. The data of the same amorphous sheet samples prepared by unidirectional solidification.

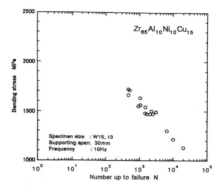

$Zr_{65}Al_{10}Ni_{10}Cu_{15}$

Specimen size : W15, t3
Supporting span: 30mm
Frequency : 10Hz

Bending stress MPa

Number up to failure N

Fig. 3 Maximum bending applied stress-cyclic numbers up to failure for
bulk amorphous $Zr_{60}Al_{10}Ni_{10}Cu_{20}$ sheets prepared by the squeeze casting.

machine and the bending fatigue test was performed at a frequency of 10 Hz with a servopulser type
testing machine. The fracture stress and displacement at the impact fracture and the impact fracture
energy were measured at room temperature with a Charpy impact fracture testing machine. The load-
displacement curve was stored in the computer and the impact fracture energy was simultaneously
calculated from the load-displacement curve. The fracture surface appearance was examined by SEM.

RESULTS AND DISCUSSION

The σ_t and Young's modulus (E) of the squeezed sample were 1800 MPa and 90 GPa, respectively.
The σ_t value is higher by 10 to 20 % than those (1550 to 1700 MPa)(19)(20) for the samples prepared by
unidirectional solidification and powder consolidation techniques. The relation between σ_t and E is
plotted in Fig. 1, where the data of other bulk amorphous alloys prepared by unidirectional
solidification(19) and powder consolidation(20) techniques and commercial crystalline alloys are also
shown for comparison. The present σ_t lies in the nearly same relation. The relation between σ_t and E
for the bulk amorphous alloys is significantly different from that for conventional crystalline alloys.
When the σ_t of the bulk amorphous alloys is compared with that of commercial crystalline alloys with the
same E values, the σ_t values for the bulk amorphous alloys are higher by about three times.

Figure 2 shows the three-point bending flexural strength-deflection curve of the cast $Zr_{55}Al_{10}Ni_5Cu_{30}$
amorphous sheet, together with the data on the $Zr_{55}Al_{10}Ni_5Cu_{30}$ amorphous sheets with the same
dimension prepared by unidirectional solidification. The σ_b and bending Young's modulus (E_b) are
evaluated by the following relations of $\sigma_b=3FL/2wh^2$ and $E_b=L^3F/4wh^3\delta(23)$, where F is the fracture load, L
is the distance between the two outer points, w is the width of the specimen, h is the height of the specimen,
and δ is the deflection of the beam when F is applied. The σ_b, E_b and δ at the final failure are 3200 MPa,
94 GPa and 2.4 mm, respectively, for the squeezed sheet, and 2200MPa, 94 GPa and 1.3 mm, respectively,
for the unidirectionally solidified sheet. It is noticed that the σ_b and δ of the squeezed sheet are about two
times higher than those for the sheets prepared by the unidirectional solidification and powder
consolidation techniques.

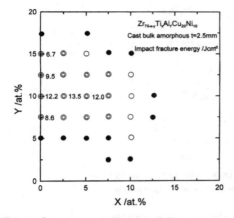

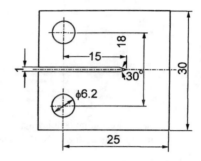

Fig. 4　Composition dependence of Charpy impact fracture energy for bulk amorphous $Zr_{70-x-y}Ti_xAl_yNi_{10}Cu_{20}$ sheets prepared by squeeze casting.

Fig. 5　Shape and size of the ASTM E-399 specimen used for the measurement of fracture toughness (K_c).

The much higher σ_b for the squeezed cast sheet indicates the possibility of a much higher fatigue strength as compared with those obtained by the other methods.　Figure 3 shows the relation between bending maximum stress and the cyclic numbers up to failure for the cast sheets prepared by squeeze casting.　When the bending fatigue strength is defined by a minimum applied stress at the constant cyclic number of 3×10^4, the σ_f is evaluated to be 1100 MPa.　Since the σ_f value after 2×10^4 cycles is 600 MPa for the unidirectionally solidified sample and 570 MPa for the consolidated sample, the σ_f of the squeezed sample is also about two times higher than those for the samples prepared by the other techniques. When the fatigue limit is defined by the ratio of σ_f/σ_b at 10^7 cycles, the ratios of all the samples are estimated to be much lower than 0.3, in agreement with the tendency for the Zr-Ti-Be-Ni-Cu amorphous alloy in which the σ_f/σ_b has been reported to be 0.04(24).

Figure 4 shows the compositional dependence of Charpy impact fracture energy for the $Zr_{70-x-y}Ti_xAl_yNi_{10}Cu_{20}$ amorphous sheets prepared by squeeze casting.　The $Zr_{55-60}Ti_{0-5}Al_{10}Ni_{10}Cu_{20}$ sheets exhibit high impact fracture energies of 120 to 135 kJ/m² which are higher by about two times in comparison with 70 kJ/m² for the unidirectionally solidified sample and 67 kJ/m² for the consolidated sample. Furthermore, by using the specimen which agrees with the dimension of ASTM E-399 shown in Fig. 5, the fracture toughness (K_c) of the $Zr_{55}Al_{10}Ni_5Cu_{30}$ amorphous sheet was measured to be 68 MPa$\sqrt{}$m.　The K_c value is higher than those (22 to 33 MPa$\sqrt{}$m) for high-strength type Al base alloys and comparable to those (50 to 90 MPa$\sqrt{}$m) for commercial Ti-based alloys(25).　Also, the Kc value is nearly equal to that (55MPa$\sqrt{}$m) for Zr-Ti-Be-Ni-Cu bulk amorphous alloy. The high K_c of the bulk amorphous alloy is supported from the fracture surface appearance in which a well-developed vein pattern is observed over the whole area including the region near the preexistent crack tip as shown in Fig. 6.

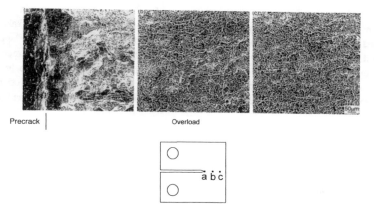

Precrack | Overload

a b c

Fig. 6　Fracture surface appearance near the preexistent crack tip for the bulk amorphous Zr$_{55}$Al$_{10}$Ni$_5$Cu$_{30}$ sheet subjected to the measurement of fracture toughness. The SEM images (a), (b) and (c) are taken the regions of a, b and c, respectively, in the illustration of the specimen.

DISCUSSION

It is noticed that all the values of σ$_b$, σ$_f$ and E$_j$ of the squeezed sample are about two times higher than those for the samples prepared by unidirectional solidification(19) and powder consolidation(20), in spite of the same alloy composition and sample shape.　The remarkable difference is thought to result from the introduction of compressive residual stress in the outer surface region where the maximum load is applied by the bending stress mode in the three types of tests.　It has been reported that the compressive residual stress (σ) generated by the maximum difference in temperature (ΔT$_{max}$) between the outer surface and central regions during sample preparation is evaluated by the relation of σ=(α E)/(1-μ)2ΔT$_{max}$/3(22). Here, the α, E and μ are the coefficient of thermal expansion, Young's modulus and Poisson's ratio, respectively, of the amorphous alloy.　The α, E, μ and ΔT$_{max}$ values used in the calculation are 21x10^{-6} K^{-1}(27), 90 GPa(19)(20), 0.34(28) and 650 K(29), respectively.　The ΔT$_{max}$ value was assumed as the difference between the squeeze temperature of molten alloy and T$_g$.　The σ value is evaluated to be 1240 MPa which corresponds to the tensile yield stress of about 1500 MPa.　The existence of the residual compressive stress of 1240 MPa seems to be the reason for the achievement of the much higher values of σ$_b$, σ$_f$ and E$_f$.　As described above, the difference in the tensile fracture strength between the squeezed sample and the other samples is as small as about 10 % because the initiation of tensile fracture occurs in the central region with tensile residual stress.　The remarkable difference in σ$_b$, σ$_f$ and E$_f$ except σ$_t$ is the same as that(24) for reinforced oxide glasses subjected to the strengthening treatment in which the oxide surface region is cooled at higher cooling rates from the supercooled liquid above T$_g$.

CONCLUSIONS

The Zr-based bulk amorphous sheets with a thickness of 3 mm were found to exhibit high values of σ$_b$,

σ_f and E_f which are about two times higher than those for the corresponding bulk amorphous sheets prepared by unidirectional solidification and powder consolidation techniques. The remarkable difference is presumed to result from the introduction of maximum compressive residual stress of approximately 1240 MPa in the outer surface region of the sheet sample which is generated by the difference in cooling rate between the outer surface and central regions during the squeeze casting. The finding of the effectiveness of the compressive residual stress for the dramatic increase in mechanical strength in the bending stress mode is expected to be widely used as a new strengthening method of bulk amorphous alloys.

REFERENCES

(1) A. Inoue, Mater. Trans., JIM, 36(1995), 866.

(2) A. Inoue, Sci. Rep. Res. Inst. Tohoku Univ., A42(1996), 1.

(3) H.W. Kui, A.L. Greer and D. Turnbull, Appl. Phys. Lett., 45(1984), 615.

(4) H.W. Kui and D. Turnbull, Appl. Phys. Lett., 47(1985), 796.

(5) R. Willnecker, K. Wittmann and G.P. Gorler, J. Non-Cryst. Solids, 156-158(1993), 450.

(6) P.L. Maitrepierre, J. Appl. Phys., 40(1969), 4826.

(7) A. Inoue, K. Ohtera, K. Kita and T. Masumoto, Jpn. J. Appl. Phys., 27(1988), L2248.

(8) A. Inoue, T. Zhang and T. Masumoto, Mater. Trans., JIM, 30(1989), 965.

(9) A. Inoue, T. Zhang and T. Masumoto, Mater. Trans., JIM, 31(1990), 177.

(10) A. Peker and W.L. Johnsosn, Appl. Phys. Lett., 63(1993), 2342.

(11) A. Inoue, T. Shibata and T. Zhang, Mater. Trans., JIM, 36(1995), 1420.

(12) A. Inoue and J.S. Gook, Mater. Trans., JIM, 36(1995), 1180.

(13) A. Inoue, N. Nishiyama and T. Matsuda, Mater. Trans., JIM, 37(1996), 181.

(14) Y. He and R.B. Schwarz, Metall. Mater. Trans., in press.

(15) A. Inoue, T. Zhang, A. Takeuchi and T. Itoi, Mater. Trans., JIM, 38(1997), 359.

(16) T. Zhang and A. Inoue, Mater. Trans., JIM, 39(1998), in press.

(17) Dunlop Catalogue, Japan Dunlop, Tokyo, (1998).

(18) Hiranuma Industrial Co., Ltd., Mito, private communication (1998).

(19) A. Inoue, Y. Yokoyama, Y. Shinohara and T. Masumoto, Mater. Trans., JIM, 35(1994), 923.

(20) A. Kato, Y. Kawamura and A. Inoue, Mater. Trans., JIM, 37(1996), 70.

(21) A. Inoue, Mater. Sci. Forum, 269-272(1998), 855.

(22) T. Inoue, Glass Handbook, ed. by Sakka, Takahashi and Sakaino, Asakurashotten, Tokyo, (1975), p.485.

(23) D.R. Askeland, Science and Engineering Materials, PWS Publishing, Boston (1994), p.144.

(24) C. J. Gilbert, J. M. Lippmann and R.O. Ritchie, Scripta Mater., 38(1998), 537.

(25) D.R. Askeland, Science and Engineering Materials, PWS Publishing, Boston, (1994), p.154.

(26) C. J. Gilbert, R. O. Ritchie and W. L. Johnson, Appl. Phys. Lett., 71(1997), 473.

(27) H.M. Kimura, A. Inoue, N. Nishiyama, K. Sasamori, O. Haruyama and T. Masumoto, Sci. Rep. Res. Inst. Tohoku Univ., A43(1997), 101.

(28) Materials Science of Amorphous Metals, Ohmu, Tokyo, (1982), p.202.

(29) A. Inoue, T. Zhang and T. Masumoto, J. Non-Cryst. Solids, 156-158(1993), 473.

Deformation Behavior of FCC Crystalline Metallic Nanowires Under High Strain Rates

Yue Qi *, Hideyuki Ikeda **, Tahir Cagin *, Konrad Samwer **, William L. Johnson **, William A. Goddard III*

*Materials and Process Simulation Center, California Institute of Technology, Pasadena, CA 91125, USA

**Keck Laboratory of Engineering Materials, California Institute of Technology, Pasadena, CA 91125, USA

ABSTRACT

We used molecular dynamics (MD) methods to study the deformation behavior of metallic alloy crystal nanowires of pure Cu, NiCu alloy and NiAu alloy, under high rates of uniaxial tensile strain, ranging from $5*10^8$/s to $5*10^{10}$/s. These nanowires are just about 2 nm thick and hence cannot sustain dislocations, instead we find that deformation proceeds through twinning and coherent slipping mechanisms. NiAu has a 13% size mismatch whereas NiCu only 2.5%. As a result the critical strain rate at which the "nanowire crystals" flow like a "liquid" is 100 times smaller for NiAu. We also calculated the elastic constants at each strain state for all strain rates to identify the relation between mechanical "shear" instability and deformation process.

INTRODUCTION

From mechanical tests and structure analyses such deformation mechanisms as, dislocations[1][2], twinning[3][4], and grain boundary sliding[5]. have been proposed and verified in various systems. The two major and competitive deformation models in single crystal are slip and twinning, increases in strain rate or decreases in temperature tend to favor twining over slip. Deformation twins are more prominent in bcc and hcp metals than in fcc, because the five independent slip systems required for general deformation may not be satisfied in bcc or hcp. Indeed twins have been observed. in fcc metals exposed to large strain rates (such as in shock waves [6]), and/or low temperature [7]. Unfortunately experiments under these conditions, tend to have many factors all playing a role simultaneously. For instance, experiments at very high strain rates lead to shear localization arising from adiabatic heat dissipation and localized thermal softening.

On the other hand , molecular dynamics (MD) simulations can isolate the direct effect of strain at various strain rates from those due to heat dissipation and the concomitant temperature increase by controlling the temperature and the stain rate. Keeping the temperature constant and can then be used to analyze the detailed mechanisms of the deformations.

SIMULATION METHODS

The periodic unit cell for the MD simulations consisted of 5*5*10 fcc cubic cells (1.78nm*1.78nm*3.56nm for NiCu) with 1000 atoms. We started with an fcc single crystal and equilibrated at 300K for 100 psec. For the alloys NiCu(50%) and NiAu(50%) we assumed a random fcc structure. We used 1-D periodic boundary condition along the <001> crystallographic direction [8], but a finite wire (about 2nm thick) in the x and y directions. The tensile strain, ε_{33}, was applied in the z direction to these infinite nanowires. To maintain the specified strain rate, the tensile strain along the <001> direction was increased by 0.5% at a time interval determined by the strain rate. To avoid the generation of stress wave, the strain was applied uniformly to all layers of atoms, which were then allowed to relax in the MD under isokinetic conditions (kinetic energy is kept constant). This isolates effects of adiabatic heat dissipation and thermal softening. At each strain state, we evaluate the stress, the elastic constants (the Born term contribution). Although our nanowire is periodic in only one dimension, we use the true atomic volume to calculate intensive variables.[9]

To obtain accurate properties for the simulations, we use the simulation is a quantum Sutten-Chen (Q-SC) many body force field.[10][11].   Q-SC leads to accurate values for the mechanical, thermodynamic, viscosity, and surface properties (for such as pure metals such as Cu and Ni and for alloys such as CuNi and NiAu). We observed Ni and NiCu lead to similar behavior but NiAu behaves very differently due to the large atomic radii difference.

RESULTS and DISCUSION

Figure 1 shows snapshots at 100% strain for different strain rates, and figure 2 displays the stress (σ_{33}) - strain (ε_{33}) curve for NiCu at T=300K for each strain rate. The stress-strain relation is linear at small strain, and the yield stress increases as the strain rate increases

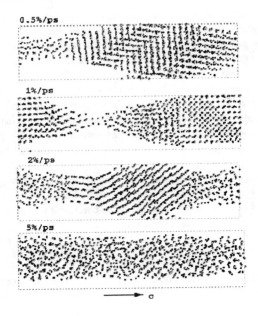

Figure-1, Snapshots of NiCu nanowire at 100% strain deformed at strain rates of
a. 0.05%/ps shows twinning and local melting in bulk
b. 1%/ps shows smaller fcc grains
c. 2%/ps shows extensive disorder
d. 5%/ps shows full amorphization to glass

At low strain rates, 0.05%/ps and 0.5%/ps, the stress drops in a repeating loading/plastic

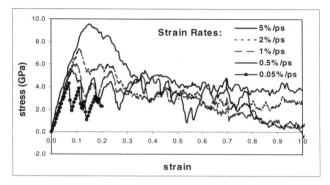

Figure-2. Stress strain curve for NiCu at different strain rates

flow cycle after passing the elastic limit. From analyzing the structures, we have been able to identify these events as coherent shearing, and the twinning formed during the shearing events. The stress drop is due to twin deformation. During the next increase of the stress, the Young's Modulus remains almost the same. Multiple coherent shearing continue to take place as tensile stress increases, but eventually necking begins around 80% strain for strain rates of 0.5%/ps and 1%/ps, and sample becomes disordered around the necking part. Eventually the stress concentration at the neck causes failure, with the stress dropping to zero. After fracture additional increments in strain cannot cause any increase in stress, however the melted neck relaxes back to fcc crystalline packing in the absence of the stress field.

But when the strain rate is increased to 5%/ps we see quite different behavior, At 5%/ps strain rate, no coherent shear bands and twins are formed within the specimen. Instead, the specimen transforms homogeneously into a disordered amorphous state, and undergoes uniform flow without twinning or necking. We reported this strain induced amorphization by analyzing the radial distribution function and elastic constants in an early work.[12]

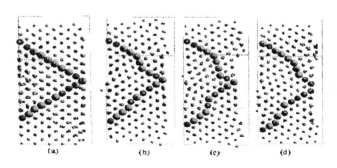

Figure-3 snapshots for NiCu during tensile tests.
(a) strain rate = 0.05%/ps strain = 7% before yielding
(b) strain rate = 0.05%/ps strain=8%, two twins formed
(c) strain rate = 0.05%/ps strain = 14%, twins grow one more layer
(d) strain rate = 0.5%/ps, strain = 12.5%/ps, three twins formed

369

To study the structural details, we show the $(01\bar{1})$ projection of the structure before and after yielding at strain rate of 0.05%/ps in figure 3. We highlight a group of neighbor atoms on (111) and $(\bar{1}11)$ planes, and keep track of them. The mismatch on (111) planes are obvious in the structure at 8% strain, right after the first load drop. There are two twins formed; one with two $(\bar{1}11)$ layers of atoms, and one only has one (111) layer of atoms. These twins are formed by cooperative shearing, and to return back to the original crystal orientation, there are two stacking faults at each side of the twin. The relative displacements for the atoms across the twin boundary is about $<\bar{2}11>$ on (111) planes, consistent with the well known (111)[112] twin system in fcc metals. The schematic of twinning in fcc metals is given in figure 4, the projection from (220) plane is included to keep the same view as the snapshots shown in figure 3. At a strain state of 14%, the second stress drop finished, and both twins increased by one more (fig3-c). Cottrell and Bilby[13] predicted that the fcc structure would produce only a monolayer stacking fault, and Venables[14] suggested a modified mechanism to allow a continuous growth of twin from a single stacking fault. Our observation agrees with this model.

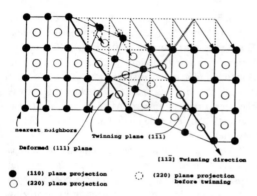

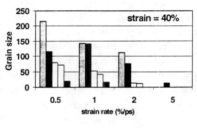

Figure-5, grain size of multi-oriented crystal after twin deformation decreases as strain rate increases.

Figure-4, schematic of twining in fcc metals

The twin deformation causes the stress drop, which in general increases with increasing temperature and the number of stacking faults. We can estimate the energy caused to form four stacking faults (there are four interface for two twins), from the stacking fault energy[15] of Ni ($150mJ/m^2$) and Cu ($80mJ/m^2$). We counted the number of atoms of the surface, and calculated the area of the surface (3.29 nm^2 and 4.19 nm^2), then use the average of stacking fault energy for Ni and Cu, $115mJ/m^2$, to obtain the stacking fault energy as $8.602*10^{-19}$ J. The elastic energy drop during the twinning process is $9.6*10^{-19}$ J. Thus the system has used the stored elastic energy to form twins. The similar deformation mode has been observed in pure Ni.

Since the system can accumulate more strain energy as the strain rate increases, it is possible that the sample can nucleate more stacking fault for twin growth. The structure of NiCu at 12.5% strain state(right after the first stress drop) under strain rate of 0.5%/ps has show in figure 3-d. Here we observe two independent monolayer stacking faults on the (111) planes. The stress drop at this strain rate is also larger than the drop at 0.05%/ps strain rate.

Figure 1 shows that after the single crystal turned to a multi-oriented crystal by continuous twinning, the grain size at the same strain state but different strain rates are different. We can show qualitatively that the grain size is decreasing as the strain rates increases. We define an atom having 12 nearest neighbors with a ABC packing sequence as a part of a perfect fcc crystal, from it we check the atoms in far nearest neighbors, and stop at the atoms, which lost the fcc nearest neighbors configuration. This method can identify a stacking fault on a twin boundary. Using this method we define the atoms inside the boundary as the grain size. The grain sizes at strain=40% states calculated by this method are given in figure 5. At strain rate 0.5%/ps, we got one grain with 215 atoms and one with 115 atoms, at strain rate 2%/ps, we still have one grain with 112 atoms, while at strain rate 5%/ps, we only found one little cluster with 13 atoms with a fcc structure. The size of grains decreases as the strain rates increases by two reasons, one is more nucleus can form, and the other is to release more stress energy, the deformed nanowire tends to form more twins, as the strain rates increases.

We can propose a deformation mechanism at high strain rates. Here the twin is considered as a unit, it's size decreases as the strain rate gets higher, after a critical strain rate, these units lose their coherent relation, and get smaller than 12, the number of nearest neighbors for fcc, a disordered system emerged. This is the amorphization behavior we observed at 5%/ps strain rate, when system flows like liquid after the transition.

This critical strain rate is a function of temperature and size difference for alloys. As AuNi, which has a large size mismatch, favors the glass formation, thus the critical strain rate to deform like liquid is much lower. We found at strain rate of 0.5%/ps is sufficient to obtain amorphous NiAu. The size mismatch destabilizes the crystalline state with respect to the amorphous state.

CONCLUSIONS

We studied the deformation behavior of Ni, CuNi and NiAu at high strain rate, from 0.05%/ps to 5%/ps. A new deformation mechanism at high strain rate is found. In systems where dislocations cannot be sustained, coherent shear of the atoms can create twins, and lower the elastic energy in the system. As the strain rate increases, the size of the twins decreases and the number of twins increases. Beyond the critical strain rate, amorphization occurs, and the specimen deforms uniformly.

ACKNOWLEDGMENTS
Financial support was provided by the Japanese Ministry of Education for HI, by DOE (DEFOG3 86ER45242) for KS and WLJ, by ARO (DAAH04-95-1-0233) for WLJ and WAG, and by NSF (CHE 95-22179 and ACR-92-17368) and DOE-ASCI for YQ, TC, and WAG. In addition, the facilities of the MSC are supported by grants from ARO-DURIP, BP Chemical, Exxon, Avery-Dennison, Owens-Corning, ARO-MURI, Asahi Chemical, Chevron, and Beckman Institute.

REFERENCES:

1. J.P. Hirth and J. Lothe, *Theory of Dislocations*, Wiley, New York, 1982

2. F.R.N. Nabarro, *Theory of Crystal Dislocations*, Oxford, Clarendon 1967

3. R.W. Cahn, Adv. Phy. **3**,363 (1954)

4. J.W. Christian and S. Mahajan, Prog. Mater. Sci. **39**, 1 (1995)

5. G.A. Ghadwick and D.A. Smith, *Grain Boundary Structure and Properties*, Academic Press, London, 1976

6. G.T. Gray III, Acta. Met. **36**, 1745 (1988)

7. T.H. Blewitt, R.R. Coltman and J.K. Redman, J. Appl. Phys. **28**, 651 (1957)

8. T. Kitamura, K. Yashiro and R. Ohtani, JSME **A 40**, 430 (1997)

9. J.R. Ray, Comput.. Phys. Rep. **8**, 109 (1988)

10. A.P. Sutton and J. Chen, Phil. Mag. Lett. **61**, 139 (1990); H. Rafii-Tabar and A.P. Sutton, Phil. Mag. Lett. **63**,217(1991)

11. Y. Kimura, T. Cagin, Y. Qi, and W. A. Goddard, III, Phys. Rev B1, submitted.

12. H. Ikeda, Y. Qi, T. Cagin, K. Samwer, W.L. Johnson and W.A. Goddard III , submitted

13. A.H. Cottrell and B.A. Bilby, Phil. Mag. **42**, 573(1951)

14. J.A. Venables, Phil. Mag. **6**, 379(1961)

15. G.E. Dieter, *Mechanical Metallurgy*, McGraw-Hill, New York, 1986

DEFORMATION BEHAVIOR IN La$_{55}$Al$_{25}$Ni$_{20}$ METALLIC GLASS

Y. KAWAMURA, T. NAKAMURA, A. INOUE
Institute for Materials Research, Tohoku University, Sendai 980-8577, Japan,
rivervil@imr.tohoku.ac.jp

ABSTRACT

We investigated the deformation behavior of the supercooled liquid in a La$_{55}$Al$_{25}$Ni$_{20}$ (at.%) metallic glass ribbon that has a wide supercooled liquid region of 65 K before crystallization. The deformation of the metallic glass was divided into three types, namely, homogeneous deformations with and without a stress overshoot and an inhomogeneous deformation. The supercooled liquid above the glass transition temperature (T_g) exhibited a Newtonian viscosity and superplastic-like behavior during isothermal tensile deformation. The metallic glass was elongated as much as 1000 % at high strain rates ranging from 10^{-2} s^{-1} to 10^0 s^{-1} without embrittlement. The maximum elongation to failure was in excess of 1800 % at a strain rate of 1.7×10^{-1} s^{-1} and at 503 K (T_g+20 K, 0.71T_m) under about 40 MPa.

INTRODUCTION

Recently, a number of metallic glasses that have a temperature range of supercooled liquid above 60 K before crystallization and high glass-forming ability have been discovered in Zr-, Pd-, La-, Mg- and Pt-based alloys [1-7]. We have previously reported that the Zr$_{65}$Al$_{10}$Ni$_{10}$Cu$_{15}$ and Pd$_{40}$Ni$_{40}$P$_{20}$ metallic glasses exhibit superplastic-like behavior at high strain rates of more than 1×10^{-2} s^{-1} in a supercooled liquid state [8,9]. The high-strain-rate superplasticity inherent in the supercooled liquid enables to produce a large-scale bulk metallic glass with full tensile strength by powder consolidation, and to work the bulk metallic glass into complex-shaped components [10-12]. In this study, we will report the deformation behavior of the supercooled liquid in a La$_{55}$Al$_{25}$Ni$_{20}$ (at.%) metallic glass that has a wide supercooled liquid region of 65 K.

EXPERIMENT

Ribbon samples that were produced by a single-roller melt-spinning method in an argon atmosphere were used for the investigation of the deformation behavior and the thermal properties. The size of the melt-spun ribbon was 1.4×0.04 mm^2. The formation of a single glassy phase was confirmed by X-ray-diffractometry. The thermal properties were measured by differential scanning calorimetry (DSC) measurements and differential thermal analysis (DTA). Tensile tests were conducted using an Instron-type tensile test machine at a constant crosshead velocity. Wide strain rates in the range of 5.0×10^{-4} to 8.3×10^{-1} s^{-1} were used. A silicone oil bath was used for heating the samples. The fluctuation of the temperature was about ± 0.2 K during the tensile tests, which is much smaller than that of an electric furnace. The gauge length of the tensile specimens was 10 mm. The tensile tests were started after allowing the sample to equilibrate for 200 s. Ductility of the samples was examined by 180 degrees bend tests at room temperature. The sample that enables to be bent through 180 degrees without fracture was defined as a ductile sample.

RESULTS

Thermal Properties

Figure 1 shows a DSC curve of the La$_{55}$Al$_{25}$Ni$_{20}$ metallic glass at a scanning rate of 0.67 K/s. The glass transition temperature (T_g), crystallization temperature (T_x) and supercooled liquid region ($\Delta T_x = T_x - T_g$) were 483 K, 548 K and 65 K, respectively. The melting temperature (T_m)

373

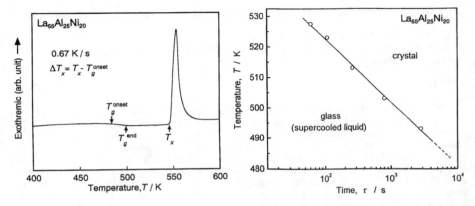

Figure 1. Differential scanning calorimetric curve of $La_{55}Al_{25}Ni_{20}$ metallic glass at a heating rate of 0.67 K/s.

Figure 2. Time-temperature-transformation diagram for the onset of the crystallization in $La_{55}Al_{25}Ni_{20}$ metallic glass.

measured by DTA was 704 K. Accordingly, the temperatures of T_g and T_x correspond to $0.68T_m$ and $0.77T_m$, respectively. Figure 2 shows the time-temperature-transformation (T-T-T) diagram for the onset of crystallization. The duration for retaining the glassy phase without decomposition at the glass transition temperature was estimated to be about 1×10^4 s. The supercooled liquid was found to have a high thermal stability.

Deformation Types

Figure 3 shows nominal stress-strain curves of the $La_{55}Al_{25}Ni_{20}$ metallic glass at a strain rate of 5.0×10^{-4} s^{-1}. The deformation mode of the metallic glass was divided into two types, namely, inhomogeneous mode and homogeneous one [13]. Moreover, there are two types of the stress-strain curves in the homogeneous deformation mode. At lower temperatures and

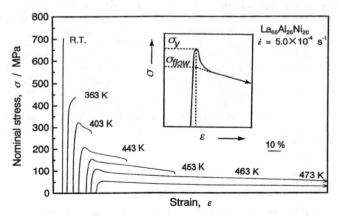

Figure 3. Nominal stress-strain curves of $La_{55}Al_{25}Ni_{20}$ metallic glass. The flow stress is defined as shown in the inset.

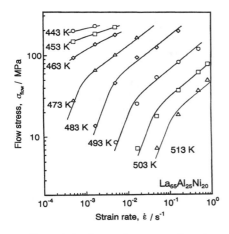 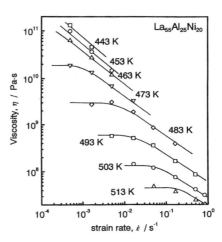

Figure 4. Strain-rate dependence of the flow stress in La$_{55}$Al$_{25}$Ni$_{20}$ metallic glass.

Figure 5. Strain-rate dependence of the viscosity in La$_{55}$Al$_{25}$Ni$_{20}$ metallic glass.

higher strain rates, the stress-strain curves were accompanied by a stress overshoot that is a transient stress-strain phenomenon [14]. After the stress overshoot, the flow reaches a steady state, namely, the plateau of stress in the stress-strain curves. The stress overshoot is dependent on stress-relaxation and strain-rate change in the course of stretch [14]. However, the steady flow stress is identical at each temperature and strain rate. On the other hand, at higher temperatures and lower strain rates, the stress transformed directly to a steady flow state without a stress overshoot.

Flow Stress and Viscosity

Figure 4 shows the relationship between the steady flow stress and strain rate for the La$_{55}$Al$_{25}$Ni$_{20}$ metallic glass. The curves exhibited a sigmoidal shape and shifted towards higher strain rates with temperature. It is clear that the supercooled liquid above the glass transition temperature enabled to deform at high strain rates above 1.0×10^{-2} s^{-1} under low flow stresses below 100 MPa. The flow stress was about 40 MPa at 503 K (T_g+20 K) and at 1.7×10^{-1} s^{-1} where the maximum elongation of 1800 % was obtained. Figure 5 shows the strain-rate dependence of the viscosity for the La$_{55}$Al$_{25}$Ni$_{20}$ metallic glass. The viscosity was simulated by the equation $\eta = \sigma_{flow} / 3\dot{\varepsilon}$ [15]. At lower strain rates, the viscosity of the supercooled liquid was essentially independent of the strain rate, showing a Newtonian flow. The higher strain rates led to a decrease in the viscosity which shows the transition from the Newtonian flow to a non-Newtonian one. This tendency was shifted towards higher strain rate and lower viscosity sides with temperature. One should not overlook that the Newtonian flow was observed even at high strain rates above 1.0×10^{-2}. The viscosity of the supercooled liquid was 7.7×10^{7} Pa·s at 503 K (T_g+20 K) and at 1.7×10^{-1} s^{-1} where the maximum elongation to failure was obtained. Strain-rate sensitivity exponent, namely, m value is usually represented by $m = \Delta \log \sigma_{flow} / \Delta \log \dot{\varepsilon}$ [16]. The m value was nearly 1.0 in the Newtonian flow, and was 0.4-0.5 in the non-Newtonian one. Figure 6 shows the plot of the Newtonian viscosity, namely, zero shear viscosity against the reciprocal of temperature. Although it is well known that viscosity of metallic glasses is represented by the Vogel-Fulcher formulation [17], the viscosity was fitted by the Arrhenius' equation. Its activation energy was estimated to be 330 kJ/mol.

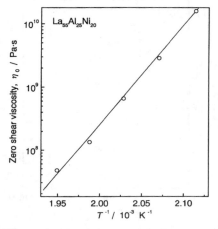

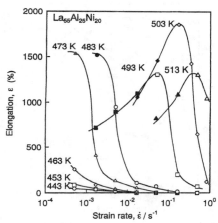

Figure 6. Newtonian flow viscosity of La$_{55}$Al$_{25}$Ni$_{20}$ metallic glass as a function of temperature.

Figure 7. Strain-rate dependence of the elongation to failure in La$_{55}$Al$_{25}$Ni$_{20}$ metallic glass. The samples indicated with the open and solid symbols exhibit the ductile and brittle nature, respectively, at ambient temperature after the tensile test.

Elongation to Failure

Figure 7 shows the strain-rate dependence of the elongation to failure at various temperatures. The metallic glass elongated significantly in the homogeneous deformation region in which the elongation was dependent on both temperature and strain rate. As the strain rate increased, the elongation increased, reached a peak and then decreased. The peak in the elongation of the supercooled liquid was shifted towards higher strain rates with temperature. The supercooled liquid exhibited elongations exceeding 1000 % over a wide range of strain rate (10^{-4} to 10^0 s^{-1}). Figure 8 shows the photograph of the La$_{55}$Al$_{25}$Ni$_{20}$ metallic glass ribbon deformed to an

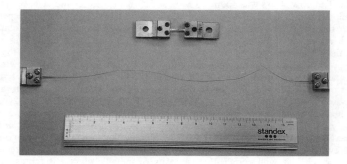

Figure 8. Photograph of La$_{55}$Al$_{25}$Ni$_{20}$ metallic glass before and after deformation of 1800 % at an initial strain rate of 1.7×10^{-1} s^{-1} and at 503 K.

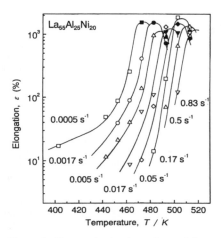 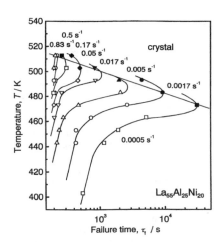

Figure 9. Changes in the elongation to failure for $La_{55}Al_{25}Ni_{20}$ metallic glass as a function of temperature. The samples indicated with the open and solid symbols exhibit the ductile and brittle nature, respectively, at ambient temperature after the tensile test.

Figure 10. Relationship between the failure time and temperature in $La_{55}Al_{25}Ni_{20}$ metallic glass. The T-T-T diagram of the crystallization shown in Fig. 2 is represented for reference. The samples indicated with the open and solid symbols exhibit the ductile and brittle nature, respectively, at ambient temperature after the tensile test.

elongation of 1800 % at an initial strain rate of 1.7×10^{-1} s^{-1} and at 503 K (T_g +20 K). The strain rate in this study is expressed as the initial strain rate. The strain rate decreased with stretching the samples because the speed of the crosshead was constant. The elongation is, therefore, expected to become much larger by controlling the crosshead speed so as to keep the constant strain rate. The samples that are indicated with an open symbol in Fig. 7 exhibited a ductile nature at room temperature after the tests. The metallic glass was elongated as much as 1000 % at high strain rates ranging from 10^{-2} to 10^{0} s^{-1} without embrittlement. This means that the metallic glass can be deformed at high-strain-rates without losing the original mechanical properties. On the other hand, the samples represented with a solid symbol in Fig. 7 had a brittle nature at room temperature after the tests. Figure 9 shows the temperature dependence of the elongation to failure. The temperature dependence of the duration to failure at various strain rates is also shown in Fig. 10. The duration was calculated by $\tau_f = \varepsilon / \dot{\varepsilon}$ + 200. The open and solid symbols in Figs. 9 and 10 represent that the fractured samples had ductile and brittle nature, respectively, at ambient temperature. The elongation decreased at higher temperature side, where the samples were brittle. The duration to failure of the embrittled samples was found in a line approximately parallel to the T-T-T line of the crystallization shown in Fig. 2. The decrease in the elongation at higher temperature or lower strain rate sides in Figs. 7 and 9, therefore, seems to be due to the crystallization during the tensile tests.

Superplasticity

Superplastic materials usually exhibit large elongations to failure in excess of 200 % for deformation in an uniaxial tension, and m values greater than 0.3 [16]. Especially, the superplasticity observed at strain rates higher than 1.0×10^{-2} s^{-1} is termed as "high-strain-rate superplasticity". The supercooled liquid of the $La_{55}Al_{25}Ni_{20}$ metallic glass exhibited the same deformation behavior as the high-strain-rate superplastic materials. Although the word of "superplasticity" is defined to be applicable to polycrystalline solids, it can be said that the metallic glasses with a wide supercooled liquid region are a new type of high-strain-rate superplastic materials [8,9]. These metallic glasses, moreover, have high strength at room temperature. The superplastic deformation of the metallic glasses, therefore, seems to promise for future development of advanced materials with high strength and good workability.

CONCLUSIONS

We have investigated the deformation behavior of the supercooled liquid in a $La_{55}Al_{25}Ni_{20}$ metallic glass with a wide supercooled liquid region of 65 K. The deformation of the metallic glass transferred from an inhomogeneous mode to a homogeneous one with a stress overshoot and then changed to a homogeneous mode without a stress overshoot with increasing temperature or decreasing strain rate. The supercooled liquid above the glass transition temperature exhibited a Newtonian flow and superplastic-like behavior during isothermal tensile deformation at strain rates in the range of 10^{-4} s^{-1} to 10^0 s^{-1}. The metallic glass, moreover, was elongated as much as 1000 % at high strain rates ranging from 10^{-2} s^{-1} to 10^0 s^{-1} without embrittlement. The maximum elongation to failure was 1800 % at a strain rate of 1.7×10^{-1} s^{-1} and at 503 K (T_g+20 K, $0.71T_m$) where the flow stress was about 40 MPa.

ACKNOWLEDGEMENTS

The author (Y.K.) is grateful to the Grant-in-Aid for scientific Research on Priority Areas of the Ministry of Education, Science and Culture, Japan for the support of this research.

REFERENCES

1. A. Inoue, T. Zhang and T. Masumoto, Mater. Trans. JIM **31**, 17 (1990).
2. A. Inoue, T. Zhang and T. Masumoto, Mater. Trans. JIM **36**, 391 (1995)..
3. A. Inoue, N. Nishiyama and T. Matsuda, Mater. Trans. JIM **37**, 181 (1996).
4. A. Inoue and J. S. Gook, Mater. Trans. JIM **36**, 1180 (1995).
5. A. Peker and W. L. Johnson, Appl. Phys. Lett. **63**, 2342 (1993).
6. A. J. Drehman, A. L. Greer and D. Turnbull, Appl. Phys. Lett. **41**, 716 (1982).
7. H. S. Chen, Acta. Metall. **22**, 1505 (1974).
8. Y.Kawamura, T.Shibata, A.Inoue and T.Masumoto, Scripta Metall. Mater. **37**, 431 (1997).
9. Y. Kawamura, T. Nakamura and A. Inoue, Scripta Metall. Mater. **39**, 301 (1998).
10. Y. Kawamura, H. Kato, A. Inoue and T. Masumoto, Appl. Phys. Lett. **67**, 2008 (1995).
11. Y. Kawamura, H. Kato, A.Inoue and T. Masumoto, Int. J. Powder Metall. **33**, 50 (1997).
12. Y. Kawamura, T. Shibata, A. Inoue and T. Masumoto, Acta Mater. **37**, 253 (1998).
13. Y. Kawamura, T. Shibata, A. Inoue and T. Masumoto, Appl. Phys. Lett. **69**, 1208 (1996).
14. Y. Kawamura, T. Shibata, A. Inoue and T. Masumoto, Appl. Phys. Lett. **71**, 779 (1997).
15. F.A. McClintock and A.S. Argon, *Mechanical Behavior of Materials*, (Addison-Wesley Reading, MA, 1966), p. 290.
16. J. Pilling and N. Ridley, *Superplasticity in Crystalline Solids*, (The Institute of Metals, London, 1989).
17. G.S. Fulcher, J. Am. Ceram. Soc. **6**, 339 (1925).

SUPERPLASTIC DEFORMATION AND VISCOUS FLOW
IN AN Zr-BASED METALLIC GLASS AT 410°C

T.G. Nieh*, J.G. Wang*, J. Wadsworth*, T. Mukai**, C.T. Liu***
*Lawrence Livermore National Laboratory, Box 808, Livermore CA 94551, nieh1@llnl.gov
**Osaka Municipal Technical Research Institute, Osaka 536, Japan
***Oak Ridge National Laboratory, P.O. Box 2008, Oak Ridge TN 37831

ABSTRACT

The thermal properties of an amorphous alloy (composition in at.%: Zr–10Al–5Ti–17.9Cu–14.6Ni), and particularly the glass transition and crystallization temperature as a function of heating rate, were characterized using Differential Scanning Calorimetry (DSC). X-ray diffraction analyses and Transmission Electron Microscopy were also conducted on samples heat-treated at different temperatures for comparison with the DSC results. Superplasticity in the alloy was studied at 410°C, a temperature within the supercooled liquid region. Both single strain rate and strain rate cycling tests in tension were carried out to investigate the deformation behavior of the alloy in the supercooled liquid region. The experimental results indicated that the alloy did not behave like a Newtonian fluid.

INTRODUCTION

Metallic glasses fabricated by rapid quenching from the melt were first discovered in 1960 [1]. Because of the high quench rate requirements ($10^4 – 10^6$ K/s), only thin ribbons and sheets with a thickness less than 0.1 mm could be fabricated. One of the most important recent developments in the synthesis of amorphous materials is the discovery that certain metallic glasses can be fabricated from the liquid state at cooling rates of the order of 10 K/s. This enables the production of bulk amorphous alloys with a thickness of ~10 mm. While advances in amorphous metallic alloy development have been impressive, they have been made largely through empirical approaches. Amorphous alloys in bulk form can be characterized, in fact, by some empirical rules [2]: (a) multi-components (at least three, and very often five components), (b) a large difference in atomic size, (c) "deep" eutectics in phase diagrams of the alloys, and (d) a large value of $\Delta T = T_x - T_g$ (T_x and T_g are the crystallization and glass transition temperatures, respectively).

Bulk amorphous alloys have many potential applications resulting from their unique properties. Among those worth mentioning are: superior strength and hardness, excellent corrosion resistance [3], shaping and forming in a viscous state [4], reduced sliding friction and improved wear resistance [5], and extremely low magnetic energy loss. These properties should lead to applications in the fields of near-shape fabrication by injection molding and die casting, coatings, joining and bonding, biomedical implants, and synthesis of nanocrystalline and composite materials [2]. Also, the development of bulk amorphous alloys with soft magnetic properties would be extremely beneficial for energy conservation.

The mechanical behavior of metallic glasses is characterized by either inhomogeneous or homogeneous deformation. Homogeneous deformation usually takes place at high temperatures ($>0.7 T_g$) [6], and the material can often exhibit significant plasticity. The transition temperature T_{tr} from inhomogeneous to homogeneous deformation (or brittle-to-ductile transition) is strongly dependent upon strain rate. For example, T_{tr} for a $Zr_{65}Al_{10}Ni_{10}Cu_{15}$ alloy is about 533K (corresponding to $0.82 T_g$) at 5×10^{-4} s^{-1}, but is 652K (corresponding to $1.0 T_g$) at 5×10^{-2} s^{-1} [4]. The strain rate dependence of T_{tr} (480–525K; 0.61–0.75T_g) has also been demonstrated in $Fe_{40}Ni_{40}B_{20}$ [7]. These results suggest that homogeneous deformation is associated with certain diffusional relaxation processes.

As early as 1980, Homer and Eberhardt [8] reported the observation of superplasticity in amorphous $Pd_{78.1}Fe_{5.1}Si_{16.8}$, in ribbon form, ($T_g$=395°C, T_x=410°C) during non-isothermal creep experiments. In their experiments, test samples were rapidly heated to the maximum temperature of 425°C under a constant load (stress range: 25–150 MPa). The resulting creep rate was rather high; for example, an applied stress of 150 MPa produced a creep rate of 0.5 s^{-1}. The strain rate sensitivity value was estimated to be about one, suggesting possible Newtonian flow. Since the test temperature of 425°C was higher than T_x, a dispersion of 0.4 µm crystalline grains in an amorphous matrix was observed in the alloy after superplastic deformation. It is of particularly importance to note that slow heating during creep testing resulted in the disappearance of superplasticity. This was apparently caused by the structural instability of this alloy, as indicated by a narrow ΔT (=15°C), and correspondingly fast crystallization kinetics.

Zelenskiy *et al.* [9] studied the formability of amorphous $Co_{68}Fe_7Ni_{13}Si_7B_5$ (T_g=563°C, T_x=583°C) at temperatures between 500 and 640°C. They observed large tensile ductility at a relatively fast strain rate of 10^{-2} s^{-1} in the 550–580°C temperature range (i.e. within the supercooled liquid region). Specifically, the maximum elongation of 180% was recorded at a corresponding minimum stress of about 150 MPa. However, the strain rate sensitivity was not measured. TEM microstructural examination indicated that annealing at 580°C resulted in the precipitation of nanometer grains (~50–70 nm) in the alloy. From these results, the authors argued that the presence of a large grain/amorphous matrix interfacial area was necessary for superplasticity. However, we want to point out that this may not be the case. In fact, several pieces of experimental evidence have indicated a reduced ductility in the presence of nanograins [4, 8]. A recent study of Busch et al [10] also showed that the viscosity of a metallic glass increases sharply once it is heated over the crystallization temperature. As expected, this increase in viscosity leads to increased resistance to plastic flow and, thus, a decrease in ductility.

To further understand superplasticity and extended plasticity in metallic glasses, Khonik and Zelenskiy [11] analyzed the available mechanical data from fifteen different metallic glasses, including both metal-metal and metal-metalloid systems. They made several important observations. First, superplasticity occurs in alloys with a large ΔT; typically, about several tens of degrees. The larger the ΔT (=T_x–T_g), the larger the tensile elongation, provided tests were conducted in the supercooled liquid region. This indicates the importance of thermal stability of a metallic glass during superplastic deformation. They also noticed that an increased heating rate usually produced an increased elongation. Apparently, this is associated with structural stability since slow heating rates result in an earlier onset of crystallization.

In studying the formability of a $La_{55}Al_{25}Ni_{20}$ alloy, Inoue *et al* [12, 13] reported that the alloy in the supercooled liquid range (480-520K) behaved like a Newtonian fluid, i.e., m=1. A tensile elongation of over 15,000% was recorded at 500K at a strain rate of 10^{-1} s^{-1}. However, a careful examination of their stress-strain rate data indicated that the strain rate sensitivity tends to decrease to less than one when the testing temperature (e.g. 510, 520K) approaches T_x. Again, this is due to a partial crystallization in the amorphous structure during testing.

Kawamura *et al* [4] recently studied the high-temperature deformation of a $Zr_{65}Al_{10}Ni_{10}Cu_{15}$ metallic glass with a wide range of ΔT (T_g=652K, T_x=757K). In the supercooled liquid region, they found that plastic flow was strongly dependent on strain rate and the strain rate sensitivity value exceeded 0.8, but was less than unity. The high strain rate sensitivity produced a corresponding high tensile elongation. For example, a tensile elongation of 340% was obtained at a strain rate of 5×10^{-2} s^{-1} and at 673K. However, true Newtonian behavior (m=1) was not observed in the alloy. As discussed before, the structure of a supercooled liquid is thermally unstable. Upon thermal exposure, and particularly under an external applied stress, the amorphous structure tends to crystallize. It is believed that the observation of an m value of less than 1 can be attributed to the fact that the deformation mode of the alloy, to a first-order approximation, is a mixture of Newtonian flow (m=1) and fine-grained superplasticity (m=0.5). Under appropriate testing conditions, both deformation modes are expected to give rise to a large tensile elongation.

To investigate further superplasticity in metallic glass systems, Kawamura *et al* [4] tested a $Pd_{40}Ni_{40}P_{20}$ alloy prepared by rapid solidification. Within the range of 560-620K, the alloy exhibited similar deformation behavior to that of $Zr_{65}Al_{10}Ni_{10}Cu_{15}$; namely, a high-strain-rate-sensitivity value accompanied by extended tensile ductility in the supercooled liquid region (T_g=578-597K, T_x=651K). In contrast to $Zr_{65}Al_{10}Ni_{10}Cu_{15}$ which is non-Newtonian, $Pd_{40}Ni_{40}P_{20}$ can behave like a true Newtonian fluid (i.e. m=1) under appropriate testing conditions. The difference may be associated with the fact that $Pd_{40}Ni_{40}P_{20}$ is more thermally stable than $Zr_{65}Al_{10}Ni_{10}Cu_{15}$ in the supercooled liquid state, as pointed out by Kawamura *et al* [14]. (ΔTs are 72 and 100K for $Zr_{65}Al_{10}Ni_{10}Cu_{15}$ and $Pd_{40}Ni_{40}P_{20}$, respectively.) Therefore, during high-temperature deformation, $Pd_{40}Ni_{40}P_{20}$ can retain its amorphous state, whereas crystallization may already have taken place in $Zr_{65}Al_{10}Ni_{10}Cu_{15}$. This is indirectly indicated by the fact that the viscosity of $Pd_{40}Ni_{40}P_{20}$ is about one order of magnitude lower than that of $Zr_{65}Al_{10}Ni_{10}Cu_{15}$.

Despite the fact that metallic alloys still have a high strain rate sensitivity at a temperature that approaches T_x, the tensile elongation reduces rapidly. This appears to be contrary to the general idea that an ultrafine grain size alloy is expected to give rise to a large elongation, presumably as a result of extensive grain boundary sliding. However, it must be pointed out that in the case of grain boundary sliding, the sliding strain must be properly accommodated either by diffusional flow or by dislocation slip (e.g., climb or glide) across neighboring grains, in order to prevent cavitation and, thus, fracture [15]. Dislocation slip in an ordered, multicomponent intermetallic compound is difficult even at temperatures near T_x (~0.8 T_m, where T_m is the melting point of the alloy). Also, at a relatively high strain rate of ~10^{-2} s^{-1}, diffusional processes are not expected to be sufficiently fast to accommodate sliding strain. This may offer an explanation for a high m value but an associated low tensile elongation in metallic glasses at temperatures in the vicinity of T_x. It is noted that, for the case of pure Newtonian flow, no such accommodation is needed.

In summary, large tensile ductilities can be obtained in the supercooled liquid region from a metallic glass with a large ΔT. The maximum ductility is expected to occur at a temperature near T_x, where the flow stress (or viscosity) is low, and at a high strain rate at which the alloy can retain its amorphous structure during deformation. The purpose of this paper is to investigate the deformation behavior of an amorphous Zr-10Al-5Ti-17.9Cu-14.6Ni alloy in the supercooled liquid region.

EXPERIMENTS

The material used in the present study has a composition of Zr-10Al-5Ti-17.9Cu-14.6Ni, and was developed at Caltech [16]. Zone-purified Zr bars (containing 12.3 appm O and 10 appm Hf), together with pure metal elements, were used as charge materials. The alloys were prepared by arc melting in inert gas, followed by drop casting into 7.0-mm-diameter by 7.2cm-long Cu molds at Oak Ridge National Laboratory. The details of fabrication of the alloy have been described previously [16].

Tensile sheet specimens were fabricated from the as-cast material by means of electrical discharge machining. They had a gage length of 4.76 mm, a thickness of 1.27 mm and a width of 1.59 mm, as shown in FIG01. Tensile tests were conducted using an Instron machine equipped with an air furnace. Due to structural instability during testing of samples at high temperatures, the heating rate must be rapid to minimize crystallization. Typically, the heating plus holding time prior to testing was about 25 minutes. For example, for test at 410°C and a constant strain rate of 10^{-2} s^{-1}, the temperature profile was: 305°C (5 min), 371°C (10 min), 397°C (15 min), 407°C (20 min), 410°C (23 min), and 410°C (24 min). Constant strain rate tests were performed with a computer-controlled machine at a strain rate of 10^{-2} s^{-1}. Strain-rate-cycling tests were also performed between 10^{-3} and 10^{-2} s^{-1} to measure strain rate sensitivity exponents.

Differential scanning calorimetry (DSC) was used to characterize the thermodynamic and kinetic properties of the amorphous alloy. Various heating rates of 10, 20, 40, and 80K per minute were used to examine the glass transition and crystallization temperatures as a function of heating rate.

FIG01 Amorphous Zr–10Al–5Ti–17.9Cu–14.6Ni alloy samples fractured in tension in the supercooled liquid region. An untested sample is also shown for comparison.

RESULTS AND DISCUSSION

Differential scanning calorimetry (DSC) curves obtained from the amorphous Zr–10Al–5Ti–17.9Cu–14.6Ni alloy at different heating rates are shown in FIG02. The temperatures for the start (T_g^s) and end (T_g^e) of glass transition, and crystallization (T_x), and supercooled range (ΔT) are listed in Table 1. As expected, these temperatures are all dependent upon the heating rate. In fact, they all increase with an increasing heating rate.

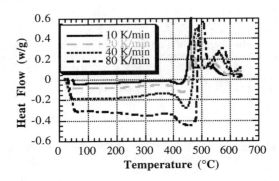

FIG02 DSC curves measured from amorphous Zr–10Al–5Ti–17.9Cu–14.6Ni alloy at different heating rates.

Table 1 Temperatures for the start (T_g^s) and end (T_g^e) of glass transition, crystallization (T_x), and supercooled range (ΔT) for amorphous Zr–10Al–5Ti–17.9Cu–14.6Ni alloy

Heating rate, (K/min)	T_g^s, K	T_g^e, K	T_x, K	ΔT, K
10	628	696	714	86
20	631	705	729	98
40	637	716	742	105
80	640	728	758	118

The stress-strain curve of the alloy at 410°C and a strain rate of 10^{-2} s^{-1} is shown in Fig. 3. After an initial yield drop, there is an apparent steady-state flow region which is followed by a sharp decrease in flow stress until the final fracture. The initial yield drop usually occurred during condition of high temperature and high strain rate deformation; this behavior has also been observed by Kawamura et al [17] who attributed it to a transient phenomenon. The fracture strain is about 1.3 which corresponds to approximately 250% tensile elongation. The fractured sample is noted to exhibit gradual necking, as shown in FIG01, almost necking down to a point. The final decrease in flow stress is, therefore, not a result of softening, but from a reduction in load bearing capacity. The fracture appearance is in contrast to that observed in a $Zr_{65}Al_{10}Ni_{10}Cu_{15}$ metallic glass [4], in which a uniform deformation was observed. This difference may be caused by the fact that the samples used by Kawamura et al were very thin (0.02 mm). As a result, the samples were subject to a plane stress condition. It is of interest to note that the maximum stress is about 700 MPa; this is considerably higher than the flow stresses reported for any existing metal or ceramic exhibiting superplasticity or extended ductility. The flow stress for a superplastic metal or ceramic is typically only about 70 MPa.

To characterize the deformation behavior, a strain-rate-cycling test was carried out to measure the strain rate sensitivity value. The result is shown in FIG04, the values of strain rate sensitivity m in equation $\sigma = Kr^m$, where r is strain rate, σ is stress, and K is a constant, were measured by strain rate cycling tests between 10^{-2} and 7×10^{-3} s^{-1}. Some observations are readily made. There is no steady state after each strain rate cycle, making it difficult to determine accurately the strain rate sensitivity. This difficulty can probably be attributed to structural instability during testing. It is noted that, despite the fact that 410°C is below the crystallization temperature, an external applied stresses can promote crystallization in amorphous alloys [18]. Thus, it is believed that some nano-scale, crystallized phase evolved during the course of the test. It is known that the presence of nanocrystalline phases can significantly affect the mechanical properties of an metallic glass. For example, Busch et al [10] recently showed that the presence of crystalline phases increases the viscosity of a $Zr_{46.75}Ti_{8.25}Cu_{7.5}Ni_{10}Be_{27.5}$ metallic glass. This observation is also consistent with the results of Kim et al [19, 20] who reported that the fracture strength of an amorphous $Al_{88}Ni_{10}Y_2$ was doubled if the alloy was crystallized and contained 5-12 nm Al particles. Therefore, in the present strain rate cycle test, a continuous strengthening is proposed to be a result of the continuous precipitation of nanocrystals in the amorphous matrix. In fact, this is also reflected by a slight increase in stress after the initial yield drop (strain >0.4) shown in FIG03.

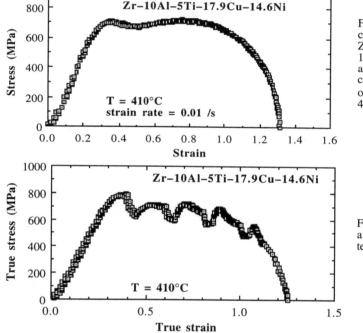

FIG03 Stress-strain curve for amorphous Zr–10Al–5Ti–17.9Cu–14.6Ni alloy tested at a constant strain rate of 10^{-2} s^{-1} and at 410°C.

FIG04 Results from a strain-rate-cycling test at 410°C.

It is noted in FIG04 that, after each strain rate decrease, except for the first one, there is no steady-state flow region. The gradual decrease in flow stress after decreasing the strain rate results from sample necking. Data from FIGs03 and 04 indicate that significant sample necking occurs at a strain of approximately 0.7-0.8, as discussed previously. The fracture strain obtained in the strain-rate-cycling test is similar to that in a constant strain rate test.

From the above results, it may be tentatively concluded that the strain-rate-cycling test, which is commonly used for measuring strain rate sensitivity values, and thus inferring deformation mechanism in superplastic polycrystalline alloys, may be an inappropriate method for metallic glasses. This is because the structure of a metallic glass is unstable compared to that of a fine-grained superplastic alloy, especially under stress.

It is worth noting that from FIG04 the average "apparent" strain rate sensitivity for the present alloy is computed to be about 0.5. Although structural instability can contribute to some variations in determining the "true" strain rate sensitivity value, its influence is not expected to be sufficiently great to imply a "true" strain rate sensitivity value of as high as one. In other words, the present alloy does *not* behave like a Newtonian fluid. The non-Newtonian behavior is also reflected by the fracture appearance of tested samples (FIG01). The deformation behavior of an ideal Newtonian fluid is, in principle, uniform and would not be expected to exhibit significant local necking.

SUMMARY

The glass transition and crystallization temperatures of an amorphous alloy (composition in at.%: Zr–10Al–5Ti–17.9Cu–14.6Ni) are dependent upon the heating rate; the faster the heating rate, the higher the temperature. The deformation behavior of the alloy at 410°C (within the supercooled liquid region) was characterized. The alloy was found to exhibit a large tensile elongation of over 250% at a high strain rate of 10^{-2} s^{-1}. Due to structural instability it is difficult to determine the "true" strain rate sensitivity value. However, preliminary experimental results indicated that the alloy does not behave like a Newtonian fluid (m=1). This is supported by the observation that tensile samples deformed non-uniformly and exhibited macroscopic necking.

ACKNOWLEDGMENT

This work was performed under the auspices of the U.S. Department of Energy by Lawrence Livermore National Laboratory under contract No. W-7405-Eng-48, and under contract DE-AC05-96OR22464 with Lockheed-Martin Energy Research Corp.

REFERENCES

1. W. Klement, R. Willens, and P. Duwez, *Nature*, **187** (1960) 869.
2. W.L. Johnson, in *Metal Handbook 10th Edition, Vol.10*, p. 804, ASM International, Metals Park, OH, 1988.
3. K. Hashimoto, in *Current Topics in Amorphous Materials: Physics and Technology*, p. 167, edited by Y. Sakurai et al, Elsevier Science Publishers B.V., 1993.
4. Y. Kawamura, T. Nakamura, and A. Inoue, *Scr. Mater.*, **39** (1998) 301.
5. D.G. Morris, in *Proc. 5th Int'l Conf. on Rapidly Quenched Metals*, p. 1775, edited by S. Steeb and H. Warlimont, Elsevier Science Publishers B.V., 1985.
6. A.S. Argon, *Acta Metall.*, **27** (1979) 47.
7. A.L. Mulder, R.J.A. Derksen, J.W. Drijver, and S. Radelaar, in *Proc. 4th International Conf. on Rapidly Quenched Metals*, p. 1345, edited by T. Masumoto and K. Suzuki, Japan Institute of Metals, Sendai, Japan, 1982.
8. C. Homer and A. Eberhardt, *Scr. Metall.*, **14** (1980) 1331.
9. V.A. Zelenskiy, A.S. Tikhonov, and A.N. Kobylkin, *Russian Metallurgy*, **4** (1985) 152.
10. R. Busch, E. Bakke, and W.L. Johnson, *Acta Mater.*, **46** (1998) 4725.
11. V.A. Khonik and V.A. Zelenskiy, *Phys. Met. Metall.*, **67** (1989) 196.
12. A. Inoue, Y.H. Kim, and T. Masumoto, *Mater. Trans., JIM*, **33** (1992) 487.
13. A. Inoue and Y. Satome, *Metals*, **3** (1993) 51.
14. Y. Kawamura, H. Kato, A. Inoue, and T. Masumoto, *Appl. Phys. Lett.*, **67** (1995) 2008.
15. T.G. Nieh, J.N. Wang, L.M. Hsiung, J. Wadsworth, and V. Sikka, *Scr. Mater.*, **37** (1997) 733.
16. X.H. Lin, W.L. Johnson, and W.K. Rhim, *Mater. Trans. JIM*, **38(5)** (1997) 473.
17. Y. Kawamura, T. Shibata, A. Inoue, and T. Masumoto, *Appl. Phys. Lett.*, **69** (1996) 1208.
18. R. Maddin and T. Masumoto, *Mater. Sci. Eng.*, **9** (1972) 153.
19. Y.H. Kim, K. Hiraga, A. Inoue, T. Masumoto, and H.H. Jo, *Mater. Trans. JIM*, **35(5)** (1994) 293.
20. Y.H. Kim, A. Inoue, and T. Masumoto, *Mater. Trans. JIM*, **31** (1990) 747.

MICROFORMING OF MEMS PARTS WITH AMORPHOUS ALLOYS

Yasunori Saotome*, Tao Zhang**, Akihisa Inoue**
* Dept of Mechanical Eng., Gunma University, Tenjin-cho 1-5-1, Kiryu, Gunma376-8515,
**Inst. for Materials Research, Tohoku University, Katahira 2-1-1, Aoba, Sendai 980-8577,
JAPAN.

ABSTRACT

Microformability of new amorphous alloys in the supercooled liquid state and microforming techniques for the materials are shown. In the supercooled liquid state, the materials reveal perfect Newtonian viscous flow characteristics and furthermore exhibit an excellent property of microformability on a submicron scale. As for microforming techniques, microforging and micro extrusion of amorphous alloys are introduced in addition to the fabrication method of micro dies of photochemically machinable glass. As a result, amorphous alloys are expected as one of the most useful materials to fabricate micromachines.

1. INTRODUCTION

There are two major requirements of micromachine materials. First, material characteristics must be exploitable on a microscopic scale in addition to conventional macroscopic scales. Second, forming, processing and machining methods to form micro geometries must be established. Methods of creating micro geometries include complex processes, such as silicone process and LIGA process, and selecting materials are two sides of the same coin. To make micromachine elements, especially structural elements, suitable materials and micromachining technology must be developed. Amorphous alloys are highly useful for realizing high-performance micro actuators and structures due to their excellent characteristics as functional or structural materials, including isotropic homogeneity free from crystalline anisotropy. The present report introduces new materials which are second-generation amorphous alloys and their microforming properties, as well as micromachine production methods for the materials.

2. CHARACTERISTIC DEFORMATION BEHAVIOR

2.1 Viscous Flow in the Supercooled Liquid State

If amorphous alloys prepared by melt quenching are reheated, the alloys usually crystallize at temperature T_{XA} after the structural relaxation at temperature T_R. However, new amorphous alloys have been developed by Inoue[1], et al. These materials reveal obvious glass transition behavior at temperature T_g below crystallization temperature T_{XB} and develop the supercooled liquid state in a wide temperature range of ΔT_X to 100K. Furthermore, the materials are characterized by their low critical cooling rates for amorphization; for example, this fact enables producing a 10mm diameter amorphous solid bar. The characteristic behavior of these amorphous alloys in the supercooled liquid state has therefore been studied with the newly developed test apparatus. The equipment has a unique method for heating the foil specimen. Inert gas (Ar) is heated and fed into the furnace, and the specimen is heated and controlled to the test temperature. In addition, the furnace has the double cylindrical structure. The specimen temperature inside the pipe can be stabilized by flowing heated air into an outer pipe. The strain of the specimen was measured under various tensile loads, and the results are shown in Fig.1, where the relationship between applied stress σ and strain rate ε is plotted.

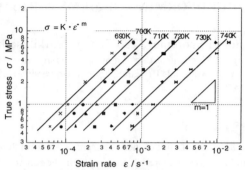

Fig.1 Characteristic flow curves of Zr55Al10Cu30Ni5 amorphous alloy in the supercooled liquid state.

In the figure, the inclination of each straight line, that is, the strain rate sensitivity exponent, is m=1 in the equation $\sigma = K\dot{\varepsilon}^m$ for deformation at high temperatures. This shows that the deformation at each temperature in the supercooled liquid region is a complete Newton viscous flow. The relation can also be shown as $\sigma = \lambda\dot{\varepsilon}$ by using a normal viscosity coefficient of λ. As for the temperature dependence of the deformation, the strain rate increases with the rise of the deformation temperature. This means the normal viscosity coefficient λ decreases as shown in Fig. 2.

As for temperature dependence in a viscous liquid, the coefficient of viscosity decreases exponentially with temperature in general. This relation is expressed by the well known Andrade equation $\lambda = Ae^{B/T}$.

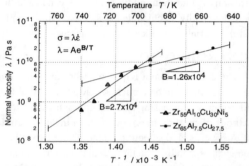

Fig.2 Temperature dependence of normal viscosity l in the supercooled liquid state.

Here, A and B are constants and T is the absolute temperature. From the results, we can show $\sigma = \lambda\dot{\varepsilon} = Ae^{B/T}\dot{\varepsilon}$. In Fig.2, result[2] in Zr65Al7.5Cu27.5 is also shown. From these experimental results, the ternary system alloy exhibited a tendency for strong temperature dependence in a Zr base amorphous alloy.

2.2 Microformability in the Supercooled Liquid State

It has been shown that an amorphous alloy exhibits a viscous flow in the supercooled liquid state. Furthermore, a superplastic alloy is known to be a polycrystalline material that exhibits strain rate dependence in its deformation behavior and large elongation under low stresses. It also has microformability properties[3]. To evaluate the microformability of superplastic materials, Backofen[4] measured the shape of a specimen which was deformed with a V-groove die 100 μm wide, and Kimura[5] measured the geometrically transferred surface roughness of a die-forged specimen using a die which had a surface dressed with abrasive paper. However, in the present study, we developed an evaluation system using a micro V-grooved die made of (100) silicon that has been processed by electron beam lithography and anisotropic etching to quantitatively evaluate the microscopic formability for these materials. The V-groove is 1 to 20 μm wide and has a base angle of 70.6 degrees. The system used to evaluate microformability is shown in Fig. 3. The materials are subjected to microforging and die forged with the micro V-grooved die. After the deformation, we measured the shape of the specimen by two methods. First, we measured the surface shape by using a three-dimensional shape measurement system with a highly sensitive back-scattered electron detector. Second, we observed the specimen cross section with a scanning electron microscope (SEM) and measured the ratio of the flow area A_f to the V-grooved area A_v and the radius of curvature ρ by image processing. From these geometrical analyses of deformed specimens, we obtained the evaluation index as a percentage of flow area $R_f = A_f/ A_v$ and curvature $\rho = 1/r$ as shown in Fig. 4. Index R_f increases with the forming time, and Fig. 4 shows the result of a specimen deformed for 600 s working time on a $Zr_{55}Al_{10}Cu_{30}Ni_5$ amorphous alloy and Al-78Zn fine grained superplastic alloy of various grain diameters. For the amorphous alloy, R_f reached 95% or more for $W_d > 5\mu m$ (W_d is the width of the V-groove). In contrast, when $W_d < 5\mu m$, R_f decreases with the decrease

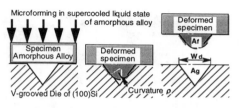

Wd : Width of die throat = 1, 2, 5, 10, 20 μm
Af : Formed and inflowed area into the V-groove
Ag : Area of V-groove

Index of microformability:
  Rf = Af/A g : Percentage of flowed area
  ρ : Curvature of a deformed specimen

Fig.3 Evaluation system of microformability for superplastic materials.

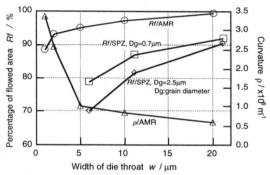

Fig.4 Microformability of $Zr_{55}Al_{10}Cu_{30}Ni_5$ amorphous alloy (AMR) in a supercooled liquid state and of Al-78Zn micrograined superplastic alloy(SPZ)
AMR : σ=10 MPa, T=720K, t=600s
SPZ : σ=10 MPa, T=520K, t=600s
σ: Applied stress, T, t : Working temperature, time

in W_d and becomes R_f=93% for W_d=2 μm. However, it can be observed in Fig. 4 that radius of curvature ρ becomes small. This means that the tip of the die forged specimen becomes acute. In addition, excellent micro formability is confirmed in Fig. 5(a, b). Similar formability results have been obtained on a nanometer scale for lanthanum(La) base amorphous alloy[8]. This excellent microformability can be understood well by comparing these results with the microformability of Al-78Zn superplastic alloy, a polycrystalline aggregate with 1.2 μm grain diameter[6,7]. In this case, the material cannot be microforged, as shown in Fig.5(c,d), due to the deformation mechanisms of grain boundary sliding and grain rotation, since the accuracy limit of microformability depends on the grain diameter as shown in Fig.6. Therefore, an ultrafine-grained microstructure is required for microformability during superplastic deformation at low stresses. However, if there is high stress or a long forming time, good formability exceeding 90% in R_f can be obtained by the deformation of each grain itself caused by conventional plastic deformation or by a diffusion process in

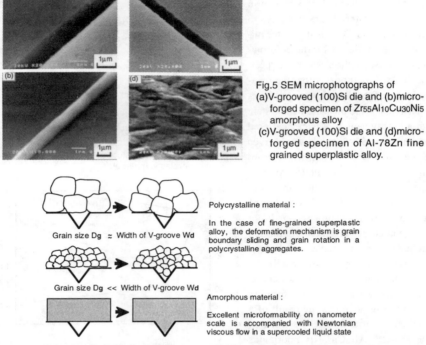

Fig.5 SEM microphotographs of
(a)V-grooved (100)Si die and (b)micro-
forged specimen of $Zr_{55}Al_{10}Cu_{30}Ni_5$
amorphous alloy
(c)V-grooved (100)Si die and (d)micro-
forged specimen of Al-78Zn fine
grained superplastic alloy.

Grain size Dg $\simeq$ Width of V-groove Wd

Polycrystalline material :

In the case of fine-grained superplastic
alloy, the deformation mechanism is grain
boundary sliding and grain rotation in a
polycrystalline aggregates.

Grain size Dg << Width of V-groove Wd

Amorphous material :

Excellent microformability on nanometer
scale is accompanied with Newtonian
viscous flow in a supercooled liquid state

Fig.6 Deformation mechanism and microformability of superplastic alloy of polycrystalline
aggregates and of amorphous alloy in the supercooled liquid state.

a polycrystalline structure. Excellent microformability should be obtained for an amorphous alloy
because it has no anisotropy and it is homogeneous with no grain boundaries or segregation.

3. DEVELOPMENT OF MICROFORMING TECHNIQUE

Microforming and fabrication processes are classified into two types.

The first is a process of transforming a micro-shape from design (virtual) space to actual
space. This process is analogous to a two-dimensional or three-dimensional positioning process
using a mechanical system or an electron-beam controller. The second is a process of imparting
the shape to materials by etching, electroplating, or milling. From the viewpoint of production
engineering, plastic forming processes offer a significant advantage in productivity and enable
mass production with controlled quality and low cost. In the forming process, the relationship
between the die and the material is based on the difference in the strength of the materials. From
this point of view, superplastic materials and amorphous alloys in the supercooled liquid state have
a great advantage in achieving deformation under very low stresses compared to conventional
plastic deformation as shown in Fig.7. Furthermore, the materials exhibit good microformability.
This excellent microformability has been applied to closed die forging and micro extrusion. In

these microforming processes, materials and fabrication processes are very important.

3.1 Micro forging

Figure 8(b) shows a micro-pyramid die forged of $Zr_{55}Al_{10}Cu_{30}Ni_5$. The pyramid base is 5 μm long, and the die of the concave pyramid was processed by anisotropic etching of (100)Si as shown in Fig.8(a). The smooth surface of the Si die and the pointed pyramid shape have been transferred, and the possibility of microforging has been confirmed[10].

3.2 Micro extrusion

The extrusion forming method is used to obtain long parts that have a large aspect ratio and the same, complex shape in cross-section. This forming method is classified as either backward extrusion or forward extrusion. Photochemically machinable glass[11] has been applied to fabricate micro dies, and the micromachining process is shown in Fig.9. A micro-gear shape is printed on a photomask by electron beam lithography[12], and the mask is projected onto photochemically machinable glass. The glass contains photosensitive, metallic ion materials. A metallic colloid is generated by the continuous exposure of ultraviolet rays by the heating processing. This becomes a nucleus, and fine (Li_2O-SiO_2) crystals grow. The solubility of these crystals in acid is up to 50 times that of the former glass, so only the exposed part is etched with weak hydrogen fluoride (HF).

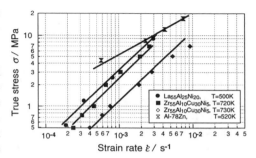

Fig.7 Characteristic flow curves for (a)Al-78Zn superplastic alloy, (b)$La_{55}Al_{25}Ni_{20}$ amorphous alloy, (c)$Zr_{55}Al_{10}Cu_{30}Ni_5$ amorphous alloy.

Fig.8 (a) (100)Si die processed by anisotropic etching and (b) micro-forged specimen of $Zr_{55}Al_{10}Cu_{30}Ni_5$ amorphous alloy

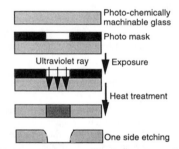

Fig. 9 Micromachining process of superplastic forming dies by using photo-chemically machinable glass

The exposure light can penetrate the photochemically machinable glass in the thickness direction. This enables micromachining in 2.5 dimensions and creating shapes with large aspect ratios. During the etching process, two-sided etching and one-sided etching are possible (Fig.9). Fig.10 shows the configuration of the developed superplastic backward extrusion machine[13]. The specimen was placed in a container, heated, and held at the working temperature then extruded with a piezoelectric actuator in a superplastic or supercooled liquid state in a vacuum or an argon gas atmosphere. The working temperature of Al-78Zn alloy was 523K in superplastic condition and that of the $La_{55}Al_{25}Ni_{20}$ amorphous Alloy was 495K in the supercooled Liquid temperature range. The results of extrusion are shown in Fig.11. The micro-gear shaft of $La_{55}Al_{25}Ni_{20}$ amorphous alloy is 50μm in gear module and that of Al-78Zn superplastic alloy is 20μm in module. In this forming process, the relation between punch displacement and punch load determines the effect of the frictional force and the existence of the high threshold punch load. In general, the working energy is classified as deformation energy of the material and the friction energy between the material and the tool (die). In micromachining, the surface roughness and lubrication of the tool affect the forming behavior.

RESULTS

Amorphous alloys in the supercooled liquid state exhibit newtonian viscous flow and an excellent microformability under very low stresses. These alloys are therefore expected to become some of the most useful materials for fabricating micromachines.

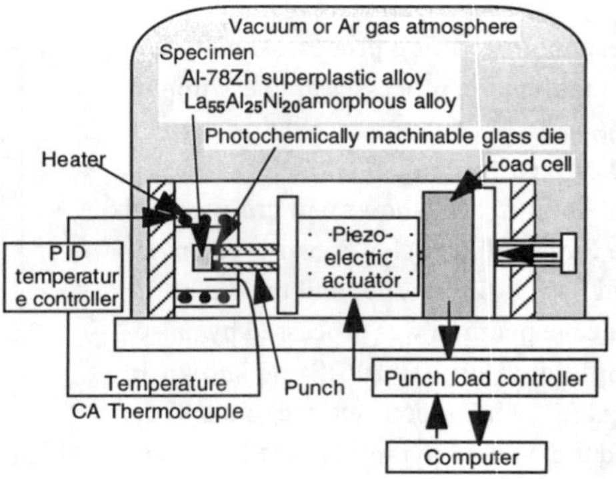

REFERENCES

1. A.Inoue, T.Zhang, T.Masumoto, Metal. Trans. JIM. **30**, 965(1989).
2. Y.Saotome, A.Inoue, Proc. the 44th Japanese Joint Conf. for the Technology of Plasticity, 445(1993).
3. M.Miyagawa, J. of the Japan Society for precision Engineering **52**, 39(1985).
4. W.A.Backofen, et.al, Metals Engineering Quarterly **10**, 1(1970).
5. M.Kimura, M.Kobayashi, Proc. the 39th Japanese Joint Conf. for the Tech. of Plasticity, 427(1988).
6. Y.Saotome, H.Satoh, et.al., Proc. the 43rd Japanese Joint Conf. for the Tec. of Plasticity, 619(1992)
7. Y.Saotome, S.Hata, et.al., Proc. of the 1992 Japanese Spring Conf. for the Tech. of Plasticity,127(1992).
8. Y.Saotome, A.Inoe, et.al., Proc. the 43th Japanese Joint Conf. for the Tech. of Plasticity, 441(1992).
9. Y.Saotome and A.Inoue, Proc. IEEE Micro Electro Mechanical Systems 343(1994).
10. H.Iwazaki, Y.Saotome, et.al., Abstract of The 117th Meeting of JIM, 337(1995).
11. Y.Saotome, H.Satoh, et.al., Proc. the 44th Japanese Joint Conf. for the Tech. of Plasticity, 437(1993).
12. H.Iwazaki, Y.Saotome, et.al. , Proc. the 45th Japanese Joint Conf. for the Tech. of Plasticity, 865(1994).
13. Y.Saotome, H.Iwazaki, et.al., Proc. of the 1996 Japanese Spring Conf. for the Tech. of Plasticity, 288(1996).

Fig.10 Schematic illustration of superplastic backward extrusion machine

(a)Al-78Zn superplastic alloy   m=20μm, Z =10, Dp =200μm, Ps = 56MPa, Tw =520K
(b)$La_{55}Al_{25}Ni_{20}$ amorphous alloy,   m=50μm, Z =10, Dp =500μm, Ps =130MPa, Tw =500K

m ;module, Z :Number of teeth, Dp :Pitch diameter
Ps : Extruding pressure loaded on punch
Tw :Working temperature

Fig.11 Extruded micro-gear shaft of (a)superplastic alloy and (b) amorphous alloy

Part VII

Mechanical and Other Properties II

Synthesis and Properties of Bulk Metallic Glass Matrix Composites

Haein Choi-Yim, Ralf Busch and William L. Johnson

W.M. Keck Laboratory of Engineering Materials, Mail Code 138-78,
California Institute of Technology, Pasadena, California 91125

ABSTRACT

Bulk metallic glass matrix composites are processed and investigated by X-ray diffraction, DSC, optical microscopy, SEM, microprobe, TEM, and mechanical testing. Ceramics such as SiC, WC, or TiC, and the metals W or Ta are introduced as reinforcements into the metallic melt. The metallic glass matrix remains amorphous after adding up to 30 vol% of particles. The thermal stability of the matrix does not deteriorate after adding the particles. ZrC layers form at the interfaces between the bulk metallic glasses and the WC or SiC particles. Si and W are released into the matrix in which Si enhanced the glass forming ability. The composites are tested in compression and tension experiments. Compressive strain to failure increases by over 300% compared to the unreinforced $Zr_{57}Nb_5Al_{10}Cu_{15.4}Ni_{12.6}$ and the energy to break of the tensile samples increases by over 50% adding 15 vol. % W.

INTRODUCTION

Bulk metallic glasses (BMG) have a high yield strength and a high elastic strain limit combined with corrosion resistance and a relatively high fracture toughness [1-4]. However, the lack of tensile ductility limits the number of applications. One way to address this problem is to reinforce the glass. Recently it was shown that bulk metallic glass matrix composites reinforced by particles and wires can be successfully processed [5-6].

Bulk metallic glass formers are excellent matrix materials for composites because of their low melting and glass transition temperatures. The low glass transition temperature helps to reduce differential thermal stresses, which develop between the reinforcement and matrix during freezing upon cooling. The low melting temperatures of metallic glass formers result in slow chemical interactions between the reinforcing particles and the liquid during processing.

In this study, it will be shown that composites combining a metallic glass with different crystalline metals or ceramics can be successfully processed. Results on the X-ray diffraction (XRD), differential scanning calorimetry (DSC), optical microscopy, scanning electron microscopy (SEM), microprobe, transmission electron microscopy (TEM) as well as compression and tension tests will be presented and discussed.

EXPERIMENTAL METHODS

Ingots of the three alloys, $Cu_{47}Ti_{34}Zr_{11}Ni_8$ (Vit101), $Zr_{52.5}Ti_5Al_{10}Cu_{17.9}Ni_{14.6}$ (Vit105) and $Zr_{57}Nb_5Al_{10}Cu_{15.4}Ni_{12.6}$ (Vit106) were prepared by arc melting a mixture of the elements [7,8]. The alloys were then mixed with SiC, WC, TiC, W or Ta particles by remelting in a water cooled copper boat under a Ti-gettered argon atmosphere. The volume fractions of particles ranged between 5 and 20 percent and the sizes of the particles varied between 12 and 50 µm. The composite ingots were remelted at temperatures ranging between 850 and 1100 °C under vacuum in a quartz tube using an induction heating coil and then injected through a nozzle into a copper mold using high purity argon at 1 atm pressure.

The composites were examined by XRD. The glass transition and crystallization of all samples was studied with a DSC. In addition, the particle distribution in the composite specimens was investigated by optical microscopy. In selected samples the interfaces between particles and matrix were examined by electron microprobe, SEM and TEM. Tensile and

Mat. Res. Soc. Symp. Proc. Vol. 554 © 1999 Materials Research Society

compression tests were performed on an Instron 4200. After the mechanical tests, the SEM was used to investigate the fracture surfaces.

RESULTS

We processed three matrices with four different reinforcement materials and studied them by XRD, optical microscopy and DSC. In Fig.1, the XRD patterns of the amorphous Vit 106 matrix, the alloy reinforced with 10% WC and pure WC particles are shown. The pattern of the composite shows the diffraction peaks of WC particles superimposed on the broad diffuse scattering maxima from the amorphous phase. No considerable amounts of other phases are detected within the sensitivity limit of X-ray diffraction. The particles maintained their crystal structure. Since no significant peak shifts of the reflexes of the particles are observed, we can assume that no considerable amount of matrix atoms were dissolved into the particles.

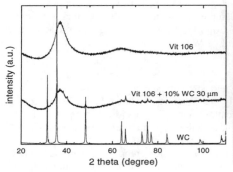

Fig. 1. X-ray diffraction patterns of the Vit 106 matrix, the alloy reinforced with 10% WC and pure WC particles. Vit 106 matrix is amorphous after processing and quenching

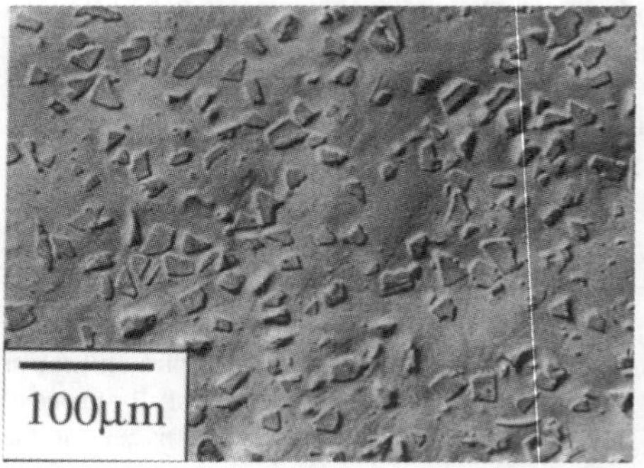

Fig. 2. Optical micrographs of polished surfaces of $Zr_{57}Nb_5Al_{10}Cu_{15.4}Ni_{12.6}$ alloy (Vit 106)/WC composites, showing a uniform distribution of particles in the Vit 106 matrix.

Figure 2 shows the optical micrographs of the uniformly distributed WC particles in the amorphous Vit 106 matrix. The matrix appears uniform and free of heterogeneities in the optical micrograph.

The glass transition and the crystallization behavior of the amorphous matrix of the composites were investigated in DSC scans. Figure 3 shows DSC scans of the pure amorphous Vit 106 and WC, SiC reinforced composites using a heating rate of 0.33 K/s. They exhibit an endothermic heat event characteristic of the glass transition followed by two characteristic exothermic heat releases indicating the transformation from the metastable undercooled liquid state into the crystalline compounds. T_g is defined as the onset of the glass transition temperature. T_{x1} is the onset temperature of the first crystallization event. The value ΔT_x, which is defined as $T_{x1}-T_g$, is referred to as the supercooled liquid region. Based on the DSC scans in Fig. 3, it is observed that the addition of WC particles into the Vit 106 produces no discernible change in either T_g or T_{x1}. Contrarily, the addition of SiC particles into the Vit 106 produces an extension of the supercooled liquid region. The same effect is observed, when SiC is introduced into Vit 101[9] and Vit 105.

Reinforcement/matrix	Vit 101	Vit 105	Vit 106		Vit 101	Vit 105	Vit 106
10% SiC 50 μm	A	A	A	10%WC 30 μm	X	X	A
20% SiC 50 μm	A	A	A	15% WC 30 μm	X	X	A
30% SiC 50 μm	A	X	X	5% W 30 μm	A	A	A
10% SiC 30 μm	A	X	A	10%W 30 μm	X	X	A
20% SiC 30 μm	A	X	X	15%W 30 μm	X	X	A
10% SiC 1 μm	X	---	---	5% W 12 μm	---	---	A
10% TiC 20 μm	A	X	A	5% Ta 30 μm	---	---	A
5% WC 30 μm	X	A	A	10% Ta 30 μm	---	---	A

Table 2.1 Lists of combinations between three different compositions of bulk metallic glasses and particles. Vit 101 is $Cu_{47}Ti_{34}Zr_{11}Ni_8$. Vit 105 is $Zr_{52.5}Ti_5Al_{10}Cu_{17.9}Ni_{14.6}$. Vit 106 is $Zr_{57}Nb_5Al_{10}Cu_{15.4}Ni_{12}$. "A" stands for an amorphous matrix after processing; "X" stands for a fully or partially crystallized matrix after processing.

Table I gives an overview of the various combinations of matrices and reinforcements which have been processed and characterized.

The interfaces between particles and matrix were investigated by scanning electron microscopy, transmission electron microscopy and electron microprobe. In general, if the particles contain C we observe (ZrTi)C layer between the matrix and the particles. As an example, the dark field (DF) TEM image of the interface between SiC particle and the Vit 106 was shown in Fig. 4. The diffraction pattern shows the (large) reflexes caused by the single SiC grain, the diffuse amorphous ring, as well as a series of rings that can be clearly identified as the reflexes corresponding to ZrC. It is worth noting that the interface between the SiC and the ZrC is mechanically very weak. At all reacted layers that were looked at, this interface was broken. The interface between the amorphous matrix and the ZrC, in contrast, appears to be rather strong.

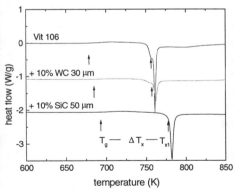

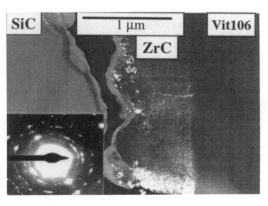

Fig. 3. DSC thermogram (heating rate of 0.33 K/s) of the Vit 106 and WC, SiC reinforced composites for Vit 106.
In no case the introduction of the particles negatively affects the thermal stability of the Vit 106 matrix.

Fig. 4. Dark field TEM of interfacial region between a SiC particle and the Vit 106 matrix. The diffraction pattern of the imaged region is also shown. A ZrC layer had formed at the interface.

In the cases where the particles are pure metals, particle atoms diffuse into the matrix. Fig. 5 shows an example where the interfaces between Vit 106 and W particles were studied. This DF TEM image also contains the diffraction patterns of the W particle, the amorphous matrix and the interface, respectively. At the interface between the matrix and the W particle crystals with a diameter of 50-100 nm have formed. According to qualitative EDAX analyses the compositions of these crystals are closed to the average matrix composition, with the exeption of the Ni. The Ni signal is reduced to about 50% compared to the matrix suggesting that the crystals are considerably depleted of Ni. Complementary microprobe analysis show that a concentration of 0.4 % W is released in the matrix.

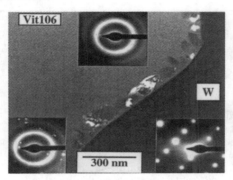

Fig. 5. TEM of interfacial region between W particle and Vit 106 matrix. The dark field TEM image of the interface is shown with the diffraction patterns of the W particle, the amorphous matrix and the interface, respectively.

Compression tests performed on Vit 106 and particulate composites containing WC, SiC, W, and Ta are shown in Fig.6. The compressive fracture takes place along the maximum shear plane, which is declined by about 45 degrees to the direction of compressive load. Pure Vit 106 exhibits only 0.5 % inelastic deformation in compression. Samples reinforced with SiC shows

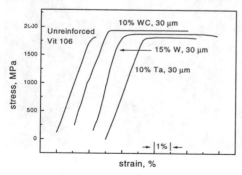

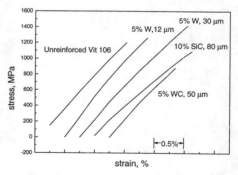

Fig. 6. Quasi-static compression stress-strain curves of Vit 106 and composites reinforced with WC, W, and Ta. The composites exhibited substantial plasticity (3-7 %) under compression.

Fig. 7. Quasi-static tensile stress-strain curves for Vit 106 and composites reinforced with WC, SiC, and W.

similar behavior like pure Vit 106. However, plastic elongation of samples reinforced with WC, W, and Ta exhibited 3 to 7 % under compression as shown in Fig. 6. The compressive strength of composites increased from 1.82 Gpa at 0 vol% reinforcement to 1.96 Gpa at 15 vol% W reinforced composites.

Figure 7 shows the stress-strain curves of Vit 106 and the composites containing WC, SiC, and W obtained by tensile tests. None of the specimens shows extensive plastic deformation as observed in the compression specimens. The composite with SiC show a smaller value of the elastic modulus than pure Vit 106. In addition, composites with 5% W, but a different particle size of 12 μm and 30 μm, respectively, are compared. They both show improved elastic modulus compared to the pure Vit 106. The energy to break (the area under the stress-strain curve), which reflects an increase in toughness, increased by over 50% in W reinforced Vit 106.

DISCUSSION

In this investigation we used either carbides or pure refractory metals as reinforcements. In the case of the carbide particulates a ZrC layer is formed at the interface during processing due to the larger heat of formation for ZrC compared to WC and SiC. The metal (W) or semiconductor (Si) atom is released into the matrix. In the case of the Ta and W particulates the metals are simply dissolved into the liquid matrix during processing. From the DSC studies it is clear that in none of the composites the thermal stability of the matrix after reheating into the supercooled liquid is reduced. The size of the observed supercooled liquid region did not change for WC, W, and Ta. In the case of the SiC particulates the glass forming ability was even improved.

As shown in Fig. 6, compressive ductility of the glass in the composite was enhanced by adding particles into the glass. Figure 8 depicts a micrograph of the side of a composite reinforced with 10 % WC. This picture reveals that multiple shear bands form in the glass. This suggests that the deformation mechanism in the composite must have changed compared to the pure glass. Evidently the constraint which the reinforcement imposes on the matrix prevents catastrophic failure of the material and leads to the formation of secondary bands parallel to the initial band.

Figure 9 shows the fracture surface of a composite around a particle. The fracture occurred

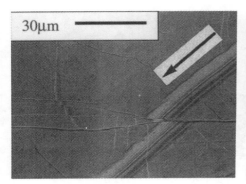
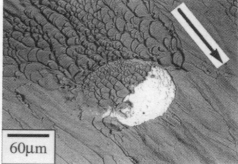

Fig. 8. SEM micrograph of the side of a 10% WC reinforced composite for Vit 106. Multiple shear bands are clearly visible and localized near the primary shear band. The arrow indicates the direction of compressive load

Fig. 9. SEM micrograph of the compressive fracture surface of a 5% W reinforced composite for Vit 106. The arrow indicates direction of shear band propagation. The propagation of shear band was slowed down by the reinforcing particles.

under compression in a specimen reinforced with 5% W. This picture reveals the characteristic vein pattern of the resulting fracture morphology. This vein-like morphology on the failure surface is attributed to localized melting during shear band failure. The direction of the veins is related to the direction of the fracture process. The slippage occurred in the direction of the arrow. The picture clearly proves that particles are restricting shear band propagation. The reduced size of the vein pattern in front of the particle indicates that the flow of material had slowed down in front of the particle. That means the propagation of the shear band was blocked by the reinforcing particles. We also observe the flow of metallic glass over the particle. This indicates that the viscosity of the glass in the shear band is low compared to the surrounding glass. It has been proposed that during the deformation of metallic glasses a fluid layer is formed in the shear plane [10,11].

SUMMARY

Bulk metallic glass forming liquids were reinforced with carbides or pure metals. The composites were characterized with respect to their structure, microstructure and thermal properties. The range of plastic deformation of the material was improved drastically under compression in the case of WC, W, and Ta particles. This is attributed to the formation of multiple shear bands in the presence of particles. The SiC, in contrast, did not improve the ductility, which is related to the fact that the interface between the SiC particles and their shell of ZrC turned out to be mechanically very weak. In tension no significant improvement of plasticity was observed. However the energy to break the material increased substantially.

ACKNOWLEDGMENT

We thank U. Köster and D. Conner for the support and fruitful discussion and C. Garland for performing TEM work. This work was supported by U.S. Army Research Office (Grant No. DAAH04-95-1-0233), the Air Force Office of Scientific research (AFS 5 F4920-97-0323) and the U.S. Department of Energy (Grant No. DEFG-03-86ER45242).

REFERENCES:

1. H.A. Bruck, T. Christman, A.J. Rosakis, and W.L. Johnson, Scripta Metall. Mater. **30**, 429 (1994).
2. H.A. Bruck, A.J. Rosakis, and W.L. Johnson, J. Mat. Res. **11**, 503 (1996).
3. C.G. Gilbert, R.O. Ritchie, and W.L. Johnson, Appl. Phys. Lett. **71**, 476 (1997).
4. D. Conner, A.J. Rosakis, and W.L. Johnson, Scripta Metall. Mater.**37**, 1373 (1997).
5. H. Choi-Yim and W.L. Johnson, Appl. Phys. Lett. **71**, 3808 (1997).
6. R.B. Dandliker, R.D. Conner and W.L. Johnson, J. Mat. Res., in press (1998).
7. X.H. Lin and W.L. Johnson, J. Appl. Phys. **78**, 6514 (1995).
8. X.H. Lin, PhD-thesis, California Institute of Technology; X.H. Lin and W.L. Johnson, Mater. Trans. JIM. **38**,475 (1997).
9. H. Choi-Yim, R. Busch, and W.L. Johnson, J. Appl. Phys. **83**, 7993 (1998).
10. H.J. Leamy, H.S. Chen and T.T. Wang, Met. Trans. **3**, 699 (1972).
11. C.A. Pampillo and A.C. Reimschuessel, J. Mat. Sci. **9**, 718 (1974).

AMORPHOUS AND DISORDERED MATERIALS – THE BASIS OF NEW INDUSTRIES

S.R. OVSHINSKY
Energy Conversion Devices, Inc., Troy, Michigan 48084

ABSTRACT

As in the past, materials will shape the new century. Dramatic changes are taking place in the fields of energy and information based on new synthetic materials. In energy, the generation of electricity by amorphous silicon alloy thin film photovoltaics; the storage of electricity in nickel metal hydride batteries which are the batteries of choice for electric and hybrid vehicles. In the information field, phase change memories based on a reversible amorphous to crystalline transformation are widely used as optical memories and are the choice for the new rewritable CDs and DVDs. The scientific and technological bases for these three fields that have become the enabling technologies are amorphous and disordered materials. We will discuss how disordered, multielemental, multiphase materials can throw new light upon metallic conductivity in both bulk and thin film materials. We will demonstrate new types of amorphous devices that have the ability to learn and adapt, making possible new concepts for computers.

INTRODUCTION

The great advances in civilization have been based on materials – the Stone Age, the Bronze Age, the Iron Age. The interaction between materials and the industries that have transformed society was the driving force for the Industrial Revolution. The twin pillars of our society today are energy and information. In a deep sense they are opposite sides of the same coin. They both must be generated, stored and transmitted. Information is structured energy that contains intelligence.

I will show how a new scientific approach to materials based upon disorder and local order can enable the development of new pollution free technology which will answer society's urgent needs to reduce its dependence upon uranium and fossil fuels, particularly oil; the latter is a causative factor not only in climate change but a root cause of war. Science and technology which can change the world's dependence on it can create new huge industries so necessary for economic growth. $30 trillion in 30 years for new electricity alone [1]. Furthermore, over 2 billion people in developing countries are without electricity. Electricity is the fundamental requirement of modern life and the common link between energy as an undifferentiated source of power and energy which can be encoded, switched and stored as information.

Devices made of amorphous and disordered materials have become the enabling technology for generating electricity through thin film photovoltaics which can be cost competitive to fossil fuels, for storage of electricity in batteries for electric and hybrid vehicles, ushering in a new, much needed transportation revolution, and high density switching and storage media based on phase change optical and electrical memories, so needed for our information society. Computers which have adaptability, can learn from experience and provide neuronal and synaptic type intelligence, are being made possible by devices described here.

How is it possible that multi-elemental disorder can be the basis for such "revolutionary" possibilities [2] when it is well-known that the great success of the 20th century, the transistor is based upon the periodicity of materials with particular emphasis on one element, silicon? Indeed, with the great success of the transistor based upon the crystal structure of germanium and silicon, we entered the historical era where achieving crystalline perfection over a very large distance became the sine qua non of materials science.

From a materials point of view, the physics that made the transistor possible was based upon the ability to utilize periodicity mathematically which permitted parts per million

Mat. Res. Soc. Symp. Proc. Vol. 554 © 1999 Materials Research Society

perturbations of the crystalline lattice by substitutional doping. But right from the beginning, the plague of their disordered surfaces prevented for a decade the fulfillment of the field effect transistor. The disorder of the surface states swamped out the transistor action. Emphasizing again Pauli's statement "God created the solids, the devil their surfaces" [3]. The irony was that the solution that made not only the field effect transistor possible but also the integrated circuit which became the basis for the information age was the utilization of amorphous silicon oxide for photolithography and for the gate oxide.

When I introduced the idea that there was a new world of interesting physics and chemistry in minimizing and removing the constraints of periodicity [4,5], one can understand the resulting consternation of the solid state physicists who had received their Ph.Ds by accepting the dogma of periodicity as being the basis of condensed matter and of the theoretical physicists to whom the control of many elements was as incomprehensible as the conundrum of many-body theory. The change from periodicity to local order permitted atomic engineering of materials in a synthetic manner by opening up new degrees of design freedom. Literally many scores of new materials could be developed and new physical phenomena could be displayed, new products made and new process dependent production technology invented.

Disorder is the common theme in the minimization and lifting of lattice constraints, (what I call the tyranny of the crystalline lattice) which permits the placing of elements in three-dimensional space where they interact in ways that were not previously available. This allowed the use of multi-elements and complex materials including metals where positional, translational and compositional disorder removed the restrictions so that new local order environments [6,7] could be generated which controlled the physical, electronic and chemical properties of the material. Just as the control of conductivity through doping was the Rosetta Stone of understanding and utilizing the transistor, the unusual bonding, orbital configurations and interactions affecting carriers, including ions, are the controlling factors in disordered materials [8-11].

The tools that I utilize for generating these configurations are hydrogen, fluorine, f- and particularly d-orbitals, and nonbonding lone-pair orbitals of the chalcogens. The latter, like the d-orbitals, can be distinguished from their cohesive bonding electrons, freeing them for varied interactions. Even in a sea there are channels and currents affected by topology and climate; in a metallic sea of electrons, we can design paths, control flow and make hospitable environments for incoming ions/protons. Since we are designing new local environments, we also utilize rapid quench technology to make for non-equilibrium configurations and offer a new degree of freedom for the production of unusual local order. Of course, rapid quench and non-equilibrium are associated with vacuum deposition, sputtering and plasma generated materials, in brief, such materials are process dependent.

The understanding of these basic premises became a design tool which we applied universally across the periodic table to build new types of semiconductors, dielectrics and metals. We showed that we can control the density of states in a band/mobility gap affecting conductivity in several ways including chemical modification and the generation of chemically reactive sites [12,13] so as to design, for example, complex, disordered, metallic electrodes of our nickel metal hydride batteries. What we mean by complex is not just that there are many elements, but that we build into a material a chemical, electronic and topological system which performs in the same material various functions such as catalysis, hydrogen diffusion paths, varying density of electrons, acceptor sites for hydrogen, etc. Such a material system in a battery must provide high energy density, power, long life and robustness [14-17].

To put into perspective the principles of disorder, what is required is a metaphor. It is helpful to continue the analogy of the sea. In ancient times, the earliest explorers stayed as close to the shore as possible; there were many things to discover that way – new people, animals, physical environment, things could be strange but understandable. However, to explore the great unknown ocean, they were filled with anxiety for out there was the end of the world and where the dragons lay. Navigational skills were needed to avoid dangerous shoals, utilize favorable currents, etc.

As I pointed out in the 1970s [18-19], amorphous tetrahedral materials to be useful would have to be as close to the four-fold coordination of their crystalline cousins as possible, otherwise, the huge density of states of dangling bonds would prohibit their use. Hydrogen and fluorine were able to act as organizers to assure sufficient four-fold coordination resulting in a low density of states so that the materials could be electronically useful, a necessity for successful photovoltaic products [20,21]. Hydrogen not only capped dangling bonds, but was also valuable as a bridging element providing the connectivity between the silicon atoms and under the proper circumstances assuring enough four-fold coordination. I chose fluorine since it is the superhalogen and provides a much stronger bond and, most importantly, expands the undercoordinated bonds of silicon and germanium so that they can have their full tetrahedral structure. Fluorine also provides useful functions in the plasma and on the surface in the growth of the film. It does its job so well not only in intermediate order but as a preferred element to make microcrystalline tetrahedral materials [22] as well as thin-film diamond-like carbon [23]. It also played a role in making superior superconducting films [24].

The most exciting physics lay in the unexplored ocean that I have been working in since 1955 with our only nautical chart the periodic table and physical intuition our compass. Even though disordered materials could not be easily categorized mathematically, one can constructively design nonequilibrium, nonstoichiometric graded and mixed phase materials to discover new phenomena [8].

In order to follow the exploration process, we will intermingle the relevant scientific and technological approaches with the materials, products, and technologies made possible by utilizing the freedom permitted by disorder to design and atomically engineer local environments.

ENERGY GENERATION –PHOTOVOLTAICS

In photovoltaics, we started our exploration relatively close to the silicon shoreline but had to push further out because contrary to conventional thinking, elemental amorphous silicon had no possible electronic use and therefore required the alloying described above so as to eliminate dangling bonds and yet to retain its four-fold coordination. I felt that it also required new technology befitting its thin film form. Rather than choosing heavy glass as a substrate, we used flexible, thin stainless steel as well as other flexible materials such as kapton.

A historical perspective is needed to show the interaction of science, technology and product. Starting in the late 1970s, we invented the materials, the products and designed and built six generations of production machines utilizing our multi-junction, continuous web technology, the most recent, our 5 MW machine shown in Fig 1. These machines, designed and built at Energy Conversion Devices (ECD), manufacture much needed, simple products using our advanced science and technology and, most importantly, this continuous web approach has shown that it is possible in larger machines to make solar energy cost competitive to conventional fuels. This, of course, would set off an enormous positive change in the use and economics of energy. The plasma physics involved in such a machine makes it possible to discard the power consuming crystal growing methods of crystalline semiconductors and the billion dollar costs now involved in building crystal wafer plants. In crystalline materials, the investment and throughput are linearly coupled, in our amorphous thin film technology, a 4 times increase of capital investment (in the millions of dollars) would yield a 20 times increase in throughput.

Fig. 1   **Fig. 2**   **Fig. 3**

Fig. 2 shows a 3 inch crystalline wafer (at bottom, center), the state of the art of semiconductors at the time, and what Gordon Moore and Bob Noyes, founders of Intel, utilized as opening humor for their talks, a huge cardboard "wafer", representing what they felt would be the electronic requirements of the 90s. On the right, made in our second generation continuous web machine in 1982 for our joint venture with Sharp Corporation, is the first half mile long, over a foot wide roll of a sophisticated multi-junction thin film semiconductor solar cell continuously deposited on thin, flexible stainless steel. This revolutionary new process was the answer to the fantasy of the cardboard wafer. Obviously the roll could be made much wider and longer. I have called our photovoltaic roll an infinite "crystal".

In order to show the extreme light weight, high energy density potential of our approach, we deposited our films on kapton and demonstrated in 1984 the highest energy density per weight of any kind of solar cell. This is shown in Fig.3 in a water pumping application at that time. These days the same approach is beginning to be utilized for telecommunication, space and satellite applications by Guha, Yang and colleagues[25]. The power density in thin film stainless steel and especially kapton which can store almost 3000 watts per kilogram is so exceptional as to become the solar cell of choice for telecommunication. Just last month, in November 1998, an amorphous silicon solar array was installed on the MIR space station. It was fabricated in Troy, Michigan by United Solar Systems Corp (United Solar), (ECD's joint venture with Canon), and assembled in Russia by Sovlux, ECD's Russian joint venture with Kvant, the developer of the original photovoltaics on MIR, and the Russian Ministry of Atomic Energy.

Fig. 4   **Fig. 5**   **Fig. 6**

Fig. 4 shows shingles made in 1980 [26] providing a new paradigm for energy generation. Paradigm shifting takes time. Figs 5 and 6 are current installations of the shingles and a standing seam roof. These products are gaining widespread approval. Fig 7 shows former Secretary of Energy Pena displaying our solar shingles in 1998.

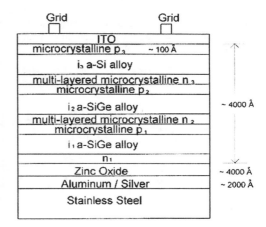

Grid		Grid
ITO		
microcrystalline p₃	~ 100 Å	
i₃ a-Si alloy		
multi-layered microcrystalline n₃		
microcrystalline p₂		
i₂ a-SiGe alloy		~ 4000 Å
multi-layered microcrystalline n₂		
microcrystalline p₁		
i₁ a-SiGe alloy		
n₁		
Zinc Oxide		~ 4000 Å
Aluminum / Silver		~ 2000 Å
Stainless Steel		

Fig. 7   **Fig. 8**

Fig 8 is a schematic of our triple-junction solar cell showing its multi-layered configuration wherein light can be absorbed in sub-cells with different bandgaps so as to utilize as much of the sun's spectrum as possible. The blue, green and red light is absorbed in layered thin films of amorphous silicon and germanium alloys containing hydrogen/fluorine. ECD and United Solar have all the world's records for efficiency, culminating in the latest world record of 10.5% stable efficiency on a one square foot module and 13% stable efficiency on a .25 square cm cell [27].

Amorphicity is crucial for several reasons. Unlike crystalline tetrahedral materials in which quantum mechanical selection rules make for indirect bandgaps and require layers of 50 to 100 microns in order to absorb the light energy, amorphous materials act as direct band gap materials and therefore the entire triple layer system is less than 1 micron in total thickness. It is important to note that in crystalline materials of different bandgaps, lattice mismatch is a serious problem and therefore such multi-layered structures could not be made in a production manner. In this case, we can see that amorphicity and the physics make possible continuous web production. Amorphous photovoltaics illustrate that when one removes the lattice constraints, atomic engineering can be merged with machine engineering to provide a new, much needed approach to energy generation.

There has been much ongoing work in photovoltaics at ECD since 1977 [28], advancing the science and technology of materials, production processes and new products. At the MRS 1998 Spring Meeting, my collaborators, Guha, Yang and coworkers at United Solar, who have made very significant contributions to our work, gave an excellent review of our recent commercialization progress [29]. The ECD-United Solar team, which also includes Masat Izu, Prem Nath, Steve Hudgens, Joe Doehler, Scott

Jones and Herb Ovshinsky among others (the latter the head of ECD's Machine Division which has designed and built our continuous web processors) has through the years made important contributions to our field and has been working on products, production and plasma technology that increase throughput.

From a materials science point of view, we note several points of importance. While amorphous tetrahedral materials are close to the crystalline shoreline in their need for low density of states and substitutional doping, being direct bandgap materials, one can make large area, thin film multi-junction devices by the decomposition of plasmas in a continuous manner. Accomplishing this, we were able to basically alter the way materials could be laid down in a continuous manner, showing the tight coupling in amorphous materials between basic science and advanced technology.

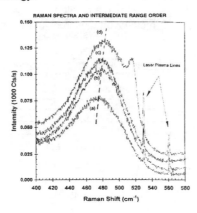

Fig. 9

Illustrating the scientific and technological richness of amorphous materials is the ability to develop intermediate range order in an amorphous matrix as we do in the intrinsic silicon alloy layer of our photovoltaic product [11,30]. (Fig. 9) It is of great interest that quasi one-dimensional ordering is accomplished without introducing grain boundaries that would interfere with electron and hole conductivity. This intermediate state is the signature of the best material. We have shown that fluorine is a great facilitator of intermediate order [31-34] and leads to crystallization. The intermediate order has important implications for the future. It is possible that the carriers have increased mobility due to the intermediate structures and could affect the important parameter of hole mobility in these materials. Microcrystalline materials can also have unique properties that bridge the gap between crystalline and disordered solids. We have incorporated them in our continuous web process by utilizing fluorine [11,21] to make under 120 angstrom micro-crystalline silicon p-layers. When one considers that this is accomplished in a continuous web, very high yield production process, it can be appreciated that atomic engineering and manufacturing are a reality.

ENERGY STORAGE – NICKEL METAL HYDRIDE BATTERIES

Nickel metal hydride batteries are in a real sense misnamed, for while nickel, by virtue of its filled d-orbital, plays an important catalytic role, there are usually seven or eight other elements that make up the alloy used in the negative electrode; certainly hydrogen is the key component. It is the smallest and simplest atom in the universe (which, by the way, in terms of actual matter is composed of over 90% hydrogen). Therefore, to attain the highest energy storage density not only for batteries but for hydrogen as a fuel is a matter of designing the highest density of reversible hydrogen sites. This can be accomplished by our principles of disorder and local order [11,14-17,35]. From 1960 on, we have demonstrated that hydrogen is not just a future source of energy but a here and now solution that, together with solar energy, offers the ultimate answer to society's need for clean, virtually inexhaustible energy.

Energy technology should be regarded as a system. We have described energy generation. Equally as important is energy storage. Now we must go much further from shore in order to discuss the basis for the materials used for energy storage and later for information which is structured energy and must also be stored.

In the energy storage area, we cannot discern the shoreline, we are in the ocean of many-body theory which has never been adequately understood. I have had a long interest and worked for many years with d-orbital materials [36]. A recent publication sums up present day advanced thinking regarding d-electrons: "d-electron-based systems, in particular, present the combined intrigue of a range of dramatic phenomena, including superconductivity and itinerant ferromagnetism, and the tendency toward inscrutability associated with the fact that key electronic states are often intermediate between the ideals of localization and itineracy which provide the starting points for most theory " [37]. By making multi-elemental, multi-phase d-orbital materials for hydrogen storage in a completely reproducible manner, it is clear that we have been able to understand how to take the mystery out of them and utilize these orbitals in new and unique ways.

We are literally at sea when we discuss metals for they are always described as being composed of a sea of electrons. Why does the sea not swamp out the background provided by local atomic environments? Paradoxically, this is because when we use many, for example, ten, different atoms, particularly those with directional d-orbitals, to make up a material such as the negative electrode in our nickel metal hydride battery, we provide through the disordered state regions of lower electronic density which have a larger probability to overlap with negative hydrogen ions in interstitial sites. To simplify, one can say that it is the s-electrons that provide the sea, the d-electrons sculpt out the channels and receptors. Obviously there is some hybridization possible in many of these materials.

It is in the multielemental f- and particularly the d-orbital material that we introduce a means of delocalizing electrons and still have them represent their parentage. This is where internal topology begins to play an important role in metallic conduction for, as noted, we need channels in the sea of electrons just as we take into account sea level/Fermi level. The different types of atoms provide the interatomic spacings for the hydrogen ions to operate in and the varying electronic density is the steering means for the ions to reach the preferred sites of low electron density. Disordered materials are therefore necessary to provide the spectrum of binding sites. In summary, large interatomic spacing and low energy density make for the optimal binding/storage of negative hydrogen ions. The electron environment surrounding the hydrogen provides the degree of negativity and coulombic repulsion is the steering means. The binding energy provided by the local environment is of such a nature as to assure reversibility so important for the rechargeable battery. While we list characteristics of individual atoms in Fig.10, it is how they act and interact in the alloy that makes for the mechanism that we have described above.

The acceptance of nickel metal hydride batteries has been very rapid. All significant manufacturers of nickel metal hydride batteries are under agreement with ECD and our Ovonic Battery Company (OBC) and over 600 million consumer batteries were sold last year with a 30% per year predicted growth rate. These high production volumes have made for a very low cost battery.

The problems of pollution, climate change and our strategic dependence on oil have provided a global impetus for the use of electric and hybrid electric vehicles [38,39]. The automotive industry has made our nickel metal hydride battery the battery of choice for these vehicles. GM, Toyota, Honda, Hyundai, Ford and Chrysler all have chosen nickel metal hydride batteries. We have a joint venture with GM, GM Ovonic, which manufactures and sells electric vehicle and hybrid electric vehicle batteries to all companies.

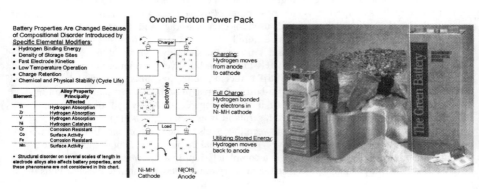

| **Fig.10** | **Fig. 11** | **Fig. 12** |

Fig. 11 shows a very simplified schematic of the nickel metal hydride battery. The battery is based on hydrogen transfer in which hydrogen is shuttled back and forth between the nickel hydroxide and metal hydride without soluble intermediates or complex phase changes. Fig. 12 shows an ingot of our materials as well as our green battery.

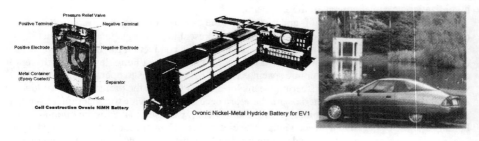

| **Fig. 13** | **Fig. 14** | **Fig. 15** |

Fig.13 is a cutaway drawing of our NiMH automotive battery illustrating its simplicity. Fig. 14 is a GM EV1 battery pack; Fig. 15 is a photo of the EV1 car whose range is between 160 and over 200 miles, its acceleration is 0 to 60 mph in less than 8 seconds, less than 15 minutes for an over 60% recharge, very robust, environmentally benign, lifetime of the car battery.

Fig. 16 shows James Worden, cofounder of Solectria, having driven his 4 passenger Solectria Sunrise from Boston to New York on the equivalent BTU energy of less than 1 gallon of gas and he had 15% energy remaining [40]. Fig. 17 gives world record ranges of EVs using

Ovonic NiMH batteries,and Fig. 18 shows the capability of nickel metal hydride to be continuously improved by atomic engineering of the materials.

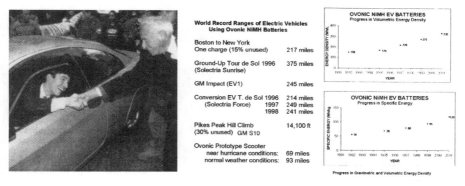

World Record Ranges of Electric Vehicles
Using Ovonic NiMH Batteries

Boston to New York One charge (15% unused)	217 miles
Ground-Up Tour de Sol 1996 (Solectria Sunrise)	375 miles
GM Impact (EV1)	245 miles
Conversion EV T. de Sol 1996	214 miles
(Solectria Force) 1997	249 miles
1998	241 miles
Pikes Peak Hill Climb (30% unused) GM S10	14,100 ft
Ovonic Prototype Scooter near hurricane conditions: normal weather conditions:	69 miles 93 miles

Fig. 16   **Fig. 17**   **Fig. 18**

A gasoline powered GM Geo Metro was tested against the same make and model car rebuilt by Solectria to be an electric vehicle with our Ovonic batteries. In New York City driving, the gasoline car provided an equivalent range of 120 miles compared to the 220 mile equivalent range of the electric vehicle with our batteries [41]. It is quite clear that scientific and technological issues are not what is holding back electric vehicles. Hybrid electric vehicles with our batteries will offer at least 80 miles per gallon in the charge sustaining mode and over 100 miles per gallon in the charge depletion mode.

INFORMATION

Disordered materials depend upon optional bonding configurations of atoms generating various kinds of new orbital relations. While boron and carbon are helpful in this regard, the elements of choice as I have shown for NiMH batteries are f- and d-orbitals. For the information side of our work, we prefer the chalcogenides characterized by Kastner as lone pair materials [42]. We utilize lone pair p-orbitals since they are not only nonbonding but have a spectrum of lone pair interactions that include various new bonding configurations [43-46]. In some respects, they have similarity to the directional d-orbitals, in other respects, they are different, for example, the empty or filled d-orbitals are very localized and do not reach out as far as the lone pair p-orbitals of the chalcogens, where two of the p-orbitals are deep in energy and serve as strong structural bonds responsible for the cohesiveness of the material. The many lone pair interactions are spread in energy throughout the mobility gap. Their similarity to d-orbitals is that neither the lone pairs nor the d-orbitals play a strong role in cohesive bonding but both are available for interesting electronic, optical and chemical interactions. However, their dissimilarity is that the lone pairs being the outer electrons (they are as far out as any valence electrons), can remain free or form weak or strong bonds, covalent or coordinate, depending on the environment. They are as far out as any valence electrons. The d-orbitals, on the other hand, form a narrow but designable band of high density states at the Fermi level and they too can act as receptors in the coordinate bond configuration. A more profound difference is that the divalency of the lone pair chalcogens allows a flexibility of structure which can be controlled by crosslinking so that we can make either an electronic Ovonic threshold switch (OTS) in which the excitation process does not affect bonding (see Fig. 19) or an Ovonic memory switch (OMS)

[4] in which the electronic processes initiate structural, that is, reversible phase change. We will concentrate here on the OMS which is now universally utilized in its optical phase change form (see Figs. 20 and 21) [47].

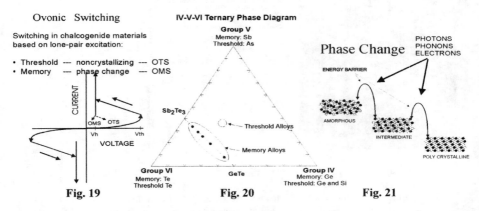

Fig. 19   Fig. 20   Fig. 21

Phase change rewritable memories have become the basis for the rapidly growing DVD rewritable market which holds so much promise for the future since it is replacing VCRs and CD-ROMs.

The energy necessary for an Ovonic optical memory material to change its atomic configuration is provided by a laser beam which couples to the non-bonding lone pairs so that the electronic energy exceeds a threshold value, causing a high atomic mobility state to occur and a change from the amorphous to the crystalline phase to take place. The same laser, but at different power, is used for recording, erasing and rewriting since the amorphous material can become crystalline again by rapid rearrangement through slight movements of atoms. The different structural phases of the material have different optical constants, so information is stored in the form of regions with different reflectivity. It is particularly favorable to use a phase congruent material for these applications and as we will show, the cycle life for phase change memories is exceptionally long. Electrical phase change memories have gone over 10^{13} cycles when testing was stopped [48,49].

Since Ovonic phase change optical memories are having great commercial success and the markets are growing rapidly, we are now entering the semiconductor memory market with devices using these materials. Conventional semiconductor memories are the basic building blocks of the information age. The data shown in Fig. 22 clearly show that conventional memories can be replaced by Ovonic semiconductor memories since a single plane of our memory can replace DRAM, SRAM and Flash. To indicate the great advantages that our multi-state memory offers in this highly competitive industry, we show a comparison with Intel's multi-level flash memory which was announced as "the Holy Grail" and which they said would have a "revolutionary" impact on the Flash market. (Figs. 23 and 24).

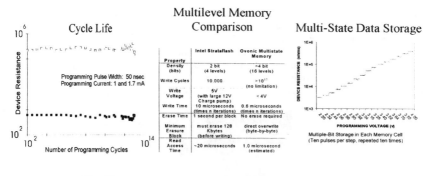

	Cycle Life	Multilevel Memory Comparison	Multi-State Data Storage

Fig. 22 **Fig. 23** **Fig. 24**

It can be appreciated that the multi-state memory is also a learning device since it adapts its electrical conductivity to the amount of information it receives. In other words, it displays what neurophysiologists refer to as plasticity as the basis for intelligence. We have also developed another version of this thin-film memory which is truly neuronal and synaptic in that it accumulates a number of sub-threshold pulses before it changes state. This device takes advantage of the fact that a small portion of the active volume of the memory will change phase upon application of every sub-threshold energy pulse. After a specified amount of energy has been deposited, a percolation path is established among the crystallized regions, and a large change in electrical resistance results. The device is reset into its virgin state in the same manner as our other semiconductor memory. This accumulation mode memory device will have important near-term applications in secure, tamper-proof information storage in smart cards and other devices.

What I have in mind for the future is an all thin-film intelligent computer. We have designed all thin-film circuits which can have logic, memory, and adaptive or intelligent memory integrated into a circuit. We have already proven that these devices can be made in three-dimensional and multi-layered circuits, can receive information from various sources, integrate it, remember it and learn from it. This is the basis of a truly cognitive machine, not artificial intelligence as we now know it, nor just a large number of parallel circuits, but a huge density of switching points, receiving and integrating information, in other words, many neurons of different thresholds and frequencies, receiving huge amounts of synaptic information, responding to it and utilizing it [50]. This is what I have wanted to build since 1955; this is what we can build now. I feel that this kind of computer is the computer of the next millenium. Combined with amorphous sensors and displays which we introduced many years ago and which have found widespread use, such a computer could perform tasks which are now beyond the ken of present "dumb" computers. All the various parts have been shown to work and they are all based on amorphous and disordered materials.

CONCLUSION

I believe that I have shown that science and technology can be utilized to build new industries that are responsive to societal problems and needs, providing jobs, educational opportunities and the chance to express the creative urge that has driven humankind since time immemorial. Fig. 25 shows a young woman climbing a mountain barefooted with her future on her back, our photovoltaics, and her future in front of her, her child, bringing our photovoltaics to a village that does not have electricity. There can be no civilization without energy and without knowledge (information). We take this picture as inspiration to continue our work.

Fig. 25

ACKNOWLEDGEMENTS

This work could not have been accomplished without the immense help and loving and fruitful collaboration of Iris Ovshinsky. Amorphous and disordered materials could not have reached its state in science without Hellmut Fritzsche. I am honored to have him as my longest scientific collaborator about whom it can be truly said that he is a giant in our field. Dave Adler's untimely death cut short our warm, close and productive friendship and collaboration. We and the amorphous field owe so much to him. I wish to thank the talented and creative teams of ECD, United Solar, and the Ovonic Battery Company. In the battery group, Subhash Dhar is an inspired multi-talented leader; Mike Fetcenko has made and is making many invaluable, critical contributions, his collaboration with me has been essential to the success of the battery company. Dennis Corrigan is owed thanks by me and the entire EV industry for his essential contributions to EV batteries; Srini Venkatesan has been a mainstay, his finger- and brain- prints are on everything that happens in batteries; Benny Reichman's great creativity has always been deeply appreciated; Art Holland's mechanical contributions have been of great importance in our battery work; John deNeufville's contributions through the years have not only been of great value to us but his latest activities contribute to the development of the amorphous silicon and germanium alloy field; we thank Paul Gifford not only for his battery talents, but for representing us in developing the new industry; particular thanks to our "young tigers" who are literally building this industry from the ground up. Our computer authority, Guy Wicker fits into that category very well. We also wish to thank Krishna Sapru for her early and current valuable work with me on hydrogen storage, batteries and atomic modeling. Our optical memory work would not be possible without the talent, commitment and hard work of Dave Strand; the electrical memory activity is dependent on the innovative talents and motivation of Wally Czubatyj and his group including Sergey Kostylev, Pat Klersy, and Boil Pashmakov; Ben Chao, the head of our analytical laboratory, is a resource beyond comparison to the entire company;

Rosa Young is a unique, extraordinarily innovative talent whose important contributions have spanned all of our areas in a remarkable manner; Steve Hudgens has contributed greatly to all areas of our work, his collaboration and advice have been extremely helpful to me and essential to the company. We are indeed fortunate to be working with Bob Stempel, the pioneer and leading figure in electric vehicles, a great engineer, leader and most appreciated partner. I am prouder of the organization and working climate that we have built than of any of my inventions.

REFERENCES

1. The Economist, p, 98, October 2, 1993.
2. U.S. Department of Energy announcement, January 18, 1994.
3. U. Hofer, Science **279**, 190 (Jan. 9, 1998).
4. S.R. Ovshinsky, Phys. Rev. Lett. **21**, 1450 (1968).
5. M.H. Cohen, H. Fritzsche and S.R. Ovshinsky, Phys. Rev. Lett. **22**, 1065 (1969).
6. S.R. Ovshinsky, Rev. Roum. Phys. **26**, 893 (1981).
7. S.R. Ovshinsky and D. Adler, Contemp. Phys. **19**, 109 (1978).
8. S.R. Ovshinsky, in *Physics of Disordered Materials*, edited by D. Adler, H. Fritzsche and S.R. Ovshinsky, (Inst. for Amorphous Studies Series, Plenum Press, New York, 1985) p. 37.
9. S.R. Ovshinsky, J. Non-Cryst. Solids **32**, 17 (1979).
10. S.R. Ovshinsky, in *Insulating and Semiconducting Glasses*, edited by P. Boolchand (World Scientific Press, Singapore, 1999).
11. For further references, see *Disordered Materials: Science and Technology, Selected papers by Stanford R. Ovshinsky*, edited by D. Adler, B.B. Schwartz and M. Silver (Institute for Amorphous Studies Series, Plenum Press, New York, 1991).
12. S.R. Ovshinsky, in *Proc. of the Seventh International Conference on Amorphous and Liquid Semiconductors*, Edinburgh, Scotland, 27 June-1 July, 1977, p. 519.
13. S.R. Ovshinsky, J. of Non-Cryst. Solids **42**, 335 (1980).
14. S.R. Ovshinsky, Presented at 1978 Gordon Research Conference on Catalysis (unpublished).
15. S.R. Ovshinsky, presented in May 1980 at Lake Angelus, MI (unpublished).
16. K. Sapru, B. Reichman, A. Reger and S.R. Ovshinsky, U.S. Pat. No. 4 633 597 (18 Nov. 1986).
17. S.R. Ovshinsky, M.A. Fetcenko and J. Ross, Science **260**, 176 (9 April 1993).
18. S.R. Ovshinsky, in *Proc. of the Sixth International Conference on Amorphous and Liquid Semiconductors*, Leningrad, USSR, 18-24 November 1975, p. 426.
19. S.R. Ovshinsky, in *Proc. of the International Topical Conference on Structure and Excitation of Amorphous Solids*, Williamsburg, Virginia, 24-27 March 1976, p. 31.
20. S.R. Ovshinsky, New Scientist **80** (1131), 674-677 (1978).
21. S.R. Ovshinsky, Solar Energy Mats. and Solar Cells **32**, 443-449 (1994).
22. S.R. Ovshinsky, in *Proc. of the International Ion Engineering Congress, ISIAT '83 & IPAT '83*, Kyoto, 12-16 September 1983, p. 817.
23. S.R. Ovshinsky and J. Flasck, U.S. Patent No. 4 770 940 (13 September 1988).
24. S.R. Ovshinsky and R.T. Young, in *Proc. of the SPIE Symposium on Modeling of Optical Thin Films II*, **1324**, San Diego, California, 12-13 July 1990, p. 32.
25. S. Guha, J. Yang, A. Banerjee, T. Glatfelter, G.J. Vendura, Jr., A. Garcia and M. Kruer, presented at the 2nd World Conference and Exhibition on Photovoltaic Solar Energy Conversion, Vienna, Austria, 6-10 July 1998.
26. J. Glorioso, Energy Management, June/July 1980, p. 45. Shown in 1980 by Stan and Iris Ovshinsky to Domtar, a Canadian roofing company and to Allside, an aluminum siding company of Akron, Ohio.

27. J. Yang, A. Banerjee, K. Lord and S. Guha, presented at the 2nd World Conference and Exhibition on Photovoltaic Solar Energy Conversion, Vienna, Austria, 6-10 July 1998.

28. S.R. Ovshinsky, presentation at the British House of Commons, July 1977.

29. S. Guha, J. Yang, A. Banerjee and S. Sugiyama, presented at the 1998 MRS Spring Meeting, San Francisco, CA, 1998 (invited).

30. D.V. Tsu, B.S. Chao, S.R. Ovshinsky, S. Guha and J. Yang, App. Phys. Lett. **71**, 1317 (1997).

31. R. Tsu, M. Izu, V. Cannella, S.R. Ovshinsky, G-J. Jan and F.H. Pollak, J. Phys. Soc. Japan Suppl. **A49**, 1249 (1980).

32. R. Tsu, M. Izu, V. Cannella, S.R. Ovshinsky and F.H. Pollak, Solid State Comm. **36**, 817 (1981).

33. R. Tsu, S.S. Chao, S.R. Ovshinsky, G-J. Jan and F.H. Pollak, J. de Physique **42**, 269 (1981).

34. R. Tsu, J. Gonzalez-Hernandez, J. Doehler and S.R. Ovshinsky, Solid State Comm. **46**, 79 (1983).

35. M.A. Fetcenko, S.J. Hudgens and S.R. Ovshinsky, Daido Journal (Denki-Seiko) **66** (2) 123-136 (April 1995). (Special Issue Electronics & Functional Materials.)

36. See for ex. 1959 Control Engineering on the Ovitron.

37. P. Kostic, Y. Okada, N.C. Collins, Z. Schlesinger, J.W. Reiner, L. Klein, A. Kapitulnik, T. H. Geballe, and M. R. Beasley, Phys. Rev. Lett. **81**, 2498, (1998).

38. S.R. Ovshinsky and R.C. Stempel, invited presentation at the 13th Electric Vehicle Symposium (EVS-13), Osaka, Japan, 13-16 October 1996.

39. R.C. Stempel, S.R. Ovshinsky, P.R. Gifford and D.A. Corrigan, IEEE Spectrum, November 1998, p. 29.

40. IEEE Spectrum, December 1997, p. 68.

41. 10th Anniversary American Tour de Sol Competition, run in New York City by Northeast Sustainable Energy Association (NESEA), (1998).

42. M. Kastner, Phys. Rev. Lett. **28**, 355 (1972).

43. S.R. Ovshinsky and K. Sapru, in *Proc. of the Fifth International Amorphous and Liquid Semiconductors*, Garmisch-Partenkirchen, Germany (1974), p. 447.

44. S.R. Ovshinsky, Phys. Rev. Lett. **36** (24), 1469-1472 (1976).

45. S.R. Ovshinsky and H. Fritzsche, *IEEE Trans. Elect. Dev.*, **ED-20** (2) 91-105 (1973).

46. M. Kastner, D. Adler and H. Fritzsche, Phys. Rev. Lett. **37**, 1504 (1976).

47. For history and early references, see S. R. Ovshinsky, "Historique du Changement de Phase" in Memoires, Optiques et Systemes, No. 127, Sept. 1994, p. 65; in the *Proc. of the Fifth Annual National Conference on Industrial Research*, Chicago, Illinois, 18-19 September 1969; Journal de Physique **42**, supplement au no. 10 October 1981.

48. S. R. Ovshinsky, presented at the 1997 International Semiconductor Conference, Sinaia, Romania , 1997.

49. S. R. Ovshinsky, presented at High Density Phase Change Optical Memories in Multi-Media Era, 9th Conference for Phase Change Media, Shizuoka, Japan, 1997.

50. S.R. Ovshinsky and I.M. Ovshinsky, Mats. Res. Bull. **5**, 681 (1970).

THERMAL STABILITY OF BULK METALLIC GLASSES. INFLUENCE ON MECHANICAL PROPERTIES

J.M. PELLETIER*, Y. JACQUEMARD*, J. PEREZ*, R. PERRIER de la BATHIE**
*GEMPPM, INSA, Bat. 502, INSA, 69621 Villeurbanne Cedex, France, pelletie@insa.insa-lyon.fr,
**CRETA, CNRS, Grenoble, France.

ABSTRACT

Two Zr-base bulk metallic glasses were investigated in the present work. DSC experiments were performed at different heating rates (dT/dt). Evolution of the characteristic temperatures, glass transition and onset of crystallisation, were determined as a function of dT/dt. Evolution of shear elastic modulus and internal friction are measured as a function of temperature and resulting microstructural evolution ; these evolutions are related to variation of the atomic mobility.

INTRODUCTION

New bulk metallic glasses are attractive due to their specific mechanical properties ; the outstanding feature is the yield strength. Values up to 1900 MPa have been reported [1-4]. The occurrence of a glass transition at a moderate temperature T_g enables a good workability above T_g. Kawamura et al [5] have published a detailed study of the deformation behaviour of a Zr-Ti-Ni-Cu-Al amorphous alloy as a function of temperature and deformation rate. In addition, recent experiments on a bulk Zr-Ti-Ni-Cu-Be alloy (at.%) have demonstrated that amorphous materials can exhibit fracture toughness as high as 55 MPa.m$^{1/2}$ [6], a value similar to that observed in aluminium alloys or in steels.

Unfortunately, thermal ability to keep the non-crystalline state is not always ensured, in contrast to polymers or oxide glasses. This thermal stability depends on composition and especially on oxygen content [7-12]. Partial crystallisation can induce beneficial effects on elastic properties, but roughness is drastically reduced by devitrification of the material. In alloys containing small crystalline particles, toughness becomes very low : about 1 MPa.m$^{1/2}$, a value which can be compared to that reported in brittle ceramics.

So, information on crystallisation kinetics and atomic mobility is required to characterise in a complete way the mechanical behaviour of the materials. As shown in previous studies calorimetry and mechanical spectroscopy [7-13] are very suitable methods to provide data on thermal stability and atomic mobility.

Both Zr-Cu-Ni-Ti-Be and Zr-Cu-Ni-Ti-Al alloys are studied in the present work.

EXPERIMENTAL PROCEDURE

Ingots of $Zr_{41.2}$-$Ti_{13.75}$-Cu_8-$Ni_{14.5}$-$Be_{22.5}$ and Zr_{50}-Ti_5-Cu_{25}-Ni_{10}-Al_{10} were prepared with electromagnetic induction in a water-cooled copper crucible in purified argon atmosphere, at CRETA (CNRS, Grenoble, France). Rapidly solidified ingots have a cylindrical shape (diameter : 5mm) and specimen for X-ray diffraction, DSC and mechanical spectroscopy are machined from these ingots.

The amorphicity of the material was examined by X-ray diffraction, using the Cu-Kα radiation. The endothermic and exothermic reactions associated with glass transition and

413

crystallisation process (respectively) were examined by differential scanning calorimetry (DSC), at different heating rates.

The shear dynamic modulus $G^*(\omega)$ was measured by a mechanical spectrometer described by Etienne et al [14]. This apparatus gives data at frequencies between 10^{-5} Hz and 1 Hz. Parallelipedal specimens ($50*4*1$ mm^3) are gripped to the oscillating system. All measurements are performed with a low nitrogen pressure atmosphere. Tan ϕ is defined as the ratio $G''(\omega)/G'(\omega)$, where G'' and G' are the loss and the storage modulus, respectively.

EXPERIMENTAL RESULTS AND DISCUSSION

$Zr_{41.2}$-$Ti_{13.75}$-Cu_8-$Ni_{14.5}$-$Be_{22.5}$ alloy

Features of this material, first developed by Johnson [1] are fairly well known and, consequently, only a few experiments have been performed on this composition. Fig. 1 shows the DSC curve at a scanning rate of 30K/min.

A single endothermic process is observed at the glass transition temperature (Tg = 368°C), while three different exothermic peaks are associated to crystallisation processes. Onset of crystallisation is observed at 443°C and the three peak temperatures are, respectively, 461°C, 480°C and 538°C. These results confirm previously reported data [4, 9, 11] and evidence the microstructural evolution at high temperature.

This alloy is faced with two problems, related to the presence of beryllium : price and toxicity of this element (especially of the BeO oxide). Therefore new materials have been developed. Two solutions were

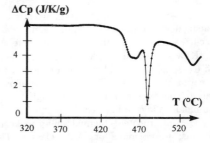

Fig. 1 : DSC curve

investigated in the literature : either suppression of beryllium (Zr-Cu-Ni-Ti alloys) or replacement of beryllium by aluminium. The second one is retained in the present work.

Zr_{50}-Ti_5-Cu_{25}-Ni_{10}-Al_{10} alloy

In this case, formation of a mainly glassy state was confirmed by X-ray diffraction.

a - DSC experiments :
Fig. 2 shows the DSC curve achieved at different scanning rates, ranging from 1 to 30 K/min. Results are similar to those observed in the Be-containing alloy : an endothermic phenomenon is associated to the glass transition (near Tg) and different crystallisation peaks (at Tx_1, Tx_2, Tx_3).

Characteristic temperatures increase with scanning rate and, in the studied temperature range, the third crystallisation peak is not observed at high scanning rate. Magnitude of the different peaks is not significantly modified by the heating rate.

This evolution can be simply related to the modification of the atomic diffusion : with a low heating rate, time allowed for diffusion is longer and hence processes are observed at lower temperature.

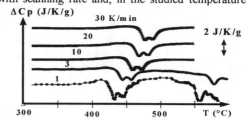

Fig. 2 : DSC at various scanning rates (curves are shifted)

Fig. 3 illustrates the influence of heating rate (dT/dt) on both glass transition and onset of crystallisation (Tg and Tx, respectively). All these parameters decrease with the scanning rate : the thermal stability is reduced by a slow heating, in agreement with results reported by Kawamura et al [5]. Consequently, the duration τ for retaining the glassy state without crystallisation depends on heating rate. Processing in the supercooled liquid region is influenced by this evolution.

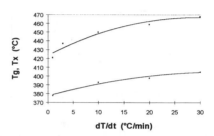

Fig. 3 : characteristic temperatures

b - Elastic and anelastic properties.

Young modulus (E) and shear elastic modulus (G) can be determined by various methods. Measurements of ultrasounds velocity is a simple and precise route. Values of E and G are simply determined from longitudinal and transverse wave propagation. Experimental results obtained at room temperature are given in table I .

	Glassy state	After crystallisation
G (GPa)	27.6	42
E (GPa)	74	110
ν (Poisson coefficient)	0.34	0.31

Table I : elastic characteristics of the Zr_{50}-Ti_5-Cu_{25}-Ni_{10}-Al_{10} alloy

Young modulus E and shear modulus G increase by as much as about 50% after crystallisation. These changes are associated to the reduction of atomic mobility due to a decrease of the number of quasi-point defect [15] and hence to an increase of the material rigidity.

To investigate in a more detailed way this evolution, either as a function of time or temperature, mechanical spectroscopy experiments have been performed.

Shear modulus and internal friction were measured at a given frequency (1 Hz) during the following temperature cycle shown in fig. 4 : step 1 and 3 correspond to, respectively, heating and cooling at a constant scanning rate (1K/min), while step 2 is an isothermal annealing (duration : 24 h).

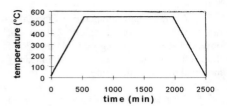

Fig. 4 : temperature cycle

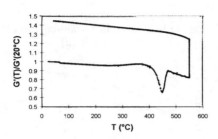

Fig. 5 : storage modulus

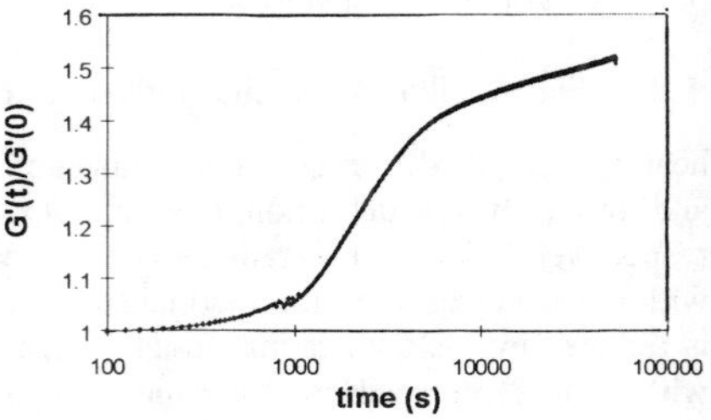

Fig. 6 : isothermal annealing

Evolution of the storage modulus G' is shown in fig. 5 : during heating, G' decreases slowly up to 380°C (i.e. Tg) ; then a sudden decrease is observed, which can be associated to the rapid evolution of elastic properties in the supercooled liquid region ; however, above about 450°C G' increases.

This increase can be associated to the formation of crystalline particles, as shown by DSC experiments (fig. 2). Isothermal annealing at 550°C (step 2) leads to a marked increase of G', due to the development of the crystallisation process. A regular decrease of G' occurs during cooling (step 3), since no microstructural evolution occurs during this stage. Fig. 6 shows that the modification of modulus during isothermal annealing is regular.

Internal friction was also measured during the same temperature cycle (Fig. 7) : a peak occurs during heating, while a decrease is observed both during isothermal annealing and cooling. In order to evidence in a different way phenomena occurring during heating, ln(tan ϕ) was plotted versus temperature during this stage.

As shown by Perez et al [15], a linear increase is associated to a simple thermal process (increase of atomic mobility with temperature), while any evolution of the microstructure induces

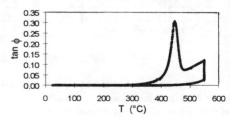

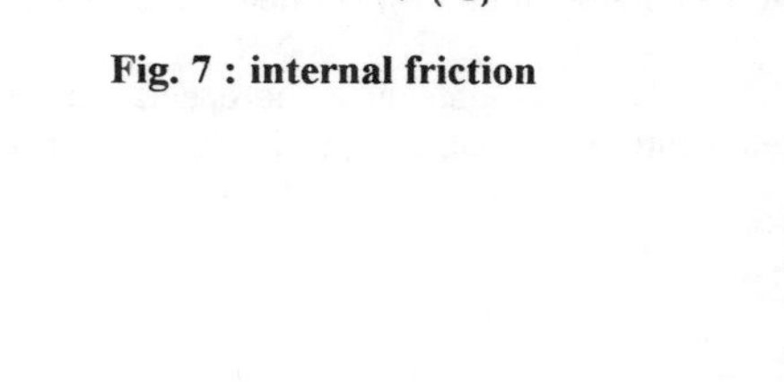

Fig. 7 : internal friction

a modification of the slope. In amorphous materials (polymeric materials, oxide glasses), it was shown that below Tg, the concentration of quasi-point defects is constant (isoconfigurational state), while above Tg this concentration is temperature dependent and induce more complex temperature variation. Thus, Tg can be determined by this way and the onset of crystallisation is also observed.

Internal friction is high in the supercooled liquid region and decreases during crystallisation. Explanation is as follows : a large atomic mobility enables large atomic movements when a stress is applied, leading to a delay of the deformation and hence to a high value of the loss modulus G'' (and therefore of the ratio tan ϕ = G ''/G'). Above Tg, mobility is high and tan ϕ is high ; in contrast, crystallisation yields a densification of the material, a reduction of the quasi-point concentration and, finally, to a large decrease of the internal friction. Vaniuk et al [16] have given the same conclusion concerning atomic mobility, using viscosity measurements.

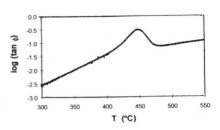

Fig. 8 : ln (internal friction)

CONCLUSION

The thermal stability of two Zr-base bulk metallic glasses has been investigated by DSC experiments and mechanical properties determination (elastic and anelastic features) ; evolution during a thermal treatment was determined. The results obtained are summarised as follows :

- During heating at a constant scanning rate, glass transition is first observed (at Tg). Then crystallisation occurs in three different steps starting from a Tx temperature. Tg and Tx can be determined either by DSC or by mechanical spectroscopy. These values increase with increasing the scanning rate.

- In the supercooled liquid region, (T>Tg), atomic mobility is large and, consequently, the glass transition is associated to a large decrease of the elastic modulus and a large increase of the internal friction.

- Crystallisation induces a reduction of the atomic mobility and hence of the internal friction. In contrast, Young modulus and shear modulus exhibit a large enhancement (up to 50%) when crystallisation phenomenon is fully achieved.

REFERENCES

1 - W.L. Johnson, Mater. Sci. Forum 225-227, p. 35 (1996).
2 - Li Ye, R.B. Schwarz, D. Mandrus and L. Jacobson, J. Non Crsyt. Sol. 205-207, p. 602 (1996).

3 - L.Q. Xing, D.M. Herlach, M. Cornet, C. Bertarand, J.P. Dallas, M.F. Trichet and J.P. Chevallier, Mat. Sci. Eng. A226-228, p. 874 (1997).

4 - L.Q. Xing, C. Bertrand, J.P. Dallas and M. Cornet, Mat. Sci. Eng. A241, P. 216 (1998).

5 - Y. Kawamura, T. Shibata, A. Inoue and T. Masumoto, Acta Mater. 46, p. 253 (1998).

6 - C.J. Gilbert, R.O. Richie and W.L. Johnson, Appl. Phys. Letters 71, p.476 (1977).

7 - A. Inoue, Mater. Sci. Forum 179-181, p. 691 (1995).

8 - A. Inoue, D. Kawase, A.P. Tsai, T. Zhang and T. Masumoto, Mat. Sci. Eng. A178, p. 255 (1994).

9 - S. Schneider, P. Thigarajan and W.L. Johnson, Appl. Phys. Letters 68, p. 493 (1996).

10 - R. Busch, S. Schneider, A. Peker and W.L. Johnson, Appl. Phys. Letters 67, p. 1544, (1995).

11 - L.Q. Xing and P. Ochin, Acta Mater. 45, p. 3765 (1997).

12 - A. Gebert, J . Eckert and L. Schultz, Acta Mater. 46, p. 5475 (1998).

13 - C. Mai, S. Etienne, J. Perez and G.P. Johari, J. Non-Cryst. Solids 74, p. 119 (1985).

14 - S. Etienne, J.Y. Cavaillé, J. Perez, R. Point and M. Salvia, Rev. Sci. Instr. 53, p. 1261 (1982).

15 - J. Perez, S. Etienne and J. Tatibouet, Phys. Stat. Sol. 121,p. 129 (1990).

16 - T.A. Vaniuk, R. Busch, A. Masuhr and W.L. Johnson, Acta Mater. 46, p. 5229 (1998).

For further information, contact :

    Jean-Marc PELLETIER
    GEMPPM, Bat. 502, INSA
    69621 Villeurbanne Cedex, France
    telephone : 33 4 72 43 83 18
    fax : 33 4 72 43 85 28
    email : pelletie@insa.insa-lyon.fr

DYNAMIC FAILURE MECHANISMS IN BERYLLIUM-BEARING BULK METALLIC GLASSES

David M. Owen*, Ares J. Rosakis*, and William L. Johnson**
*Graduate Aeronautical Laboratories, MC 105-50, **Department of Materials Science, MC 138-78, California Institute of Technology, Pasadena, CA 91125

ABSTRACT

The understanding of dynamic failure mechanisms in bulk metallic glasses is important for the application of this class of materials to a variety of engineering problems. This is true not only for design environments in which components are subject to high loading rates, but also when components are subjected to quasi-static loading conditions where observations have been made of damage propagation occurring in an unstable, highly dynamic manner. This paper presents preliminary results of a study of the phenomena of dynamic crack initiation and growth as well as the phenomenon of dynamic localization (shear band formation) in a beryllium-bearing bulk metallic glass, $Zr_{41.25}Ti_{13.75}Ni_{10}Cu_{12.75}Be_{22.5}$. Pre-notched and pre-fatigued plate specimens were subjected to quasi-static and dynamic three-point bend loading to investigate crack initiation and propagation. Asymmetric impact loading with a gas gun was used to induce dynamic shear band growth. The mechanical fields in the vicinity of the dynamically loaded crack or notch tip were characterized using high-speed optical diagnostic techniques. The results demonstrated a dramatic increase in the crack initiation toughness with loading rate and subsequent crack tip speeds approaching 1000 m s^{-1}. Dynamic crack tip branching was also observed under certain conditions. Shear bands formed readily under asymmetric impact loading. The shear bands traveled at speeds of approximately 1300 m s^{-1} and were accompanied by intense localized heating measured using high-speed full-field infrared imaging. The maximum temperatures recorded across the shear bands were in excess of 1500 K.

INTRODUCTION

The earliest studies on the mechanical properties of metallic glasses revealed their extraordinary characteristics, most notably their relatively high strength and low elastic moduli [1]. However, until recently these alloys could only be fabricated as thin sheets or rods with diameters of less than 2 mm, thereby limiting the extent of detailed experimental characterization of their mechanical properties using conventional testing techniques. The development of bulk metallic glasses by Inoue and co-workers [2,3] and Peker & Johnson [4] has heightened interest in this class of materials for a variety of engineering applications. Furthermore, the larger ingot sizes have facilitated the study of a broad range of material properties.

Although several compositions of bulk metallic glasses have successfully been processed, one of the more promising and widely studied alloys is $Zr_{41.25}Ti_{13.75}Ni_{10}Cu_{12.75}Be_{22.5}$, also known by its commercial name, Vitreloy 1. Studies of the static and dynamic constitutive response of this alloy by Rosakis, Johnson and their research group have demonstrated essentially elastic-perfectly plastic behavior with a strain rate-independent tensile / compressive flow stress of approximately 2 GPa [5,6]. Failure in these specimens occurs in a highly unstable manner along planes inclined by 45° with respect to the loading axis as a result of shear localization. Using high-speed infrared diagnostics Bruck, Rosakis & Johnson [6] observed that the temperatures accompanying localized shear failure due to adiabatic heating approach 775 K, more than 80% of the melting temperature.

Direct measurement of the static mode-I fracture toughness, K_{IC} of Vitreloy 1 have been made by Conner et al. [7] and Gilbert et al. [8] with K_{IC} on the order of 55 MPa m$^{1/2}$, a value similar to the toughnesses of typical high strength polycrystalline metallic alloys. However, in contrast to many steels, aluminum alloys and titanium alloys, the onset of failure is followed by highly unstable, dynamic crack propagation. A dramatic reduction in toughness has been reported when a significant fraction of crystalline phases are introduced through heating treatment [8,9], resulting materials with $K_{IC} \sim 2$ MPa m$^{1/2}$. Also Lowhaphandu & Lewandowski [10] have observed a value of $K_{IC} \sim 20$ MPa m$^{1/2}$ on a nominally amorphous material of similar, but not identical composition to Vitreloy 1. It was suggested that this apparently lower toughness might have resulted from differences in specimen geometry or preparation or slight variations in composition.

In the studies on bulk metallic glasses cited above, dynamically propagating cracks or shear bands have been reported to occur consistently under either quasi-static or dynamic loading conditions. Although such observations have never been studied systematically thus far, these early reports indicate the importance of developing a detailed understanding of dynamic failure mechanisms in bulk metallic glasses, as well as of the transitions in failure behavior from tensile cracking to shear banding. The dynamic fracture behavior of any material may be broadly classified into one of two types of problems. First, the response of a stationary crack to a relatively high or dynamic loading rate (dynamic crack initiation). Second, the propagation of failure (cracks or shear bands) at relatively high speeds as compared to the characteristic wave speeds in the material (dynamic failure propagation).

This paper describes the first results of a comprehensive study on dynamic failure in $Zr_{41.25}Ti_{13.75}Ni_{10}Cu_{12.75}Be_{22.5}$. High speed optical and infrared experiments capturing the details of the dynamic crack initiation and dynamic failure propagation phenomena in this are presented. Specifically, the goals of the current program are to determine the variation of the mode-I dynamic crack initiation fracture toughness, K_{IC}^d as a function of loading rate ($\dot{K}_I^d$) and the dynamic crack propagation toughness, K_D as a function of crack tip speed. In addition, the possibility of a transition in failure mechanism from tensile cracking to dynamic, adiabatic shear banding under shear-dominated (mode-II) loading conditions is investigated. Furthermore, the investigation of the complex phenomenon of adiabatic heat generation during dynamic failure during both fracture and shear banding is of great interest.

EXPERIMENTAL

Plate specimens of $Zr_{41.25}Ti_{13.75}Ni_{10}Cu_{12.75}Be_{22.5}$ were provided by Amorphous Technologies International, Inc. Two different batches of material were supplied that were fabricated using varying processing parameters, the details of which are proprietary. The two batches will be designated as Lots 41 and 46, respectively. The dilatational and shear wave speeds of the two different batches were measured. The wave speeds, along with the measured densities were used to calculate the elastic constants of the material.

The nominal dimensions of the as-received plates of were 75 x 150 x 4 mm. These large plate specimens were used for the dynamic fracture and shear banding experiments described below. A limited number of plates were cut into smaller specimens having planar dimensions of 25 x 50 mm for fracture testing under quasi-static conditions which are also described below. A wire EDM notch was fabricated into the center of the longest side of the specimen, having a

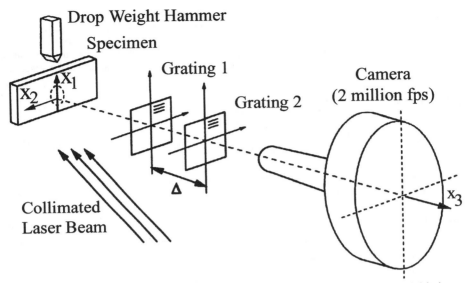

Figure 1 Schematic illustration of experimental setup for recording dynamic crack initiation and propagation using high-speed photography in conjunction with CGS interferometry.

depth of approximately one-third of the specimen width. A fatigue crack was grown from the end of the notch using K_{min}=4 MPa m$^{1/2}$ and K_{max}=8 MPa m$^{1/2}$ at a frequency of 10 Hz. The extension of the fatigue crack from the end of the notch was approximately 1 mm for the smaller specimens and 2 mm for the larger specimens. The specimens were lapped and polished to provide a flat and specularly reflective surface for optical interferometry resulting in final specimen thicknesses of 3.2±0.1 mm.

In the quasi-static range, $\dot{K}_I^d < 10^4$ MPa m$^{1/2}$ s^{-1}, the fracture experiments were conducted in a three-point bend configuration using a hydraulic Materials Testing System. The displacement rate was varied systematically, yielding a broad range of loading rates. The time history of the stress intensity factor, K_I^d was calculated directly from the time varying load and the specimen geometry. The fracture toughness was taken as the peak value of K_I^d, whereas the quasi-static loading rate was determined from the slope of the K_I^d versus time curve, expressed as $\dot{K}_I^d$.

In the dynamic range, $\dot{K}_I^d > 10^4$ MPa m$^{1/2}$ s^{-1}, a different loading configuration and diagnostic methods were used. A schematic illustration of the experimental configuration for these experiments is shown in Fig. 1. A Dynatup drop weight tower was used to load specimens in dynamic three-point bend at impact velocities ranging from 2 to 6 m s^{-1}. The mechanical fields in the vicinity of the dynamically loaded crack tip were recorded using a Cordin 330 high-speed camera in conjunction with optical interferometry. The interferometric technique employed was Coherent Gradient Sensing (CGS) which has been described in great detail elsewhere [11]. Briefly, a collimated laser beam of 50 mm in diameter was incident on the reflective specimen surface. The reflected beam, carrying information on specimen surface slopes, was processed by two gratings, G1 and G2 separated by a distance Δ. The first diffraction order was filtered using an aperture and recorded using the high speed camera. The resulting interferograms are comprised of fringes of constant in-plane gradients of the out-of-plane displacement, u_3. For example in Fig. 1, the crack line is parallel to the x_1 direction and the grating lines are parallel to the x_2 direction: such a configuration would yield fringes of constant $\partial u_3/\partial x_1$. Under conditions of K_I^d dominance and plane stress, the fringe fields in polar coordinates r and θ are related to K_I^d using

$$K_I^d(t) = -\left(\frac{mp}{2\Delta}\right)\left(\frac{E}{vh}\right)\frac{2\sqrt{2\pi}r^{3/2}}{F(v)\cos(3\theta/2)} \quad m = 0,\pm1,\pm2,....$$ (1)

where m is the fringe order, p is the pitch of the gratings (0.0254 mm), E and v are the Young's modulus and Poisson's ratio, respectively, h is the specimen thickness, and $F(v)$ is a known function of crack tip speed [11]. For a stationary, dynamically loaded crack, expression (1) is still valid for $v=0$. From a series of high-speed interferograms, the time history of K_I^d and the crack tip motion history can be measured. Using this information, the values of K_I^d at crack initiation, (K_{IC}^d) and the propagation toughness, K_D, associated with a given speed can be determined.

In addition to fracture experiments, unfatigued, EDM-notched specimens were loaded in an asymmetric impact configuration as shown in Fig. 2. Such a configuration was used to investigate the possibility of the formation of a single shear band under predominately shear loading conditions. For these experiments, dynamic loading was provided by a 50 mm diameter projectile fired from a gas gun in the speed range of 25 to 35 m s^{-1}. In a manner similar to the fracture experiments, high-speed photography in conjunction with CGS interferometry was utilized. An infrared detector array was employed to record the transient temperature fields in the vicinity of the dynamically loaded notch tip. The square array consists of 64 HgCdTe elements that were mapped onto an area of approximately 1.3 mm^2 on the surface of the specimen not used for optics. The infrared images can be captured at rates up to 1 million frames per second. Details of the system and infrared temperature measurement are given elsewhere [12].

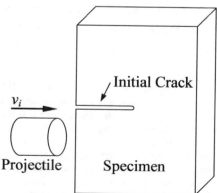

Figure 2 Schematic illustration of asymmetric impact configuration.

RESULTS

The wave speeds, density and elastic constants of the two batches of material are given in Table 1. It can be seen from the table that the elastic constants are essentially similar for the two materials, however Lot 46 has slightly faster wave speeds and correspondingly higher moduli. The results of dynamic fracture initiation, crack propagation and shear band formation are described in the sections that follow. The values of the elastic constants are necessary for accurate evaluation of the stress intensity factors using equation (1).

Table 1 Wave speeds and computed elastic constants for the two batches of Vitreloy 1.

	Lot 41	Lot 46
Shear wave speed, c_s (m s^{-1})	2596	2612
Dilatational wave speed, c_l (m s^{-1})	5143	5167
Density, ρ (g cm^{-3})	5.89	5.94
Shear modulus, G (GPa)	39.7	40.5
Young's modulus, E (GPa)	105.5	107.7
Poisson's ration, ν	0.33	0.33

Dynamic Crack Initiation

A selected sequence of high-speed interferograms recorded from a Lot 41 specimen impacted at 4 m s^{-1} in the drop weight tower is shown in Fig. 3. The sequence shows four images: three prior to crack initiation, including the image immediately prior to initiation and a

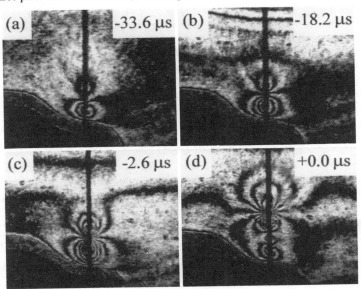

Figure 3 Selected sequence of four high-speed interferograms prior to dynamic crack initiation in a Lot 41 specimen loaded in the drop weight tower at an impact speed of 4 m s^{-1}.

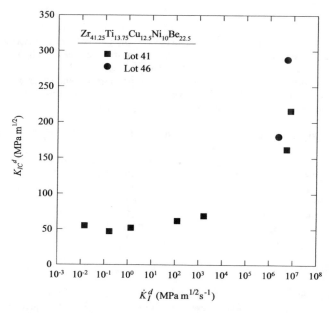

Figure 4 Variation of dynamic fracture toughness with loading rate.

images recorded during the 34 μs time interval noted. Time t=0 corresponds to crack initiation, while negative times correspond to interferograms recorded prior to initiation. Comparison of the first three images reveals an increase in the density of fringes near the crack tip, located at the center of the symmetric pattern, indicating an increase in the stress intensity factor with time. The value of K_I^d for any given time can be determined by measuring fringe positions relative to a polar coordinate system centered at the crack tip and performing a least squares fit to the data using equation (1). In this manner, the loading rate can be calculated from the series of images prior to crack initiation and the dynamic fracture toughness from the last image in which the crack is stationary. For the experiment shown in Fig. 3, $\dot{K}_I^d = 9 \times 10^6$ MPa m$^{1/2}$ s^{-1} and K_{IC}^d=225 MPa m$^{1/2}$.

The variation of K_{IC}^d with $\dot{K}_I^d$ is shown in Fig. 4, presenting data obtained over 9 orders of magnitude in loading rate. A majority of the data shown in Fig. 4 is for the Lot 41 material, with limited dynamic fracture toughness data for experiments conducted on the Lot 46 material. Several interesting features are apparent in Fig. 4. First, the values of K_{IC}^d at the highest loading rates are 4 to 6 times greater than those obtained using the hydraulic loading frame when $\dot{K}_I^d < 10^4$ MPa m$^{1/2}$ s^{-1}. Second, in addition to the dramatic increase in dynamic toughness at the highest loading rates, there is measurable increase K_{IC}^d at intermediate rates from ~50 to 70 MPa m$^{1/2}$ as $\dot{K}_I^d$ approaches 10^4 MPa m$^{1/2}$ s^{-1}. Third, the Lot 46 material exhibits greater dynamic toughness than the Lot 41 material, such that a given loading rate the toughness is approximately 1.8 times greater in the Lot 46 material.

Typically, the variation in toughness with loading rate can be rationalized by considering the interplay of (i) strain rate hardening, (ii) thermal softening that occurs as a result of adiabatic heating at the highest loading rates and (iii) inertia. In the case of Vitreloy 1, previous studies have revealed no strain rate dependence of the flow stress at strain rates up to 5000 s^{-1} [6]. Therefore, the dramatic increase in toughness with loading rate can be attributed primarily to the effect of inertia and thermal softening on the dynamic crack initiation process. Qualitatively similar variations of K_{IC}^d with $\dot{K}_I^d$ have been reported for a variety of other materials: for one such example, see [13].

Dynamic Crack Propagation

Figures 5 and 6 are sequences of three selected images showing crack propagation in the Lot 41 and 46 materials, respectively. The images in Fig. 5 are a continuation of the sequence shown in Fig. 3. The two sequences illustrate the marked differences in the propagation behavior of the two batches of materials. In Fig. 5, a single crack is seen to propagate from the bottom to top of the field of view with the fringe field clearly discernable around the traveling crack. However, the sequence recorded during an experiment in which a Lot 46 specimen was impacted at 5 m s^{-1} exhibits almost immediate crack branching and the emission of intense release waves from the crack tip at the onset of crack propagation. The emitted waves make it difficult to resolve CGS fringes near the crack tip, although the crack tip position can still be identified.

Figure 7 plots the computed crack tip speed as a function of time after initiation for the two experiments from which the images shown in Figs. 5 and 6 were taken. The data for the Lot 46 material corresponds to the propagation of the left branch in Fig. 6. The plot illustrates the significantly greater velocities observed for the branched crack, which approaches almost 1000 m s^{-1} as compared to ~800 m s^{-1} for the single crack. The left-hand axis in Fig. 7 gives the crack tip speed normalized by the shear wave speed, c_s, revealing that the maximum normalized velocities are almost 40% of c_s. It is interesting to note that these values of v/c_s are similar to those observed in highly brittle materials, such as ceramics, and greater than those observed for high strength metallic alloys. For example, normalized crack tip velocities in

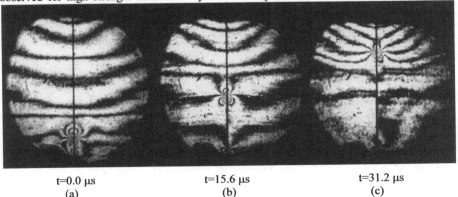

t=0.0 μs   t=15.6 μs   t=31.2 μs
(a)   (b)   (c)

Figure 5 Selected sequence of three high-speed interferograms showing dynamic crack propagation in a Lot 41 specimen loaded at an impact speed of 4 m s^{-1}.

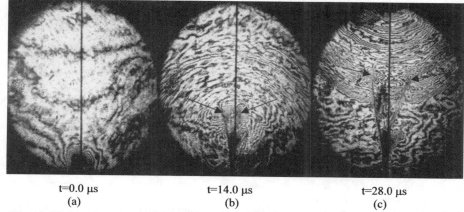

t=0.0 µs   t=14.0 µs   t=28.0 µs

(a)   (b)   (c)

Figure 6 Selected sequence of three high-speed interferograms showing dynamic crack propagation in a Lot 46 specimen loaded at an impact speed of 5 m s^{-1} (arrows indicate position of branched crack tips).

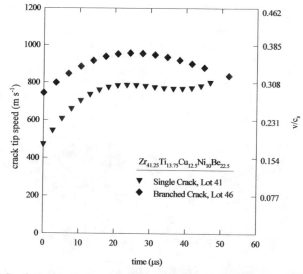

Figure 7 Crack tip speed versus time after crack initiation for a single crack and a branched crack.

ductile 2024-T3 aluminum [13], high strength AISI 4340 steel [14] and brittle soda-lime glass [15] have been reported as ~0.05, 0.32, and 0.45 respectively.

Experiments were conducted on materials from both batches of material over a range of impact speeds. It was observed that crack branching occurred in the Lot 46 material at impact speeds as low as 2 m s^{-1}. Conversely, crack branching could not be induced in the Lot 41 material at impact speeds as high as 6 m s^{-1}. The observations of branching may simply be a

426

consequence of the much higher dynamic initiation fracture toughness observed in the Lot 46 material noted in the previous section. The higher dynamic toughness results in the storage of significantly more elastic energy prior to initiation. Upon crack initiation, the initial speed is much greater in the tougher material, as shown in Fig. 7. Subsequent to the initiation of a single "fast" crack, the formation of two cracks becomes energetically favorable in Lot 46 material. A goal of the ongoing research on these two materials will be an investigation of the specific criteria for crack branching in each material.

<u>Shear Band Formation</u>

Previous observations of micro-shear band formation during uni-axial deformation experiments on bulk metallic glasses has illustrated the importance of this mechanism in the understanding of failure in this class of materials. Figure 8 shows a sequence of 4 high-speed interferograms recorded when a Lot 46 specimen was loaded asymmetrically using a gas gun firing a 50 mm diameter steel projectile at 32 m s^{-1}. The specimen was not pre-fatigued prior to testing and was loaded below the notch from the left in reference to the figure. The sequence of images shows the propagation of the shear band, as a dark shadow region, from the middle left to the lower right of the field of view. Any interference fringes near the propagating shear band are not visible as a result of the severe deformation near the band causing the deviation of the initially parallel laser light outside of the optical system. An estimate of the shear band speed was made by tracking the leading edge of the propagating shadow spot, revealing an average speed of approximately 1300 m s^{-1}.

Figure 9 shows sequence high-speed infrared images from the same experiment depicted in Fig. 8. The sequence shows six out of 30 images recorded over the elapsed time of

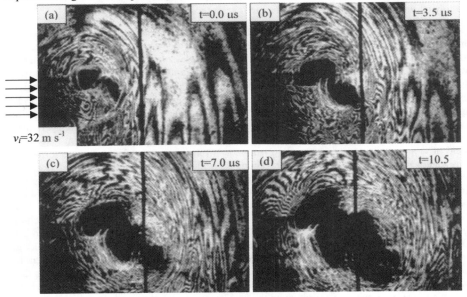

Figure 8 Selected sequence of four high-speed interferograms of shear band propagation initiation in a Lot 46 specimen impacted asymmetrically at a speed of 32 m s^{-1}.

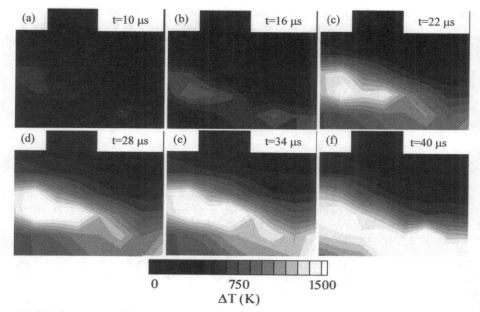

Figure 9 Sequence of high-speed infrared images ahead of an asymmetrically loaded notch showing intense localized heating leading to macroscopic shear band formation.

30 μs. The initial notch was located just to the left and in the middle of the field of view which maps 1.3 mm² on the specimen surface. The sequence of thermal images reveals features of shear band formation and propagation in metallic glasses. First, the extensive plastic deformation evident in the high-speed optical interferograms is accompanied by a dramatic temperature increase. By the time the shear band is fully formed (Fig. 9f) a region approximately 300 μm wide exhibits temperatures in excess of 1500 K, which is significantly greater than the melting temperature of 936 K [4]. Second, shear band formation appears to occur by the development and subsequent linking of regions where relatively high temperatures were recorded. Thus, in Fig. 9d, three regions of elevated temperature are apparent across the field of view, one of which is emanating from the lower right corner. These regions grow and link together, resulting in a shear band traversing the width of the field of view.

Investigation of the specimens following the asymmetric impact experiment brought to light several other interesting phenomena. The surfaces of the shear band were extremely shiny and rough in comparison to surfaces of specimens that had failed by tensile cracking. The shininess of the surfaces may have resulted from localized melting in the shear band. The shear bands propagated over limited distances in the specimen on the order of 20 mm, before a failure mode transition to tensile crack occurred. After the onset of failure by tensile cracking, numerous crack bifurcations were apparent, resulting in an extremely fragmented specimen. In addition, inspection of the specimen surface showed a large region around the initial notch that was affected by the large temperature rises associated with the shear band. This region was extended over a diameter of approximately 25 to 30 mm centered around the initial notch and was evidenced by a clear change in surface color and morphology.

CONCLUSIONS

The preliminary experimental studies of dynamic failure in amorphous $Zr_{41.25}Ti_{13.75}Ni_{10}Cu_{12.75}Be_{22.5}$ presented here, recorded for the first time many distinct characteristics of dynamic failure in this class of materials. Experiments to determine the nature of dynamic crack initiation, dynamic crack propagation and shear band formation and propagation were conducted using two high-speed, high resolution optical and infrared imaging techniques. Two materials of nominally the same composition, but fabricated using different processing techniques, were tested.

The dynamic crack initiation fracture toughness was investigated over a range of 9 orders of magnitude in loading rate. The average quasi-static fracture toughness was 50 ± 3 MPa $m^{1/2}$. At the highest loading rates of $\sim10^7$ MPa $m^{1/2}$ s^{-1}, the toughness was 4 to 6 times greater than that measured at the lowest rates, with a slight increase in K_{IC}^d at intermediate loading rates. Comparison of the two different batches of material at the same dynamic loading rate showed a difference in toughness by a factor of ~1.8, demonstrating the important role of processing in the subsequent failure behavior. The relatively high quasi-static and dynamic toughness values in $Zr_{41.25}Ti_{13.75}Ni_{10}Cu_{12.75}Be_{22.5}$ are similar to those observed in many metallic alloys such as steels, titanium alloys and aluminum alloys.

Dynamic crack propagation speeds were measured using high speed interferograms and were found to be typically in the range of 30 to 40% of the shear wave speed of the material. The batch of material exhibiting a higher dynamic toughness, also exhibited a tendency for cracks to branch under all the impact conditions investigated (impact velocities greater than 2 m s^{-1}). In contrast, the material having lower dynamic toughness did not exhibit crack branching at impact speeds up to 6 m s^{-1}. The branched crack traveled with higher velocities, approaching 1000 m s^{-1}, whereas the single cracks reached maximum velocities of ~800 m s^{-1}. The high normalized velocities are comparable to those measured in highly brittle materials such as glass or engineering ceramics.

Shear band formation and propagation was studied under asymmetric impact conditions. Shear band speeds were in excess of those measured for tensile cracks, with speeds of approximately 1300 m s^{-1}. The concurrent, full-field infrared temperature measurements revealed the formation of shear bands resulting from the nucleation, growth and linkage of areas of highly localized temperature increase ahead of the notch tip. Once fully formed, the maximum shear band temperature exceeded 1500 K and the apparent shear band width was ~300 μm.

ACKNOWLEDGEMENTS

The authors gratefully acknowledge support from the Air Force Office of Scientific Research under contract number F49620-97-1-0323. The contribution of the material used in this study by Amorphous Technologies, Inc. is appreciated. The development of the high-speed infrared imaging system was made possible through grant DE-FG03-95ER14560 from the Department of Energy. Pradeep R. Guduru and David D. Anderson are acknowledged for their assistance in conducting some of the experiments described herein.

REFERENCES

1. C. Pampillo, J. Mater. Sci. **10**, 1194 (1975).

2. A. Inoue, T. Zhang, and T. Masumoto, Mat. Trans. Jpn Inst. Metals. **31**, p. 425 (1990).

3. T. Zhang, A. Inoue, and T. Masumoto, Mat. Trans. Jpn Inst. Metals. **32**, p. 1005 (1991).

4. A. Peker and W.L. Johnson, Appl. Phys. Lett. **63**, p. 2342 (1993).

5. H.A. Bruck, T. Christman, A.J. Rosakis, and W.L. Johnson, Scripta Metall. Mater. **30**, p. 429 (1994).

6. H.A. Bruck, A.J. Rosakis, and W.L. Johnson, J. Mater. Res. **11**, p. 503 (1996).

7. R.D. Conner, A.J. Rosakis, W.L. Johnson, and D.M. Owen, Scripta Mater. **37** p. 1373 (1997).

8. C.J. Gilbert, R.O. Ritchie, and W.L. Johnson, Appl. Phys. Lett. **71**, p. 476 (1997).

9. C.J. Gilbert, J.M. Lippmann, and R.O. Ritchie, Scripta Mater. **38**, p. 537 (1998).

10. P. Lowhaphandu and J.J. Lewandowski, Scripta Mater. **38**, p. 1811 (1998).

11. A.J. Rosakis, in *Experimental Techniques in Fracture,* edited by J.S. Epstein (VCH Publishers, Inc., New York, NY 1993) p. 327-425.

12. P. Guduru, A.T. Zehnder, A.J. Rosakis, and G. Ravichandran, Caltech GALCIT Solid Mechanics Report, SM 98-16, presented at the ASME Winter Meeting, Anaheim, CA, November, 1998.

13. D.M. Owen, S.M. Zhuang, A.J. Rosakis and G. Ravichandran, to appear in Inter. J. Fracture (1999).

14. A.T. Zehnder and A.J. Rosakis, Inter. J. Fracture **43**, p. 271 (1990).

15. D.M. Owen, A.J. Rosakis, and G. Ravichandran, in preparation (1999).

RESIDUAL STRESSES IN BULK METALLIC GLASSES DUE TO DIFFERENTIAL COOLING OR THERMAL TEMPERING

E. USTUNDAG[†], B. CLAUSEN*, J. C. HANAN[†], M. A. M. BOURKE*, A. WINHOLTZ[‡‡]
and A. PEKER[†]**
[†]California Institute of Technology, Department of Materials Science, M/C 138-78, Pasadena, CA 91125;
ustundag@hyperfine.caltech.edu
*Los Alamos National Laboratory, LANSCE, MS H805, Los Alamos, NM 87545
[‡‡]Research Reactor Center, Research Park, University of Missouri, Columbia, MO 65211
**Amorphous Technologies, 27722 El Lazo, Laguna Niguel, CA 92677

ABSTRACT

Due to their very low thermal conductivities and large thermal expansion values, bulk metallic glasses (BMGs) undergo differential cooling during processing. Large thermal gradients are generated across a specimen leading to residual stress buildup. A thin surface layer contains compressive stresses balanced by tension in the middle. Such stresses can not only influence the mechanical behavior of BMGs, but they can also lead to problems during manufacturing of large or intricate components. Analytical and finite element modeling was used to predict the values and distribution of such stresses as a function of processing conditions. Neutron diffraction measurements were then performed on model specimens which included crystalline phases as "strain gages". It was shown that significant stresses, on the order of several hundred MPa, can be generated in BMGs. Modeling and diffraction results are presented and their implications discussed.

INTRODUCTION

Although metallic glasses have been made since 1960s, specimen dimensions were previously limited to tens of μm due to the very fast cooling rates (about 10^6 K/s) needed in order to prevent crystallization in most systems. Recently, multicomponent alloys have been developed with exceptional glass formation ability that allow the processing of *bulk* specimens. One of the most successful bulk metallic glass (BMG) alloy series (Zr-Ti-Cu-Ni-Be) has been developed at Caltech [1]. These alloys form metallic glasses at cooling rates as low as 1 K/s allowing the casting of specimens up to 5 cm in diameter while still retaining the glassy structure. The ability to prepare large specimens has permitted the bulk characterization of these materials using more "traditional" techniques. The unique properties of BMGs potentially place them among significant engineering materials: very high strength (up to 1.9 GPa) and initiation fracture toughness (40-55 MPa.m$^{1/2}$), a near theoretical specific strength, excellent wear and corrosion resistance, high elastic strain limit (up to 2%), and so on [2, 3].

The properties of BMGs are influenced by, among other parameters, residual stresses. A potentially major source of such stresses is differential cooling. The cooling rates currently used to process large specimens are still fast enough to lead to large thermal gradients inside BMGs. This is magnified by their extremely poor thermal conductivity ($k \approx 4$ W/m-K, compared to ~400 W/m-K for Cu) which leads to a rapid cooling of the outside while the middle is still well above the glass transition temperature. Additional factors such as rapid variation of viscosity [4] and coefficient of thermal expansion (CTE) [5] with temperature can also add to this effect leading to potentially very high residual stresses [6, 7]. The result is an effect called 'thermal tempering' as is seen in silica-based glasses where a surface layer with compressive stresses is balanced with a tensile stress in the middle [8, 9, 10, 11]. Such a stress state significantly increases the fracture and impact resistance of silica glass due to inhibition of surface cracks.

BMGs have, in principle, a higher potential in tempering due to their higher strength and CTE values. Indeed, as it will be shown by our preliminary calculations and measurements, residual stresses of up to 900 MPa can be generated due to this process (silica glass would usually be tempered with surface stresses around 100-150 MPa [8]). Such high stresses are expected to have profound effects on material properties as well as processing conditions. This paper presents the preliminary results of a systematic study to identify mechanisms that influence tempering stress generation and relaxation in BMGs, to measure these stresses using neutron diffraction and compare the results to analytical and finite element calculations.

BACKGROUND

Residual stress generation due to rapid cooling, sometimes referred to as *thermal tempering*, is a well known phenomenon in metals and silica glasses. Despite its high potential for significant stress buildup, however, this topic has received little attention in metallic glasses. Almost all studies noted in literature investigated the tempering stresses in magnetic metallic glass ribbons (see for example [12, 13]). The thrust of this work was the determination of the effect of internal stresses on the magnetic domain structure of these materials due to the magnetoelastic effect. Among the noteworthy results of these studies is the measurement of residual compressive stresses of about 100 MPa despite the use of very thin (30-40 μm) ribbons. This is another demonstration of the potential of internal stress generation due to rapid cooling in metallic glasses. On the other hand, no systematic tempering stress studies are known on *bulk* metallic glasses where these stresses are potentially higher due to large specimen dimensions.

The tempering of silica glass, in turn, is an extensively studied topic. It has been known since the demonstration of the so-called Prince Rupert drops in the 17th century, but more systematically since 1891 when the Schott's process was introduced, that glass can be treated to achieve high strength and impact resistance [14]. There are a number of methods to strengthen silica glass, e.g., via chemical means [14], but here only the thermal methods will be reviewed briefly due to their direct relevance to the metallic glass case.

The tempering of silica glass involves rapid quenching from a temperature above its strain point. During this process, surface cools more rapidly than the interior, and in a few seconds a large thermal gradient is attained. Then the interior cools more rapidly than the surface until room temperature is reached. Initially, therefore, the thermal contraction of the surface is greater than that of the midplane. This differential contraction tends to produce tensile stresses on the surface and compressive stresses in the interior. In an elastic solid these stresses would be cancelled by stresses of opposite sign during the later stage of quenching, in which the cooling rate of the midplane exceeds that of the surface. However, since glass is a viscoelastic material with rapidly varying viscosity as a function of temperature, the initial stresses are relaxed at high temperatures. On the other hand, the stresses generated later, by the contraction of the middle, are not relaxed leading to residual compressive stresses balanced by interior tension [8].

THEORETICAL CALCULATIONS OF TEMPERING STRESSES

Analytical Calculation

A simple calculation was performed adapting an analytical model developed for silica glass tempering [15] called the 'Instant Freezing Model'. The model was later modified by Indenbom and Vidro [16]. It assumes that the material is a perfect fluid above the glass transition temperature, T_g, incapable of supporting any stresses and a perfectly elastic solid with no stress relaxation below it. Obviously this is a gross oversimplification considering the continuous albeit fast variation of material properties around this temperature. However, it has been shown that predictions of the model are reasonably close to measured stress values although information about transient effects is lost [8].

In this calculation [6] the room-temperature values of BMG elastic constants ($E = 90$ GPa, $v = 0.35$) were used as well as $T_g = 625$ K, final temperature, $T_a = 298$ K and $k \approx 4$ W/m-K. Since specimens are usually quenched in water after casting, heat transfer coefficients were chosen to correspond to laminar and turbulent water flow over sample surface: $h = 2000$ W/m².K for the former and $h = 10000$

W/m^2.K for the latter [17]. In the 'Instant Freezing Model' midplane tension, σ_M, of an *infinite plate* is given by [15, 16]:

$$\sigma_M = \frac{\beta E}{1 - v}(T_g - T_a)(1 - \frac{\sin\delta_1}{\delta_1})$$
(1)

where, β is the CTE and δ_l is the first root of $\delta \tan(\delta) = ht/k$ (Biot number), and t is the half thickness of the plate. The surface stress is then calculated from $\sigma_S = -\sigma_M f(t)$ where $f(t)$ is a function of the thickness and the Biot number and takes values between 2 and 5.

The calculated surface compressive stresses for each cooling case are shown in Fig. 1. There is about 200 MPa difference in compressive stress values between the laminar and turbulent flow cases for a given half thickness. In either case, it is seen that very high compressive stress values can be obtained due to tempering, e.g. up to 900 MPa for a half thickness of 2 cm, a dimension easily accessible with the current BMG processing techniques.

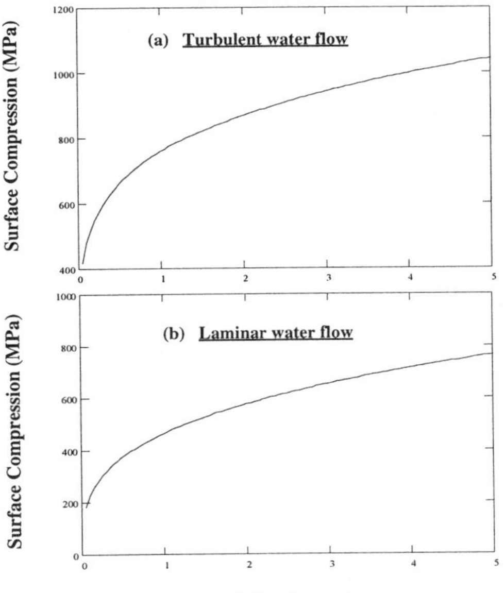

Figure 1. Surface stresses in an infinite plate predicted by the "Instant Freezing Model'.

Finite Element Modeling

The finite element (FE) calculations of the two geometries chosen as model specimens (and described later) were made using ABAQUS (version 5.7) in the standard non-linear implicit mode. The full details of the calculation will be presented elsewhere [6]; here only one of the specimens is discussed for brevity. The FE model is axis-symmetric; hence, only a quarter of a plane along the cylinder axis was taken into account. A heat flux through the top and right hand side of the model was applied to simulate the cooling in water. As a typical assumption used in FE models, the interfaces were regarded as perfect, which in this case is a realistic assumption. Coupled temperature and displacement calculations were used to determine the transient stresses and strains in the sample. The element size was biased so the smallest elements were at the top surface and at the interfaces.

The stainless steel (SS) used in the model specimens was assumed to be an elastic-plastic material with linear strain hardening, while the BMG was modeled as a viscoelastic material. Some approximations were made for the viscoelastic properties, as these properties are not yet available for BMG. Namely, the shape of the curve that describes the relaxation of stresses as a function of time at a given temperature (in the form of a Prony series in ABAQUS [18]) was taken to be that of silica glass [11]. It is difficult to estimate the errors introduced by using these adapted material data. However, the functional form of the relaxation curve is the same for all viscoelastic materials which should ensure that only the numerical values of the stresses and strains calculated using these parameters are subjected to errors, but the general trends should be well described by the model. The cooling rate was assumed to be that of laminar water flow, similar to the case with the analytical model. Another important variable in tempering, the initial temperature, was taken to be around 670 K (close to the glass transition temperature of BMG) although the casting of the model specimens was done at 1173 K. Since a higher initial temperature is predicted to lead to larger tempering stresses (based on the silica glass data [8]), the current version of the FE model should be viewed as an underestimate. Furthermore, the thermal conductivity ($k \approx 16$ W/m-K) of type 314 SS used is much lower than that of Cu normally used in the casting molds for BMG. For this reason, the model specimen shown in Fig. 2 is not optimized in terms of tempering stresses.

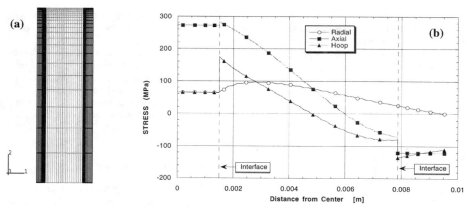

Figure 2. Finite element model (a), and calculated residual stress distribution due to tempering in a model specimen (b). The dark lines in (a) indicate the stainless steel regions; the rest is BMG.

Nevertheless, the stresses predicted are still substantial. The stress distribution shown in Fig. 2(b) follows the predictions of the tempering theory. The middle of the sample is under high tension and the surface is compressed. In the FE model, three different calculation schemes were assumed (details are reported elsewhere [6]). The FE model predictions for each case are shown in Fig. 3.

Scenario 1: only CTE mismatch between SS and BMG (both materials are elastic);
Scenario 2: Scenario 1 plus viscoelasticity in BMG;
Scenario 3: all the above plus plasticity in SS.

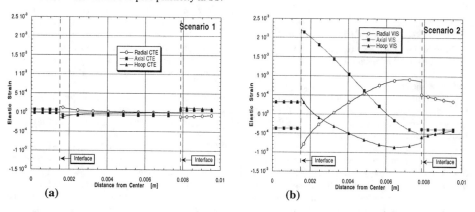

Figure 3. Finite element model predictions of *elastic strain* components for each calculation scenario. The specimen geometry is depicted in Fig. 2(a).

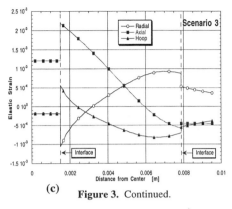

(c) **Figure 3.** Continued.

It is seen that only the introduction of viscoelasticity in BMG (Scenario 2) leads to a substantial, and more 'realistic' change in the strain state. The average residual strains in *steel* are shown in Table I. Despite some relaxation by yielding in this region, the residual stresses predicted are still quite appreciable.

EXPERIMENTAL

Sample Preparation

The specimen shown in Fig. 2 was prepared by casting a BMG alloy (Vitreloy 1: $Zr_{41.2}Ti_{13.8}Cu_{12.5}Ni_{10}Be_{22.5}$) in a long, type 314 stainless steel (SS) tube (19.0 mm outer diam., 15.6 mm inner diam.) at 1173 K after which it was dropped in water at room temperature for quenching. Other details of specimen preparation were presented elsewhere [1]. The tube had a welded pin (3.2 mm diam.) in its middle intended to provide 'contrast' in the strain state due to tempering. Both interfaces were observed to be intact after processing. A 50 mm long section of the BMG/SS composite structure was cut for strain measurements. An identical tube (with an attached pin) was also heat treated under the same conditions, but without a BMG core, to be used as a stress-free reference. The primary function of the SS components was to serve as 'strain gages' in neutron diffraction strain measurements.

Table I: Calculated and measured *average* elastic strain components in steel in the model sample (Fig. 2)

Strain Component	Measured	Calculated, scenario 1	Calculated, scenario 2	Calculated, scenario 3
Axial, pin	1420×10^{-6}	72×10^{-6}	-385×10^{-6}	1210×10^{-6}
Axial, tube	-137×10^{-6}	72×10^{-6}	-375×10^{-6}	-454×10^{-6}
Radial, tube	847×10^{-6}	-92×10^{-6}	314×10^{-6}	500×10^{-6}

Neutron Diffraction Strain Measurements

Some preliminary strain measurements in the model sample were performed at the Missouri University Research Reactor (MURR). These were made using a monochromatic neutron beam of 1.478 Å on the 2XD powder diffractometer. The (311) stainless steel peak at 86.6° (2θ) was employed. The gage volume for the measurements, defined by a 4 inch long 1x8 mm boron nitride incident beam collimator and a 1 mm cadmium slit on the diffracted beam, was almost a perfect rectangular box (a 1x1x8 mm "match stick").

DISCUSSION

The neutron strain measurement results and their comparison to the three different finite element models are shown in Table I. There is a very close correspondence between experimental data and the model predictions by the third scenario. This is encouraging although the model is still very preliminary and has not been optimized [6]. The occurrence of tempering can be concluded, for instance, from a comparison between the first and third scenarios. If only the CTE mismatch were the source of residual stresses, then the first case should have applied. However, the presence of the viscoelasticity in BMG leads to tempering. At this stage, these residual stresses are not optimized. The relatively poor thermal conductivity of SS prevented the realization of faster cooling. In any case, it should be noted that the SS/BMG assembly is already a tempered product with appreciable compressive stresses in the SS tube.

SUMMARY

The viscoelastic nature of BMGs, their poor thermal conductivity and the fast cooling utilized in their processing lead to thermal tempering; namely, compressive stresses on the surface and tension in the interior. Both a simple analytical calculation and finite element models predict substantial residual stress generation in BMGs due to this effect (up to 900 MPa on the surface). Preliminary neutron diffraction strain measurements on model specimens support this and point to a potential of significant property manipulation via such residual stresses. A systematic study is underway that involves modeling and experimentation to investigate stress generation/relaxation mechanisms in BMGs and to quantify the effect of these stresses on their properties.

REFERENCES

1. A. Peker and W. L. Johnson, *Appl. Phys. Lett.*, **63**, 2342 (1993).
2. C. J. Gilbert, R. O. Ritchie and W. L. Johnson, *Appl. Phys. Lett.*, **71** (4), 476 (1997).
3. H. A. Bruck, T. Christman, A. J. Rosakis and W. L. Johnson, *Scripta Metall.* **30**, 429 (1994).
4. E. Bakke, R. Busch and W. L. Johnson, *Appl. Phys. Lett.*, **67** (22), 3260 (1995).
5. K. Ohsaka, S. K. Chung, W. K. Rhim, A. Peker, D. Scruggs and W. L. Johnson, *Appl. Phys. Lett.*, **70** (6), 726 (1997).
6. B. Clausen, E. Üstündag, J. C. Hanan and M. A. M. Bourke, *in preparation* (1998).
7. B. Clausen, E. Üstündag and M. A. M. Bourke, *in preparation* (1998).
8. R. Gardon in *Glass Science and Technology, vol. 5: Elasticity and Strength in Glasses,* edited by D. R. Uhlmann and N. J. Kreidl, pp. 146-216, Academic Press, New York, 1980.
9. O. S. Narayanaswamy and R. Gardon, *J. Am. Ceram. Soc.*, **52** (10), 554 (1969).
10. O. S. Narayanaswamy *J. Am. Ceram. Soc.*, **54** (10), 491 (1971).
11. H. Carre and L. Daudeville, *J. de Phys. IV*, vol. **6**, 175, (1996).
12. M. de Jong, J. Sietsma, M. Th. Rekveldt and A. van den Beukel, *Mat. Sci. and Eng.*, **A179-A180**, 341 (1994).
13. M. Tejedor, J. A. Garcia, J. Carrizo and L. Elbaile, *J. Mater. Sci.*, **32**, 2337 (1997).
14. R. F. Bartholomew and H. M. Garfinkel, in *Glass Science and Technology, vol. 5: Elasticity and Strength in Glasses,* edited by D. R. Uhlmann and N. J. Kreidl, pp. 217-270, Academic Press, New York, 1980.
15. Z. M. Bartenev, *Zh. Tekhn. Fiz.*, **19** (12), 1423 (1949).
16. V. L. Indenbom and L. I. Vidro, *Sov. Phys. Solid State*, **6** (4), 767 (1964).
17. S. Yanniotis and D. Kolokotsa, *J. of Food Eng.*, **30** (3-4) 313 (1996).
18. ABAQUS Theory Manual, Hibbitt-Karlsson-Sorensen, Inc., Pawtucket, RI, 1995.

TRIBOLOGICAL CHARACTERISTICS OF $Zr_{41.2}Ti_{13.8}Cu_{12.5}Ni_{10.0}Be_{22.5}$ BULK METALLIC GLASS

XI-YONG FU, D. A. RIGNEY
Materials Science and Engineering, The Ohio State University, Columbus, OH 43210

ABSTRACT

The sliding characteristics of $Zr_{41.2}Ti_{13.8}Cu_{12.5}Ni_{10.0}Be_{22.5}$ bulk metallic glass have been examined in vacuum and in air using a pin/disk geometry without lubrication. The counterface material was either the same metallic glass (self-mated) or 52100 steel. Normal load was 0.1-1 kgf. The test system was equipped for continuous measurement of friction force. In addition, a Kelvin probe allowed continuous monitoring of changes in the structure and chemistry of the disk surface. Post-test characterization included optical and electron microscopy, X-ray diffraction and EDS of worn surfaces and debris. Friction coefficients in both vacuum and air were typical of those expected for ductile materials, with values ranging from 0.4 to 0.9 (higher values with lower load). Wear rates and average debris size increased with load. Wear rates were larger in air than in vacuum. Wear surface appearance and chemical composition were influenced by plastic deformation, material transfer and environmental interactions. After tests in air, the wear tracks and debris had a granular appearance and oxygen concentrations were high. Changes in the Kelvin probe signal correlated well with visual observations and with concentrations of oxygen detected by EDS.

INTRODUCTION

Applications of bulk metallic glasses include those which will require suitable tribological characteristics. Therefore it is important to determine the friction and wear behavior of such materials under different experimental conditions, e.g., different applied loads, counterface materials and test environments.

The literature on the tribological behavior of metallic glasses is not large [1-13]. Most of these papers reported work on Fe, Ni or Co based alloys containing Si, B and sometimes C. In most cases, the test environment was laboratory air and the counterface material was a hardened alloy steel such as 52100. Imura et al. [12] and Lee and Evetts [13] used counterfaces of metallic glass. Except for the work in those two papers, tribological test specimens were only in the form of thin films or relatively thin sections because bulk metallic glasses were not available. Imura et al. [12] reported that test environment affected fatigue results, but did not report effects of environment on sliding behavior. Sudarsan et al. [5] reported that friction was higher in the presence of water and suggested that this was related to oxidation.

The work presented in references 1-13 is intriguing but raises many questions. In particular, the following need further attention: load dependence, the extent of plastic deformation, and effects of structural and chemical changes, including phase transformations and environmental effects. The present paper is an initial report in which these topics are considered for the case of unlubricated sliding involving $Zr_{41.2}Ti_{13.8}Cu_{12.5}Ni_{10.0}Be_{22.5}$ bulk metallic glass.

EXPERIMENT

Unlubricated sliding tests involved a pin pressed against a rotating disk. The axis of rotation was horizontal, allowing convenient collection of wear debris. Loads were applied through a pulley and lever system. Normal loads ranged from 0.10 kgf to 1.20 kgf. Disks 31mm in diameter and 3 mm thick were made of $Zr_{41.2}Ti_{13.8}Cu_{12.5}Ni_{10.0}Be_{22.5}$ bulk metallic glass (MG) provided by Howmet Corp. Pins were either 52100 bearing steel balls of diameter 6 mm or rods of MG, each 3 mm x 3 mm x 12 mm. Contacting surfaces of the MG pins were of two types, one flat and the other with one end machined to a curvature of 15 mm about one of the transverse axes. The sliding speed was kept low (0.05 m·s^{-1}) to control frictional heating. Friction force was monitored with the aid of a full bridge arrangement of four strain gauges mounted on the pin holder arm. The entire pin/disk system was contained within a bell jar to allow control of test environment. A turbomolecular pump allowed tests in vacuum with typical pressures of about 10^{-5} torr (10^{-3} Pa). A special feature of the sliding test system was a Kelvin probe, described elsewhere [14]. The Kelvin probe signal (KPS) was sensitive to structural and chemical changes on the wear track of the disk. This feature provided information which aided interpretation of differences in sliding behavior when test conditions were varied. Together with friction force data, the KPS was also used to determine when to stop a given test.

Disk surfaces were prepared by standard metallographic procedures, with final polishing using 1 μm diamond paste. X-ray diffraction (XRD) confirmed that the polished surface re-mained amorphous. Disks and pins were ultrasonically cleaned sequentially in acetone, methanol and distilled water. Worn surfaces and debris were characterized by optical microscopy, scanning electron microscopy (SEM), energy dispersive spectroscopy (EDS), XRD, differential scanning calorimetry (DSC), transmission electron microscopy (TEM), X-ray photoelectron spectroscopy (XPS) and microhardness. SEM and EDS were done with a Philips XL-FEG instrument. XRD was done with a Scintag PAD-V system using Cu Kα radiation, and TEM was performed with Philips CM-200 and CM-300FEG (high resolution) instruments. DSC was done with a Seiko DSC320, and XPS used a VG ESCALABII with a Mg Kα source.

RESULTS AND DISCUSSION

The as-received MG was first examined with the aid of several characterization tech-niques. Average Vickers microhardness values on four different disks were 538±6, 516±1, 526±6 and 530±7. X-ray diffraction patterns were typical of those expected for an amorphous material, and no peaks from crystalline phases were detected. Conventional TEM bright field images and diffraction patterns were also as expected for an amorphous material. High resolution TEM showed no evidence of crystalline phases in most areas selected for viewing. One region did contain isolated features which resembled defective crystals a few nm in diameter, as shown in Figure 1a. During observation, there was some sharpening of these features, presumably related to the 300 keV electron beam, as shown in Figure 1b. Digital diffraction patterns obtained from the image of Figure 1b also suggested the presence of small crystals, but most of the material remained amorphous.

DSC analysis of the same material showed three prominent exothermic peaks, as shown in Figure 2. The process associated with the first peak started at 419°C, and the process associated with the second peak started before the first process was complete. Both processes were essentially complete at 460°C. Peaks like these have been reported previously (15,16). Johnson et al. (15) associated the first peak with phase separation and primary crystallization and the second peak with secondary crystallization. Gebert et al. (16), working with a related Zr-Al-Cu-Ni alloy, reported that the first peak was absent when oxygen in the alloy was below 0.3 at. %. They also suggested that the oxygen effect was related to the formation of a metastable phase which aided the nucleation of other phases.

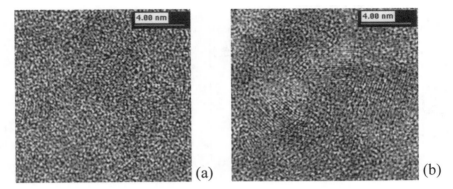

Fig. 1: High resolution TEM images of as received $Zr_{41.2}Ti_{13.8}Cu_{12.5}Ni_{10.0}Be_{22.5}$ bulk metallic glass (MG): (a.) Soon after focusing. Some regions resemble defective crystals a few nm in diameter. (b.) A different area, after observing for 30 minutes. Some sharpening of crystalline regions is apparent.

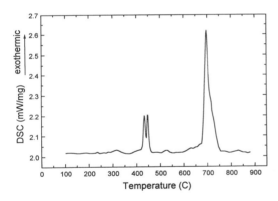

Fig. 2: DSC trace for as received MG, showing three peaks associated with exothermic processes during heating. Heating rate was 5°C/min. with argon flow of 100 ml/min.

XRD of material annealed at 500°C showed that devitrification had occurred. The hardness of the devitrified material was 673±4. Figure 2 shows that a third exothermic process began at 677°C. The process associated with this peak has not yet been identified. XRD after annealing the material at 900°C showed the same peaks as after the 500°C annealing but also several new peaks.

Steady-state friction coefficients ranged from 0.4 to 0.9, depending on normal load and test environment (air, vacuum). These are typical friction coefficients for unlubricated sliding of ductile materials. Figure 3 shows that friction decreased as load increased. The same figure shows that, for a given load, the friction was typically higher in vacuum than in air.

Wear rates in both environments increased with normal load, as shown in Figure 4, with the wear rate higher in air than in vacuum. Wear rates tended to be higher during early stages of sliding. They decreased to lower steady-state values after extended sliding. An example is shown in Figure 5.

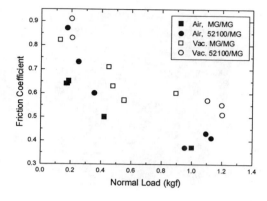

Fig. 3: Friction coefficient vs. normal load for sliding of MG disks against pins of MG and 52100 steel in air and vacuum.

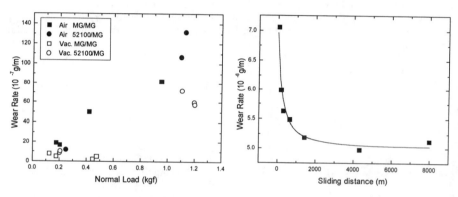

Fig. 4: Wear rate of MG disk vs. normal load for unlubricated sliding against pins of MG and 52100 steel in air and vacuum.

Fig. 5: Wear rate of MG disk vs. sliding distance with MG pin in air (normal load 0.50 kgf).

Wear tracks on MG disks were quite different after sliding in air and vacuum. Fine granular material was visible on the wear track after tests in air but not after tests in vacuum. Plastic deformation of surface material was evident in both cases. However, the Vickers hardness of the wear tracks after sliding of MG vs. MG and MG vs. 52100 in vacuum remained the same as in the as-received material. This absence of work hardening suggests behavior like that of a model elastic/perfectly plastic material (17). Disks tested in air showed slight increases in hardness (550-560) in most areas of the wear tracks. As the load increased, local patches having much higher hardness (638±14) appeared. As the load increased further, the area fraction of these patches increased. The worn surfaces of the corresponding pins were somewhat harder (~580) than most areas of the worn disks. Oxygen content of non-patch regions decreased as load increased. XPS was used to detect the presence of ZrO_2, TiO_2, NiO, CuO, Cu_2O and BeO (and the respective alloy elements in metallic form) on the wear tracks of samples tested in air. All of these oxides would be expected if oxidation occurred at low temperatures where the more noble components could not diffuse away from metal oxide interfaces (18). Samples tested in vacuum did not have significant amounts of oxygen on either the wear tracks or the debris. Wear debris particles from tests in vacuum ranged in size from 20 to 200 µm, while particle sizes from tests in air ranged from 1 to 100 µm. Particle sizes increased with increased load.

The use of 52100 steel pins allowed detection of transfer of MG from disk to pin. This transfer meant that sliding became effectively MG on MG, justifying the inclusion of data from MG/MG and 52100/MG tests on the same plots, as in Figures 3 and 4. In these cases, wear track widths on the disks were determined by the dimensions of the transferred material. Some transfer of 52100 also occurred locally on the disk during testing in air, as shown by the Fe peaks in EDS spectra. The corresponding debris, which occurred as loose clusters of fine particles, also contained Fe.

Debris from MG on MG tests in air and vacuum showed XRD patterns similar to those obtained from as-received material. Some of these patterns also contained a few small, broadened peaks superimposed on the dominant amorphous pattern. TEM revealed that these debris contained a small fraction of crystalline islands about 100 nm in diameter. When MG on MG debris were annealed at 900°C the resulting XRD patterns of sharp lines were similar to each other, but quite different from the patterns mentioned earlier for unworn material annealed at the same temperature. The various phases in these specimens have not yet been identified.

Figure 6 shows DSC results for MG on MG wear debris generated in air and vacuum using different loads. The results shown earlier in Fig.2 are included for comparison. Differences include the disappearance of both lower temperature peaks for all the debris, changes in shape for the higher temperature peak, and small shifts with both load and environment. The simplest explanation would be that sliding causes more thorough homogenization of the alloy via mechanical mixing processes, thus eliminating prime sites for nucleation of new phases. Alternatively, changes in the distribution of oxygen may be responsible. However, evidence to support either hypothesis is not yet available.

A few MG disks and pins were annealed at 500°C for 120 minutes. XRD showed that these samples were crystallized. Sliding tests in vacuum and air with 0.20 kgf were then done with these annealed materials. XRD of the resulting debris gave patterns similar to those of the as-received material, but with decreased ratios of peak/width , especially for the air case.

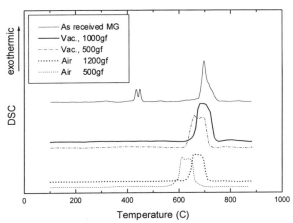

Fig. 6: DSC traces for as received MG and debris generated in vacuum and air with different loads. All peaks are exothermic. Heating rate was 5°C/min. with argon flow of 100ml/min.

SUMMARY AND CONCLUSIONS

1. Unlubricated sliding behavior of $Zr_{41.2}Ti_{13.8}Cu_{12.5}Ni_{10.0}Be_{22.5}$ depends on normal load, counterface material and test environment.
2. Friction coefficients are similar to those reported for unlubricated sliding of ductile materials.
3. Friction coefficients were higher with lower loads and higher in air than in vacuum.
4. Wear rates and debris size increased with load. Wear rates were larger in air than in vacuum.
5. Sliding can cause amorphous material to become partly crystalline and annealed (crystalline) material to become amorphous.

ACKNOWLEDGMENTS

The authors are grateful to D. Larsen and the Howmet Corp. for support of this project and for providing the $Zr_{41.2}Ti_{13.8}Cu_{12.5}Ni_{10.0}Be_{22.5}$ Vitreloy® material. We also acknowledge support by the U. S. Civilian Research and Development Foundation (CRDF) and the help of our CRDF collaborator, A.Zharin.

REFERENCES

1. P.G. Boswell, J. Mater. Sci. **14**, p. 1505(1979).
2. J.K.A. Amuzu, J. Phys. **D13**, p. L127(1980).
3. S. H. Whang and B.C. Giessen, in *Rapidly Solidified Amorphous and Crystalline Alloys*, eds. B.H. Kear, B.C. Giessen and M. Cohen, North Holland (1982), p. 301.
4. K. Miyoshi and D.H. Buckley, NASA Technical Reports No. 1990, 2001 and 2140.
5. U. Sudarsan, K. Chattopadhyay and C. Kishore, in *Rapidly Quenched Metals*, Volumes 1 and 2, eds. S. Steeb and H. Warlimont (Elsevier, Oxford, 1985), p. 1439.
6. N. Dolezal and G. Hausch, p. 1767, in Procs. cited in ref. 5 above.
7. D.H. Buckley, *Surface Effects in Adhesion, Friction, Wear and Lubrication*, Elsevier, 1981.
8. D.G. Morris, J. Mater. Sci. **17**, p. 1789(1982).
9. D.G. Morris, pp. 1775-1778, in Procs. cited in ref. 5 above.
10. K. H. Zum Gahr, Z. Metallkunde **73**, p. 267(1982).
11. C. Kishore, N. Chattopadhyay and K. Sudarsan, Acta Metall.**35**, p. 1463(1987).
12. T. Imura, K. Hasegawa, M. Moori, T. Nishiwaki, M. Takagi and Y. Kawamura, Mater. Sci. & Engin. **A133**, p. 332(1991).
13. D.-H. Lee and J. E. Evetts, Acta Metall. **32**, p. 1035(1984).
14. T.Kasai, D.A.Rigney and A.Zharin, Scripta Metall. **39**, p. 561(1998).
15. T.A.Waniuk, R.Busch, A.Masuhr and W.L.Johnson, Acta Mater. **46**, p. 5229(1998).
16. A.Gebert, J.Eckeert and L. Schultz, Acta Mater. **46**, p. 5475(1998).
17. A.Leonhard, L.Q.Xing, M.Heilmaier, A.Gebert, J.Eckert, L.Schultz, submitted to Nano-structured Materials (1998).
18. X.Sun, S.Schneider, U.Geyer, W.L.Johnson and M.-A.Nicolet, J.Mater.Res. **11**, p. 2738 (1996).

AUTHOR INDEX

SUBJECT INDEX

SUBJECT INDEX